AF344532

Plant Responses to Hypoxia

Plant Responses to Hypoxia

Editors

Elena Loreti
Gustavo Striker

MDPI • Basel • Beijing • Wuhan • Barcelona • Belgrade • Manchester • Tokyo • Cluj • Tianjin

Editors
Elena Loreti
National Research Council
Italy

Gustavo Striker
University of Buenos Aires &
National Scientific and Technical
Research Council
Argentina

Editorial Office
MDPI
St. Alban-Anlage 66
4052 Basel, Switzerland

This is a reprint of articles from the Special Issue published online in the open access journal *Plants* (ISSN 2223-7747) (available at: https://www.mdpi.com/journal/plants/special_issues/plant_hypoxia).

For citation purposes, cite each article independently as indicated on the article page online and as indicated below:

LastName, A.A.; LastName, B.B.; LastName, C.C. Article Title. *Journal Name* **Year**, *Volume Number*, Page Range.

ISBN 978-3-0365-0148-2 (Hbk)
ISBN 978-3-0365-0149-9 (PDF)

Contents

About the Editors

Elena Loreti PhD in Plant Physiology, holds a position as Researcher at the National Research Council (Institute of Biology and Agricultural Biotechnology). Elena Loreti obtained her degree in Biological Science from the University of Pisa; she then joined the Scuola Superiore Sant'Anna of Pisa as a PhD student in Plant Physiology. Research activities include the physiological and molecular response of plants to low oxygen stress, using Arabidopsis thaliana as a model plant. The investigation methods used are gene expression techniques by using q-PCR or transcriptomic analysis by microarray and molecular approaches (sense/antisense transgenesis).

Gustavo Striker (Doctor in Agricultural Sciences), an Argentinean national, holds a position as Assistant Professor of Plant Physiology at the University of Buenos Aires (UBA) and as Researcher at the Argentine National Council for Scientific Research. Gustavo obtained his degree of Agronomical Engineer, and later his Doctorate in Agricultural Sciences, both at UBA. Gustavo investigates waterlogging and submergence responses of forage (grasses and legumes) and crop species through ecophysiological approaches.

Editorial

Plant Responses to Hypoxia: Signaling and Adaptation

Elena Loreti [1],* and Gustavo G. Striker [2],*

[1]   Institute of Agricultural Biology and Biotechnology, CNR, National Research Council, Via Moruzzi, 56124 Pisa, Italy

[2]   IFEVA, CONICET, Cátedra de Fisiología Vegetal, Facultad de Agronomía, Universidad de Buenos Aires, Av. San Martín 4453, Buenos Aires C1417DSE, Argentina

*   Correspondence: elena.loreti@ibba.cnr.it (E.L.); striker@agro.uba.ar (G.G.S.)

Received: 25 November 2020; Accepted: 30 November 2020; Published: 3 December 2020

1. Introduction

Molecular oxygen deficiency leads to altered cellular metabolism and can dramatically reduce crop productivity. Nearly all crops are negatively affected by lack of oxygen (hypoxia) due to adverse environmental conditions such as excessive rain and soil waterlogging. Extensive efforts to fully understand how plants sense oxygen deficiency and their ability to respond using different strategies are crucial to increase hypoxia tolerance. It was estimated that 57% of crop losses are due to floods [1]. Progress in our understanding has been significant in the last years. This topic deserved more attention from the academic community; therefore, we have compiled a Special Issue including four reviews and thirteen research articles reflecting the advancements made thus far.

2. Advances in Hypoxia Sensing and Responses

Oryza sativa (rice) is an important crop widely used in areas prone to suffer waterlogging and submergence. This Special Issue has contributions that address essential aspects related to its tolerance to the lack of oxygen. Publications that address rice aspects include reviews on the molecular regulatory pathways and the metabolic adaptation in the seed germination and early seedling growth [2,3]. A first review examines in detail aspects about the coleoptile elongation under submergence, anaerobic gene regulation in rice coleoptile, chromosomal regions regulating coleoptile elongation under oxygen shortage, starch degradation during anaerobic germination, and the hormonal regulation of anaerobic rice coleoptile elongation [2]. A second review focuses on the recent advances underlying anaerobic germination and coleoptile elongation and highlights the prospect of introducing quantitative trait loci (QTL) for anaerobic germination into rice mega varieties [3]. In addition, interesting experimental information—indicating that chlorophyll retention, content, low hydrogen peroxide accumulation, and catalase activity are related to better performance under submergence in seedlings of five japonica rice varieties—is also shown [4]. Conversely, another paper examined the potassium efflux and cytosol acidification as primary anoxia-induced events in wheat and rice seedlings and found that rice responses were more distinct and reversible upon reoxygenation when compared with sensitive wheat [5].

Root aeration is essential to withstand water excess scenarios such as waterlogging [6]. The formation of aerenchyma in roots is critical to enable the diffusive oxygen transport to reach the root tips. Additionally, this longitudinal transport of oxygen towards root tips can be constrained if there is an excessive radial loss of oxygen towards the rhizosphere (ROL). Hence, the presence of a barrier preventing ROL is a desirable trait for more efficient root aeration. In this regard, it is shown for rice that auxin-mediated signaling contributes to ethylene-dependent inducible aerenchyma formation in roots [7]. It was demonstrated that an auxin transport inhibitor stopped aerenchyma formation under oxygen-deficient conditions and reduced the expression of genes encoding ethylene biosynthesis enzymes [7]. Complementarily, another contribution assessed the formation of barrier to oxygen loss of four genotypes from two wild rice species (*Oriza glumaepatula* and *O. rufipogon*) and found that the three *O. glumaepatula* accessions formed a

ROL barrier constitutively, while the accession of *O. rufipogon* accession did not [8]. Therefore, these wild relatives' selected accessions might be crossed with elite commercial rice materials to incorporate or improve this root trait aiming at better root aeration when waterlogged [8].

The traits aiding to the recovery from submergence-induced hypoxia have been less examined and identified than those conferring tolerance during the stress period. A detailed study in the legume *Lotus japonicus* showed genotypic variation in the recovery ability (RGR) from short-term complete submergence and a trade-off between growth during vs. after the stress. In addition, an inverse relationship between growth after submergence and the shoot to root ratio (SR) was identified, where genotypes with low values of SR were able to maintain high stomatal conductance, a better leaf water status, and chlorophyll retention [9].

Among the consequences of flooded plants are the involvement of hormone and metabolic responses as well as enhanced production of reactive oxygen species (ROS). To achieve progress in the mechanisms underlying anoxia tolerance is crucial to develop a broader view considering interactions between different signaling pathways. An exciting overview of the hypoxia field's classical and recent findings is reported in this Special Issue [10]. The review summarizes various aspects of low oxygen stress: (i) hypoxia sensing; (ii) adaptation to hypoxia at the cellular level; (iii) environmental hypoxia; and finally (iv) developmental hypoxia representing a physiologically relevant condition for the functionality of specific plant tissues [10].

The molecular mechanism of oxygen perception has been revealed in plants where proteins belonging to group VII Ethylene Response Factors (EFR-VIIs) play a pivotal role in becoming stable under hypoxia and degraded by proteasome machinery under aerobic conditions [11,12]. A new contribution provides information about the function of the *Jatropha curcas* ERFVIIs and the consequent N-terminal modification that stabilized the protein under low oxygen availability [13]. It was shown that JcERFVII2 is an N-end rule regulated waterlogging-responsive transcription factor that modulates gene expression under different stress-responsive conditions, including low-oxygen, oxidative, and pathogen response [13].

The involvement of hormones during hypoxia stress demonstrated that in root apical meristem, crosstalk between hypoxia and JA signaling occurs [14]. The jasmonate synthesis is initially enhanced but later decreased probably due to lack of oxygen and a consequent energy crisis [14]. Previous research has shown that when low oxygen occurs, ethylene signal drives hypoxia responses and improves survival in Arabidopsis [15,16]. A new paper showed that the hypoxia response triggered by ethylene is conserved in *Solanum* species, and, as it occurs in *Arabidopsis*, it enhances hypoxia tolerance [17]. One of ethylene signaling effects is hydrogen peroxide (H_2O_2) production that acts as a second messenger. To clarify the relationship between ethylene and H_2O_2 under low oxygen availability the rbohD/ein2-5 double mutant plants was analyzed under hypoxic stress [18]. The results demonstrated a synergistic interaction between ethylene and H_2O_2 signaling in modulating seed germination, seedling root growth, leaf chlorophyll content, and hypoxia-inducible gene expression [18].

In *Medicago truncatula*, the metabolic response differs between shoots and roots tissue [19]. Analyzing the composition of phloem exudate sap, they demonstrated that roots and leaves have distinct metabolic responses. Overall, the metabolomic data suggest that the decrease in sugar import in waterlogged roots increases the phloem sugar pool, which exerts a negative feedback regulation on sugar metabolism in shoots tissue [19].

In recent years, a group of non-coding RNA molecules, microRNAs (miRNAs), were identified. They play a pivotal role in different cellular processes. Among them, it was proposed to have a role in response to environmental stresses triggering the repression of target genes [20]. Taking advantage of high-throughput small RNA sequencing, a group of micoRNAs was identified in maize and teosinte under two crucial environmental stresses, submergence and drought and alternated stresses. Therefore, the identified miRNAs are a good starting point to establish the roles of miRNAs in stress response and could be useful for improving stress tolerance [21].

Finally, some practical aspects are also approached by articles on this Special Issue for other agricultural species. For instance, in *Physalis peruviana* (cape gooseberry), it was reported that foliar glycine betaine or hydrogen peroxide sprays ameliorate waterlogging stress, and that these can be potentially used as tools in managing waterlogging in this horticultural Andean fruit crop species [22]. Additionally, in *Pisum sativum* (pea) it was proved that the application of a hypoxic treatment decreases the physiological action of the herbicide imazamox due to an amelioration of the effects on total soluble sugars, starch accumulation, and changes in some amino acids [23]. This allows the authors to suggest that fermentation might constitute a plant defense mechanism that decreases the herbicidal effect [23]. Finally, a complete revision of the state of the art is provided regarding the findings that explain the traits conferring tolerance to root hypoxia in woody fruit species [24]. Special attention is given to the strategies for managing the energy crisis in *Prunus* species, and less explored topics in recovery and stress memory in woody fruit trees are pointed out.

3. Future Perspectives

The studies briefly summarized above provide an advance in our understanding of the molecular basis, the ecophysiological traits, and some of the genetic diversity of the model species *Arabidopsis thaliana*, *Lotus japonicus*, and *Medicago truncatula*. They also increase our understanding of some agricultural species (rice and its wild relatives, wheat, cape gooseberry, alfalfa, *Prunus* spp.) in response to waterlogging and submergence. These advances could provide opportunities to breed crops tolerant to oxygen deficiency. Despite the advances in knowledge gained over the last years, some challenges need to be addressed: (i) the study of genes indicative of ethylene-mediated hypoxia acclimation must be deepened, exploring the universality of discovered mechanisms of tolerance across species; (ii) physiological traits associated with plant recovery from submergence and their regulation must be identified; and (iii) trait-to-gene-to-field approaches (i.e., "translational research"), warranting the development of strategies to cope with oxygen deficiency stress aiming to stabilize crop yields, must be promoted to resolve food insecurity in the future. The coordinated effort of research groups working at the different organization levels (e.g., molecular, plant and field) will increase the chances of success of the "translational research" as the main goal, implying the translation of basic scientific discovery into improved agricultural productivity.

Funding: This research received no external funding.

Acknowledgments: We would like to thank Sylvia Guo for the guidance and support throughout the entire process of this Special Issue. We also would like to thank the numerous reviewers and authors who contributed to this challenge with their science and expertise.

Conflicts of Interest: The authors declare no conflict of interest.

References

1.	FAO. Damage and Losses from Climate-Related Disasters in Agricultural Sectors; Food and Agricultural Organization of United States. Retrieved from FAO, 2016 I6486EN/1/11.16. Available online: http://www.fao.org/3/a-i6486e.pdf (accessed on 2 December 2020).
2.	Pucciariello, C. Molecular mechanisms supporting rice germination and coleoptile elongation under low oxygen. *Plants* **2020**, *9*, 1037. [CrossRef]
3.	Ma, M.; Cen, W.; Li, R.; Wang, S.; Luo, J. The molecular regulatory pathways and metabolic adaptation in the seed germination and early seedling growth of rice in response to low O_2 stress. *Plants* **2020**, *9*, 1363. [CrossRef] [PubMed]
4.	Li, Y.-S.; Ou, S.-L.; Yang, C.-Y. The seedlings of different japonica rice varieties exhibit differ physiological properties to modulate plant survival rates under submergence stress. *Plants* **2020**, *9*, 982. [CrossRef] [PubMed]
5.	Yemelyanov, V.V.; Chirkova, T.V.; Shishova, M.F.; Lindberg, S.M. Potassium efflux and cytosol acidification as primary anoxia-induced events in wheat and rice seedlings. *Plants* **2020**, *9*, 1216. [CrossRef] [PubMed]

6. Yamauchi, T.; Colmer, T.D.; Pedersen, O.; Nakazono, M. Regulation of root traits for internal aeration and tolerance to soil waterlogging-flooding stress. *Plant Physiol.* **2018**, *176*, 1118–1130. [CrossRef]

7. Yamauchi, T.; Tanaka, A.; Tsutsumi, N.; Inukai, Y.; Nakazono, M. A role for auxin in ethylene-dependent inducible aerenchyma formation in rice roots. *Plants* **2020**, *9*, 610. [CrossRef]

8. Ejiri, M.; Sawazaki, Y.; Shiono, K. Some accessions of amazonian wild rice (*Oryza glumaepatula*) constitutively form a barrier to radial oxygen loss along adventitious roots under aerated conditions. *Plants* **2020**, *9*, 880. [CrossRef]

9. Buraschi, F.B.; Mollard, F.P.O.; Grimoldi, A.A.; Striker, G.G. Eco-physiological traits related to recovery from complete submergence in the model legume *Lotus japonicus*. *Plants* **2020**, *9*, 538. [CrossRef]

10. Loreti, E.; Perata, P. The many facets of hypoxia in plants. *Plants* **2020**, *9*, 745. [CrossRef]

11. Licausi, F.; Kosmacz, M.; Weits, D.A.; Giuntoli, B.; Giorgi, F.M.; Voesenek, L.A.C.J.; Perata, P.; Van Dongen, J.T. Oxygen sensing in plants is mediated by an N-end rule pathway for protein destabilization. *Nature* **2011**, *479*, 419–422. [CrossRef]

12. Gibbs, D.J.; Lee, S.C.; Md Isa, N.; Gramuglia, S.; Fukao, T.; Bassel, G.W.; Correia, C.S.; Corbineau, F.; Theodoulou, F.L.; Bailey-Serres, J.; et al. Homeostatic response to hypoxia is regulated by the N-end rule pathway in plants. *Nature* **2011**, *479*, 415–418. [CrossRef] [PubMed]

13. Juntawong, P.; Butsayawarapat, P.; Songserm, P.; Pimjan, R.; Vuttipongchaikij, S. Overexpression of *Jatropha curcas* ERFVII2 transcription factor confers low oxygen tolerance in transgenic *Arabidopsis* by modulating expression of metabolic enzymes and multiple stress-responsive genes. *Plants* **2020**, *9*, 1068. [CrossRef] [PubMed]

14. Shukla, V.; Lombardi, L.; Pencik, A.; Novak, O.; Weits, D.A.; Loreti, E.; Perata, P.; Giuntoli, B.; Licausi, F. Jasmonate signalling contributes to primary root inhibition upon oxygen deficiency in *Arabidopsis thaliana*. *Plants* **2020**, *9*, 1046. [CrossRef] [PubMed]

15. Voesenek, L.A.C.J.; Sasidharan, R. Ethylene–and oxygen signalling–drive plant survival during flooding. *Plant Biol.* **2013**, *15*, 426–435. [CrossRef]

16. Sasidharan, R.; Voesenek, L.A.C.J. Ethylene-mediated acclimations to flooding stress. *Plant Physiol.* **2018**, *169*, 3–12. [CrossRef]

17. Hartman, S.; van Dongen, N.; Renneberg, D.M.; Welschen-Evertman, R.A.; Kociemba, J.; Sasidharan, R.; Voesenek, L.A.C.J. Ethylene differentially modulates hypoxia responses and tolerance across *Solanum* species. *Plants* **2020**, *9*, 1022. [CrossRef]

18. Hong, C.P.; Wang, M.C.; Yang, C.Y. NADPH oxidase RbohD and ethylene signaling are involved in modulating seedling growth and survival under submergence stress. *Plants* **2020**, *9*, 471. [CrossRef]

19. Lothier, J.; Diab, H.; Cukier, C.; Limami, A.M.; Tcherkez, G. Metabolic responses to waterlogging differ between roots and shoots and reflect phloem transport alteration in *Medicago truncatula*. *Plants* **2020**, *9*, 1373. [CrossRef]

20. Song, X.; Li, Y.; Cao, X.; Qi, Y. MicroRNAs and their regulatory roles in plant–environment interactions. *Ann. Rev. Plant Biol.* **2019**, *70*, 489–525. [CrossRef]

21. Sepúlveda-García, E.B.; Pulido-Barajas, J.F.; Huerta-Heredia, A.A.; Peña-Castro, J.M.; Liu, R.; Barrera-Figueroa, B.E. Differential expression of maize and teosinte microRNAs under submergence, drought, and alternated stress. *Plants* **2020**, *9*, 1367. [CrossRef]

22. Castro-Duque, N.E.; Chávez-Arias, C.C.; Restrepo-Díaz, H. Foliar glycine betaine or hydrogen peroxide sprays ameliorate waterlogging stress in cape gooseberry. *Plants* **2020**, *9*, 644. [CrossRef] [PubMed]

23. Gil-Monreal, M.; Royuela, M.; Zabalza, A. Hypoxic treatment decreases the physiological action of the herbicide IMAZAMOX on *Pisum sativum* roots. *Plants* **2020**, *9*, 981. [CrossRef] [PubMed]

24. Salvatierra, A.; Toro, G.; Mateluna, P.; Opazo, I.; Ortiz, M.; Pimentel, P. Keep calm and survive: Adaptation strategies to energy crisis in fruit trees under root hypoxia. *Plants* **2020**, *9*, 1108. [CrossRef] [PubMed]

Publisher's Note: MDPI stays neutral with regard to jurisdictional claims in published maps and institutional affiliations.

Review

The Many Facets of Hypoxia in Plants

Elena Loreti [1,*] **and Pierdomenico Perata** [2,*]

1   Institute of Agricultural Biology and Biotechnology, CNR, National Research Council, Via Moruzzi, 56124 Pisa, Italy
2   PlantLab, Institute of Life Sciences, Scuola Superiore Sant'Anna, Via Giudiccioni 10, 56010 San Giuliano Terme, 56124 Pisa, Italy
*   Correspondence: loreti@ibba.cnr.it (E.L.); p.perata@santannapisa.it (P.P.)

Received: 5 June 2020; Accepted: 12 June 2020; Published: 12 June 2020

Abstract: Plants are aerobic organisms that require oxygen for their respiration. Hypoxia arises due to the insufficient availability of oxygen, and is sensed by plants, which adapt their growth and metabolism accordingly. Plant hypoxia can occur as a result of excessive rain and soil waterlogging, thus constraining plant growth. Increasing research on hypoxia has led to the discovery of the mechanisms that enable rice to be productive even when partly submerged. The identification of Ethylene Response Factors (ERFs) as the transcription factors that enable rice to survive submergence has paved the way to the discovery of oxygen sensing in plants. This, in turn has extended the study of hypoxia to plant development and plant–microbe interaction. In this review, we highlight the many facets of plant hypoxia, encompassing stress physiology, developmental biology and plant pathology.

Keywords: anaerobiosis; anoxia; Arabidopsis; flooding; hypoxia; rice; submergence; waterlogging; development

---

1. Hypoxia and Its Sensing

Oxygen availability is a pre-requisite for life in several living organisms. In plants, when the oxygen supply is insufficient, most cellular functions are compromised, which can lead to death [1]. In order to keep the level of oxygen under control, cells sense the oxygen levels and react to insufficient oxygen (hypoxia) by adopting survival strategies ranging from gene regulation to morphological adaptive responses [2].

Hypoxia occurs when the oxygen level limits aerobic respiration (usually between 1% and 5%), while anoxia takes place when oxygen is absent in the environment [3]. It is also important to highlight the difference between acute and chronic hypoxia [4]. The term acute hypoxia is used when the drop in oxygen availability is transitory in plants, which can be due to adverse environmental conditions such as flooding events or when unusual, transient increases in oxygen consumption occur in a plant tissue. Acute hypoxia is perceived by the plants as stressful. On the other hand, chronic hypoxia is a constitutive and, usually, non-stressful condition where the oxygen level is maintained low only in a given group of cells and not in the entire plant. In other words, chronic hypoxia can be a physiological condition in specific plant tissues.

The molecular mechanism of oxygen perception has been revealed both in animals and plants. In mammals, the Hypoxia Inducible Factor (HIF-1) is responsible for low-oxygen sensing, while this molecular activity is controlled by the Cys branch of the N-degron pathway in plants [5,6].

HIF-1 is formed by two subunits: HIF-1β is constitutively expressed, while HIF-1α is regulated by oxygen through post-translational modifications [7–9]. During normoxia, HIF-1α is degraded via the ubiquitin–proteasome pathway, whereas during hypoxia, it is stabilized and protected from proteolysis, allowing its migration into the nucleus [10]. Here, the complex HIF1α and HIF-1β is reconstituted [7], which induces the transcription of hypoxic genes.

The oxygen-sensing mechanism in plants (Figure 1) was revealed several years later after the discovery of HIF-1 [5,11,12]. In plants, proteins belonging to group VII of the Ethylene Response Factors (ERF-VIIs) are destabilized by oxygen and become stable only under hypoxia, with a mechanism conceptually similar to the one that regulating HIF-1. ERF-VIIs are transcription factors characterized by a Cys residue at their N-terminal [13]. Under normoxia, the Cys residue is constitutively oxidized by a class of plant enzymes named Plant Cysteine Oxidases (PCOs), which target the ERF-VII proteins to degradation via the proteasome, following the Cys branch of the N-degron pathway [14,15]. Under hypoxia, on the other hand, ERF-VIIs are stable, given that the absence of oxygen prevents the oxidization of the Cys residue by PCOs. The stabilized ERF-VIIs can translocate to the nucleus where they activate the transcription of anaerobic genes by binding to the Hypoxia-Responsive Promotor Element (HRPE) present in the promoter of anaerobic genes [16].

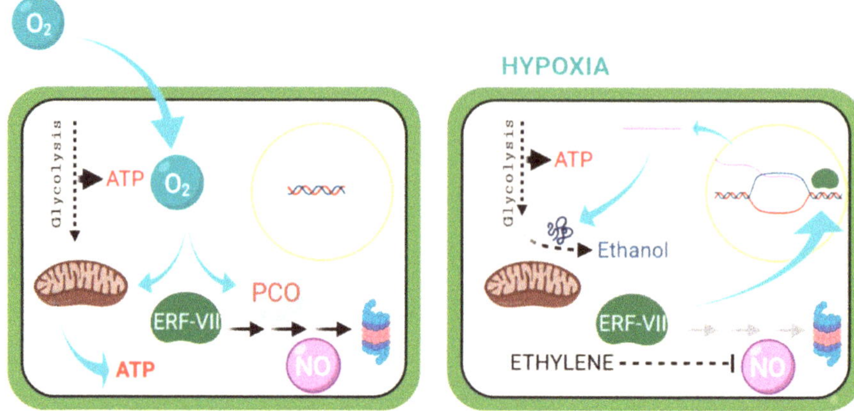

Figure 1. How plants sense oxygen. Under aerobic conditions (left), aerobic respiration in the mitochondria provides most of the energy (ATP) required for the cell metabolism. The ERF-VII transcription factor genes are constitutively expressed, but their stability is compromised by the activity of PCOs, which, in a process requiring oxygen, oxidize the N-terminal Cys residue, channeling the ERF-VII proteins to the proteasome, in a process also requiring nitric oxide (NO). Under hypoxia (right), the respiration in the mitochondria is drastically reduced, and ATP production can only occur because of enhanced glycolytic activity. The ERF-VII proteins are stabilized because of the absence of oxygen and also thanks to ethylene production, which dampens the presence of NO in the cell. The stable ERF-VII proteins migrate to the nucleus where they activate the transcription of Hypoxia-Responsive Genes (HRGs), including genes encoding proteins required for alcoholic fermentation. This figure was created using BioRender [17].

Once oxygen deficiency has been perceived, plants change their metabolism, which is also achieved by a modulation in gene expression [18]. In Arabidopsis, 49 anaerobic core hypoxic genes are induced in all plant organs during a hypoxic event [18].

Although oxygen-sensing mechanisms in mammals and plants are different, an overlapping sensing machinery between plants and the animal kingdom was recently discovered [19]. Mammals possess a cysteamine dioxygenase enzyme, named ADO, which is similar to the PCOs of plants and operates similarly on proteins possessing a Cys residue at the N-terminus [19]. With their differences and similarities [12], oxygen-sensing mechanisms in animals and plants represent the first steps in the cascade of events following hypoxia.

2. Adaptation to Hypoxia at the Cellular Level

Aerobic organisms have developed a variety of adaptive responses to hypoxia at the cellular, tissue and organism levels [20]. When the oxygen supply is adequate, the mitochondria produce enough ATP for survival, while under hypoxia alcoholic fermentation replaces mitochondrial respiration [21]. The aim of the fermentative metabolism is to allow ATP production through the glycolytic pathway by recycling NAD+ via the action of two key enzymes, pyruvate decarboxylase (PDC) and alcohol dehydrogenase (ADH) [21]. These two enzymes belong to the class of anaerobic polypeptides (ANPs), which are induced and produced under hypoxia [21]. These are produced through the action of ERF-VIIs, which activate the transcription of the Hypoxia-Responsive Genes (HRG) that encode the ANPs.

In addition to those involved in the fermentative pathway, ANPs also include proteins linked to aerenchyma formation, cytoplasmatic pH and carbohydrate metabolism [2,22]. Carbohydrate degradation through glycolysis coupled with the fermentative pathway leads to 2 moles of ATP instead of the 36 normally produced during aerobic respiration [21]. Although limited, ATP production through fermentation is important for hypoxia tolerance [1]. Moreover, mutants defective in alcoholic fermentation are intolerant to hypoxia, demonstrating that this pathway contributes significantly to hypoxia tolerance [23]. In order to be efficient at ATP production, glycolysis coupled to fermentation under low oxygen availability requires an adequate glucose supply [1,24]. In this context, the role of starch as a source of sugars to be used during prolonged hypoxia has been demonstrated, both in cereals and in Arabidopsis [1,24]. Of the cereals, only rice is able to germinate under anoxia due to its ability to exploit the starchy reserves present in the caryopses, a consequence of the successful induction of α-amylases even in the absence of oxygen [25,26].

These key enzymes, which are required for starch degradation, are not produced in other cereals, which results in their inability to germinate under hypoxia [27,28]. Although Arabidopsis tolerance to hypoxia depends on distinct mechanisms that enable rice seeds to germinate, starch is also an essential component of the anaerobic response in this species. Arabidopsis adult plants require starch for their tolerance to submergence and, interestingly, a relation between sugar starvation and the plant's ability to induce HRG transcription has been observed [29]. Low-oxygen conditions weaken the anaerobic response at the transcriptional level, indicating the existence of a homeostatic mechanism linking oxygen sensing with sugar sensing, which controls the intensity of the induction of HRGs, so that this matches the available carbon resources [29].

3. Environmental Hypoxia

Hypoxia can be caused by specific environmental conditions, which we refer to as "environmental hypoxia". In humans, a lack of oxygen can be due to environmental hypoxia at high altitudes, where the partial oxygen pressure (pO_2) is decreased in proportion to the lower ambient pressure [30]. Besides environmental hypoxia, pathological conditions such as chronic obstructive pulmonary disease [31], obstructive sleep apnea [32] and anemia [33] can also lead to hypoxia. Similarly, during intense exercise, hypoxia can also be generated by the intense respiratory metabolism [34].

In plants, hypoxia often originates during flooding, leading to waterlogging or submergence [1] (Figure 2). Waterlogging is the saturation of soil with water and occurs when roots cannot respire due to a water excess, whereas the term submergence means that, in addition to the roots, the aerial part of the plant is under water [3]. In both cases, the final effect is a lack of oxygen.

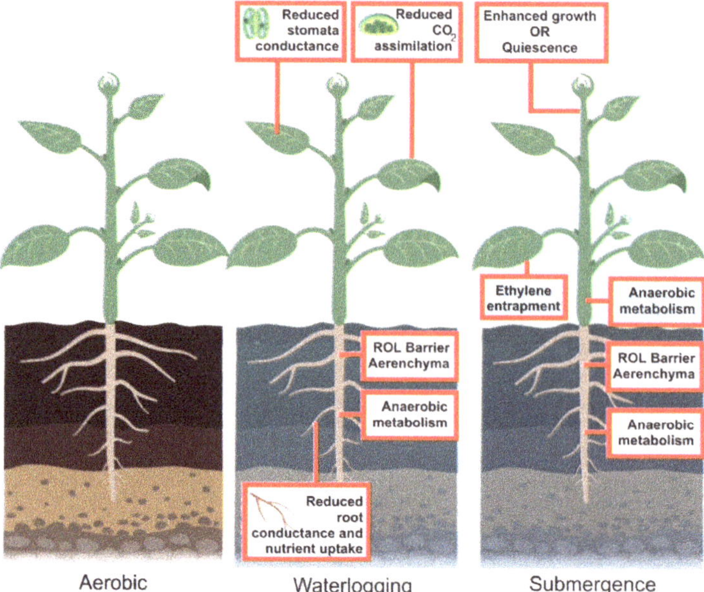

Figure 2. Environmental hypoxia is generated by soil waterlogging or by plant submergence. An excess of water in the soil reduces aerobic respiration, which is replaced by anaerobic metabolism. The uptake of nutrients is reduced. A radial oxygen loss (ROL) barrier and aerenchyma eventually develops in the root system to provide oxygen to the submerged roots. The aerial part of the plant, although not submerged, is also affected by waterlogging, with reduced stomata conductance and CO_2 assimilation. Complete or partial plant submergence may trigger quiescence or escape strategies, characterized by reduced or enhanced elongation, respectively. Ethylene entrapment in the submerged plant plays an important role in defining the overall plant response to submergence. Aerenchyma develops and the metabolism is mostly anaerobic, with the exception of tissues and organs that receive sufficient oxygen through the aerenchyma. This figure was created using BioRender [17].

Table 1. Traits affected by waterlogging and submergence in plants.

Trait	Function	Plant Species	References
Aerenchyma	Improvement in internal gas diffusion	*Zea mays; Oryza sativa; Pisum sativum; Triticum aestivum; Arabidopsis thaliana*	[2,35–40]
Hypertrophic lenticels	Facilitating O_2 diffusion; venting ethylene and CO_2	Woody plant species	[41]
Radial oxygen loss barrier	Barrier impermeable to radial O_2 loss	*Oryza sativa; Phragmites australis; Phalaris aquatica*	[37,41,42]
Increased specific leaf area (indicating a large surface area relative to mass)	CO_2 enters the mesophyll cells via diffusion through the epidermis and not via stomata	*Rumex palustris* and other amphibious species	[2]
Petiole elongation	Reaching water surface	*Rumex palustris*	[2,43]
Reorientation of petioles in upright position	Reaching water surface	*Rumex palustris*	[2,43]
Coleoptile elongation	Reaching water surface	*Oryza sativa*	[26,44,45]

Table 1. *Cont.*

Trait	Function	Plant Species	References
Fast stem elongation	Reaching water surface	*Oryza sativa* (deep water rice)	[46,47]
Inhibition of stem elongation	Reducing growth-associated costs (quiescence strategy)	*Oryza sativa*	[48,49]
Root architecture	Minimize the distance between the aerial surface and the flooded root tips	*Oryza sativa; Zea mais; Triticum aestivum*	[1,38]
Adventitious roots production	Replace primary root systems; roots at surface of water; enhance supply of water and minerals	*Zea mais; Solanum lycopersicon; Solanum dulcamara*	[50,51]

How plants survive flooding has been studied for decades and, besides the molecular responses described above, survival also entails morpho-physiological modifications such as aerenchyma development, formation of barriers to radial oxygen loss, production of adventitious roots, elongation of stems, and leaf petioles (Table 1).

Overall, plants respond to environmental hypoxia through different strategies, aimed at prolonging their life even under these unfavorable conditions [2]. Some plant species respond to submergence by attempting to escape from the low-oxygen environment by elongating their stems, petioles or leaves so that at least part of the plant is above water. This facilitates the transport of oxygen to the submerged organs, by means of the aerenchyma, whose development is usually enhanced by submergence [2]. The production of ethylene, entrapped by the water surrounding the submerged plant, plays an important role and regulates several aspects of the enhanced growth, enabling the plant to escape submergence [52]. Plant species that adopt the escape strategy include wild species, such as *Rumex palustris*, and crop plants, such as some rice varieties [2].

The escape strategy is energetically costly. It is only advantageous if the rate of elongation is fast enough to keep part of the plant above the water surface. If, instead, growth is enhanced but is insufficient for the plant to reach the surface of the submerging water, then this strategy is detrimental to survival, since the plant will use most of its reserve carbohydrates to fuel growth and may be starved of carbon when the water recedes and normal plant physiology should be re-established. Some plant species are very well adapted to prevent the starvation syndrome arising from submergence, and, instead of enhancing growth, they adopt a quiescence strategy based on very limited growth when the plant is submerged.

Rice is a wetland species that has evolved in very distinct environments, very often characterized by phases of growth suffering from flooding events. Some rice species (known as deep-water rice or floating rice), have adapted well to environments where submergence is prolonged and deep [46,53]. Deep-water rice survives due to a set of genes, named *Snorkel*, which are ethylene-responsive and trigger a very fast elongation response [53]. Other flooding-resistant rice varieties, also known as "Scuba-rice", can tolerate short-term complete submergence by adopting a quiescence strategy made possible by the *Sub1A* gene. *Sub1A* is ethylene-responsive but, unlike the *Snorkel* genes, triggers reduced elongation, thus allowing the plant to preserve its carbohydrate reserves, which are required to ensure vigorous re-growth when the water recedes [48,49].

The regulation of gene expression is thus important during environmental hypoxia because it controls the induction of genes that are required for tolerance. The cellular events previously described occur during environmental hypoxia in all those tissues which, because of submergence, do not receive an adequate oxygen supply, but not in tissues that become aerated because of oxygen transport via the aerenchyma to the submerged organs.

4. Developmental Hypoxia

In humans, hypoxia can arise as a consequence of impaired blood flow. This can significantly damage organ structure and function, resulting, for example, in a stroke (cerebral ischemia) or heart infarction (myocardial ischemia). Hypoxia also regulates tumor growth and metastasis [54].

Although, in plants, hypoxia traditionally refers to the stress due to flooding events, it can also occur in tissues and organs under normal oxygen availability. In this case, hypoxia represents a physiological status that is experienced by the plant throughout its life or during a specific stage of development. Oxygen diffuses from the air (21% [*v/v*] oxygen) into the plant body across apertures in the epidermis and through intracellular air spaces within a tissue [36]; however, unlike in animals, it is still unknown if plants have an efficient mechanism that distributes oxygen to cells. As a consequence, plants have to deal with oxygen gradients in tissues and organs [4].

The anatomy of specific tissues or organs such as in seeds, fruits and roots may limit oxygen diffusion, leading to hypoxia [55–57]. In bulky organs, the oxygen concentration can be very low in the inner part of the organ [58], thus limiting oxygen for respiration. For example, oxygen concentration in the center of growing potato tubers can decrease to around 5%, thus making this tissue hypoxic [59].

With the notable exception of bulky fruits and organs, most plant organs possess a relatively high surface-to-volume ratio that should allow oxygen diffusion, thus preventing the establishment of hypoxia [55]. Even tissues in which oxygen diffusion is not problematic may become hypoxic when there is high metabolic activity, especially if there are no intercellular air spaces. High local rates of oxygen consumption may take place in tissues such as the phloem, which has been shown to be hypoxic [55]. Furthermore, as discussed by van Dongen et al. [55], oxygen access might be restricted because phloem is a dense tissue with few intercellular spaces. Measurements of oxygen profiles across stems of *Ricinus communis* plants showed that oxygen levels can be as low as approximately 7% (*v/v*) in the vascular regions, even when plants are grown under fully aerobic conditions [55].

Interestingly, alcohol dehydrogenase activities, as well as the presence of ethanol, have been observed in the vascular cambium of trees [60], supporting the existence of hypoxic tissues near the phloem. Physiological hypoxia usually occurs in heterotrophic tissues, given that autotrophic tissue produces photosynthetic oxygen and can therefore more easily prevent the occurrence of hypoxia. During germination, reduced oxygen availability in soil plays a positive role in the survival and establishment of seedlings following darkness, through the enhanced stability of ERF-VIIs, which enhance dark-activated development and repress light-activated development [61].

Environment-independent hypoxia can be limited to a small cluster of cells or to entire organs. In animals, the formation of a hypoxic microenvironment occurs when the cells need to be kept in an undifferentiated state, such as in stem cells [62,63]

Interest in hypoxic microenvironments is also growing in plant sciences. Pollen can be hypoxic [64], and, in fact, given that hypoxia arises naturally within growing anther tissue, it acts as a positional cue to establish germ cell fate [65].

The concept of hypoxia as a developmental, positional cue was recently strengthened (Figure 3), and hypoxic niches were shown to be present in shoot apical meristems (SAM; [66]). Using a micro-scale oxygen electrode, Weits et al. [66] measured the oxygen concentration inside the shoot apical meristems of Arabidopsis (which are very small) and that of *Solanum lycopersicum* (which are larger). In both cases, the oxygen concentration inside the meristem was below 5%, highlighting that hypoxia niches are a conserved occurrence in plants. Interestingly, Weits et al. [66] demonstrated that low oxygen in the SAM is not only a non-stressful condition, but is actually required for the production of new leaves [66]. This is because, in SAM-related proteins, a novel substrate of the Cys branch of the N-degron pathway, namely, LITTLE ZIPPER 2 (ZPR2), due to its hypoxia-dependent stabilization, drives leaf organogenesis in a hypoxia-dependent manner by interacting with its target class III homeodomain leucine zipper (HD-ZIP III) proteins [66]. The hypoxic niche at the shoot meristem not only promotes the stability of ZPR2, but also of VERNALIZATION 2 (VRN2; [67]), which requires low oxygen to be stable and thus regulates reproductive and vegetative development by acting on its target FLOWERING LOCUS C (FLC) [67].

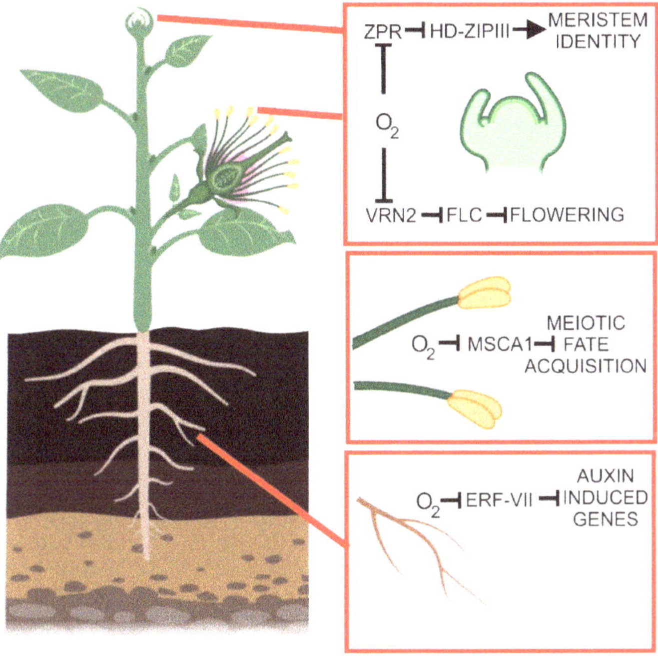

Figure 3. Hypoxic niches affecting developmental processes. In the roots, hypoxia in the lateral root primordia affects auxin-regulated genes through the action of ERF-VII proteins. In the anthers, hypoxia represents a developmental signal influencing meiotic fate acquisition (MSCA: male sterile converted anther 1). In the shoot apical meristem, a hypoxic niche controls the stability of ZPR2 and VRN2, affecting meristem identity and flowering, respectively. This figure was created using BioRender [17].

Hypoxia in the SAM triggers the stabilization of the ERF-VII transcription factors, which drive the expression of the core hypoxic genes, including those involved in the fermentative pathway such as ALCOHOL DEHYDROGENASE (ADH) and PYRUVATE DECARBOXYLASE (PDC). These genes are induced not only in SAM but also in the Lateral Root Primordia (LRP) of Arabidopsis, suggesting that also LRP are hypoxic [68]. Low oxygen influences root architecture, such as the formation of adventitious roots [69], the promotion of root bending and the inhibition of primary roots [70]. Physiological hypoxia occurs during LR development irrespectively of the oxygen concentration in the environment [68]. The consequence of the formation of hypoxic niches during lateral root primordia development is the stabilization of ERF-VII transcription factors, which induce the set of hypoxic genes and, at the same time, the downregulation of crucial auxin-induced genes that promote lateral root production [68].

Roots are, in fact, the organs most affected by limited oxygen supply when the plants are waterlogged or submerged. Hypoxia causes the primary roots to grow sideways, possibly to escape soil patches with reduced oxygen availability via ERF-VII activity and mediated by auxin signaling in the root tips [70].

It is still unclear how hypoxic conditions are established in the SAM and in root primordia. While the slow oxygen diffusion rate through plant tissues explains why bulky plant organs are hypoxic, it is less obvious why oxygen levels drop to very low levels in plant meristems. One hypothesis is the possibly very active respiratory metabolism in the SAM, which makes the rate of oxygen diffusion insufficient to cope with oxygen consumption in the cells. Alternatively, the presence, yet to be demonstrated, of barriers to oxygen diffusion may keep the SAM hypoxic.

5. Hypoxia in Plant–Microbe Interactions

In legumes, the capacity to establish a symbiotic relationship with endosymbiotic dinitrogen-fixing rhizobia requires the establishment of hypoxic conditions, given that the activity of the nitrogenase enzyme, which can fix N_2, is inactivated by free O_2. Nodules are able to maintain an internal low-O_2 environment, including a nodule O_2 diffusion barrier, and by expressing O_2-carrying symbiotic plant hemoglobins (reviewed by Pucciariello et al. [71]).

In other plant species, hypoxia-responsive genes have been reported to be induced in roots infected by *Plasmodiophora brassicae* [72] or in stems infected by *Agrobacterium tumefaciens* [73], suggesting that hypoxic conditions are generated at the sites of infection by these bacteria. Slower O_2 diffusion rates in tumorigenic tissues, such as clubroot, and in crown galls, may be responsible for the establishment of hypoxia in these conditions [74]. Alternatively, intracellular competition for O_2 between the host and the pathogen can reduce the oxygen available to the plant and lead to hypoxia [72]. In the case of crown galls, the increased respiration rate in the plant cells, which is required to fuel rapid cell proliferation rates, may be the cause of the hypoxic conditions [73].

The evidence of a possible role of ERF-VII proteins, which require hypoxia to be stable, in pathogen resistance has raised the question as to whether hypoxia is established during pathogen infection [72,75–77]. Low-oxygen conditions may be established during pathogen infection, as described above, or by modulation of nitric oxide (NO) levels, as NO is another ERF-VII de-stabilizing molecule [72,78]. Given that ethylene is often involved in plant–pathogen interaction, the NO-scavenging activity of the ethylene-induced PHYTOGLOBIN1 (PGB1) may also play a role which has not yet been studied [79]. In fact, hypoxia is present in leaves that are infected with *Botrytis cinerea* [80]. The hypoxic conditions are established locally, at the exact sites of *B. cinerea* infection, where an oxygen concentration of less than 1% has been measured. Given that leaves are usually fully aerobic tissues, also due to their ability to produce oxygen through photosynthesis, the establishment of hypoxia suggests that competition of oxygen at the sites of infection may be very strong. Although increased oxygen consumption was actually measured at the sites of *B. cinerea* infection, it is still not clear whether it is fungal or plant respiration that is enhanced to a level that leads to the rapid depletion of oxygen availability [80].

In addition to *B. cinerea*, infection by *Alternaria brassicicola*, but not *Pseudomonas syringae*, also leads to the establishment of hypoxia in leaves. Besides playing a role in enhanced respiration in the infected tissue, hypoxia arising from infection by necrotrophic fungi can also be the consequence of tissue waterlogging in the necrotic area [80]. An important question is thus whether hypoxia contributes to plant tolerance to *B. cinerea* infection or is actually an advantage for the fungus. Given that ERF-VII proteins can contribute to *B. cinerea* resistance in Arabidopsis [75] and that these proteins require hypoxia to be stable and active, it is tempting to speculate that the low-oxygen environment that is established at the site of infection is advantageous for the plant. The stability of RAP2.3 (an ERF-VII) makes it possible to interact with OCTADECANOID-RESPONSIVE ARABIDOPSIS 59 (ORA59) [76], thus enhancing tolerance to the fungus.

The plant's response to fungal infection includes the production of elicitor molecules that are aimed at boosting the plant's response to the pathogen. Oligogalacturonides (OGs) are important elicitors in the response to *B. cinerea* infection. Interestingly, the fate of OGs as elicitors might depend on the establishment of local hypoxia at the site of *B. cinerea* infection [80]. OGs can be oxidized by Arabidopsis berberine bridge enzyme-like proteins (OXOGs) [81]. Oxidized OGs are less able to trigger the immune responses in the plant and are less hydrolysable by fungal polygalacturonases [81]. Hypoxia could therefore be beneficial for the fungus because it prevents the oxidation of OGs that would otherwise be oxidized and are no longer metabolically useful for the fungus. On the other hand, the plant benefits from hypoxia, which prevents oxidation of OGs making them less effective as elicitors.

The overexpression of OXOGs results in enhanced resistance to *B. cinerea* [81], suggesting that the fungus takes advantage of OGs as a metabolic source of carbon. Establishing whether it is the fungus or the plant that benefits from hypoxia at the site of infection thus remains an open question (Figure 4).

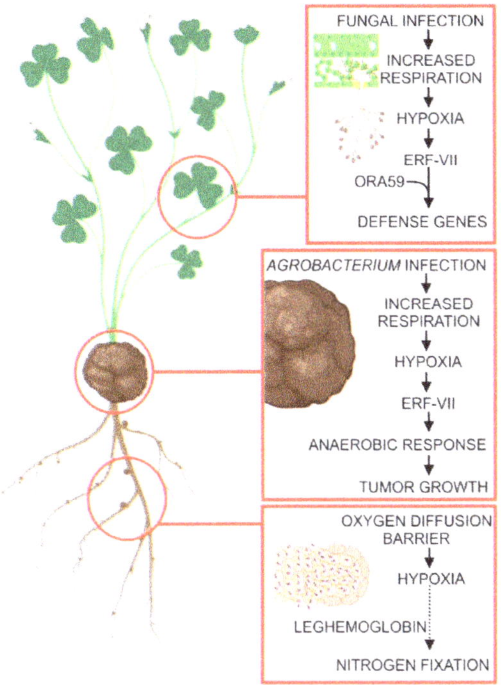

Figure 4. Local hypoxia establishment during plant–microbe interactions. Leaf infection by necrotrophic fungi results in enhanced respiration, which eventually leads to hypoxia. Under these conditions, ERF-VII proteins are stable and, through interaction with ORA59, they influence the efficacy of a plant's response to the pathogen. Similarly, agrobacterium infection enhances respiration in the infected tissue, which leads to hypoxia. Stable ERF-VII activates the anaerobic response that favors tumor growth. In legumes, rhizobia requires hypoxic conditions, given that the activity of nitrogenase enzyme, which can fix N_2, is inactivated by oxygen. Nodules maintain an inner low-O_2 environment thanks to the presence of a nodule O_2 diffusion barrier and by expressing O_2-carrying symbiotic plant hemoglobins (leghemoglobin). This figure was created using BioRender [17].

6. Concluding Remarks

The hypoxia field of study in animal physiology is quite broad, ranging from sensing at the cellular level to its role in pathogenesis. In plants, for decades, research on hypoxia has been mostly restricted to how plants respond to adverse environmental conditions such as excessive rain, resulting in soil waterlogging and plant submergence. Although hypoxia is of great relevance for agriculture, also due to the challenges of climate change, plant hypoxia has been erroneously considered a minor abiotic stress compared to, for example, drought.

Understanding the mechanisms enabling rice plants to survive submergence, involving ERF-VII transcription factors, has paved the way to the discovery of oxygen sensing in plants. Several papers in this research area have recently extended hypoxia research to other fields, including developmental biology and plant pathology.

During plant–microbe interactions, hypoxia is required for generating an environment that is compatible with oxygen-sensitive enzymatic activities, such as during nitrogen fixation in leguminous plants. On the other hand, hypoxia at the site of infection of necrotrophic pathogens may be required to activate oxygen-sensitive plant resistance pathways, or it may alter the production and stability of elicitors in a hypoxia-dependent manner.

The identification of new targets of the oxygen-dependent N-degron pathway unequivocally demonstrates that the absence of oxygen has profound effects on the physiology and developmental biology of the plant, well beyond the metabolic adaptive response.

The existence of hypoxic niches in an otherwise aerobic plant body raises a number of issues. Firstly, we need to know how hypoxia is achieved in hypoxic niches. Secondly, it is of importance to assess how many enzymatic activities are affected in hypoxic niches, including those involved in the synthesis of plant hormones, with consequences for the hormone concentration gradients. Thirdly, we need to assess the energy status of hypoxic niches and the impact of energy sensing on developmental processes. How is hypoxia achieved in hypoxic niches? The limited diffusion of oxygen though tissues and intense respiratory activity are good candidates for explaining developmental hypoxia; however, experimental evidence is missing. The existence of a gene family of PCOs with different affinities for oxygen and, possibly, distinct expression patterns across different cell types is another fascinating aspect of oxygen sensing in plants, which may provide evidence of a much more complex regulatory network for hypoxia-modulated processes in plants. Besides PCOs, limited oxygen availability will inevitably influence the activity of several other enzymes, including those involved in the synthesis of plant hormones, but also as-yet-unidentified oxygen-requiring enzymes whose activity will be restricted in hypoxic niches. The identification and evaluation of these enzymes is of importance, as they may represent ancillary oxygen-sensing mechanisms in specific plant tissues. Finally, how is metabolism adjusted so that energy production is not compromised during chronic hypoxia? Is fermentation sufficient? All these questions await experimental evidence.

Author Contributions: Both authors collected literature data, wrote the manuscript, and All authors read and agreed to the published version of the manuscript.

Funding: This research received no external funding.

References

1. Loreti, E.; van Veen, H.; Perata, P. Plant responses to flooding stress. *Curr. Opin. Plant Biol.* **2016**, *33*, 64–71. [CrossRef]

2. Bailey-Serres, J.; Voesenek, L.A.C.J. Flooding Stress: Acclimations and Genetic Diversity. *Annu. Rev. Plant Biol.* **2008**, *59*, 313–339. [CrossRef]

3. Sasidharan, R.; Bailey-Serres, J.; Ashikari, M.; Atwell, B.J.; Colmer, T.D.; Fagerstedt, K.; Fukao, T.; Geigenberger, P.; Hebelstrup, K.H.; Hill, R.D.; et al. Community recommendations on terminology and procedures used in flooding and low oxygen stress research. *New Phytol.* **2017**, *214*, 1403–1407. [CrossRef]

4. Weits, D.A.; van Dongen, J.T.; Licausi, F. Molecular oxygen as a signaling component in plant development. *New Phytol.* **2020**. [CrossRef]

5. Licausi, F.; Kosmacz, M.; Weits, D.A.; Giuntoli, B.; Giorgi, F.M.; Voesenek, L.A.C.J.; Perata, P.; Van Dongen, J.T. Oxygen sensing in plants is mediated by an N-end rule pathway for protein destabilization. *Nature* **2011**, *479*, 419–422. [CrossRef] [PubMed]

6. Gibbs, D.J.; Holdsworth, M.J. Every Breath You Take: New Insights into Plant and Animal Oxygen Sensing. *Cell* **2020**, *180*, 22–24. [CrossRef]

7. Wang, G.L.; Jiang, B.H.; Rue, E.A.; Semenza, G.L. Hypoxia-inducible factor 1 is a basic-helix-loop-helix-PAS heterodimer regulated by cellular O2 tension. *Proc. Natl. Acad. Sci. USA* **1995**, *92*, 5510–5514. [CrossRef] [PubMed]

8. Iliopoulos, O.; Levy, A.P.; Jiang, C.; Kaelin, W.G.; Goldberg, M.A. Negative regulation of hypoxia-inducible genes by the von Hippel-Lindau protein. *Proc. Natl. Acad. Sci. USA* **1996**, *93*, 10595–10599. [CrossRef] [PubMed]

9. Maxwell, P.H.; Wiesener, M.S.; Chang, G.W.; Clifford, S.C.; Vaux, E.C.; Cockman, M.E.; Wykoff, C.C.; Pugh, C.W.; Maher, E.R.; Ratcliffe, P.J. The tumour suppressor protein VHL targets hypoxia-inducible factors for oxygen-dependent proteolysis. *Nature* **1999**, *399*, 271–275. [CrossRef]

10. Depping, R.; Steinhoff, A.; Schindler, S.G.; Friedrich, B.; Fagerlund, R.; Metzen, E.; Hartmann, E.; Köhler, M. Nuclear translocation of hypoxia-inducible factors (HIFs): Involvement of the classical importin α/β pathway. *Biochim. Biophys. Acta—Mol. Cell Res.* **2008**, *1783*, 394–404. [CrossRef]

11. Gibbs, D.J.; Lee, S.C.; Md Isa, N.; Gramuglia, S.; Fukao, T.; Bassel, G.W.; Correia, C.S.; Corbineau, F.; Theodoulou, F.L.; Bailey-Serres, J.; et al. Homeostatic response to hypoxia is regulated by the N-end rule pathway in plants. *Nature* **2011**, *479*, 415–418. [CrossRef] [PubMed]

12. Licausi, F.; Giuntoli, B.; Perata, P. Similar and Yet Different: Oxygen Sensing in Animals and Plants. *Trends Plant Sci.* **2020**, *25*, 6–9. [CrossRef] [PubMed]

13. Giuntoli, B.; Perata, P. Group vii ethylene response factors in arabidopsis: Regulation and physiological roles. *Plant Physiol.* **2018**, *176*, 1143–1155. [CrossRef] [PubMed]

14. Weits, D.A.; Giuntoli, B.; Kosmacz, M.; Parlanti, S.; Hubberten, H.M.; Riegler, H.; Hoefgen, R.; Perata, P.; Van Dongen, J.T.; Licausi, F. Plant cysteine oxidases control the oxygen-dependent branch of the N-end-rule pathway. *Nat. Commun.* **2014**, *5*, 1–10. [CrossRef] [PubMed]

15. White, M.D.; Klecker, M.; Hopkinson, R.J.; Weits, D.A.; Mueller, C.; Naumann, C.; O'Neill, R.; Wickens, J.; Yang, J.; Brooks-Bartlett, J.C.; et al. Plant cysteine oxidases are dioxygenases that directly enable arginyl transferase-catalysed arginylation of N-end rule targets. *Nat. Commun.* **2017**, *8*, 14690. [CrossRef] [PubMed]

16. Gasch, P.; Fundinger, M.; Müller, J.T.; Lee, T.; Bailey-Serres, J.; Mustropha, A. Redundant ERF-VII transcription factors bind to an evolutionarily conserved cis-motif to regulate hypoxia-responsive gene expression in arabidopsis. *Plant Cell* **2016**, *28*, 160–180. [CrossRef]

17. BioRender.com. Available online: https://biorender.com/ (accessed on 27 May 2020).

18. Mustroph, A.; Lee, S.C.; Oosumi, T.; Zanetti, M.E.; Yang, H.; Ma, K.; Yaghoubi-Masihi, A.; Fukao, T.; Bailey-Serres, J. Cross-Kingdom comparison of transcriptomic adjustments to low-oxygen stress highlights conserved and plant-specific responses. *Plant Physiol.* **2010**, *152*, 1484–1500. [CrossRef]

19. Masson, N.; Keeley, T.P.; Giuntoli, B.; White, M.D.; Lavilla Puerta, M.; Perata, P.; Hopkinson, R.J.; Flashman, E.; Licausi, F.; Ratcliffe, P.J. Conserved N-terminal cysteine dioxygenases transduce responses to hypoxia in animals and plants. *Science* **2019**, *364*, 65–69. [CrossRef]

20. Bailey-Serres, J.; Fukao, T.; Gibbs, D.J.; Holdsworth, M.J.; Lee, S.C.; Licausi, F.; Perata, P.; Voesenek, L.A.C.J.; van Dongen, J.T. Making sense of low oxygen sensing. *Trends Plant Sci.* **2012**, *17*, 129–138. [CrossRef]

21. Perata, P.; Alpi, A. Plant responses to anaerobiosis. *Plant Sci.* **1993**, *93*, 1–17. [CrossRef]

22. Vartapetian, B.B. Plant anaerobic stress as a novel trend in ecological physiology, biochemistry, and molecular biology: 2. Further development of the problem. *Russ. J. Plant Physiol.* **2006**, *53*, 711–738. [CrossRef]

23. Jacobs, M.; Dolferus, R.; Van Den Bossche, D. Isolation and biochemical analysis of ethyl methanesulfonate-induced alcohol dehydrogenase null mutants of Arabidopsis thaliana (L.) Heynh. *Biochem. Genet.* **1988**, *26*, 105–122. [CrossRef] [PubMed]

24. Cho, H.; Loreti, E.; Shih, M.; Perata, P. Energy and sugar signaling during hypoxia. *New Phytol.* **2019**, nph.16326. [CrossRef] [PubMed]

25. Perata, P.; Geshi, N.; Yamaguchi, J.; Akazawa, T. Effect of anoxia on the induction of α-amylase in cereal seeds. *Planta* **1993**, *191*, 402–408. [CrossRef]

26. Yu, S.; Lee, H.; Lo, S.; Ho, T.D. How does rice cope with too little oxygen during its early life? *New Phytol.* **2020**, nph.16395. [CrossRef]

27. Guglielminetti, L.; Yamaguchi, J.; Perata, P.; Alpi, A. Amylolytic activities in cereal seeds under aerobic and anaerobic conditions. *Plant Physiol.* **1995**, *109*, 1069–1076. [CrossRef]

28. Perata, P.; Guglielminetti, L.; Alpi, A. Mobilization of endosperm reserves in cereal seeds under anoxia. *Ann. Bot.* **1997**, *79*, 49–56. [CrossRef]

29. Loreti, E.; Valeri, M.C.; Novi, G.; Perata, P. Gene regulation and survival under hypoxia requires starch availability and Metabolism. *Plant Physiol.* **2018**, *176*, 1286–1298. [CrossRef]

30. Burtscher, M.; Gatterer, H.; Burtscher, J.; Mairbäurl, H. Extreme terrestrial environments: Life in thermal stress and hypoxia. A narrative review. *Front. Physiol.* **2018**, *9*, 572. [CrossRef]

31. Baldi, S.; Aquilani, R.; Pinna, G.D.; Poggi, P.; de Martini, A.; Bruschi, C. Fat-free mass change after nutritional rehabilitation in weight losing COPD: Role of insulin, C-reactive protein and tissue hypoxia. *Int. J. COPD* **2010**, *5*, 29–39. [CrossRef]

32. Garvey, J.F.; Taylor, C.T.; McNicholas, W.T. Cardiovascular disease in obstructive sleep apnoea syndrome: The role of intermittent hypoxia and inflammation. *Eur. Respir. J.* **2009**, *33*, 1195–1205. [CrossRef] [PubMed]

33. Grocott, M.; Montgomery, H.; Vercueil, A. High-altitude physiology and pathophysiology: Implications and relevance for intensive care medicine. *Crit. Care* **2007**, *11*, 1–5. [CrossRef] [PubMed]

34. Ameln, H.; Gustafsson, T.; Sundberg, C.J.; Okamoto, K.; Jansson, E.; Poellinger, L.; Makino, Y. Physiological activation of hypoxia inducible factor-1 in human skeletal muscle. *FASEB J.* **2005**, *19*, 1009–1011. [CrossRef] [PubMed]

35. Evans, D.E. Aerenchyma formation. *New Phytol.* **2004**, *161*, 35–49. [CrossRef]

36. Drew, M.C.; He, C.J.; Morgan, P.W. Programmed cell death and aerenchyma formation in roots. *Trends Plant Sci.* **2000**, *5*, 123–127. [CrossRef]

37. Takahashi, H.; Yamauchi, T.; Colmer, T.D.; Nakazono, M. Aerenchyma formation in plants. *Plant Cell Monogr.* **2014**, *21*, 247–265. [CrossRef]

38. Pedersen, O.; Sauter, M.; Colmer, T.D.; Nakazono, M. Regulation of root adaptive anatomical and morphological traits during low soil oxygen. *New Phytol.* **2020**, nph.16375. [CrossRef]

39. Yamauchi, T.; Shimamura, S.; Nakazono, M.; Mochizuki, T. Aerenchyma formation in crop species: A review. *Field Crop. Res.* **2013**, *152*, 8–16. [CrossRef]

40. Yamauchi, T.; Tanaka, A.; Tsutsumi, N.; Inukai, Y.; Nakazono, M. A Role for Auxin in Ethylene-Dependent Inducible Aerenchyma Formation in Rice Roots. *Plants* **2020**, *9*, 610. [CrossRef]

41. Yamamoto, F.; Sakata, T.; Terazawa, K. Physiological, morphological and anatomical responses of Fraxinus mandshurica seedlings to flooding. *Tree Physiol.* **1995**, *15*, 713–719. [CrossRef]

42. Colmer, T.D. Long-distance transport of gases in plants: A perspective on internal aeration and radial oxygen loss from roots. *Plant Cell Environ.* **2003**, *26*, 17–36. [CrossRef]

43. Voesenek, L.A.C.J.; Benschop, J.J.; Bou, J.; Cox, M.C.H.; Groeneveld, H.W.; Millenaar, F.F.; Vreeburg, R.A.M.; Peeters, A.J.M. Interactions Between Plant Hormones Regulate Submergence-induced Shoot Elongation in the Flooding-tolerant Dicot Rumex palustris. *Ann. Bot.* **2003**, *91*, 205–211. [CrossRef] [PubMed]

44. Alpi, A.; Beevers, H. Effects of O2 Concentration on Rice Seedlings. *Plant Physiol.* **1983**, *71*, 30–34. [CrossRef]

45. Kretzschmar, T.; Pelayo, M.A.F.; Trijatmiko, K.R.; Gabunada, L.F.M.; Alam, R.; Jimenez, R.; Mendioro, M.S.; Slamet-Loedin, I.H.; Sreenivasulu, N.; Bailey-Serres, J.; et al. A trehalose-6-phosphate phosphatase enhances anaerobic germination tolerance in rice. *Nat. Plants* **2015**, *1*, 1–5. [CrossRef] [PubMed]

46. Hattori, Y.; Nagai, K.; Ashikari, M. Rice growth adapting to deepwater. *Curr. Opin. Plant Biol.* **2011**, *14*, 100–105. [CrossRef]

47. Voesenek, L.A.C.J.; Bailey-Serres, J. Plant biology: Genetics of high-rise rice. *Nature* **2009**, *460*, 959–960. [CrossRef]

48. Xu, K.; Xu, X.; Fukao, T.; Canlas, P.; Maghirang-Rodriguez, R.; Heuer, S.; Ismail, A.M.; Bailey-Serres, J.; Ronald, P.C.; Mackill, D.J. Sub1A is an ethylene-response-factor-like gene that confers submergence tolerance to rice. *Nature* **2006**, *442*, 705–708. [CrossRef]

49. Perata, P.; Voesenek, L.A.C.J. Submergence tolerance in rice requires Sub1A, an ethylene-response-factor-like gene. *Trends Plant Sci.* **2007**, *12*, 43–46. [CrossRef]

50. Dawood, T.; Rieu, I.; Wolters-Arts, M.; Derksen, E.B.; Mariani, C.; Visser, E.J.W. Rapid flooding-induced adventitious root development from preformed primordia in Solanum dulcamara. *AoB Plants* **2014**, *6*. [CrossRef]

51. Steffens, B.; Rasmussen, A. The physiology of adventitious roots. *Plant Physiol.* **2016**, *170*, 603–617. [CrossRef]

52. Hartman, S.; Sasidharan, R.; Voesenek, L.A.C.J. The role of ethylene in metabolic acclimations to low oxygen. *New Phytol.* **2020**, nph.16378. [CrossRef] [PubMed]

53. Hattori, Y.; Nagai, K.; Furukawa, S.; Song, X.J.; Kawano, R.; Sakakibara, H.; Wu, J.; Matsumoto, T.; Yoshimura, A.; Kitano, H.; et al. The ethylene response factors SNORKEL1 and SNORKEL2 allow rice to adapt to deep water. *Nature* **2009**, *460*, 1026–1030. [CrossRef] [PubMed]

54. Michiels, C. Physiological and pathological responses to hypoxia. *Am. J. Pathol.* **2004**, *164*, 1875–1882. [CrossRef]

55. Van Dongen, J.T.; Schurr, U.; Pfister, M.; Geigenberger, P. Phloem metabolism and function have to cope with low internal oxygen. *Plant Physiol.* **2003**, *131*, 1529–1543. [CrossRef]

56. Borisjuk, L.; Rolletschek, H. The oxygen status of the developing seed. *New Phytol.* **2009**, *182*, 17–30. [CrossRef]

57. Ho, Q.T.; Verboven, P.; Verlinden, B.E.; Schenk, A.; Delele, M.A.; Rolletschek, H.; Vercammen, J.; Nicolaï, B.M. Genotype effects on internal gas gradients in apple fruit. *J. Exp. Bot.* **2010**, *61*, 2745–2755. [CrossRef]

58. Armstrong, W.; Armstrong, J. Plant internal oxygen transport (Diffusion and convection) and measuring and modelling oxygen gradients. *Plant Cell Monogr.* **2014**, *21*, 267–297. [CrossRef]

59. Geigenberger, P.; Fernie, A.R.; Gibon, Y.; Christ, M.; Stitt, M. Metabolic activity decreases as an adaptive response to low internal oxygen in growing potato tubers. *Biol. Chem.* **2000**, *381*, 723–740. [CrossRef]

60. Kimmerer, T.W.; Stringer, M.A. Alcohol Dehydrogenase and Ethanol in the Stems of Trees. *Plant Physiol.* **1988**, *87*, 693–697. [CrossRef]

61. Abbas, M.; Berckhan, S.; Rooney, D.J.; Gibbs, D.J.; Vicente Conde, J.; Sousa Correia, C.; Bassel, G.W.; Marín-De La Rosa, N.; León, J.; Alabadí, D.; et al. Oxygen sensing coordinates photomorphogenesis to facilitate seedling survival. *Curr. Biol.* **2015**, *25*, 1483–1488. [CrossRef]

62. Dunwoodie, S.L. The Role of Hypoxia in Development of the Mammalian Embryo. *Dev. Cell* **2009**, *17*, 755–773. [CrossRef] [PubMed]

63. Mohyeldin, A.; Garzón-Muvdi, T.; Quiñones-Hinojosa, A. Oxygen in stem cell biology: A critical component of the stem cell niche. *Cell Stem Cell* **2010**, *7*, 150–161. [CrossRef] [PubMed]

64. Tadege, M.; Kuhlemeier, C. Aerobic fermentation during tobacco pollen development. *Plant Mol. Biol.* **1997**, *35*, 343–354. [CrossRef] [PubMed]

65. Kelliher, T.; Walbot, V. Maize germinal cell initials accommodate hypoxia and precociously express meiotic genes. *Plant J.* **2014**, *77*, 639–652. [CrossRef]

66. Weits, D.; Kunkowska, A.; Kamps, N.; Portz, K.M.S.; Packbier, N.K.; Venza, Z.N.; Gaillochet, C.; Lohmann, J.U.; Pedersen, O.; van Dongen, J.T.; et al. An apical hypoxic niche sets the pace of shoot meristem activity. *Nature* **2019**, *569*, 714–717. [CrossRef] [PubMed]

67. Gibbs, D.J.; Tedds, H.M.; Labandera, A.M.; Bailey, M.; White, M.D.; Hartman, S.; Sprigg, C.; Mogg, S.L.; Osborne, R.; Dambire, C.; et al. Oxygen-dependent proteolysis regulates the stability of angiosperm polycomb repressive complex 2 subunit VERNALIZATION 2. *Nat. Commun.* **2018**, *9*, 1–11. [CrossRef]

68. Shukla, V.; Lombardi, L.; Iacopino, S.; Pencik, A.; Novak, O.; Perata, P.; Giuntoli, B.; Licausi, F. Endogenous Hypoxia in Lateral Root Primordia Controls Root Architecture by Antagonizing Auxin Signaling in Arabidopsis. *Mol. Plant* **2019**, *12*, 538–551. [CrossRef]

69. Vidoz, M.L.; Loreti, E.; Mensuali, A.; Alpi, A.; Perata, P. Hormonal interplay during adventitious root formation in flooded tomato plants. *Plant J.* **2010**, *63*, 551–562. [CrossRef]

70. Eysholdt-Derzsó, E.; Sauter, M. Root bending is antagonistically affected by hypoxia and ERF-mediated transcription via auxin signaling. *Plant Physiol.* **2017**, *175*, 412–423. [CrossRef]

71. Pucciariello, C.; Boscari, A.; Tagliani, A.; Brouquisse, R.; Perata, P. Exploring legume-rhizobia symbiotic models for waterlogging tolerance. *Front. Plant Sci.* **2019**, *10*, 578. [CrossRef]

72. Gravot, A.; Richard, G.; Lime, T.; Lemarié, S.; Jubault, M.; Lariagon, C.; Lemoine, J.; Vicente, J.; Robert-Seilaniantz, A.; Holdsworth, M.J.; et al. Hypoxia response in Arabidopsis roots infected by Plasmodiophora brassicae supports the development of clubroot. *BMC Plant Biol.* **2016**, *16*, 251. [CrossRef] [PubMed]

73. Kerpen, L.; Niccolini, L.; Licausi, F.; van Dongen, J.T.; Weits, D.A. Hypoxic Conditions in Crown Galls Induce Plant Anaerobic Responses That Support Tumor Proliferation. *Front. Plant Sci.* **2019**, *10*, 56. [CrossRef] [PubMed]

74. Jubault, M.; Lariagon, C.; Taconnat, L.; Renou, J.P.; Gravot, A.; Delourme, R.; Manzanares-Dauleux, M.J. Partial resistance to clubroot in Arabidopsis is based on changes in the host primary metabolism and targeted cell division and expansion capacity. *Funct. Integr. Genom.* **2013**, *13*, 191–205. [CrossRef] [PubMed]

75. Zhao, Y.; Wei, T.; Yin, K.-Q.; Chen, Z.; Gu, H.; Qu, L.-J.; Qin, G. Arabidopsis RAP2.2 plays an important role in plant resistance to Botrytis cinerea and ethylene responses. *New Phytol.* **2012**, *195*, 450–460. [CrossRef]

76. Kim, N.Y.; Jang, Y.J.; Park, O.K. AP2/ERF Family Transcription Factors ORA59 and RAP2.3 Interact in the Nucleus and Function Together in Ethylene Responses. *Front. Plant Sci.* **2018**, *9*, 1675. [CrossRef]

77. Vicente, J.; Mendiondo, G.M.; Pauwels, J.; Pastor, V.; Izquierdo, Y.; Naumann, C.; Movahedi, M.; Rooney, D.; Gibbs, D.J.; Smart, K.; et al. Distinct branches of the N-end rule pathway modulate the plant immune response. *New Phytol.* **2019**, *221*, 988–1000. [CrossRef]

78. Gibbs, D.J.; MdIsa, N.; Movahedi, M.; Lozano-Juste, J.; Mendiondo, G.M.; Berckhan, S.; Marín-delaRosa, N.; VicenteConde, J.; SousaCorreia, C.; Pearce, S.P.; et al. Nitric Oxide Sensing in Plants Is Mediated by Proteolytic Control of Group VII ERF Transcription Factors. *Mol. Cell* **2014**, *53*, 369–379. [CrossRef]

79. Hartman, S.; Liu, Z.; van Veen, H.; Vicente, J.; Reinen, E.; Martopawiro, S.; Zhang, H.; van Dongen, N.; Bosman, F.; Bassel, G.W.; et al. Ethylene-mediated nitric oxide depletion pre-adapts plants to hypoxia stress. *Nat. Commun.* **2019**, *10*, 1–9. [CrossRef]

80. Valeri, M.C.; Novi, G.; Weits, D.A.; Mensuali, A.; Perata, P.; Loreti, E. *Botrytis cinerea* induces local hypoxia in Arabidopsis leaves. *New Phytol.* **2020**, nph.16513. [CrossRef]

81. Benedetti, M.; Verrascina, I.; Pontiggia, D.; Locci, F.; Mattei, B.; De Lorenzo, G.; Cervone, F. Four Arabidopsis berberine bridge enzyme-like proteins are specific oxidases that inactivate the elicitor-active oligogalacturonides. *Plant J.* **2018**, *94*, 260–273. [CrossRef]

Review

Molecular Mechanisms Supporting Rice Germination and Coleoptile Elongation under Low Oxygen

Chiara Pucciariello

PlantLab, Institute of Life Sciences, Scuola Superiore Sant'Anna, 56124 Pisa, Italy; chiara.pucciariello@santannapisa.it

Received: 7 July 2020; Accepted: 10 August 2020; Published: 15 August 2020

Abstract: Rice germinates under submergence by exploiting the starch available in the endosperm and translocating sugars from source to sink organs. The availability of fermentable sugar under water allows germination with the protrusion of the coleoptile, which elongates rapidly and functions as a snorkel toward the air above. Depending on the variety, rice can produce a short or a long coleoptile. Longer length entails the involvement of a functional transport of auxin along the coleoptile. This paper is an overview of rice coleoptiles and the studies undertaken to understand its functioning and role under submergence.

Keywords: anoxia; coleoptile; flooding; hypoxia; rice; submergence

1. Introduction

Submergence stress is one of the most critical constraints on crop production, with a significant impact on food security. Higher plants die rapidly when oxygen (O_2) is limited due to water submergence. However, plants adapted to semi-aquatic environments are able to survive weeks of complete submergence by modifying their metabolism and morphology. Rice is known for its capacity to grow in paddy fields. However, in countries where rice is the predominant food, prolonged flooding can affect yield, since some varieties can be sensitive to complete submergence [1]. The rise of extreme weather events—with inundation and flash flooding—has caused considerable production losses.

In the last two decades, important adaptations have been identified with a significant impact on rice cultivation. The identification of the *SUBMERGENCE1* (*SUB1*) locus, responsible for limiting elongation during submergence in order to conserve energy, and its introgression into rice popular varieties [2,3], led to the development of SUB1 mega-varieties, which provide submergence tolerance with no effects on grain production and quality [4].

Rice seeds can germinate and elongate the coleoptile in the absence of O_2. Other cereals such as wheat, barley, oat and rye lack this trait, and early experiments found it to be related to the capacity of rice to hydrolyse starch into readily fermentable sugars to generate ATP [5,6]. The energy produced by anaerobic fermentation is much lower to that produced in air. However, the reoxidation of NADH to NAD^+ by ethanolic fermentation for use in glycolysis supports plant survival under O_2 shortage.

The availability of α-amylase enzymes, which degrade starch under low O_2, is among the key determinant of the germination of this species under submergence. ALPHA AMYLASE 3 (RAMY3D) is the only isoform significantly modulated under anoxia at the transcriptional level [7] and is regulated by sugar starvation [8], a condition occurring at the beginning of germination under O_2-limiting conditions. This pathway enables rice grains to use the starchy reserves available in the endosperm in order to fuel the sink organs.

During germination, rice elongates the coleoptile. The role of the coleoptile is to protect the true leaves during emergence from the soil and likely provide nutrients for the developing tissues [9]. Under submergence, the coleoptile acts like a snorkel that reaches the water surface and makes connection with the air [10]. The hollow structure allows a flow of O_2 to travel down to underwater organs in order to establish aerobic respiration and fulfill the energy requirements of the developing leaves and roots.

Together with coleoptile elongation, a delay in radicle emergence is observed under O_2 shortage. This suggests that the development program is different under low O_2 than in air. Under O_2 shortage, the energy is preferentially allocated to fuel cell extension in the epigeal part of the plant and not to the root, which instead rapidly grow in air [10].

This article reviews how research into anaerobic rice coleoptiles has evolved and thus highlights the important role of this organ in rice seedling establishment under water. This trait may be of interest in areas particularly prone to flooding. Moreover, direct seeding associated with early flooding tolerance helps in suppressing weed growth and reducing the costs of mechanical and chemical weeding.

2. The Coleoptile Elongation under Submergence

During aerobic germination, the rice coleoptile is initially white and turns green after one day [11]. The conical structure of the coleoptile is interrupted at the apex, where a vertical crack results in the following extension of true leaves [11]. The coleoptile generally has two vascular bundles running in parallel longitudinally. Some cell layers of the outer coleoptile epidermis develop chloroplasts during maturation, while the inner regions contain large amyloplasts [11]. Amyloplasts degrade rapidly in air with a complete decomposition at day 3, although the plastid membrane persists longer [12]. Along with the split in the coleoptile, aerenchyma is formed and senescence sets in [13]. In air, the energy supply from aerobic respiration, along with starch degradation in the endosperm, enables cell division and primary root development [14].

Under O_2 shortage, there is no evidence of cell death events leading to coleoptile opening, aerenchyma formation and senescence [13]. Coleoptile maturation and cell death are blocked, and the coleoptile rapidly elongates while the root growth is dampened. The extension of the coleoptile under water occurs mainly through the cell expansion of cells preformed in the embryo [15], and the expansion increases with the time underwater [16]. On the contrary, the percentage of cells in mitotic phase decreases [16].

Indeed, both the length and the number of cells determine the final length of a coleoptile. However, cell division requires more energy than cell elongation, thus elongation is likely facilitated under O_2 shortage [17]. Cell length increases from the base of the coleoptile with a peak one third of the way up [10]. The basal region of the hypoxic rice coleoptile seems to be the most intense for cell elongation and division.

When the submerged coleoptile reaches the water surface, the hollow structure of the coleoptile enables O_2 to flow from the surface to the submerged parts. Coleoptile elongation is thus a strategy of the rice plant to avoid submergence stress and is referred to as the "snorkel effect" [18]. After reoxygenation, the rice plant shows a phenotype similar to that in air, since it will develop the primary root and leaf after coleoptile degeneration [14].

Under submergence, the rice coleoptile can grow several days without apparent senescence; it then reaches a plateau where cell extension is likely at a maximum [19,20]. At this plateau stage, different rice varieties show variability in coleoptile length, which can be a trait of genetic origin (Figure 1). Indeed, this condition can be considered specific for low O_2 state, since aerobic coleoptiles degenerate after a few days. In this context, different water regimes in cultivation practices may have influenced the selection of genotypes with the capacity to reach a certain final coleoptile length as an adaptation.

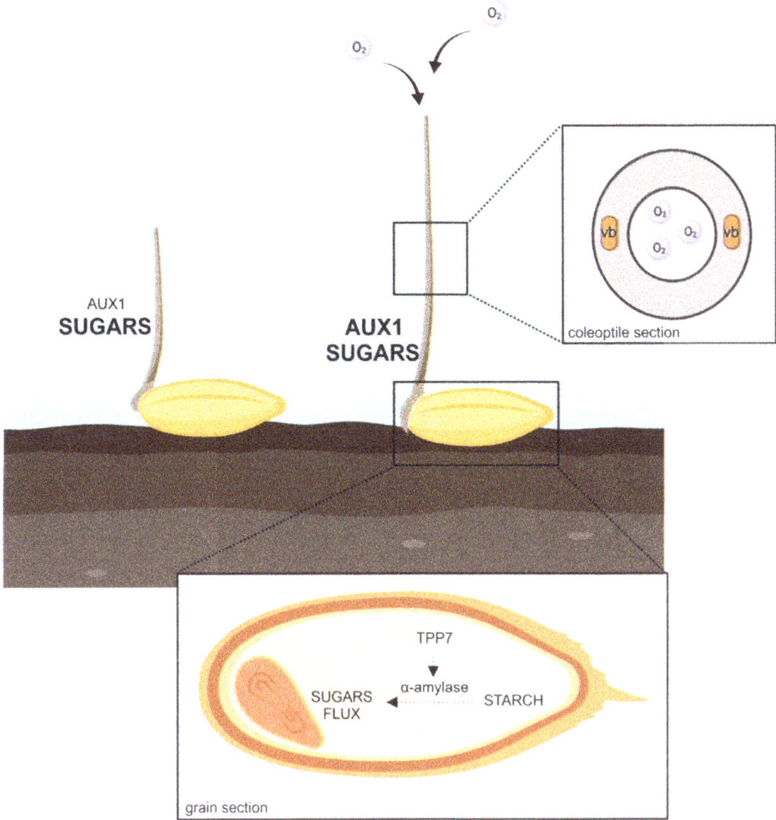

Figure 1. Rice elongates the coleoptile under water. Treahalose 6 phosphate phosphatase 7 (TPP7) availability in rice genotypes enables a better use of sugar, which is transferred from the source zone (endosperm) to the sink (embryo and coleoptile). Starch mobilization to fermentable sugars to fuel elongation takes place thanks to a battery of α-amylase whose transcription is activated by sugar starvation and oxygen shortage-dependent messages. Increased auxin transport by the influx carrier AUX1 contributes to a long coleoptile.

3. Anaerobic Gene Regulation in Rice Coleoptile

Analysis conducted on coleoptiles grown under anoxia until four days after germination, using the Nipponbare variety, revealed the regulation of the expression of several genes belonging to different families [7]. A detailed picture of this phenomenon in comparison to air was drawn, showing the upregulation of genes involved in the biochemical steps of pyruvate metabolism, glycolysis and fermentation. Interestingly, *RAMY3D* is strongly expressed in anoxic rice coleoptiles, besides the known role in the endosperm. The significance of this phenomenon is not known, but has been suggested to be a futile cycle of starch synthesis and degradation to fill amyloplasts with starch for the gravitropic response [7]. Another possibility is that starch is used as a provisional store derived from the degeneration of endosperm with the coleoptile representing a temporary sink of resources for the subsequent fast growth.

In parallel, the upregulation of *EXPA7* and *EXPA12* transcripts was observed. In this context, a subsequent study provided insights into the difference in expansin expression in rice varieties showing long (Arborio) and short (Lamone) coleoptile length [21]. No correlation was found between the transcriptional regulation of expansins and coleoptile elongation, suggesting that the differential elongation observed in the two varieties is not related to these genes.

Genes coding for heat shock proteins (HSP) were also found to be induced in anoxic coleoptiles [7], in a mechanism that is devoted to the protection of plant cells under stress and that overlaps with the response to heat stress [22].

A general reduction in enzymes, whose activity requires O_2, was observed [7], likely in an attempt to limit the energy spent on enzymes that are destined to be non-functional. In parallel, a possible sugar signalling role in genes induced under anoxia was observed, which the authors suggested was due to a moderate sugar starvation experience by anoxic rice coleoptiles.

The study of cis-element enrichment in promoters of genes regulated in rice coleoptiles under anoxia related to glycolysis and fermentation [7] revealed the association with the transcription factors of ARF, ERF, MYB, WRKY and bZIP families [23]. In that study, the analysis of cis-element enrichment revealed the potential link with hormonal signalling, e.g., GA, IAA, ABA and ethylene.

Dissimilarities in gene expression have been observed when comparing the basal with the tip coleoptile sections under anoxia and hypoxia [10]. These regions are characterised by a longitudinal steep gradient of growth, where the basal region is the most intense in terms of cell elongation and division. Some genes were found to be specifically expressed in the tip rather than the coleoptile base. A specific transcriptional program was also suggested to occur under anoxia and hypoxia. In anoxic coleoptiles, a specific expression of α-amylase genes has been identified at the base of the coleoptile in comparison to the tip [10].

Significant changes were observed in the epigenomic state of the coleoptile, with changes correlated to cell elongation during anaerobic conditions [14]. Very interestingly, the analysis in the DNA methylation observed in the reoxygenation phase after anaerobiosis showed a pattern similar to dry seeds rather than the pattern observed during exposure to air, suggesting that the plant's internal clock is reset to react rapidly to molecular changes involved in root and true leaf elongation [14].

Hsu and Tung [24] conducted an RNA seq analysis of rice genotypes that have varying capacities to elongate the coleoptile under submergence. They assigned those genes that were significantly responsive in genotypes with rapid coleoptile growth, above all, to ethylene signalling and cell wall modification related pathways.

4. Chromosomal Regions Regulating Coleoptile Elongation under Oxygen Shortage

Several works have recently been conducted to identify genomic regions associated with coleoptile elongation under water (Table 1). Screenings were performed considering rice panels composed of different varietal groups. Rapid coleoptile growth of submerged rice was studied in 153 rice accessions from *japonica* and *indica* varietal groups, measuring the difference in coleoptile length between control and submerged plants [25]. The authors used a Genome Wide Association Study (GWAS) and identified several genomic regions significantly associated with anaerobic germination. They also used a recombinant inbred line population obtained from a cross between *indica* and *japonica* varieties and identified a single significant Quantitative Trait Locus (QTL) region on chromosome 1 [25].

Table 1. Chromosomal regions identified in genetic studies focused on rice germination and coleoptile elongation traits.

Trait of Study	Genotypes	Major Chromosomal Region	Reference
Tolerance to flooding during germination	Khao Hlan On backcross population with IR64 recurrent parent	Chr 9 *qAG-9-1* and *qAG-9-2* QTLs	Angaji et al., 2010 [26]
Tolerance to anaerobic conditions during germination	Population derived from a cross between Ma-Zhan Red and IR42	Chr 7 *qAG-7-1* QTL	Septiningsih et al., 2013 [2]
Coleoptile elongation under anaerobic germination	Recombinant inbred line population derived from a cross between *japonica* and *indica* varieties	Chr1 QTL	Hsu and Tung, 2015 [24]
Coleoptile length during germination under flooding	Panel of 432 *indica* rice varieties	Chr 6 MTAs	Zhang et al., 2017 [27]
Coleoptile length under dark submergence	Panel of 273 *japonica* rice accessions	Chr1, Chr5 MTAs	Nghi et al., 2019 [20]
Coleoptile length in anaerobic solution	39 chromosome segment substitution lines derived from a cross between Koshihikari and IR64 rice varieties	Chr3 *qACE3.1*	Noshimura et al., 2020 [28]

Zhang et al. investigated a pool of 432 *indica* varieties in terms of variations in the length of flooded coleoptiles, and used GWAS to detect significant single nucleotide polymorphisms (SNPs) [27]. One of the haplotypes of a candidate gene, coding for a DUF domain containing protein, contributed significantly to coleoptile elongation.

A screening of over 8000 rice accessions performed at the International Rice Research Institute (IRRI) identified a few genotypes tolerant to anaerobic germination (AG) [26]. The QTL mapping of the backcross population obtained by the tolerant Khao Hlan On and the sensitive IR-64 identified five putative QTL on chromosome 1 (*qAG-3-1*), 3 (*qAG-3-1*), 7 (*qAG-7-2*) and 9 (*qAG-9-1* and *2*), which explained the phenotypic variation from 17.9 to 33.5%. The *OsTTP7* gene responsible for enhanced anaerobic germination was subsequently identified on *qAG-9-2* [29].

A further QTL mapping conducted using a population obtained from a cross between the sensitive cultivar IR42 and the tolerant Ma-Zhan Red identified six significant QTL on chromosome 2, 5, 6, and 7, with the largest effect on anaerobic germination played by *qAG-7.1* [2].

Our laboratory screened a panel of 273 *japonica* rice accessions for coleoptile length [20]. There were wide differences in coleoptile length, with a maximum length being reached at day 8 of dark submergence. This limited growth of the coleoptile can be due to energy restrictions by starch availability in the endosperm and release of soluble sugar or restrained capacity to sustain cell elongation over a certain limit. The interesting aspect of this panel was the high homogeneity [30], which led to the identification of new regions probably masked by high diversity in other panels. In this work, two marker trait associations (MTAs) were highly significant on chromosomes 1 and 5, with the identification of a subgroup of genes likely related to the coleoptile length trait. Of those genes, some related to auxin transport and sensing were found to be differentially expressed in long versus short coleoptile-harbouring varieties. In particular, the auxin transporter *AUX1* was found to be more expressed in rice varieties having a low ratio between the coleoptile length at day 8 on day 4, thus elongating very rapidly and showing a long coleoptile already at day 4. This gene was also more expressed in the group of rice varieties showing a long coleoptile at day 8.

Recently, chromosome segment substitution lines (CSSLs) obtained crossing Koshihikari variety, characterised by a long coleoptile, with IR 64 background, characterised by a short coleoptile, were evaluated for coleoptile elongation in anaerobic solution and tested in paddy field [28]. A novel QTL was identified on chromosome 3, referred to as *qACE3.1*, likely affecting the expression of genes involved in fermentative metabolism.

5. Starch Degradation during Anaerobic Rice Germination

The difference that makes rice stand out from other cereals is the use of the starch contained in the endosperm even under O_2 shortage. Alpha amylase enzymes enable rice grains to degrade starch, which thereby allows the production of soluble sugars for sink organs. In air, these enzymes function through gibberellins (GA), abscisic acid (ABA) and sugar-demand mediated activation. In aerobic conditions, the α-amylases subfamily *AMY1* and *AMY2* are principally regulated by phytohormones GA and ABA [31]. The expression of α-amylase in air and anoxia differs in relation to where they are located. Under anoxia, the expression of total α-amylases transcripts decreases in aleurone and increases in embryo. In particular, the anoxia-dependent reduction of *AMY1* was observed with a simultaneous increase in *AMY3* [31].

Loreti et al. [8] investigated the *Tan-ginbozu* rice mutant, impaired in GA biosynthesis, and revealed that, under anoxia, the expression of α-amylase genes is independent of GA. Under O_2 shortage, the subfamily *AMY3* is predominant [31], whose transcriptional regulation is mediated by sugar starvation and low O_2 signalling [8]. Other cereals fail to degrade starch in low O_2 conditions, due to the absence of these isoforms [6]. As a consequence, they cannot exploit the starchy reserves to activate the fermentation pathway and produce energy for growth.

Several studies have contributed to understanding the cascade effect that culminates with starch degradation by α-amylase. Rice plants mutated in a CALCINEURIN B-LIKE (CBL) INTERACTING PROTEIN KINASE 15 (CIPK15) *cipk15* were identified to be extremely sensitive to submergence during the early phase of development [32]. The *cipk15* mutant fails to elongate the coleoptile under low O_2 and to express anaerobic genes (e.g., *ADH* and *AMY3*) [32]. In addition, a sucrose supply restored the *cipk15* phenotype under submergence, confirming the importance of sugar availability for this trait.

CIPK proteins are involved in decoding the Ca^{2+} signal sensed by CBL proteins, constituting a dual component for the Ca^{2+} sensing-responding system [33]. Although it is not clear how the mechanism functions under O_2 deprivation, CBL4 was found to interact with CIPK15 [34] to positively regulate downstream events. On the contrary, CBL10 has been found to be a negative regulator of the CIPK15-dependent pathway since the maintenance of a low *CBL10* expression level in some rice varieties allows a higher expression of *AMY3* [35]. In the activation cascade, CIPK15 was shown to regulate the major sugar regulatory kinase SUCROSE NONFERMETNING 1-RELATED PROTEIN KINASE SnRK1A that subsequently activates the MYELOBLASTOSIS SUCROSE 1 (MYBS1) transcription factor involved in the transcriptional regulation of α-amylase genes [36–38]. Under sugar starvation, MYBS1 binds to the sugar response element located on the α-amylase promoter [37,38].

In the context of sugar availability, a panel of 141 Italian and 23 Sri Lankan rice varieties was tested for coleoptile length under anoxia [39]. Expansins were shown not to be involved in the coleoptile length difference among varieties [21], while the long coleoptile varieties showed a higher ethanol production, suggesting a better performance in fermentation in comparison to the short coleoptile varieties [39]. The *rice alcohol dehydrogenase 1-deficient* (*rad*) mutant showed a shorter coleoptile under low O_2 in comparison to the wild-type [16]. This was mainly related to a reduced longitudinal cell length and repression of cell division in the coleoptile, together with ADH essentiality for fermentation in embryo and endosperm [40].

The TREHALOSE 6 PHOSPHATE PHOSPHATASE 7 (TPP7) gene, the genetic determinant of the major QTL *qAG-9-2* identified in the rice *japonica* variety Khao Hlan On [26], plays a key role in the mobilization of starch under water in terms of coleoptile elongation and embryo development [29]. This gene is absent in the variety IR64, which barely germinates and whose coleoptile does not elongate significantly under submergence. When *TPP7* gene is available, the dampening of the starch degradation due to a balance between trehalose 6 phosphate and sucrose likely decreases, and α-amylases can work toward the production of sugars for sink organs. The consumption of T6P with the production of trehalose by TPP7 removes the dampening effect of T6P on SnRK1A, activating the starch catabolism. TPP7 availability has been shown to contribute substantially to the elongation of coleoptiles. The isolation of the near isogenic line NIL-AG1 in the IR64 background showed an increase in coleoptile length and α-amylase activity [29]. The anaerobic germination sensitivity of IR64 was rescued by a sucrose supply, while it was not influenced by ABA or GA. The presence of haplotypes in a rice *japonica* panel characterised by the systematic presence of *TPP7* was shown not to influence coleoptile length [20].

6. Hormonal Regulation of Anaerobic Rice Coleoptile Elongation

The role of auxin in coleoptile development under submergence has been questioned for several years. There are contrasting data on the availability and role of auxin in anaerobic coleoptiles [41–46]. Early works reported that the treatment of IAA inhibitors in water resulted in dampened elongation growth [41,42]. During anaerobic treatment, an increase in the IAA level was found, which was suggested to be translocated from the endosperm [43]. However, the addition of IAA to coleoptiles under anoxia did not affect the final coleoptile length, unlike air where IAA enhances coleoptile elongation [44]. Subsequently, experiments in Heller medium flasks showed that IAA addition has an initial positive influence on the elongation of coleoptile segments, while the second step of elongation depends on ethylene [45]. The level of IAA was found to be lower in underwater rice segments than in air [46].

More recently, the submergence-dependent inhibition of *miR393* expression has been shown to reduce the degradation of the mRNA *TRANSPORT INHIBITOR RESPONSE 1* (*TIR1*), activating the auxin-dependent signalling pathway in hypoxic rice [47]. Under submergence, the auxin-dependent pathway induces coleoptile elongation and likely stomata development. *EXPA7* expression was found to be positively regulated by target-mimic lines of miR393, and negatively by miR393 overexpressing lines. Subsequently, *EXPA7* expression has been found to be significantly regulated in *japonica* rice varieties that have a long coleoptile [19]. A role for auxin availability and transport mediated by AUX1 has been identified in long coleoptile harbouring varieties [19]. In these varieties, the coleoptile tip likely induces a further production or redistribution of auxin along the coleoptile longitudinal axis that culminates in an increased length.

There is a higher rate of ethylene production in rice genotypes that germinate and grow rapidly under low O_2 [48]. Indeed, ethylene is a primary signal under submergence [49] and ethylene treatment increases the coleoptile length [44,50]. In rice, ethylene drives the expression of the SUB1A gene and the SNORKELs genes, which belong to the Ethylene Responsive Factor of group VII (ERF-VII) and help adult plants to tolerate submergence. In coleoptiles under anoxia, more ERFs are upregulated than in air, such as ERF60, ERF67, and ERF68 [7], which belong to the ERF-VII group, like SUB1 and SNORKEL genes. Group VII ERFs function as O_2 sensors in Arabidopsis, and ERF66 and ERF67 have been identified as a target of SUB1A and, in contrast to SUB1A [51], of the N-end rule pathway [52]. In fact, a characteristic of ERF-VII is the conservation of the N-terminus, which promotes degradation in presence of O_2 and stabilisation under hypoxia. This means that, under O_2 shortage, ERF66 and ERF67 escape proteolysis and are stabilised for downstream transcriptional regulation of anaerobic genes [52].

Ethylene interacts with auxin in order to inhibit root elongation in rice seedling development [53] and it may extend the auxin-dependent elongation of rice coleoptiles under submergence [54]. It will be interesting to investigate how ERF-VIIs function in anaerobic coleoptiles and what gene targets they have in this organ.

7. Conclusions

The ability to produce coleoptiles under hypoxia and anoxia, rather than in air, is a unique feature of rice. Other cereals fail to germinate under submergence. The involvement of sugar in this important trait has long been studied. Rice can hydrolase and use the starchy reserves available in the endosperm in times when O_2 is absent. The cascade, which culminates with the activation of starch hydrolysing enzyme α-amylase, is not regulated by GA and ABA, like in air, but is modulated by sugar availability and low O_2. The pathway has been explored in detail with the identification of key components, such as CIPK15, MYBS1 and SnRK1A.

Rice varieties that have the *TPP7* gene are more effective than other varieties in moving sugars from source (endosperm) to sink (embryo and coleoptile). Of the phytohormones, the role of auxin in coleoptile length grown under water has long been controversial. Indeed, the discovery of miR393 regulation of *TIR1* transcription under submergence and the key role of auxin in long coleoptile open again the possibility that this hormone is involved in the trait. Under submergence, O_2 is going down very rapidly [20]; however, a certain level is maintained in the first days and may support tryptophan-dependent auxin biosynthesis in germinating rice. Some authors hypothesised that the rice endosperm releases auxin during low O_2. Rice grain contains IAA [19], which may be translocated to the elongating coleoptile. A key role in auxin transport has been proved by the phenotype of *osaux1* mutants, which shows a reduced coleoptile length under submergence in comparison to the background [19]. This result is very interesting also in relation to the possible auxin interaction with ethylene, as previously suggested, and the positive effect of ethylene on coleoptile growth. We know that the ethylene level increases under water and governs rice adaptation strategies to submergence.

In this framework, many questions are still open: i) is the endosperm an auxin source in coleoptile elongation under submergence? ii) what is the auxin gradient in the submerged coleoptile? iii) is auxin involved in the delay of radicle emergence observed in submerged rice? iv) is there any interaction between auxin and ethylene in the regulation of coleoptile length under water?

The answer to these questions should provide a comprehensive vision of the mechanisms involved in coleoptile elongation under water. This would help to find ways to develop coleoptile elongation in rice varieties that are needed for direct seeding or are cultivated in areas exposed to unexpected flooding.

Funding: This research received no external funding.

Acknowledgments: I would like to thank Pierdomenico Perata for critically reading this manuscript.

Conflicts of Interest: The author declares no conflict of interest.

References

1. Ismail, A.M.; Singh, U.S.; Singh, S.; Dar, M.H.; Mackill, D.J. The contribution of submergence-tolerant (Sub1) rice varieties to food security in flood-prone rainfed lowland areas in Asia. *Field Crop. Res.* **2013**, *152*, 83–93. [CrossRef]

2. Septiningsih, E.M.; Ignacio, J.C.I.; Sendon, P.M.D.; Sanchez, D.L.; Ismail, A.M.; Mackill, D.J. QTL mapping and confirmation for tolerance of anaerobic conditions during germination derived from the rice landrace Ma-Zhan Red. *Theor. Appl. Genet.* **2013**, *126*, 1357–1366. [CrossRef]

3. Singh, S.; Mackill, D.J.; Ismail, A.M. Responses of SUB1 rice introgression lines to submergence in the field: Yield and grain quality. *Field Crop. Res.* **2009**, *113*, 12–23. [CrossRef]

4. Bailey-Serres, J.; Fukao, T.; Ronald, P.; Ismail, A.; Heuer, S.; Mackill, D. Submergence tolerant rice: SUB1's journey from landrace to modern cultivar. *Rice* **2010**, *3*, 138–147. [CrossRef]

5. Perata, P.; Pozueta-Romero, J.; Akazawa, T.; Yamaguchi, J. Effect of anoxia on starch breakdown in rice and wheat seeds. *Planta* **1992**, *188*, 611–618. [CrossRef] [PubMed]

6. Guglielminetti, L.; Perata, P.; Alpi, A. Effect of anoxia on carbohydrate metabolism in rice seedlings. *Plant Physiol.* **1995**, *108*, 735–741. [CrossRef] [PubMed]

7. Lasanthi-Kudahettige, R.; Magneschi, L.; Loreti, E.; Gonzali, S.; Licausi, F.; Novi, G.; Beretta, O.; Vitulli, F.; Alpi, A.; Perata, P. Transcript profiling of the anoxic rice coleoptile. *Plant Physiol.* **2007**, *144*, 218–231. [CrossRef]

8. Loreti, E.; Alpi, A.; Perata, P. α-amylase expression under anoxia in rice seedlings: An update. *Russ. J. Plant Physiol.* **2003**, *50*, 737–742. [CrossRef]

9. Fröhlich, M.; Kutschera, U. Changes in soluble sugars and proteins during development of rye coleoptiles. *J. Plant Physiol.* **1995**, *146*, 121–125. [CrossRef]

10. Narsai, R.; Edwards, J.M.; Roberts, T.H.; Whelan, J.; Joss, G.H.; Atwell, B.J. Mechanisms of growth and patterns of gene expression in oxygen-deprived rice coleoptiles. *Plant J.* **2015**, *82*, 25–40. [CrossRef]

11. Inada, N.; Sakai, A.; Kuroiwa, H.; Kuroiwa, T. Three-dimensional progression of programmed death in the rice coleoptile. *Int. Rev. Cytol.* **2002**, *218*, 221–258. [CrossRef] [PubMed]

12. Inada, N.; Sakai, A.; Kuroiwa, H.; Kuroiwa, T. Senescence in the nongreening region of the rice (*Oryza sativa*) coleoptile. *Protoplasma* **2000**, *214*, 180–193. [CrossRef]

13. Kawai, M.; Uchimiya, H. Coleoptile senescence in rice (*Oryza sativa* L.). *Ann. Bot.* **2000**, *86*, 405–414. [CrossRef]

14. Narsai, R.; Secco, D.; Schultz, M.D.; Ecker, J.R.; Lister, R.; Whelan, J. Dynamic and rapid changes in the transcriptome and epigenome during germination and in developing rice (*Oryza sativa*) coleoptiles under anoxia and re-oxygenation. *Plant J.* **2017**, *89*, 805–824. [CrossRef] [PubMed]

15. Jones, T.J.; Rost, T.L. The developmental anatomy and ultrastructure of somatic embryos from rice (*Oryza sativa* L.) scutellum epithelial cells. *Bot. Gaz.* **1989**, *150*, 41–49. [CrossRef]

16. Takahashi, H.; Saika, H.; Matsumura, H.; Nagamura, Y.; Tsutsumi, N.; Nishizawa, N.K.; Nakazono, M. Cell division and cell elongation in the coleoptile of rice alcohol dehydrogenase 1-deficient mutant are reduced under complete submergence. *Ann. Bot.* **2011**, *108*, 253–261. [CrossRef]

17. Atwell, B.J.; Waters, I.; Greenway, H. The effect of oxygen and turbulence on elongation of coleoptiles of submergence-tolerant and -intolerant rice cultivars. *J. Exp. Bot.* **1982**, *33*, 1030–1044. [CrossRef]

18. Kordan, H.A. Patterns of shoot and root growth in rice seedlings germinating under water. *J. Appl. Ecol.* **1974**, *11*, 685. [CrossRef]

19. Nghi, K.N.; Tagliani, A.; Mariotti, L.; Weits, D.A.; Perata, P.; Pucciariello, C. Auxin is required for the long coleoptile trait in *japonica* rice under submergence. *New Phytol.* **2020**. [CrossRef]

20. Nghi, K.N.; Tondelli, A.; Valè, G.; Tagliani, A.; Marè, C.; Perata, P.; Pucciariello, C. Dissection of coleoptile elongation in japonica rice under submergence through integrated genome-wide association mapping and transcriptional analyses. *Plant Cell Environ.* **2019**, *42*, 1832–1846. [CrossRef]

21. Magneschi, L.; Kudahettige, R.L.; Alpi, A.; Perata, P. Expansin gene expression and anoxic coleoptile elongation in rice cultivars. *J. Plant Physiol.* **2009**, *166*, 1576–1580. [CrossRef] [PubMed]

22. Pucciariello, C.; Banti, V.; Perata, P. ROS signaling as common element in low oxygen and heat stresses. *Plant Physiol. Biochem.* **2012**, *59*, 3–10. [CrossRef] [PubMed]

23. Mohanty, B.; Herath, V.; Wijaya, E.; Yeo, H.C.; de los Reyes, B.G.; Lee, D.Y. Patterns of cis-element enrichment reveal potential regulatory modules involved in the transcriptional regulation of anoxia response of *japonica* rice. *Gene* **2012**, *511*, 235–242. [CrossRef] [PubMed]

24. Hsu, S.-K.; Tung, C.-W. RNA-Seq analysis of diverse rice genotypes to identify the genes controlling coleoptile growth during submerged germination. *Front. Plant Sci.* **2017**, *8*, 762. [CrossRef]

25. Hsu, S.K.; Tung, C.W. Genetic mapping of anaerobic germination-associated QTLs controlling coleoptile elongation in rice. *Rice* **2015**, *8*, 38. [CrossRef]

26. Angaji, S.A.; Septiningsih, E.M.; Mackill, D.J.; Ismail, A.M. QTLs associated with tolerance of flooding during germination in rice (*Oryza sativa* L.). *Euphytica* **2010**, *172*, 159–168. [CrossRef]

27. Zhang, M.; Lu, Q.; Wu, W.; Niu, X.; Wang, C.; Feng, Y.; Xu, Q.; Wang, S.; Yuan, X.; Yu, H.; et al. Association mapping reveals novel genetic loci contributing to flooding tolerance during germination in *indica* rice. *Front. Plant Sci.* **2017**, *8*, 678. [CrossRef]

28. Nishimura, T.; Sasaki, K.; Yamaguchi, T.; Takahashi, H.; Junko Yamagishi, J.; Kato, Y. Detection and characterization of quantitative trait loci for coleoptile elongation under anaerobic conditions in rice. *Plant Prod. Sci.* **2020**, *23*, 374–383. [CrossRef]

29. Kretzschmar, T.; Pelayo, M.A.F.; Trijatmiko, K.R.; Gabunada, L.F.M.; Alam, R.; Jimenez, R.; Mendioro, M.S.; Slamet-Loedin, I.H.; Sreenivasulu, N.; Bailey-Serres, J.; et al. A trehalose-6-phosphate phosphatase enhances anaerobic germination tolerance in rice. *Nat. Plants* **2015**, *1*, 15124. [CrossRef]

30. Biscarini, F.; Cozzi, P.; Casella, L.; Riccardi, P.; Vattari, A.; Orasen, G.; Perrini, R.; Tacconi, G.; Tondelli, A.; Biselli, C.; et al. Genome-wide association study for traits related to plant and grain morphology, and root architecture in temperate rice accessions. *PLoS ONE* **2016**, *11*, e0155425. [CrossRef]

31. Hwang, Y.S.; Thomas, B.R.; Rodriguez, R.L. Differential expression of rice α-amylase genes during seedling development under anoxia. *Plant Mol. Biol.* **1999**, *40*, 911–920. [CrossRef] [PubMed]

32. Lee, K.; Chen, P.; Lu, C.; Chen, S.; Ho, T.D.; Yu, S. Coordinated responses to oxygen and sugar deficiency allow rice seedlings to tolerate flooding. *Sci. Signal.* **2009**, *2*, ra61. [CrossRef] [PubMed]

33. Weinl, S.; Kudla, J. The CBL-CIPK Ca^{2+}-decoding signaling network: Function and perspectives. *New Phytol.* **2009**, *184*, 517–528. [CrossRef] [PubMed]

34. Ho, V.T.; Tran, A.N.; Cardarelli, F.; Perata, P.; Pucciariello, C. A calcineurin B-like protein participates in low oxygen signalling in rice. *Funct. Plant Biol.* **2017**, *44*, 917–928. [CrossRef]

35. Ye, N.H.; Wang, F.Z.; Shi, L.; Chen, M.X.; Cao, Y.Y.; Zhu, F.Y.; Wu, Y.Z.; Xie, L.J.; Liu, T.Y.; Su, Z.Z.; et al. Natural variation in the promoter of rice calcineurin B-like protein10 (OsCBL10) affects flooding tolerance during seed germination among rice subspecies. *Plant J.* **2018**, *94*, 612–625. [CrossRef]

36. Lu, C.; Lim, E.; Yu, S. Sugar response sequence in the promoter of a rice a-Amylase gene. *J. Biolo. Chem.* **1998**, *273*, 10120–10131. [CrossRef]

37. Lu, C.A.; Ho, T.H.; Ho, S.L.; Yu, S.M. Three novel MYB proteins with one DNA binding repeat mediate sugar and hormone regulation of alpha-amylase gene expression. *Plant Cell* **2002**, *14*, 1963–1980. [CrossRef]

38. Lu, C.-A.; Lin, C.-C.; Lee, K.-W.; Chen, J.-L.; Huang, L.-F.; Ho, S.-L.; Liu, H.-J.; Hsing, Y.-I.; Yu, S.-M. The SnRK1A protein kinase plays a key role in sugar signaling during germination and seedling growth of rice. *Plant Cell* **2007**, *19*, 2484–2499. [CrossRef]

39. Magneschi, L.; Kudahettige, R.L.; Alpi, A.; Perata, P. Comparative analysis of anoxic coleoptile elongation in rice varieties: Relationship between coleoptile length and carbohydrate levels, fermentative metabolism and anaerobic gene expression. *Plant Biol.* **2009**, *11*, 561–573. [CrossRef]

40. Takahashi, H.; Greenway, H.; Matsumura, H.; Tsutsumi, N.; Nakazono, M. Rice alcohol dehydrogenase 1 promotes survival and has a major impact on carbohydrate metabolism in the embryo and endosperm when seeds are germinated in partially oxygenated water. *Ann. Bot.* **2014**, *113*, 851–859. [CrossRef]

41. Nagao, M.; Ohwaki, Y. The action of trans-cinnamic and 2,3,5-triiodobenzoic acids in the rice seedling. *Sci. Rep. Tohoku Univ.* **1955**, *21*, 96–108.

42. Kefford, N.P. Auxin-gibberellin interaction in rice coleoptile elongation. *Plant Physiol.* **1962**, *37*, 380–386. [CrossRef] [PubMed]

43. Pegoraro, R.; Mapelli, S.; Torti, G.; Bertani, A. Indole-3-acetic acid and rice coleoptile elongation under anoxia. *J. Plant Growth Regul.* **1988**, *7*, 85–94. [CrossRef]

44. Horton, R.F. The effect of ethylene and other regulators on coleoptile growth of rice under anoxia. *Plant Sci.* **1991**, *79*, 57–62. [CrossRef]

45. Breviario, D.; Giani, S.; Di Vietri, P.; Coraggio, I. Auxin and growth regulation of rice coleoptile segments: Molecular analysis. *Plant Physiol.* **1992**, *98*, 488–495. [CrossRef]

46. Hoson, T.; Masuda, Y.; Pilet, P.E. Auxin content in air and water grown rice coleoptiles. *J. Plant Physiol.* **1992**, *139*, 685–689. [CrossRef]

47. Guo, F.; Han, N.; Xie, Y.; Fang, K.; Yang, Y.; Zhu, M.; Wang, J.; Bian, H. The miR393a/target module regulates seed germination and seedling establishment under submergence in rice (*Oryza sativa* L.). *Plant Cell Environ.* **2016**, *39*, 2288–2302. [CrossRef]

48. Ismail, A.M.; Ella, E.S.; Vergara, G.V.; Mackill, D.J. Mechanisms associated with tolerance to flooding during germination and early seedling growth in rice (*Oryza sativa*). *Ann. Bot.* **2009**, *103*, 197–209. [CrossRef]

49. Hartman, S.; Liu, Z.; van Veen, H.; Vicente, J.; Reinen, E.; Martopawiro, S.; Zhang, H.; van Dongen, N.; Bosman, F.; Bassel, G.W.; et al. Ethylene-mediated nitric oxide depletion pre-adapts plants to hypoxia stress. *Nat. Commun.* **2019**, *10*, 4020. [CrossRef]

50. Yang, C.; Ma, B.; He, S.J.; Xiong, Q.; Duan, K.X.; Yin, C.C.; Chen, H.; Lu, X.; Chen, S.Y.; Zhang, J.S. MAOHUZI6/ETHYLENE INSENSITIVE3-LIKE1 and ETHYLENE INSENSITIVE3-LIKE2 regulate ethylene response of roots and coleoptiles and negatively affect salt tolerance in rice. *Plant Physiol.* **2015**, *169*, 148–165. [CrossRef]

51. Gibbs, D.J.; Lee, S.C.; Md Isa, N.; Gramuglia, S.; Fukao, T.; Bassel, G.W.; Correia, C.S.; Corbineau, F.; Theodoulou, F.L.; Bailey-Serres, J.; et al. Homeostatic response to hypoxia is regulated by the N-end rule pathway in plants. *Nature* **2011**, *479*, 415–418. [CrossRef] [PubMed]

52. Lin, C.C.; Chao, Y.T.; Chen, W.C.; Ho, H.Y.; Chou, M.Y.; Li, Y.R.; Wu, Y.L.; Yang, H.A.; Hsieh, H.; Lin, C.S.; et al. Regulatory cascade involving transcriptional and N-end rule pathways in rice under submergence. *Proc. Natl. Acad. Sci. USA* **2019**, *116*, 3300–3309. [CrossRef] [PubMed]

53. Qin, H.; Zhang, Z.; Wang, J.; Chen, X.; Wei, P.; Huang, R. The activation of OsEIL1 on YUC8M transcription and auxin biosynthesis is required for ethylene-inhibited root elongation in rice early seedling development. *PLoS Genet.* **2017**, *13*, e1006955. [CrossRef]

54. Ishizawa, K.; Esashi, Y. Osmoregulation in rice coleoptile elongation as promoted by cooperation between IAA and ethylene. *Plant Cell Physiol.* **1984**, *25*, 495–504. [CrossRef]

Review

The Molecular Regulatory Pathways and Metabolic Adaptation in the Seed Germination and Early Seedling Growth of Rice in Response to Low O_2 Stress

Mingqing Ma [1], Weijian Cen [1], Rongbai Li [1,2], Shaokui Wang [3] and Jijing Luo [1,*]

[1] College of Life Science and Technology (State Key Laboratory for Conservation and Utilization of Subtropical Agro-Bioresources), Guangxi University, Nanning 530004, China; mqma@st.gxu.edu.cn (M.M.); cweijian@gxu.edu.cn (W.C.); lirongbai@126.com (R.L.)

[2] Agriculture College, Guangxi University, Nanning 530004, China

[3] Agriculture College, South China Agricultural University, Guangzhou 510642, China; shaokuiwang@scau.edu.cn

* Correspondence: jjluo@gxu.edu.cn; Tel.: +86-180-7779-2389

Received: 11 September 2020; Accepted: 12 October 2020; Published: 14 October 2020

Abstract: As sessile organisms, flooding/submergence is one of the major abiotic stresses for higher plants, with deleterious effects on their growth and survival. Therefore, flooding/submergence is a large challenge for agriculture in lowland areas worldwide. Long-term flooding/submergence can cause severe hypoxia stress to crop plants and can result in substantial yield loss. Rice has evolved distinct adaptive strategies in response to low oxygen (O_2) stress caused by flooding/submergence circumstances. Recently, direct seeding practice has been increasing in popularity due to its advantages of reducing cultivation cost and labor. However, establishment and growth of the seedlings from seed germination under the submergence condition are large obstacles for rice in direct seeding practice. The physiological and molecular regulatory mechanisms underlying tolerant and sensitive phenotypes in rice have been extensively investigated. Here, this review focuses on the progress of recent advances in the studies of the molecular mechanisms and metabolic adaptions underlying anaerobic germination (AG) and coleoptile elongation. Further, we highlight the prospect of introducing quantitative trait loci (QTL) for AG into rice mega varieties to ensure the compatibility of flooding/submergence tolerance traits and yield stability, thereby advancing the direct seeding practice and facilitating future breeding improvement.

Keywords: submergence; direct seeding; anaerobic germination; low O_2 stress; regulatory mechanism; metabolic adaptation

---

1. Introduction

Rice (*Oryza sativa* L.) is a staple food crop feeding more than half of the world's population [1]. The traditional rice production system in Asian countries commonly involves transplanting seedlings from the nursery into a paddy field. This production pattern is labor-, water-, and energy-consuming and is becoming less profitable [2]. Therefore, in recent years, direct seeding has been receiving much attention worldwide, especially in Asian countries, and farmers have shifted to direct seeding of rice due to its low cost and labor-saving strategy [2–4]. There are three major methods for rice direct seeding: dry seeding (sowing dry seeds into dry soil), wet seeding (sowing pregerminated seeds on wet puddled soils), and water seeding (sowing seeds into standing water) [5]. However, poor crop stand establishment has become a key obstacle that prevents the subsequent growth, development, and yield in the direct seeding of rice [5]. In particular, water seeding severely affects the establishment of rice seedlings due to low water oxygen under the submergence condition.

For the early emerging seedlings, submergence reduces air diffusion, which limits O_2 and carbon dioxide (CO_2) availability in the submerged tissues, resulting in a hypoxic (<21% O_2) environment, with an O_2 concentration below that under normoxic conditions [6–9]. At low O_2 supply (hypoxia) conditions, O_2 concentration can affect the respiration in plants, and respiration can still active although at a strong reduced rate when O_2 concentration is below the critical oxygen pressure for respiration (COPR), which is the lowest oxygen partial pressure to support maximum respiration rate [10,11]. Under O_2 deficiency conditions, the accumulation of several phytotoxic substances such as reduced iron (Fe^{2+}), manganese (Mn^{2+}), hydrogen sulfide (H_2S), oxygen radicals, and the products of fermentation, cause severe damage to plants. The ability of rice seeds to avoid the damages caused by these toxins is helpful to prevent further injury [4,12]. Although most modern rice varieties have low ability to germinate under water, rice has developed various adaptive mechanisms morphologically and physiologically in order to adapt to a wide range of hydrological environments. There are genetic variations exist among rice varieties, and these variations lay the base for the possibility of studying the molecular regulatory mechanism of AG-related traits and facilitate breeding rice varieties that are suitable for direct seeding systems [4]. In direct seeding practice, varieties that are suitable for water seeding have characteristics of high anaerobic germination tolerance, early vigorous seedling growth, fast root growth, early tillering, and lodging resistance [13]. Therefore, the breeding of rice varieties that are capable of surviving and accommodating submergence conditions during germination and early growth stages improves the success of direct-seeded rice [14]. Thus, prior to breeding applications, it is of great importance to isolate related genes for AG tolerance from tolerant genotypes and to understand the genetic basis and the molecular regulatory mechanisms underlying anaerobic germination and submerged seedling growth.

Despite the sophisticated molecular mechanisms involved in the regulation of the anaerobic germination and seedling growth in rice, considerable progress on understanding the genetic, molecular and physiological basis of rice in response to submergence stress at the germination stage, has been achieved in the last few decades [15–23]. Here, we mainly focus on reviewing the molecular genetic mechanisms that regulate the metabolic adaptation of rice to anaerobic germination stress; we could not cover all the related advances due to space limitations. The advances reviewed here are crucial for understanding the fundamental mechanisms and for the breeding design in direct seeding practice. Last but not least, we proposed future research prospects and premised the issues raised that remain for further investigation.

2. Strategies are Adopted in Rice under Submerged Germination

Anaerobic germination (AG) tolerance is an important trait required for rice to successfully germinate in a direct seeding system. Rice has the ability to perceive low O_2 stress and regulate germination and early seedling growth under hypoxic conditions [24,25]. There are two distinct adaptive responses or survival strategies for coping with submergence stress in plants: one is low O_2 escape strategy (LOES), and the other is low O_2 quiescence strategy (LOQS) [26–29]. In the case of the established rice plants, the two adaptive strategies for flooding tolerance are *SUBMERGENCE 1A* (*SUB1A*)-dependent quiescence strategy and *SNORKEL 1* and *SNORKEL 2* (*SK1/2*)-dependent escape strategy [30]. Overexpression of *SUB1A* stimulated the expression of *alcohol dehydrogenase* (*ADH*) genes for ethanolic fermentation and conferred submergence tolerance at the vegetative stage in rice by up-regulating the expression of *slender rice-1* (*SLR1*) and *SLR1 like-1*(*SLRL1*), two negative regulators for gibberellin (GA) signaling, thereby inhibiting rice elongation, reducing the carbohydrate consumption, and exhibiting a strong submergence tolerance in rice plants. The rice varieties with *SUB1A* gene could survive up to two weeks of the complete submergence [31–33]. For *SNORKEL 1* and *SNORKEL 2* (*SK1/2*)-dependent escape strategy, *SK1/2* promotes the internode elongation via the stimulation of GA biosynthesis in deep-water rice under flooding stress, and thereby enabling rice grows upward to the water surface for air exchange [34].

During germination and early seeding stages of rice, the LOES is characterized by promoting the elongation of the mesocotyl and coleoptile to allow plants to reach the water surface to increase air exchange between the aerial and submerged tissues. In rice, varieties that have adopted LOES in response to the submergence stress during germination are considered stress-tolerant [24]. Conversely, LOQS is a strategy that constrains the elongation of the mesocotyl and coleoptile and preserves energy for prolonged submergence stress. The varieties that exhibit slow coleoptile growth under low O_2 conditions are sensitive to submergence stress, thereby impairing the subsequent growth and development of the seedlings (Figure 1). Therefore, coleoptile elongation is a main index for selection of anerobic germination-tolerant rice varieties.

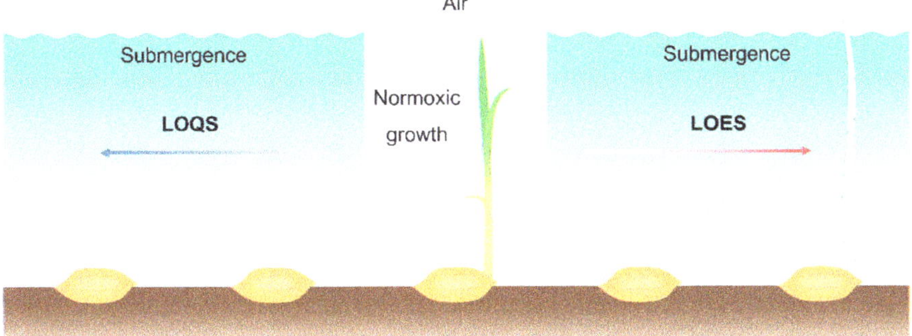

Figure 1. Low O_2 escape strategy (LOES) and low O_2 quiescence strategy (LOQS) strategy of rice germination under submergence conditions. Oxygen is one of the important factors that affects the germination and early seedling growth of rice. Under normoxic conditions, the surrounding oxygen ensures normal seed germination and early growth of the seedlings. Under the submergence condition, the formation of roots is inhibited in rice seedlings after germination. Sensitive cultivars adopt the quiescence strategy that inhibits germination and coleoptile elongation under low O_2 stress, while tolerant varieties show phenotypes with faster coleoptile elongation capacity to reach the water surface for air exchange.

Previous studies have revealed that cell elongation contributes to the rapid increase in the length of the coleoptile. During the early stage of germination, cell division is activated to increase cell numbers for coleoptile growth, and the subsequent cell expansion contributes to elongation of the coleoptile under anerobic conditions [25,35,36]. Some specific expansins were reported to contribute to coleoptile elongation in rice under anaerobic conditions [25,37,38]; for example, *EXPA2, EXPA4, EXPA1, EXPB11*, and *EXPB17* have been reported to be highly expressed in the coleoptiles under submergence stress [36,37,39]. To adapt to the submergence environment during the germination and early stages of growth, rice plants have evolved specific regulatory mechanisms to modulate metabolic shifts in response to low O_2 stress [40,41], which will be described in more detail in the following sections.

3. Metabolic Adaption to Anaerobic Germination in Rice

In the submerged germination of rice, aerobic respiration is severely inhibited due to water O_2 deficiency, leading to differential expression of the genes that contribute to activating the essential mechanisms and, in turn, to shifting the metabolic processes to anerobic respiration, which is related to energy production and utilization in the germinating seeds [42]. These alternative pathways to aerobic respiration are known as fermentation, including alcoholic fermentation, lactic acid fermentation, and the alanine pathway [43–45]. Meanwhile, tricarboxylic acid (TCA) cycle and oxidative phosphorylation processes are inhibited, and the production of ATP is shifted from electron transport chains (ETCs) to glycolysis and ethanol fermentation. Consequently, the net energy yield from anaerobic fermentation (2 mol ATP per mol of glucose) is lower than that from aerobic respiration

(30–32 mol ATP per mol of glucose) (Figure 2). Although limited amounts of energy are produced from ethanol fermentation, it is extremely important for the germination and seedling growth under submergence condition. The energy is allocated to critical bioprocesses, especially protein synthesis in the cells of the coleoptile, contributing to submergence tolerance and success in the establishment of seedlings [46]. Likewise, the reoxidization of NADH to NAD$^+$ in anaerobic fermentation is also required for continuous glycolysis [6,47]. A recent report suggested that the AG-tolerant genotypes of rice adopt a strategy that strongly increases starch degradation in the endosperm into metabolizable sugars, thereby supplying substrates for subsequent glycolysis and alcohol fermentation to produce the energy required for enhancing the germination vitality and for the rapid outgrowth of the coleoptile [15].

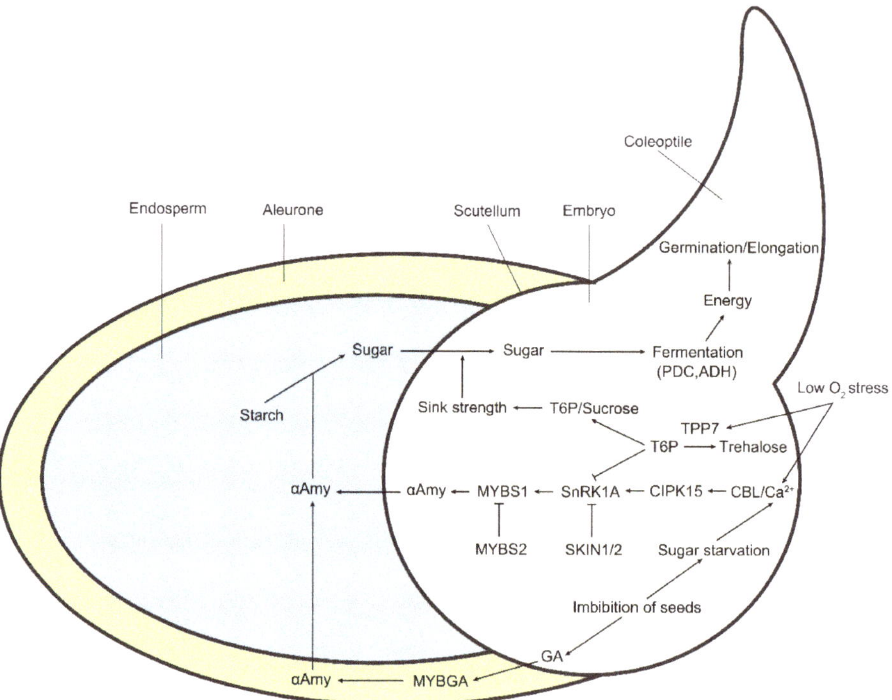

Figure 2. The molecular regulatory pathways and metabolic pathways of rice under hypoxic germination and early seedling growth. Sugar starvation caused by low O_2 stress under the submergence condition is the key upstream signal affecting metabolic regulation pathways. The Ca^{2+} signal acts as a secondary messenger to mediate the downstream responses. The calcineurin B-like (CBL) proteins bind to Ca^{2+} and then interact with CIPK15, leading to the activation of its kinase activity. Subsequently, the activated CIPK15 physically interacts with SnRK1A, an upstream protein kinase of the transcription factor MYBS1, and activates its activity, consequently elevating the activity of αAmy for seed stored starch degradation. *SKIN1/2* and *MYBS2* negatively regulate SnRK1A and MYBS1, respectively, thereby repressing *αAmy* expression. Additionally, seed imbibition also results in the biosynthesis of gibberellin (GA) in the embryo. GA induces the expression of *MYBGA* in cereal aleurone cells, thereby upregulating the expression of the *αAmy* gene. Meanwhile, low O_2-induced *OsTPP7* desuppresses the SnRK1A activity that is inhibited by trehalose 6-phosphate (T6P). OsTPP7 increases the sink strength of the embryo axis–coleoptile by the perception of low sugar availability by enhancing the conversion of T6P to trehalose and leads to a decrease in the T6P/sucrose ratio, thus enhancing starch mobilization for energy production to promote coleoptile elongation. In addition, low O_2 results in the shift of aerobic respiration to anaerobic fermentation, therefore inducing the expression of essential components, including *PDC* and *ADH*.

In the starch metabolic pathways, α-amylase (αAmy) (1,4-α-D-glucan maltohydrolase) is one of the most abundant hydrolases in rice and is involved in mobilization of the stored starch [48,49]; it catalyzes the hydrolysis of α-1,4-glucosidic bonds of starch, yielding α-glucose, and α-maltose [50]. In rice, a total of 10 distinct *αAmy* genes are classified into three subfamilies: subfamilies RAmy1 (A, B, and C), RAmy2A, and RAmy3 (A, B, C, D, E, and F). Most of the *αAmy* genes belong to the RAmy3 subfamily [51,52], in which only one gene (LOC_Os08g36910) is induced by sugar starvation and O_2 deficiency [53]. Thus, RAmy3D is considered an important enzyme in the fermentative metabolism pathway for metabolic shift regulation and in the production of energy in response to submergence stress [19] (Table 1). Regarding the alcoholic fermentation process, ADH, pyruvate decarboxylase (PDC), and aldehyde dehydrogenase (ALDH) are key enzymes that are involved in the reduction of pyruvate into alcohol. ADH catalyzes the rate-limiting step of ethanol metabolism and is considered to be essential for carbohydrate metabolism that is critical for the germination and growth of rice under low O_2 stress [54,55]. It has been reported that a mutation in rice *ADH1* leads to a reduction in the level of ADH protein, and thus compromises the tolerance to anaerobic stress of the *adh*-null mutant, suggesting that a functional ADH protein is required for enhancing tolerance to submergence during rice germination [20,56] (Figure 2). Moreover, pyruvate is a final product of glycolysis, which is converted to lactate by lactate dehydrogenase (LDH). The reduction in cytoplasmic pH inhibits the activity of LDH and activates the activity of PDC, leading to the reduction of intracellular lactate concentration. Therefore, the regulation of cytoplasmic pH, enables rice to avoid acidosis and improves the survival under low O_2 conditions [4,25,40]. Therefore, enhancement of the activity of major enzymes for either starch hydrolysis or ethanol synthesis is in favor of promoting the elongation of rice coleoptile under low O_2 stress [24,39,54,57].

Table 1. Subfamilies of the rice αAmy genes.

Gene	Subfamily	Chromosome	Locus Name	CDS Coordinates (5′-3′)	Regulatory Function
RAmy1 A	RAmy1	2	LOC_Os02g52700	32243146-32245056	High temperature in developing seeds [58]
RAmy1 B		2	LOC_Os02g52710	32250180-32248279	Chemical inhibition [59]
RAmy1 C		1	LOC_Os01g25510	14459951-14461849	High temperature in developing seeds [58]
RAmy2 A	RAmy2	6	LOC_Os06g49970	30262778-30266915	Unknown
RAmy3 A	RAmy3	9	LOC_Os09g28400	17288993-17291295	High temperature in developing seeds [58]
RAmy3 B		9	LOC_Os09g28420	17296166-17305076	Chemical inhibition [59]
RAmy3 C		8	LOC_Os08g36900	23335165-23337151	Unknown
RAmy3 D		8	LOC_Os08g36910	23340676-23343533	High temperature in developing seeds [58] Sugar starvation [53,60] Calcium signaling [21]
RAmy3 E		4	LOC_Os04g33040	20006128-2000927	High temperature in developing seeds [58] Chemical inhibition [59]
RAmy3 F		1	LOC_Os01g51754	29760719-29770037	Unknown

Upon imbibition of cereal grains, the rapid consumption of soluble carbohydrates during the germination and seedling growth stages causes sugar starvation and the activation of αAmy biosynthesis in the scutellum, which is mediated by the CIPK15-SnRK1A-MYBS1 signaling pathway [61]. Among major cereals, only rice produces a complete set of enzymes to degrade starch under AG [19]. With a higher αAmy activity for the degradation of starch to produce energy in germinating seeds, rice promotes coleoptile elongation for surviving the anerobic stress [15,19,53].

4. The Molecular Regulatory Pathway of Anaerobic Germination in Rice

As mentioned above, rice submergence-tolerant varieties adopt LOES in response to low O_2 stress [24]. In submerged germination, O_2 deficiency and sugar starvation are the key signals for rice responsiveness. O_2 is one of the key components for the important biochemical reactions in rice cells. When rice germinates under submergence conditions, it senses hypoxic stress and induces the responsive regulatory pathways to adapt to the adverse environment. Calcium (Ca^{2+}) signal, acting as a secondary messenger, is involved in multiple plant abiotic/biotic stress responsive signaling pathways [62]. The calcineurin B-like (CBL) proteins, a family of Ca^{2+}-binding proteins and one of the major calcium ion sensors in plants, interact with CBL protein-interacting protein kinases (CIPKs) to form the CBL–CIPK complex, thereby decoding Ca^{2+} signaling in response to various abiotic stresses [63–65]. CIPKs is a group of plant-specific Ser/Thr protein kinases that contain an N-terminal kinase catalytic domain and a C-terminal regulatory NAF (Asn-Ala-Phe) domain. The NAF domain of CIPKs is sufficient for interaction with CBL protein and leads to the activation of the CIPKs [66].

Recent studies have revealed that the CBL/*CIPK15-SnRK1A-MYBS1*-mediated sugar-signaling cascade plays an important role in low-O_2 tolerance during the seed germination stage of rice [15,17]. CIPK15 is essential for regulating the expression of genes encoding αAmy and ADH [15] (Figure 2). CBL4 has been reported to be induced by low-oxygen conditions and strongly interacts with the CIPK15 protein. The silence of *CBL4* results in downregulation of the expression of α*Amy3* (*RAmy3 D*). These findings suggest that the CBL4/CIPK15 complex plays a critical role in the sugar starvation signaling pathway and may contribute to the germination of rice under the hypoxia condition [21,67]. OsCBL10, another member of the rice CBL family, has been reported to be involved in flooding stress response during rice germination by comparing flooding-tolerant (Low88) with flooding-intolerant (Up221) cultivars. The natural variation in the *OsCBL10* promoter is associated with the tolerance of anaerobic germination and seedling growth among rice subspecies [16]; according to the sequence variation, its promoter can be characterized into two types, including a flooding-tolerant type (T-type) that only exists in *japonica* lowland cultivars and a flooding-intolerant type (I-type) that exists in *japonica* upland-, and *indica* upland/lowland-, cultivars. The flooding-intolerant cultivars that contain an I-type promoter can downregulate CIPK15 protein accumulation by inducing the expression of *OsCBL10*. In contrast, the cultivars containing a T-type promoter have downregulated *OsCBL10* transcription, higher αAmy activity, and increased CIPK15 protein accumulation [16]. Furthermore, however, the *OsCBL10*-overexpressing line is more sensitive to flooding stress than the wild type, suggesting that *CBL10* negatively regulates the *CIPK15-SnRK1A-MYBS1*-mediated pathway [16].

Rice sucrose nonfermenting-1 (SNF1) homolog SNF1-related protein kinase-1 (SnRK1) is encoded by *SnRK1A* and *SnRK1B*, and SnRK1A is structurally and functionally analogous to SNF1 in yeast and AMP-activated protein kinase (AMPK) in mammals [17,68]. SnRK1A, SNF1, and AMPK are considered as metabolic sensors for monitoring cellular carbohydrate status and/or AMP/ATP levels, play an important role in the regulation of carbon and nitrogen metabolism, and are essential for the proper growth and development of higher eukaryotes [68–70]. In the downstream of CBL/CIPK15 complex, SnRK1A is involved in sugar starvation and O_2 deficiency signaling pathways and acts as a conserved energy and stress sensor under submergence condition [15]. Additionally, SnRK1A-interacting negative regulators (SKINs) physically interact with SnRK1A and negatively regulate its activity to modulate the nutrient starvation signaling and its downstream pathways, including inhibition of the expression of *MYBS1* and α*Amy 3*, consequently resulting in compromising anaerobic germination and seedling growth under submergence stress in rice [22] (Figure 2).

Myeloblastosis sugar response complex 1 (MYBS1) and MYBGA are two crucial transcription factors that can integrate diverse nutrient starvation and GA signaling pathways in downstream of SnRK1A in rice [71]. Under sugar-depleted conditions, MYBS1 specifically binds to the TATCCA *cis*-acting element (the TA-box) that is located on the sugar response complex (SRC) of the α*Amy* promoter, and consequently elevates the activity of αAmy for hydrolysis of starch in the endosperm [17,72,73]. SnRK1A mediates the interaction between MYBS1 and α*Amy 3* SRC, and thus regulates the seed germination and seedling growth of rice [17] (Figure 2). In addition, the phytohormone GA controls diverse aspects of plant growth and development, including seed germination and stem elongation [74]. Endogenous GA is synthesized in embryo and is then transported to the aleurone cells to promote the secretion of αAmy [75–77]. Furthermore, MYBGA is a GA-inducible MYB transcription factor that is expressed in cereal aleurone cells [78]. With GA induction, MYBGA upregulates the expression of *αAmy* by binding to the GA-responsive element (GARE) of the *αAmy 3* promoter [75,79,80]. The formation of a stable bipartite MYB (MYBS1-MYBGA)-DNA (GARE-TA box) complex in the cytoplasm coordinates the expression of *αAmy* genes [71]. A recent study revealed that the transcription factor MYBS2 competes with MYBS1 for the same *cis*-element of the *αAmy* promoter and represses *αAmy* expression (Figure 2). Via promoting or inhibiting the expression level of *αAmy*, the competition between MYBS1 and MYBS2 can regulate and maintain sugar within an appropriate range during germination and early seedling growth under sugar starvation stress [23].

Moreover, trehalose-6-phosphate (T6P) metabolism is modulated in response to submergence stress. T6P phosphatase catalyzes the conversion of T6P to trehalose and changes the T6P:sucrose ratio [18,81]. The elevated level of T6P represses the activity of SnRK1A [82] (Figure 2). Therefore, T6P, is considered a sensor of sucrose effectiveness under low oxygen stress and is critical for allocating carbon from source to sink tissues and for maintaining sucrose homeostasis, which influences many metabolic and developmental processes in plants [83,84].

In addition, phytohormones also play an important role in rice under submergence. Ethylene has a regulatory role in the upstream of both the *SUB1A*-dependent quiescence strategy and *SK1/2*-dependent escape strategy [28,85]. GA plays negative roles in *SUB1A*-dependent pathway mediating submergence tolerance and plays positive roles in the *SK1/2*-dependent pathway, triggering rapid internode elongation. GA also promote the elongation of rice coleoptile in the *CIPK15*-dependent pathways under submergence [15,28,40,86].

Thus, *CIPK15* is induced by sugar starvation/low O_2 signals caused by submergence and then activates the accumulation of SnRK1A protein. The accumulation of SnRK1A triggers a SnRK1A-dependent signaling cascade in response submergence stress. The downstream *SnRK1A-MYBS1-αAmy 3* sugar signaling cascade promotes starch degradation and sugar mobilization, indicating that the *CIPK15-SnRK1A-MYBS1*-mediated sugar sensing pathway plays a key role in the submerged germination of rice seeds [15,17].

5. Identification of QTLs/Genes for AG Tolerance

Many QTLs and candidate genes for AG tolerance have been reported in recent years. Coleoptile elongation and seedling survival rate under submerged conditions have been widely used as major indicator traits for tolerance phenotyping in QTL mapping and genome-wide association study (GWAS) approach (Table 2). Among these, five QTLs associated with tolerance of submergence during germination were mapped on chromosomes 1 (*qAG-1-2*), 3 (*qAG-3-1*), 7(*qAG-7-2*), and 9 (*qAG-9-1* and *qAG 9 2*), respectively [87]; four QTLs (*qAG7.1*, *qAG7.2*, *qAG7.3*, and *qAG3*), which derived from *aus* landrace Kharsu 80A for AG tolerance were identified in a recent study [88]. Of which, only one QTL (*qAG-9-2*) has been cloned. *OsTPP7*, a gene encoding T6P phosphatase, was determined to be a target of the QTL *qAG-9-2*. OsTPP7 is involved in the regulation of sugar signaling and in the linking of trehalose metabolism to starch mobilization. The modulation of the local T6P: sucrose ratio by OsTPP7 promotes the flux of sugar from the source (endosperm reserve) toward the sink (embryo axis–coleoptile growth). These processes are partly mediated via the upregulation of MYBS1 and CIPK15 under low-oxygen stress, thereby increasing carbohydrate availability and consequently enhancing AG tolerance [18] (Figure 2).

GWAS is an effective and popular strategy for dissecting complex traits, which can be able to capture complex trait variations by utilizing both evolutionary as well as historical recombination events at the population level [97]. In recent studies, 15 AG tolerance loci were detected through GWAS and one candidate gene, a DUF domain-containing protein (LOC_0s06g03520) was identified and found to be highly induced by AG [94]. Additionally, two candidate genes (LOC_Os03g31550 and LOC_Os12g31350) which encode *xanthine dehydrogenase 1* (OsXDH1) and *SSXT* family protein, respectively, have been identified using GWAS approach [96]. Nonetheless, the candidate genes currently isolated from rice seedlings for AG do not always overlap in the results of QTL mapping and GWAS. These may result from the different principle of two strategies. QTL mapping is based on linkage analysis of molecular markers using artificially constructed populations, while GWAS is performed based on the association analysis between single nucleotide polymorphisms (SNPs) and phenotypes and is able to uncover a large number of associated loci in natural populations. Although many QTLs/associated loci have been mapped using molecular marker assisted selection and GWAS, respectively; however, it is required for the identification of more AG tolerance genes in order to insightfully elucidate the molecular mechanisms underlying AG tolerance and provide major loci for breeding the AG tolerant varieties via molecular breeding.

Table 2. The identified quantitative trait loci (QTLs)/candidate genes for anaerobic germination (AG) tolerance.

QTLs/Candidate Genes	Chromosome	Traits	Description & Reference
qAG-1-2	1		
qAG-3-1	3	survival rate	[87]
qAG-7-2	7		
qAG-9-1, qAG-9-2	9		
qAG7.1	7	survival rate	[89]
qAG7	7		
qAG11	11	survival rate	[90]
qAG2.1	2		
qAG1a, qAG1b	1	survival rate	[91]
qAG8	8		
qAG7.1, qAG7.2, qAG7.3	7	survival rate	[88]
qAG3	3		
qSUR6–1	6	survival rateseedling height	[92]
qSH1–1	1		
OsTPP7	9	coleoptile length	Enhancing germination and coleoptile elongation [18]
HXK6	1	coleoptile length	Encoded a hexokinase [93]
LOC_Os06g03520	6	coleoptile length	DUF domain containing protein [94]
TIR1	5		F-Box auxin receptor protein [95]
AUX1	1	coleoptile length	Control lateral root initiation in rice [95]
COI1a	1		Codes for an E3 ubiquitin ligase complex component [95]
ABC1-2	2		The ATP binding cassette transporter [95]
LOC_Os03g31550	3	survival rate and coleoptile	encodes for enzyme *xanthine dehydrogenase 1(OsXDH1)* [96]
LOC_Os12g31350	12		encodes for *SSXT* family protein [96]

6. Breeding Applications Using QTLs/Genes Underlying AG Tolerance

Introgression of the loci conferring submergence tolerance into rice mega varieties using marker-assisted selection has been extensively explored recently. The results showed that introgression has significantly enhanced submergence tolerance and did not exhibit negative impacts on rice development, yield, or grain quality [98–101].

Under field trials, the *qAG-9-2*-containing near-isogenic line (NIL-*AG1*), which carries a small Khao Hlan On (KHO) introgression in an elite cultivar (IR64) background, confers AG tolerance in the field and exhibits similar grain quality as IR64 [18]. The introgression of *AG1* into Ciherang-Sub1 significantly increased AG tolerance [101]. For rice direct seeding in the field, *AG1* and *AG2* introgression lines show no negative impact on the early seedling growth and exhibit yield stability under flooding condition [100], suggesting that the development of high-yielding varieties with submergence tolerance is feasible and imperative for the direct seeding practice.

In further studies, the more commonly detected QTLs and their candidate genes could provide promising perspectives for molecular genetic characterization of AG tolerance and for the rapid development of cultivars with enhanced submergence tolerance. The identified QTLs/genes are in favor of developing a more effective and efficient breeding strategy for direct seeding practice around the world [100,102].

7. Conclusions and Perspectives

The complex trait AG is influenced by many factors, including intrinsic genetic factors and seed quality factors, such as dormancy, maturity, and storage of nutrients, and environmental factors, such as light intensity, water oxygen content, pH, salinity, temperature, and soil physical state. Considering the major obstacle encountered in the direct seeding practice, poor seedling establishment under anerobic germination, it is critical for rice varieties to possess the ability to elongate the coleoptile faster to reach the water surface and escape the low O_2 condition to ensure normal early growth [103]. Rice with faster coleoptile elongation can be applied in direct seeding and breeding improvement [104,105]. Compared with AG-tolerant varieties (LOES), AG-intolerant varieties (LOQS) severely inhibit growth and have lower survival rate under submergence; however, AG-intolerant varieties that have the characteristics of avoiding growth under submergence conditions also seem to be meaningful for short-term flooding in direct seeding applications in the paddy fields.

Here, we highlighted the important genes and regulators that underlie the essential mechanisms for AG tolerance. Among the stress-responsive pathways, the *CIPK15*-mediated O_2-deficiency signaling regulatory pathway plays a key role in controlling sugar and energy production during AG. *CIPK15* regulates the *SnRK1A*-dependent sugar sensing, thereby regulating the abundance of αAmy and ADH for the metabolic shift and energy production to adapt to submergence conditions [15,40]. The upregulation of the key enzyme *αAmy 3* is effective in mobilizing starch to produce energy and shifts ATP production from aerobic to anaerobic respiration through signaling cascades and metabolic regulation under low O_2 stress.

Recent progress allows us to better understand the stress-related regulatory mechanisms and the field trials. However, more extensive investigations are required to resolve global food security challenges in the future. It is necessary to highlight the significance of trait-to-gene-to-field approaches to ensure that appropriate coping strategies are adopted to enable staple food crops that are more tolerant to various abiotic stresses to maintain yield stability in the field [106,107].

Author Contributions: Conceptualization, M.M. and J.L.; writing—original draft preparation, M.M. and J.L.; writing—review and editing, M.M., W.C., R.L., S.W., and J.L.; visualization, M.M.; supervision, J.L. All authors have read and agreed to the published version of the manuscript.

Funding: This work was supported by the grants from National Natural Science Foundation of China (CN) (31671646), Guangxi Hundred-Talents Program (2015), Guangxi innovation-driven development special funding project (Guike-AA17204070), and the State Key Laboratory for Conservation and Utilization of Subtropical Agro-Bioresources (SKLCUSA-a201907, -a201918).

Conflicts of Interest: The authors declare no conflict of interest.

References

1. Khush, G. Productivity Improvements in Rice. *Nutr. Rev.* **2003**, *61* (Suppl. 6), S114–S116. [CrossRef]
2. Kumar, V.; Ladha, J.K. Direct Seeding of Rice. Recent Developments and Future Research Needs. *Adv. Agron.* **2011**, *111*, 297–413.
3. Pandey, S.; Velasco, M.L. Economics of direct seeding in Asia: Patterns of adoption and research priorities. *IRRI* **1999**, *24*, 6–11.
4. Miro, B.; Ismail, A.M. Tolerance of anaerobic conditions caused by flooding during germination and early growth in rice (*Oryza sativa* L.). *Front. Plant Sci.* **2013**, *4*, 269. [CrossRef]
5. Farooq, M.; Siddique, K.H.M.; Rehman, H.; Aziz, T.; Lee, D.-J.; Wahid, A. Rice direct seeding: Experiences, challenges and opportunities. *Soil Till. Res.* **2011**, *111*, 87–98. [CrossRef]
6. Gibbs, J.; Greenway, H. Review: Mechanisms of anoxia tolerance in plants. I. Growth, survival and anaerobic catabolism. *Funct. Plant Biol.* **2003**, *30*, 353. [CrossRef] [PubMed]
7. Sasidharan, R.; Bailey-Serres, J.; Ashikari, M.; Atwell, B.J.; Colmer, T.D.; Fagerstedt, K.; Fukao, T.; Geigenberger, P.; Hebelstrup, K.H.; Hill, R.D.; et al. Community recommendations on terminology and procedures used in flooding and low oxygen stress research. *New Phytol.* **2017**, *214*, 1403–1407. [CrossRef] [PubMed]
8. Sasidharan, R.; Hartman, S.; Liu, Z.; Martopawiro, S.; Sajeev, N.; van Veen, H.; Yeung, E.; Voesenek, L. Signal Dynamics and Interactions during Flooding Stress. *Plant Physiol.* **2018**, *176*, 1106–1117. [CrossRef] [PubMed]
9. Voesenek, L.A.; Sasidharan, R.; Visser, E.J.; Bailey-Serres, J. Flooding stress signaling through perturbations in oxygen, ethylene, nitric oxide and light. *New Phytol.* **2016**, *209*, 39–43. [CrossRef]

10. Wegner, L.H. Oxygen Transport in Waterlogged Plants. In *Waterlogging Signalling and Tolerance in Plants*; Springer: Berlin, Heidelberg, 2010; pp. 3–22.

11. Armstrong, W.; Webb, T.; Darwent, M.; Beckett, P. Measuring and interpreting respiratory critical oxygen pressures in roots. *Ann. Bot.* **2008**, *103*, 281–293. [CrossRef]

12. Drew, M.; Lynch, J. Soil Anaerobiosis, Microorganisms, and Root Function. *Annu. Rev. Phytopathol.* **1980**, *18*, 37–66. [CrossRef]

13. Hafeez-ur-Rehman; Nawaz, A.; Awan, M.I.; Ijaz, M.; Hussain, M.; Ahmad, S.; Farooq, M. Direct Seeding in Rice: Problems and Prospects. In *Agronomic Crops: Volume 1: Production Technologies*; Hasanuzzaman, M., Ed.; Springer: Singapore, 2019; pp. 199–222.

14. Pathak, D.S.; Tewari, A.; Sankhyan, S.; Dubey, D.; Mina, U.; Kumar, V.; Jain, N.; Bhatia, A. Direct-seeded rice: Potential, performance and problems-A review. *Curr. Adv. Agric. Sci.* **2011**, *3*, 77–88.

15. Lee, K.-W.; Chen, P.-W.; Lu, C.-A.; Chen, S.; Ho, T.-H.; Yu, S.-M. Coordinated Responses to Oxygen and Sugar Deficiency Allow Rice Seedlings to Tolerate Flooding. *Sci. Signal.* **2009**, *2*, ra61. [CrossRef] [PubMed]

16. Ye, N.H.; Wang, F.Z.; Shi, L.; Chen, M.X.; Cao, Y.Y.; Zhu, F.Y.; Wu, Y.Z.; Xie, L.J.; Liu, T.Y.; Su, Z.Z.; et al. Natural variation in the promoter of rice calcineurin B-like protein10 (OsCBL10) affects flooding tolerance during seed germination among rice subspecies. *Plant J.* **2018**, *94*, 612–625. [CrossRef] [PubMed]

17. Lu, C.A.; Lin, C.C.; Lee, K.W.; Chen, J.L.; Huang, L.F.; Ho, S.L.; Liu, H.J.; Hsing, Y.I.; Yu, S.M. The SnRK1A protein kinase plays a key role in sugar signaling during germination and seedling growth of rice. *Plant Cell* **2007**, *19*, 2484–2499. [CrossRef]

18. Kretzschmar, T.; Pelayo, M.A.; Trijatmiko, K.R.; Gabunada, L.F.; Alam, R.; Jimenez, R.; Mendioro, M.S.; Slamet-Loedin, I.H.; Sreenivasulu, N.; Bailey-Serres, J.; et al. A trehalose-6-phosphate phosphatase enhances anaerobic germination tolerance in rice. *Nat. Plants* **2015**, *1*, 15124. [CrossRef]

19. Guglielminetti, L.; Yamaguchi, J.; Perata, P.; Alpi, A. Amylolytic Activities in Cereal Seeds under Aerobic and Anaerobic Conditions. *Plant Physiol.* **1995**, *109*, 1069–1076. [CrossRef]

20. Matsumura, H.; Takano, T.; Yoshida, K.; Takeda, G. A Rice Mutant Lacking Alcohol Dehydrogenase. *Jpn. J. Breed.* **1995**, *45*, 365–367. [CrossRef]

21. Ho, V.T.; Tran, A.; Cardarelli, F.; Perata, P.; Pucciariello, C. A calcineurin B-like protein participates in low oxygen signalling in rice. *Funct. Plant Biol.* **2017**, *44*, 917–928. [CrossRef]

22. Lin, C.R.; Lee, K.W.; Chen, C.Y.; Hong, Y.F.; Chen, J.L.; Lu, C.A.; Chen, K.T.; Ho, T.H.; Yu, S.M. SnRK1A-interacting negative regulators modulate the nutrient starvation signaling sensor SnRK1 in source-sink communication in cereal seedlings under abiotic stress. *Plant Cell* **2014**, *26*, 808–827. [CrossRef]

23. Chen, Y.-S.; David Ho, T.-H.; Liu, L.; Lee, D.H.; Lee, C.-H.; Chen, Y.-R.; Lin, S.-Y.; Lu, C.-A.; Yu, S.-M. Sugar starvation-regulated MYBS2 and 14-3-3 protein interactions enhance plant growth, stress tolerance, and grain weight in rice. *Proc. Natl. Acad. Sci. USA* **2019**, *116*, 21925. [CrossRef] [PubMed]

24. Ismail, A.M.; Ella, E.S.; Vergara, G.V.; Mackill, D.J. Mechanisms associated with tolerance to flooding during germination and early seedling growth in rice (*Oryza sativa*). *Ann. Bot.* **2009**, *103*, 197–209. [CrossRef] [PubMed]

25. Magneschi, L.; Perata, P. Rice germination and seedling growth in the absence of oxygen. *Ann. Bot.* **2009**, *103*, 181–196. [CrossRef]

26. Voesenek, L.A.; Bailey-Serres, J. Flood adaptive traits and processes: An overview. *New Phytol.* **2015**, *206*, 57–73. [CrossRef] [PubMed]

27. Bailey-Serres, J.; Voesenek, L.A. Flooding Stress: Acclimations and Genetic Diversity. *Annu. Rev. Plant Biol.* **2008**, *59*, 313–339. [CrossRef]

28. Bailey-Serres, J.; Voesenek, L.A.C.J. Life in the balance: A signaling network controlling survival of flooding. *Curr. Opin. Plant Biol.* **2010**, *13*, 489–494. [CrossRef]

29. Voesenek, L.A.C.J.; Bailey-Serres, J. Flooding tolerance: O_2 sensing and survival strategies. *Curr. Opin. Plant Biol.* **2013**, *16*, 647–653. [CrossRef]

30. Bailey-Serres, J.; Lee, S.C.; Brinton, E. Waterproofing crops: Effective flooding survival strategies. *Plant Physiol.* **2012**, *160*, 1698–1709. [CrossRef]

31. Xu, K.; Xu, X.; Fukao, T.; Canlas, P.; Maghirang-Rodriguez, R.; Heuer, S.; Ismail, A.M.; Bailey-Serres, J.; Ronald, P.C.; Mackill, D.J. Sub1A is an ethylene-response-factor-like gene that confers submergence tolerance to rice. *Nature* **2006**, *442*, 705–708. [CrossRef]

32. Fukao, T.; Xu, K.; Ronald, P.C.; Bailey-Serres, J. A variable cluster of ethylene response factor-like genes regulates metabolic and developmental acclimation responses to submergence in rice. *Plant Cell* **2006**, *18*, 2021–2034. [CrossRef]

33. Fukao, T.; Bailey-Serres, J. Submergence tolerance conferred by Sub1A is mediated by SLR1 and SLRL1 restriction of gibberellin responses in rice. *Proc. Natl. Acad. Sci. USA* **2008**, *105*, 16814–16819. [CrossRef] [PubMed]

34. Hattori, Y.; Nagai, K.; Furukawa, S.; Song, X.-J.; Kawano, R.; Sakakibara, H.; Wu, J.; Matsumoto, T.; Yoshimura, A.; Kitano, H.; et al. The ethylene response factors SNORKEL1 and SNORKEL2 allow rice to adapt to deep water. *Nature* **2009**, *460*, 1026–1030. [CrossRef] [PubMed]

35. Atwell, B.J.; Waters, I.; Greenway, H. The Effect of Oxygen and Turbulence on Elongation of Coleoptiles of Submergence-Tolerant and -Intolerant Rice Cultivars. *J. Exp. Bot.* **1982**, *33*, 1030–1044. [CrossRef]

36. Takahashi, H.; Saika, H.; Matsumura, H.; Nagamura, Y.; Tsutsumi, N.; Nishizawa, N.K.; Nakazono, M. Cell division and cell elongation in the coleoptile of rice alcohol dehydrogenase 1-deficient mutant are reduced under complete submergence. *Ann. Bot.* **2011**, *108*, 253–261. [CrossRef]

37. Huang, J.; Takano, T.; Akita, S. Expression of α-expansin genes in young seedlings of rice (*Oryza sativa* L.). *Planta* **2000**, *211*, 467–473. [CrossRef] [PubMed]

38. Choi, D.; Lee, Y.; Cho, H.T.; Kende, H. Regulation of expansin gene expression affects growth and development in transgenic rice plants. *Plant Cell* **2003**, *15*, 1386–1398. [CrossRef]

39. Lasanthi-Kudahettige, R.; Magneschi, L.; Loreti, E.; Gonzali, S.; Licausi, F.; Novi, G.; Beretta, O.; Vitulli, F.; Alpi, A.; Perata, P. Transcript Profiling of the Anoxic Rice Coleoptile. *Plant Physiol.* **2007**, *144*, 218–231. [CrossRef]

40. Lee, K.W.; Chen, P.W.; Yu, S.M. Metabolic adaptation to sugar/O_2 deficiency for anaerobic germination and seedling growth in rice. *Plant Cell Environ.* **2014**, *37*, 2234–2244.

41. Kuroha, T.; Ashikari, M. Molecular mechanisms and future improvement of submergence tolerance in rice. *Mol. Breed.* **2020**, *40*, 41. [CrossRef]

42. Hsu, S.K.; Tung, C.W. RNA-Seq Analysis of Diverse Rice Genotypes to Identify the Genes Controlling Coleoptile Growth during Submerged Germination. *Front. Plant Sci.* **2017**, *8*, 762. [CrossRef]

43. Greenway, H.; Gibbs, J. Mechanisms of anoxia tolerance in plants. II. Energy requirements for maintenance and energy distribution to essential processes. *Funct. Plant Biol.* **2003**, *30*, 999–1036. [CrossRef] [PubMed]

44. Sousa, C.A.F.D.; Sodek, L. The metabolic response of plants to oxygen deficiency. *Braz. J. Plant Physiol.* **2002**, *14*, 83–94. [CrossRef]

45. Tadege, M.; Dupuis, I.; Kuhlemeier, C. Ethanolic fermentation: New functions for an old pathway. *Trends Plant Sci.* **1999**, *4*, 320–325. [CrossRef]

46. Edwards, J.M.; Roberts, T.H.; Atwell, B.J. Quantifying ATP turnover in anoxic coleoptiles of rice (Oryza sativa) demonstrates preferential allocation of energy to protein synthesis. *J. Exp. Bot.* **2012**, *63*, 4389–4402. [CrossRef]

47. Miro, B.; Longkumer, T.; Entila, F.D.; Kohli, A.; Ismail, A.M. Rice Seed Germination Underwater: Morpho-Physiological Responses and the Bases of Differential Expression of Alcoholic Fermentation Enzymes. *Front. Plant Sci.* **2017**, *8*, 1857. [CrossRef]

48. Damaris, R.N.; Lin, Z.; Yang, P.; He, D. The Rice Alpha-Amylase, Conserved Regulator of Seed Maturation and Germination. *Int. J. Mol. Sci* **2019**, *20*, 450. [CrossRef]

49. Senapati, S.; Kuanar, S.; Sarkar, R. Anaerobic Germination Potential in Rice (*Oryza sativa* L.): Role of Amylases, Alcohol deydrogenase and Ethylene. *J. Stress Physiol. Biochem.* **2019**, *15*, 39–52.

50. Pujadas, G.; Palau, J. Evolution of alpha-amylases: Architectural features and key residues in the stabilization of the (beta/alpha)(8) scaffold. *Mol. Biol. Evol.* **2001**, *18*, 38–54. [CrossRef]

51. Huang, N.; Sutliff, T.D.; Litts, J.C.; Rodriguez, R.L. Classification and characterization of the rice alpha-amylase multigene family. *Plant Mol. Biol.* **1990**, *14*, 655–668. [CrossRef]

52. Huang, N.; Stebbins, G.; Rodriguez, R. Classification and evolution of α-amylase genes in plants. *Proc. Natl. Acad. Sci. USA* **1992**, *89*, 7526–7530. [CrossRef]

53. Hwang, Y.S.; Thomas, B.R.; Rodriguez, R.L. Differential expression of rice α-amylase genes during seedling development under anoxia. *Plant Mol. Biol.* **1999**, *40*, 911–920. [CrossRef] [PubMed]

54. Takahashi, H.; Greenway, H.; Matsumura, H.; Tsutsumi, N.; Nakazono, M. Rice alcohol dehydrogenase 1 promotes survival and has a major impact on carbohydrate metabolism in the embryo and endosperm when seeds are germinated in partially oxygenated water. *Ann. Bot.* **2014**, *113*, 851–859. [CrossRef] [PubMed]

55. Vijayan, J.; Senapati, S.; Ray, S.; Chakraborty, K.; Molla, K.; Basak, N.; Pradhan, B.; Yeasmin, L.; Chattopadhyay, K.; Sarkar, R. Transcriptomic and physiological studies identify cues for germination stage oxygen deficiency tolerance in rice. *Environ. Exp. Bot.* **2017**, *147*, 234–248. [CrossRef]

56. Saika, H.; Matsumura, H.; Takano, T.; Tsutsumi, N.; Nakazono, M. A Point Mutation of Adh1 Gene is Involved in the Repression of Coleoptile Elongation under Submergence in Rice. *Breed. Sci.* **2006**, *56*, 69–74. [CrossRef]

57. Kato-Noguchi, H.; Morokuma, M. Ethanolic fermentation and anoxia tolerance in four rice cultivars. *J. Plant Physiol.* **2007**, *164*, 168–173. [CrossRef]

58. Hakata, M.; Kuroda, M.; Miyashita, T.; Yamaguchi, T.; Kojima, M.; Sakakibara, H.; Mitsui, T.; Yamakawa, H. Suppression of α-amylase genes improves quality of rice grain ripened under high temperature. *Plant Biotechnol. J.* **2012**, *10*, 1110–1117. [CrossRef]

59. Hu, Q.; Fu, Y.; Guan, Y.; Lin, C.; Cao, D.; Hu, W.; Sheteiwy, M.; Hu, J. Inhibitory effect of chemical combinations on seed germination and pre-harvest sprouting in hybrid rice. *Plant Growth Regul.* **2016**, *80*, 281–289. [CrossRef]

60. Hwang, Y.S.; Karrer, E.E.; Thomas, B.R.; Chen, L.; Rodriguez, R.L. Three cis-elements required for rice α-amylase Amy3D expression during sugar starvation. *Plant Mol. Biol.* **1998**, *36*, 331–341. [CrossRef]

61. Yu, S.M.; Lee, Y.C.; Fang, S.C.; Chan, M.T.; Hwa, S.F.; Liu, L.F. Sugars act as signal molecules and osmotica to regulate the expression of alpha-amylase genes and metabolic activities in germinating cereal grains. *Plant Mol. Biol.* **1996**, *30*, 1277–1289. [CrossRef]

62. Knight, H.; Knight, M. Abiotic stress signaling pathways: Specificity and cross-talk. *Trends Plant Sci.* **2001**, *6*, 1360–1385. [CrossRef]

63. Batistic, O.; Kudla, J. Integration and channeling of calcium signaling through the CBL calcium sensor/CIPK protein kinase network. *Planta* **2004**, *219*, 915–924. [CrossRef]

64. Das, R.; Pandey, G.K. Expressional analysis and role of calcium regulated kinases in abiotic stress signaling. *Curr. Genom.* **2010**, *11*, 2–13. [CrossRef] [PubMed]

65. Luan, S. The CBL-CIPK network in plant calcium signaling. *Trends Plant Sci.* **2009**, *14*, 37–42. [CrossRef] [PubMed]

66. Albrecht, V.; Ritz, O.; Linder, S.; Harter, K.; Kudla, J. The NAF domain defines a novel protein-protein interaction module conserved in Ca^{2+}-regulated kinases. *EMBO J.* **2001**, *20*, 1051–1063. [CrossRef] [PubMed]

67. Kurusu, T.; Hamada, J.; Nokajima, H.; Kitagawa, Y.; Kiyoduka, M.; Takahashi, A.; Hanamata, S.; Ohno, R.; Hayashi, T.; Okada, K.; et al. Regulation of Microbe-Associated Molecular Pattern-Induced Hypersensitive Cell Death, Phytoalexin Production, and Defense Gene Expression by Calcineurin B-Like Protein-Interacting Protein Kinases, OsCIPK14/15, in Rice Cultured Cells. *Plant Physiol.* **2010**, *153*, 678–692. [CrossRef]

68. Halford, N.G.; Hey, S.; Jhurreea, D.; Laurie, S.; McKibbin, R.S.; Paul, M.; Zhang, Y. Metabolic signalling and carbon partitioning: Role of Snf1-related (SnRK1) protein kinase. *J. Exp. Bot.* **2003**, *54*, 467–475. [CrossRef] [PubMed]

69. Rolland, F.; Baena-Gonzalez, E.; Sheen, J. Sugar sensing and signaling in plants: Conserved and novel mechanisms. *Annu. Rev. Plant Biol.* **2006**, *57*, 675–709. [CrossRef] [PubMed]

70. Ghillebert, R.; Swinnen, E.; Wen, J.; Vandesteene, L.; Ramon, M.; Norga, K.; Rolland, F.; Winderickx, J. The AMPK/SNF1/SnRK1 fuel gauge and energy regulator: Structure, function and regulation. *FEBS J.* **2011**, *278*, 3978–3990. [CrossRef]

71. Hong, Y.F.; Ho, T.H.; Wu, C.F.; Ho, S.L.; Yeh, R.H.; Lu, C.A.; Chen, P.W.; Yu, L.C.; Chao, A.; Yu, S.M. Convergent starvation signals and hormone crosstalk in regulating nutrient mobilization upon germination in cereals. *Plant Cell* **2012**, *24*, 2857–2873. [CrossRef]

72. Lu, C.A.; Ho, T.H.; Ho, S.L.; Yu, S.M. Three novel MYB proteins with one DNA binding repeat mediate sugar and hormone regulation of alpha-amylase gene expression. *Plant Cell* **2002**, *14*, 1963–1980. [CrossRef]

73. Lu, C.A.; Lim, E.K.; Yu, S.M. Sugar response sequence in the promoter of a rice α-amylase gene serves as a transcriptional enhancer. *J. Biol. Chem.* **1998**, *273*, 10120–10131. [CrossRef] [PubMed]

74. Yamaguchi, S. Gibberellin metabolism and its regulation. *Annu. Rev. Plant Biol.* **2008**, *59*, 225–251. [CrossRef] [PubMed]

75. Chen, P.W.; Chiang, C.M.; Tseng, T.H.; Yu, S.M. Interaction between rice MYBGA and the gibberellin response element controls tissue-specific sugar sensitivity of alpha-amylase genes. *Plant Cell* **2006**, *18*, 2326–2340. [CrossRef] [PubMed]

76. Chen, P.W.; Lu, C.A.; Yu, T.S.; Tseng, T.H.; Wang, C.S.; Yu, S.M. Rice alpha-amylase transcriptional enhancers direct multiple mode regulation of promoters in transgenic rice. *J. Biol. Chem.* **2002**, *277*, 13641–13649. [CrossRef]

77. Kaneko, M.; Itoh, H.; Ueguchi-Tanaka, M.; Ashikari, M.; Matsuoka, M. The alpha-amylase induction in endosperm during rice seed germination is caused by gibberellin synthesized in epithelium. *Plant Physiol.* **2002**, *128*, 1264–1270. [CrossRef]

78. Tsuji, H.; Aya, K.; Ueguchi-Tanaka, M.; Shimada, Y.; Nakazono, M.; Watanabe, R.; Nishizawa, N.K.; Gomi, K.; Shimada, A.; Kitano, H.; et al. GAMYB controls different sets of genes and is differentially regulated by microRNA in aleurone cells and anthers. *Plant J.* **2006**, *47*, 427–444. [CrossRef]

79. Gubler, F.; Kalla, R.; Roberts, J.K.; Jacobsen, J.V. Gibberellin-Regulated Expression of a myb Gene in Barley Aleurone Cells: Evidence for Myb Transactivation of a High-pI a-Amylase Gene Promoter. *Plant Cell* **1995**, *7*, 1879. [CrossRef]

80. Gubler, F.; Raventos, D.; Keys, M.; Watts, R.; Mundy, J.; Jacobsen, J. Target genes and regulatory domains of the GAMYB transcription activator in cereal aleurone. *Plant J.* **1999**, *17*, 1–9. [CrossRef]

81. Lunn, J.; Delorge, I.; Figueroa, C.; Van Dijck, P.; Stitt, M. Trehalose metabolism in plants. *Plant J.* **2014**, *79*, 544–567. [CrossRef]

82. Zhang, Y.; Primavesi, L.F.; Jhurreea, D.; Andralojc, P.J.; Mitchell, R.A.; Powers, S.J.; Schluepmann, H.; Delatte, T.; Wingler, A.; Paul, M.J. Inhibition of SNF1-related protein kinase1 activity and regulation of metabolic pathways by trehalose-6-phosphate. *Plant Physiol.* **2009**, *149*, 1860–1871. [CrossRef]

83. Figueroa, C.M.; Lunn, J.E. A Tale of Two Sugars: Trehalose 6-Phosphate and Sucrose. *Plant Physiol.* **2016**, *172*, 7–27. [CrossRef] [PubMed]

84. Yadav, U.P.; Ivakov, A.; Feil, R.; Duan, G.Y.; Walther, D.; Giavalisco, P.; Piques, M.; Carillo, P.; Hubberten, H.M.; Stitt, M.; et al. The sucrose-trehalose 6-phosphate (Tre6P) nexus: Specificity and mechanisms of sucrose signalling by Tre6P. *J. Exp. Bot.* **2014**, *65*, 1051–1068. [CrossRef] [PubMed]

85. Fukao, T.; Xiong, L. Genetic mechanisms conferring adaptation to submergence and drought in rice: Simple or complex? *Curr. Opin. Plant Biol.* **2013**, *16*, 196–204. [CrossRef] [PubMed]

86. Nagai, K.; Hattori, Y.; Ashikari, M. Stunt or elongate? Two opposite strategies by which rice adapts to floods. *J. Plant Res.* **2010**, *123*, 303–309. [CrossRef] [PubMed]

87. Angaji, S.A.; Septiningsih, E.M.; Mackill, D.J.; Ismail, A.M. QTLs associated with tolerance of flooding during germination in rice (*Oryza sativa* L.). *Euphytica* **2010**, *172*, 159–168. [CrossRef]

88. Baltazar, M.; Ignacio, J.C.; Thomson, M.; Ismail, A.; Mendioro, M.; Septiningsih, E. QTL mapping for tolerance to anaerobic germination in rice from IR64 and the *aus* landrace Kharsu 80A. *Breed. Sci.* **2019**, *69*, 227–233. [CrossRef]

89. Septiningsih, E.M.; Ignacio, J.C.; Sendon, P.M.; Sanchez, D.L.; Ismail, A.M.; Mackill, D.J. QTL mapping and confirmation for tolerance of anaerobic conditions during germination derived from the rice landrace Ma-Zhan Red. *Theor. Appl. Genet.* **2013**, *126*, 1357–1366 [CrossRef]

90. Baltazar, M.D.; Ignacio, J.C.I.; Thomson, M.J.; Ismail, A.M.; Mendioro, M.S.; Septiningsih, E.M. QTL mapping for tolerance of anaerobic germination from IR64 and the aus landrace Nanhi using SNP genotyping. *Euphytica* **2014**, *197*, 251–260. [CrossRef]

91. Kim, S.-M.; Reinke, R.F. Identification of QTLs for tolerance to hypoxia during germination in rice. *Euphytica* **2018**, *214*, 160. [CrossRef]

92. Ghosal, S.; Casal, C.; Quilloy, F.; Septiningsih, E.; Mendioro, M.; Dixit, S. Deciphering Genetics Underlying Stable Anaerobic Germination in Rice: Phenotyping, QTL Identification, and Interaction Analysis. *Rice* **2019**, *12*, 50. [CrossRef]

93. Hsu, S.K.; Tung, C.W. Genetic Mapping of Anaerobic Germination-Associated QTLs Controlling Coleoptile Elongation in Rice. *Rice* **2015**, *8*, 38. [CrossRef] [PubMed]

94. Zhang, M.; Lu, Q.; Wu, W.; Niu, X.; Wang, C.; Feng, Y.; Xu, Q.; Wang, S.; Yuan, X.; Yu, H.; et al. Association Mapping Reveals Novel Genetic Loci Contributing to Flooding Tolerance during Germination in Indica Rice. *Front. Plant Sci.* **2017**, *8*, 678. [CrossRef] [PubMed]

95. Nghi, K.N.; Tondelli, A.; Valè, G.; Tagliani, A.; Marè, C.; Perata, P.; Pucciariello, C. Dissection of coleoptile elongation in japonica rice under submergence through integrated genome-wide association mapping and transcriptional analyses. *Plant Cell Environ.* **2019**, *42*, 1832–1846. [CrossRef] [PubMed]

96. Rohilla, M.; Singh, N.; Mazumder, A.; Sen, P.; Roy, P.; Chowdhury, D.; Singh, N.K.; Mondal, T.K. Genome-wide association studies using 50 K rice genic SNP chip unveil genetic architecture for anaerobic germination of deep-water rice population of Assam, India. *Mol. Genet. Genom.* **2020**, *295*, 1211–1226. [CrossRef] [PubMed]

97. Nordborg, M.; Tavaré, S. Linkage disequilibrium: What history has to tell us Trends Genet. *Trends Genet. TIG* **2002**, *18*, 83–90. [CrossRef]

98. Singh, S.; Mackill, D.; Ismail, A. Responses of SUB1 rice introgression lines to submergence in the field: Yield and grain quality. *Field Crops Res.* **2009**, *113*, 12–23. [CrossRef]

99. Mackill, D.; Ismail, A.; Singh, U.; Labios, R.; Paris, T. Development and Rapid Adoption of Submergence-Tolerant (Sub1) Rice Varieties. *Adv. Agron.* **2012**, *115*, 299–352.

100. Mondal, S.; Khan, M.; Dixit, S.; Cruz, P.C.S.; Septiningsih, E.; Ismail, A. Growth, productivity and grain quality of AG1 and AG2 QTLs introgression lines under flooding in direct-seeded rice system. *Field Crops Res.* **2020**, *248*, 107713. [CrossRef]

101. Toledo, A.M.U.; Ignacio, J.C.I.; Casal, C.; Gonzaga, Z.J.; Mendioro, M.S.; Septiningsih, E.M. Development of Improved Ciherang-Sub1 Having Tolerance to Anaerobic Germination Conditions. *Plant Breed. Biotechnol.* **2015**, *3*, 77–87. [CrossRef]

102. Ismail, A.; Singh, U.; Singh, S.; Dar, M.; Mackill, D. The contribution of submergence-tolerant (Sub1) rice varieties to food security in flood-prone rainfed lowland areas in Asia. *Field Crops Res.* **2013**, *152*, 83–93. [CrossRef]

103. Yu, S.M.; Lee, H.T.; Lo, S.F.; Ho, T.D. How does rice cope with too little oxygen during its early life? *New Phytol.* **2019**. [CrossRef] [PubMed]

104. Ray, S.; Vijayan, J.; Sarkar, R.K. Germination Stage Oxygen Deficiency (GSOD): An Emerging Stress in the Era of Changing Trends in Climate and Rice Cultivation Practice. *Front. Plant Sci.* **2016**, *7*, 671. [CrossRef] [PubMed]

105. Mahender, A.; Anandan, A.; Pradhan, S.K. Early seedling vigour, an imperative trait for direct-seeded rice: An overview on physio-morphological parameters and molecular markers. *Planta* **2015**, *241*, 1027–1050. [CrossRef] [PubMed]

106. Bailey-Serres, J.; Parker, J.E.; Ainsworth, E.A.; Oldroyd, G.E.D.; Schroeder, J.I. Genetic strategies for improving crop yields. *Nature* **2019**, *575*, 109–118. [CrossRef]

107. Mickelbart, M.V.; Hasegawa, P.M.; Bailey-Serres, J. Genetic mechanisms of abiotic stress tolerance that translate to crop yield stability. *Nat. Rev. Genet.* **2015**, *16*, 237–251. [CrossRef]

Publisher's Note: MDPI stays neutral with regard to jurisdictional claims in published maps and institutional affiliations.

Article

Jasmonate Signalling Contributes to Primary Root Inhibition Upon Oxygen Deficiency in *Arabidopsis thaliana*

Vinay Shukla [1], Lara Lombardi [2], Ales Pencik [3], Ondrej Novak [3], Daan A. Weits [1], Elena Loreti [4], Pierdomenico Perata [1], Beatrice Giuntoli [1,2] and Francesco Licausi [2,5,*]

[1] Plantlab, Institute of Life Sciences, Scuola Superiore Sant'Anna, 56127 Pisa, Italy; vinay.shukla@unige.ch (V.S.); d.weits@santannapisa.it (D.A.W.); p.perata@santannapisa.it (P.P.); beatrice.giuntoli@unipi.it (B.G.)

[2] Department of Biology, University of Pisa, 56126 Pisa, Italy; lara.lombardi@unipi.it

[3] Laboratory of Growth Regulators, Faculty of Science, Palacký University & Institute of Experimental Botany, The Czech Academy of Sciences, CZ-783 71 Olomouc, Czech Republic; alespencik@seznam.cz (A.P.); ondrej.novak@upol.cz (O.N.)

[4] The Institute of Agricultural Biology and Biotechnology, National Research Council, 20133 Milan, Italy; loreti@ibba.cnr.it

[5] Department of Plant Sciences, University of Oxford, Oxford OX1 3RB, UK

[*] Correspondence: francesco.licausi@unipi.it

Received: 15 June 2020; Accepted: 11 August 2020; Published: 17 August 2020

Abstract: Plants, including most crops, are intolerant to waterlogging, a stressful condition that limits the oxygen available for roots, thereby inhibiting their growth and functionality. Whether root growth inhibition represents a preventive measure to save energy or is rather a consequence of reduced metabolic rates has yet to be elucidated. In the present study, we gathered evidence for hypoxic repression of root meristem regulators that leads to root growth inhibition. We also explored the contribution of the hormone jasmonic acid (JA) to this process in *Arabidopsis thaliana*. Analysis of transcriptomic profiles, visualisation of fluorescent reporters and direct hormone quantification confirmed the activation of JA signalling under hypoxia in the roots. Further, root growth assessment in JA-related mutants in aerobic and anaerobic conditions indicated that JA signalling components contribute to active root inhibition under hypoxia. Finally, we show that the oxygen-sensing transcription factor (TF) RAP2.12 can directly induce Jasmonate Zinc-finger proteins (JAZs), repressors of JA signalling, to establish feedback inhibition. In summary, our study sheds new light on active root growth restriction under hypoxic conditions and on the involvement of the JA hormone in this process and its cross talk with the oxygen sensing machinery of higher plants.

Keywords: root hypoxia; oxygen sensing; jasmonate; root meristem

---

1. Introduction

Heavy soil structure and intense rainfalls often result in waterlogging events that impose severe hypoxic conditions to the root system of plants due to the poor diffusivity of gases in water [1]. Establishment of anaerobic conditions triggers an energy crisis in root cells due to the fact that oxygen is required by the mitochondrial electron transport chain as a terminal acceptor. Indeed, plant tissues that experience hypoxia mainly rely on glycolysis supported by fermentation for ATP production, although this is considerably lower than that obtained via oxidative phosphorylation [2]. Additionally, restricted oxygen supply affects a wide variety of biochemical processes in which this molecule participates as a substrate, including the production of reactive oxygen species (ROS), fatty acid desaturation and synthesis of phytohormones [3,4]. Therefore, plants developed an array of strategies to save energy and to selectively dedicate the little amount produced to structural maintenance and stress endurance [5].

In higher plants, several of these metabolic adaptations are transcriptionally controlled by constitutively expressed members of the group VII of Ethylene Response Factors (ERF-VII) [6–8]. For example, the Arabidopsis ERF-VII RAP2.12 was shown to re-localise from the plasma membrane into the nucleus to induce the expression of enzymes required for fermentative pathways in response to a drop in oxygen availability [9,10]. Indeed, oxygen levels determine ERF-VII stability by acting upon the amino terminal sequence of these transcriptional regulators and by addressing them to proteolysis via the N-end rule pathway [11,12]. This regulatory function has been demonstrated to be catalysed by plant cysteine oxidases, enzymes that use molecular oxygen as a substrate to oxidize an exposed cysteine residue to its sulfinated or possibly sulfonated form [13–15]. N-terminal oxidized cysteines are subjected to the subsequent activity of Arginyl-transferases (ATEs) that conjugate an arginine, thereby marking the protein for recognition by the E3 ubiquitin ligase Proteolysis 6 (PRT6) and subsequent proteasomal degradation [16]. Under hypoxia, this chain of events is inhibited at the initial oxidation step, leading to ERF-VII accumulation in the nucleus, where they reprogram the cell transcriptome to accommodate anaerobic metabolism [17,18].

Notwithstanding these metabolic adjustments, growth and developmental penalties are evident in plants that face waterlogging stress [19]. In soybean roots, rhizosphere hypoxia has been shown to mimic the morphological responses caused by waterlogging [20], indicating that the low oxygen component of this stress condition holds major potential to direct root architectural variation [21–23]. In general, adaptation to abiotic stresses in roots is guided by genetically controlled post-embryonic root development that allows phenotypic plasticity, fundamentally by the determination of cell division in the apical and lateral root meristems and cell expansion in the elongation zone [24–26]. Anatomical root adaptation to hypoxia is species specific. In Rumex and tomato, for example, it involves a reduction in root extension, accompanied by divergence from its regular gravitropic habitus and inhibition of lateral root production in favour of above-water adventitious roots [27,28]. Additionally, some species, including rice, are shown to develop aerenchyma and barriers to radial oxygen losses to enhance oxygen channelling from above-water tissues to the submerged organs [29,30].

Previously, microarray studies have been carried out to investigate the reprogramming of gene expression and metabolism in Arabidopsis roots and to identify a highly specific set of genes induced at moderate levels of hypoxia [31]. Cell-specific genetic regulation across individual cell types of Arabidopsis roots under hypoxia has been studied further at the translatome level [32]. Crosstalk of hypoxia responsive factors and hormonal signalling has been described for ethylene, gibberellic acid and abscisic acid signalling pathways. Ethylene, accumulated under submergence, plays a crucial role in the adaptation to low oxygen environment as well as developmental control in concert with gibberellic acid (GA) and abscisic acid (ABA) signalling. Ethylene promotes shoot elongation in an escape strategy of submergence tolerance in rice [33,34]. The interplay of ethylene with GA and ABA is known to control adventitious root and aerenchyma formation in rice [35,36]. In tomato, the ethylene and auxin signalling pathways crosstalk to ensure adventitious root formation in flooded tomato plants [24]. Instead, the impact of hypoxia on jasmonic acid (JA) signalling pathways is yet to be studied in depth.

JA plays crucial roles in both plant defence and developmental processes [37], including root growth [38], tuberization, tendril coiling [39], senescence [40] and fertility [41]. The contribution of JA to root development has been object of several studies that shed light on a dual role for this hormone. In fact, JA and auxin signalling modules are connected by the transcription factors MYC2 [42] and PLETHORA (PLT1 and PLT2), crucial regulators of root meristem activity and stem cell maintenance depending on auxin gradients. MYC2 has been shown to repress PLT1/2 expression by directly binding to their promoters [43], de facto counteracting the positive contribution to auxin biosynthesis by JA, which occurs via the enzymatic step catalysed by Anthranilate Synthase A1 (ASA1) [44].

In the current study, we investigated the crosstalk of hypoxia on JA signalling in the root apical meristem to test whether root growth was hindered by oxygen deficiency not only as a consequence of reduced metabolic activity but also due to active repressive signalling.

2. Results

2.1. Primary Root Growth Is Transiently Inhibited under Hypoxia

To examine the extent to which reduced oxygen levels affect root growth, we compared the primary root length of Arabidopsis seedlings treated with either normoxic (21% O_2) or hypoxic conditions (1% v/v O_2) for four days after one week of aerobic growth. Plants were supplemented with 1% sucrose to prevent possible carbon starvation. As expected and reported previously [28], the length of the primary root was significantly reduced in plants subjected to a hypoxic atmosphere (Figure 1A,B), raising the question of whether this phenomenon was due to transient inactivation of root meristem or rather due to stem cell death. To test these alternative hypotheses, we repeated the experiment, this time including root length measurements also at the onset of hypoxic stress and four days after its end. We observed that primary roots actually maintained elongation under hypoxia and grew further in the post-hypoxia phase, although to a significantly reduced rate when compared to the controls maintained under constant aerobiosis (Figure 1C), indicating that the meristematic activity at the root apex was not lost. Remarkably, root growth inhibition was maintained in the post-hypoxic phase (Figure 1C, Supplementary Figure S1), suggesting that active repression of root growth was established by hypoxia rather than by being a simple metabolic consequence of reduced respiratory metabolism.

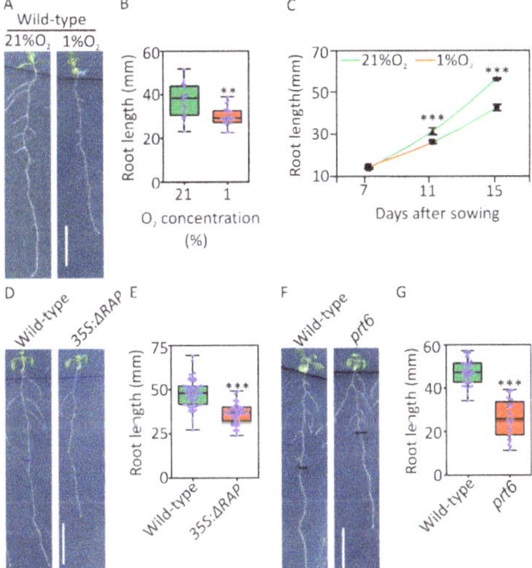

Figure 1. Root length restriction by hypoxia and ERF-VII stabilisation: (**A**) Root development of 7-day-old plants treated with 1% or 21% v/v O_2 for 4 days; (**B**) primary root length of 7-day-old plants treated with 1% or 21% v/v O_2 for four days ($n \geq 20$); and (**C**) primary root length of plants grown on vertical plates 7, 11 and 15 days after sowing. Circles represent averages of plants subjected to 4-day-long hypoxia (1% O_2 v/v) and subsequently to 4 days of recovery. Error bars represent the standard deviation of $n \geq 20$. (**D**) Root phenotype of 11-day-old Col-0 and Δ13RAP plants under normoxic conditions; (**E**) root length of wildtype and Δ13RAP plants after 11 days of growth on vertical plates under normoxic conditions ($n \geq 40$); (**F**) root phenotype of wildtype and prt6 mutant plants after 11 days of growth on vertical plates under normoxic conditions; and (**G**) root length of wildtype and prt6 plants after 11 days of growth on vertical plates under normoxic conditions ($n \geq 25$): Asterisks indicate statistical significance (** p-value ≤ 0.01, *** p-value ≤ 0.001, t-test).

Indeed, reduction of root growth under hypoxic conditions can be explained as a consequence of ATP shortage although one could also speculate that repressive signalling may be set in action to save energy and resources and possibly to reorient root growth towards areas with more favourable oxygen conditions. Therefore, we tested whether the regulators of the molecular response to anaerobiosis, the ERF-VII transcription factors [8], may also play a role in regulating root growth under hypoxic conditions. We compared root growth of wildtype and transgenic genotypes in which ERF-VII signalling is constitutively active due to constitutive stabilisation of these TFs independently of the actual oxygen availability. More specifically, we exploited the ΔRAP line that expresses an N-terminally deleted version of the ERF-VII RAP2.12 that makes it insensitive to oxygen [18] and the *prt6* mutant, which bears a nonfunctional version of the E3 ligase responsible for recognition and ubiquitination of ERF-VII proteins [16]. At the transcriptional level, both genotypes exhibit an active anaerobic response even when their mitochondria are perfectly able to carry out oxidative phosphorylation [17]. After 11 days of growth on vertical agar plates, we noticed that the primary root length of the ΔRAP line was significantly reduced when compared to Col-0 plants (Figure 1D,E). Similarly, we observed reduced root growth in the prt6 mutant as compared to wildtype plants (Figure 1F,G). Since hypoxia and ERF-VII stabilisation under aerobic conditions lead to inhibition of primary root growth to a similar extent, we hypothesized that the observed phenotype is at least partly ascribable to active signalling rather than exclusive metabolic arrest.

2.2. Root Meristem Regulators Are Repressed by Hypoxia

Root elongation is ensured by the production of new cells at the apex. Here, basally channelled auxin and apically diffused (CLAVATA3/Embryo Surrounding Region-Related) (CLE)-like peptides restrict the expression of the WUSCHEL-related homeobox 5 (WOX5) transcriptional regulator to the quiescent centre (QC), which in turn promotes cell proliferation in the surrounding meristematic neighbours in a cell nonautonomous manner through the PLETHORA (PLT) family [45]. mRNA quantification of PLTs and WOX5 gene expression in root apex samples did not reveal significant alterations in response to hypoxia, although PLT4 and WOX5 expression was significantly reduced in ΔRAP plants as compared with wildtype ones (Figure 2A). Instead, in PLT2 and PLT3, accumulation was reduced in the meristem as a consequence of 12 h of exposure to 1% oxygen, as revealed by the respective pPLT:PLT-YFP translational reporter lines [44] (Figure 2B,C, Supplementary Figure S2). Similarly, GFP expression was strongly inhibited in the QC when its coding sequence was placed under control of the WOX5 promoter (Figure 2D). We therefore speculated that hypoxia drove posttranscriptional repression of PLT2 and 3 expression, whereas low oxygen-induced stabilisation of RAP2.12 caused WOX5 transcriptional inhibition in the QC. Since the GFP reporter is controlled by the whole genomic region upstream of WOX5 coding sequence, posttranscriptional regulation impacting on the 5′UTR could also not be excluded. The overall reduction in transcripts and proteins involved in root meristem activity under hypoxia can potentially explain the reduction in root growth observed in the previous experiments (Figure 1A).

2.3. Assessment of Jasmonate Levels and Activity in Hypoxic Roots

Primary root growth inhibition was one of the first physiological effects attributed to JA [46]. Therefore, we decided to investigate whether hypoxia could impact JA signalling in roots. First, we quantified the levels of free JA, the bio-active JA-isoleucine conjugate (JA-Ile) [47] and the JA-precursor cis-12-oxo-phytodienoic acid (cis-OPDA) [48] in roots of wildtype plants grown under aerobic and hypoxic conditions. We sampled the roots at the end of two hypoxia treatments, a short period of 6 h and a prolonged one of 4 d, both starting after seven days of normoxic growth. After 6 h hypoxia, we could not observe statistically significant changes in endogenous JA and cis-OPDA levels while the active form JA-Ile was increased (Figure 3A–C). Prolonged exposure to hypoxia, instead, led to significant reduction in JA, JA-Ile and cis-OPDA when compared to aerobic roots (Figure 3A–C).

This latter observation is not unexpected, since JA biosynthesis involves several biochemical steps that use oxygen as a co-substrate, including cycles of beta-oxidation in the peroxisomes [49].

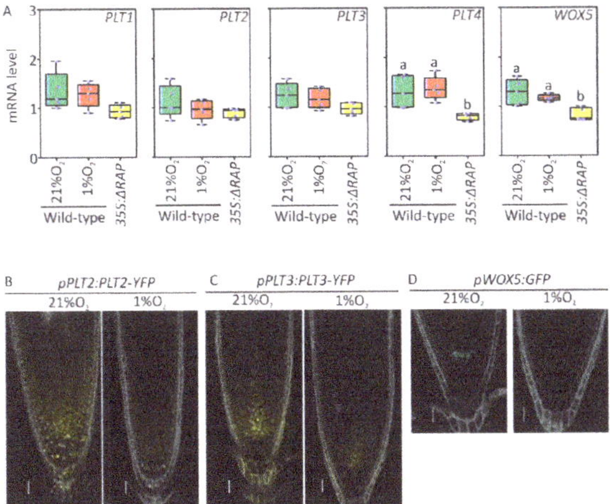

Figure 2. ERF-VII-dependent regulation of transcription factors involved in meristem activity: (**A**) Expression profiles of PLT1-4 and WOX5 in wildtype roots under aerobic and hypoxic conditions and in 35S:ΔRAP plants grown in normoxia. mRNA level of each gene is normalised to the house-keeping gene (UBQ10-AT4G05320). Data shown in the graphs are the re-normalised ΔΔCt values to one of the biological replicates of the wildtype sample at 21% O_2. The letters indicate statistically significant difference calculated by 1-way ANOVA and Tukey post hoc test ($p \leq 0.05$). (**B-C**)The YFP signal in the root apex of pPLT2:PLT2-YFP (**B**) and pPLT3:PLT3-YFP (**C**) plants maintained 12 h under aerobic (21% O_2) and hypoxic (1% O_2) conditions (**D**) GFP signal in the quiescent centre (QC) of pWOX5:GFP plants treated 12 h under normoxia (21% O_2) or hypoxia. (1% O_2). The yellow colour is associated with YFP signal, the green colour is assigned to the GFP signal, while grey is used for propidium iodide staining of or cell walls (Scale bar = 20 μm).

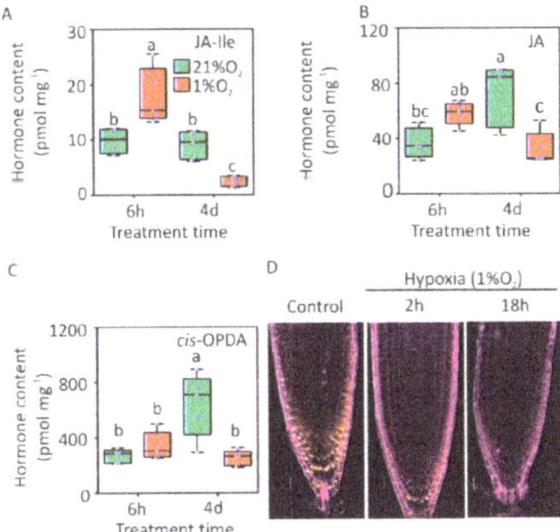

Figure 3. Active jasmonates are produced in roots under hypoxia. Levels of JA (**A**), JA-Ile (**B**) and cis-12-oxo-phytodienoic acid (cis-OPDA) (**C**) in the roots of Arabidopsis plants kept at 21% O_2 (green boxes) or 1% O_2 (red boxes) ($n = 4$). The letters in A-C indicate statistical significance compared with untreated controls ($p \leq 0.05$, Two-way ANOVA, Tukey's post hoc test). (**D**) Jas9-VENUS fluorescent protein abundance in root apical meristems under aerobic and hypoxic conditions: Yellow associates to the YFP signal, and purple associates to propidium iodide dye (Scale bar = 20 μm).

To understand whether the observed increase in active JA levels in Arabidopsis roots subjected to hypoxia actually impact on JA signalling, we exploited a genetically encoded reporter for JA activity in plants, consisting of the JA-labile JAS domain of the JAZ9 protein fused to the VENUS yellow fluorescent protein, of which degradation relates directly to the amount of active JA [50]. Indeed, treatment with 1% O_2, a level of hypoxia that we previously proved to be able to promote a physiological response in roots (Figure 1A), decreased the fluorescent signal in the entire meristematic region (Figure 3D). The signal was decreased as early as 2 h after the onset of hypoxia and did not reappear in the following 16 h (Figure 3D), confirming JA signalling being activated under low oxygen conditions.

2.4. JA-Responsive Genes Are Affected in Hypoxic Roots

In order to understand the contribution of JA signalling to the hypoxic response in Arabidopsis, we first analysed a microarray dataset that interrogates the transcriptional effect of exogenous methyl-JA (MeJA) treatment on cultured Arabidopsis cells for 2 and 6 h [51]. Using the Genevestigator software [52], we selected 70 genes that exhibited the strongest up- (43) or downregulation (27) in response to Me-JA treatments, and associated their corresponding fold change in a hypoxic dataset where cell-specific translatomes from aerobic and hypoxic root samples were compared (Supplementary Table S2) [32]. This analysis returned 22 hypoxia-inducible hormone-responsive mRNAs, characterized by a hypoxic fold change (FC) larger than 2 and 7 in which abundance was at least halved in one root cell type (Figure 4A, Supplementary Table S3). Among the upregulated genes, six coded for JAZ proteins (JAZ1/2/5/6/8/10), established markers of JA responses. Analysing the impact of RAP2.12 stabilisation under aerobic conditions [18] on the same set of genes, we observed a remarkable overlap: 9 of the 11 upregulated genes were induced, and 8 of the 13 downregulated ones were repressed by RAP2.12 (Figure 4A). This last observation indicated the capacity of ERF-VII TFs to regulate JA-responsive genes, either, in principle, by direct transcriptional control or rather by promotion of active JA signalling. In the first instance, we identified candidate ERF-VII target genes in those containing the Hypoxia Responsive Promoter Element (HRPE), bound by ERF-VII proteins [9], in the 1 kb 5′ region upstream of their coding sequence. Since we could detect this DNA motif in 3 of the 24 selected genes only (Supplementary Table S3), we rather favoured the hypothesis of indirect regulation exerted by these transcription factors on JA-responsive genes.

Among the 11 upregulated genes, we identified JAZ1 (jasmonate-ZIM Domain 1), a repressor of JA signalling [53]. An in-depth analysis of the whole JAZ gene family revealed that also JAZ8 is induced by constitutive expression of the oxygen-insensitive RAP2.12 variant (Supplementary Figure S3).

To demonstrate the involvement of the ERF-VII-mediated oxygen sensing pathway, on one side, and of the MYC-dependent jasmonate signalling, on the other, we analysed the root tissues of wildtype, erf-VII and myc234 triple mutant plants and looked for alterations imposed by hypoxia to the expression of selected differentially expressed genes (DEGs). The five genes (three upregulated and two downregulated) chosen as markers displayed a rather heterogeneous behaviour (Figure 4B). GRX480 maintained hypoxic inducibility irrespective of the genotype, whereas JAZ1 and At2g47950 lost it in one or in both knockout backgrounds, respectively (Figure 4B). Like GRX480, the regulation of each selected repressed gene (MAN7 and NLM2) was unaffected by multiple gene knockout (Figure 4B). Altogether, these results supported a complex picture of JA-mediated gene expression under hypoxia, only partially involving canonical JA or hypoxia signalling and probably requiring additional components.

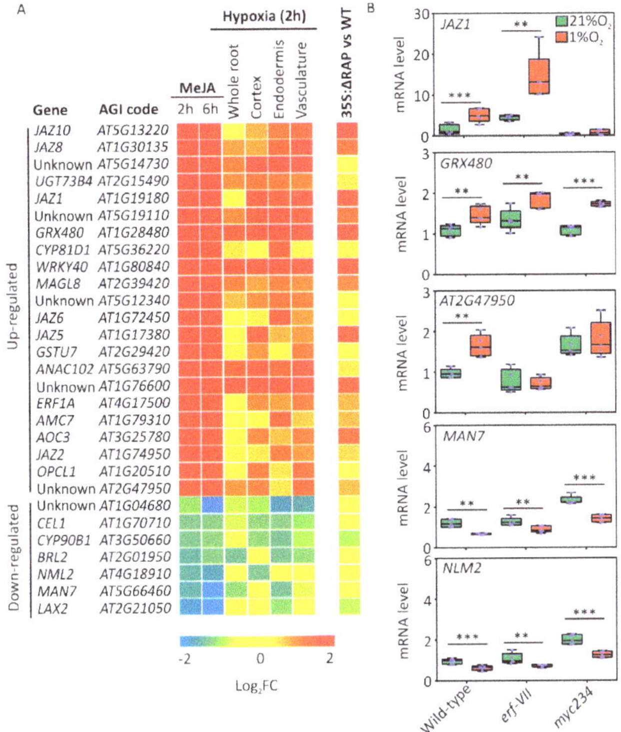

Figure 4. Jasmonic acid (JA)-responsive genes are affected by hypoxia. (**A**) Comparative expression profiles of highly MeJA-responsive genes [51] across root cell types kept in a hypoxic environment [32] or in a ΔRAP2.12 leaf dataset: the colours depict relative expression in each condition, with red indicating upregulation and blue indicating downregulation according to a logarithmic scale. (**B**) Expression of five selected hypoxia and MeJA-responsive genes in wildtype, erf-VII and myc234 mutants under aerobic and hypoxic conditions: mRNA level of each gene is normalised to the house-keeping gene (UBQ10-AT4G05320). Data shown in the graphs are the re-normalised ΔΔCt values to one of the biological replicates of the wildtype sample at 21% O_2. Data are mean ± SD of five biological replicates. Asterisks mark statistically significant differences in the indicated pairwise comparisons (**$p \leq 0.01$, *** $p \leq 0.001$, t- test).

2.5. JA Signaling Is Active under Hypoxia and Contributes to Root Growth Restriction

The observations collected in the previous experiments encouraged us to test the effect of exogenous jasmonates on root growth under aerobic and hypoxic conditions. This could reveal whether the repression of primary root elongation imposed by hypoxia is saturated or can be further enhanced by JA. Root length was significantly reduced in the wildtype by a 4-day-long treatment with hypoxia (1% O_2 *v/v*), as observed before, and further inhibited by the presence of 1 µM JA (Figure 5A,B). A two-way ANOVA analysis did not detect significant interaction between the two treatments. We tested the additional repressive capacity of exogenous JA treatment also in seedings with constitutively active hypoxia signalling. prt6 but not ΔRAP roots showed Ja-sensitivity, although a stronger treatment (5 µM Methyl-JA) was able to repress growth even in the latter genotype (Figure 5C). Together, these observations indicated that ectopic expression of RAP2.12 saturates the repressive effect of moderate JA signalling on primary root growth, although its limited induction in hypoxia and the prt6 background is not sufficient for it. To further study the JA-related signalling cascade involved in this regulatory process, we assessed the extent of root growth repression in JA-insensitive mutants under hypoxia. We observed that primary root growth was reduced in myc234 and jaz1 mutants, and exposure to 4 d hypoxia further inhibited it to a similar extent as in wildtype plants (Figure 5A,B,D,E). Instead, inactivation of the JA-Ile conjugation enzyme JAR1 caused reduced primary root growth under aerobic conditions but did not show further decrease under hypoxia, suggesting that in this genotype active root growth repression by JA signalling was lost (Figure 5D–E). We therefore concluded that hypoxia triggers a JA-mediated signalling pathway that is independent of the release of JAZ-MYC2 repressive interaction.

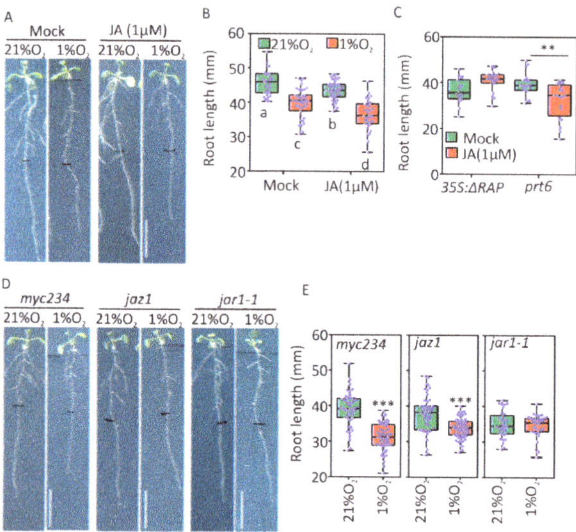

Figure 5. Involvement of JA signalling components in root growth repression under hypoxia: (**A**) Root phenotype of 11-day-old Col-0 plants treated for four days with 1% and 21% O_2 atmosphere in the presence or absence of 1 µM jasmonic acid for four days (size bar: 10 mm); (**B**) root length of 11-day-old Col-0 plants treated with 1% and 21% O_2 in the presence or absence of 1 µM JA for four days ($n \geq 30$) ($p < 0.05$, Two-way ANOVA, Tukey's post hoc test); (**C**) root length of 11-day-old Δ13RAP and prt6 plants treated with 1 µM JA ($n \geq 20$) ($p < 0.05$, T-test); (**D**) root phenotype of myc234, jaz1 and jar1-1 11-day-old plants treated for four days in 1% and 21% O_2 (size bar: 10 mm); and (**E**) root length of 11-day-old myc234, jaz1 and jar1-1 plants treated for four days in 1% and 21% O_2 ($n \geq 30$). The wildtype control is presented in 5A and 5B. Asterisks indicate statistical significance compared with untreated controls (**$p \leq 0.01$, ***$p \leq 0.001$ t-test) (size bar: 10mm).

3. Discussion

Throughout their lifespan, plants accumulate a limited amount of resources that can be dedicated to either growth and reproduction or enduring biotic and abiotic stresses. Therefore, when plants face adverse environmental conditions, growth and development are necessarily affected [4,5]. In the frame of this study, we focused on the inhibition of primary root elongation that occurs under hypoxia. We confirmed that elongation of the primary root of Arabidopsis plants grown on agarised plates was significantly hampered by imposition of a hypoxic atmosphere (Figure 1A,D,F). The slanting tropism reported by Eysholdt-Derzso and Sauter [54] could still be observed, although to a moderate extent (Figure 1A). This can be explained by the fact that, differently from the original report of this phenomenon, hypoxic treatments were applied to plants maintained under photoperiodic regime in order to avoid interference with light signalling and possible starvation due to prolonged darkness. Therefore, the photo-oxygenic activity likely raised the oxygen concentration inside the plate when oxygen was exchanged with the external atmosphere through the stomata. Nevertheless, continuous flushing of an atmosphere containing 1% O_2 was sufficient to restrict root elongation (Figure 1A,B).

Remarkably, primary root growth was maintained at a reduced rate four days after the end of the hypoxic treatment (Figure 1C, Supplementary Figure S1), which led us to speculate about the persistence of an inhibiting signalling after reoxygenation. It should be noted, however, that Arabidopsis root elongation in vertical plates is expected to increase with plant age, as reported by Yazdanbakhsh and Fisahn [55]. The comparable growth rate observed for seedlings grown permanently under aerobic conditions for 11 days and 15-day-old ones treated for 4 days with hypoxia (Supplementary Figure S1) can be considered as a simple delay imposed by active repression under hypoxia and fully relieved by restoration of aerobiosis.

The involvement of JA signalling in the active restriction of root growth that arose from the current study (Figure 5) is rather controversial, considering the requirement of several oxygen molecules for JA biosynthesis [56]. However, it is possible that, under naturally occurring hypoxia, such as waterlogging, the actual oxygen available in the non-submerged organs sustains JA biosynthesis, which is subsequently transported basipetally to restrict root meristem activity. Indeed, we could only observe significant increase of the Ile-conjugated form of JA, whereas the unconjugated version or cis-OPDA precursor were unaffected (Figure 3), suggesting that it is the last step of JA activation to be stimulated by short-term hypoxia. Recently, JA influx and efflux transporters have been identified in Arabidopsis [57]. The transient nature of Ja-Ile accumulation under hypoxia could thus also be likely explained by the energy requirement of AtABCG16/jasmonate transporter 1 for its activity [58]. Moreover, the observation of considerably lower levels of cis-OPDA over prolonged hypoxia when compared to aerobic roots suggested that, in the long run, oxygen and ATP became limiting factors in the synthesis of this hormone. In addition to substrate shortage, it is also possible that the JA biosynthetic pathway is actively repressed at the transcriptional or posttranscriptional levels by hypoxia. For instance, a reduction in JA-biosynthetic gene expression and consequently of JA levels in the seedlings of the N-end rule pathway mutant of Arabidopsis ate1ate2 has been reported previously [59]. Additionally, moderate repression of the JA-biosynthetic genes AOC2, OAS, VSP1 and JAR1 by hypoxia was also observed in root cell types of Arabidopsis [32].

The role of JA in root growth inhibition has been extensively characterized in the past (reviewed in [60]). This regulatory activity has been shown to be mediated by different transcriptional regulators belonging to the basic helix-loop-helix (bHLH) and ethylene insensitive 3 (EIN3) families [61]. The observed root inhibition under hypoxia does not seem to be mediated by the bHLH MYC2, 3 and 4, as indicated by maintenance of repression in a myc234 mutant (Figure 5). In fact, whereas MYC2 has been shown to repress PLT1 and PLT2, thereby inhibiting room meristem activity [43], hypoxia and possibly its JA-signalling component seem to act post-transcriptionally on PLT2 and PLT3 (Figure 2). Transcriptional repression was instead observed for WOX5, which therefore suggests a parallel regulatory pathway to be in action (Figure 2D). Since significant, although slight, WOX5 repression could be measured in a ΔRAP genotype (Figure 2A), it is tempting to speculate about a repressive role for the ERF-VII transcription factors in the root apical meristem. Further analyses in plants impaired in jasmonate synthesis or signalling will be required to test this hypothesis. Their involvement seems to be indirect, since we could not identify the HRPE motif, preferentially recognized by ERF-VII, in the promoters of most JA responsive genes (Supplementary Table S4). This role might rather be ascribed to transcriptional repressors activated by ERF-VIIs under hypoxia, such as the Hypoxia Responsive Attenuator 1 or Lateral Organ Boundaries (LOB) domain-containing protein 41 [10,62]. Further analyses are therefore required to shed light on the role of these transcription factors in limiting root growth under low oxygen environment.

In summary, with this study, we provided novel evidence that, under hypoxic conditions, restriction of root growth relies at least partly on JA-signalling pathways that negatively affect PLT2/3 and WOX5 activity. The synthesis of jasmonate is initially enhanced under such stress in the root, although it later decreases, probably due to oxygen and energy deficiency. We also provided the first evidence that the main known mechanism accounting for plant transcriptional responses to low oxygen, which is based on ERF-VII stabilisation, might be intertwined with this process. These observations will need to be corroborated in the future with similar analyses using single and high-order level erf-VII mutants. This study paves the way to understanding adaptive root growth responses under localised hypoxic conditions and therefore suggests previously unrecognized targets for breeding crops with improved waterlogging tolerance.

4. Materials and Methods

4.1. Plant Material and Growth Conditions

The Columbia-0 (Col-0) ecotype of *Arabidopsis thaliana* was used as the wildtype background in all experiments. The transgenic line expressing a stabilised RAP2.12 factor (35S:Δ13RAP #10) used for experiments has been previously described in [12]. The ERF-VII mutant [50] was provided by Michael Holdsworth (University of Nottingham, UK). *prt6-5* (N684039), *jar1-1* (N8072) and *jaz1* (N2107784) seeds were obtained from the European Arabidopsis Stock Centre (NASC). myc234 mutant seeds were previously described in [63] and provided by Roberto Solano (Spanish National Research Council, Madrid, Spain). Plethora reporter lines, pPLT1:PLT1-YFP, pPLT2:PLT2-YFP, pPLT3:PLT3-YFP, and pPLT4:PLT4-YFP and WOX5 reporter line WOX5:GFP, were previously described in [64] and provided by Ben Scheres (University of Wageningen). A jasmonate responsive reporter, Jas9-VENUS, has been previously described in [65] and was provided by Laurent Laplaze (Institute of Research for Development, Marseille). Seeds were grown on vertical plates on agarised medium composed of half-strength Murashige and Skoog (Duchefa) basal salt mixture, 0.8% plant agar, supplemented with 1% sucrose (Sigma-Aldrich). Seeds were stratified at 4 °C in the dark for 48 h and germinated at 22 °C day/18 °C night with a photoperiod of 12 h light and 12 h dark. For root development evaluation, treatments were applied as described in [31] to plants grown in air containing 21% oxygen (normoxia) or 1% O_2 as hypoxia for the time indicated in the figure legends and while keeping plants under unchanged growth photoperiod.

4.2. Measurement of Primary Root Length

Transparent square plates (12 cm side) containing the Arabidopsis plants were scanned at the end of the treatments, using a Perfection V700 Photo scanner (Seiko Epson) at the resolution of 300 DPI. The software EZ–Rhizo [66] was used for root length measurement, while the number of emerged lateral roots was calculated manually.

4.3. Confocal Imaging

For GFP and YFP visualisation in the primary root, plants were grown for eleven days on vertical square plates. Roots were observed under a FluoView1000 (Olympus) inverted confocal laser scanning microscope with a 20× objective lens. GFP fluorescence was excited with 488-nm laser light (4% laser transmissivity, PMT voltage 805 V) and collected with a 497–554 nm long-pass emission filter. YFP fluorescence was excited with 515-nm laser light (10% laser transmissivity, PMT voltage 890 V) and collected with a 520–560 nm long-pass emission filter. Propidium iodide (PI) fluorescence was excited with 515-nm laser light (10% laser transmissivity, PMT voltage 905 V) and collected at 590–680 nm. Scanner and detector settings were optimized and kept unchanged for all the experiments. Images were analysed with the FluoView FV1000 software (Olympus).

4.4. In Silico Analysis of Differentially Expressed Genes

Differentially expressed genes (DEGs) upon jasmonate treatment were selected from the microarray experiment by [67] (−0.5 < log2FC ≥ 2, corresponding to 70 DEGs in that data set) and compared to the hypoxic root microarray profiles produced by [32], as retrieved from the Genevestigator platform [65].

JA-responsive genes for which mRNA association with ribosome and/or overall abundance is affected by hypoxia were selected by filtering those coherently up- or downregulated in at least one root cell-type.

4.5. RNA Extraction and Gene Expression Analysis by Real-Time qRT-PCR

Roots from 11-day-old Arabidopsis plants grown on vertical agar plates were collected and immediately frozen in liquid nitrogen. RNA extraction was performed and processed to cDNA as described in [68]. An ABI Prism 7300 sequence detection system (Applied Biosystems) was used to carry out real-time PCR, using iQSYBR Green Supermix (Biorad). Ubiquitin10 (At4g05320) was used as the housekeeping gene. Relative expression of each gene was calculated using the $2^{-\Delta\Delta Ct}$ method [69]. A full list of the qPCR primers used is provided in Supplementary Table S1.

4.6. Hormone Quantification

Quantification of JA, cis-OPDA and JA-Ile was performed according to the method described in [70]. Samples were homogenized, and hormones were extracted using an aqueous solution of 10% MeOH/H_2O, *v/v*. The extracts were purified using Oasis HLB columns (30 mg/L mL, Waters), and hormones were eluted with 80% MeOH. Eluent was evaporated to dryness under a stream of nitrogen. Jasmonates levels were determined by ultra-high performance liquid chromatography-electrospray tandem mass spectrometry (UHPLC-MS/MS) using Acquity UPLC® System (Waters) equipped with an Acquity UPLC BEH C18 column (100 × 2.1 mm, 1.7 μm; Waters) coupled to triple quadrupole mass spectrometer Xevo™ TQ-S MS (Waters).

4.7. Statistical Analysis

Sigmaplot 14 (Systat Software) was used to carry out statistical analyses. Comparisons between treatments and genotypes were performed, after testing the normality of data distribution, using t-test or Analysis of Variance (ANOVA) depending on the number of averages. In the case of ANOVA, a Tukey post hoc test was applied for pairwise comparison. Results shown in figures are representative of those obtained from experiments repeated twice with qualitatively equal relative output.

Supplementary Materials: The following are available online at http://www.mdpi.com/2223-7747/9/8/1046/s1. Figure S1. Primary root growth rate in aerobic, hypoxic and post-hypoxic recovery conditions. Figure S2. Effect of hypoxia on PLT1 and PLT4 protein level in the root apex. Figure S3. Effect of hypoxia on the expression JAZ genes in the roots of wildtype and oxygen or jasmonate sensing mutants. Figure S4. Effect of 5 μM methyl-jasmonate (MeJA) treatment on Δ13RAP root growth. Table S1. List of oligonucleotides used for real-time qPCR in this study. Table S2. Expression levels of up- or downregulated genes (log fold change |1|) in response to Me-JA treatments [51] and the corresponding fold change in ribosome association caused by short-term exposure to hypoxia in root tissues [32]. Table S3. Numerical values of the expression levels of 24 JA-responsive genes shown as a heatmap in Figure 1. Table S4. Presence of HRPE motif in the promoter of JA regulated genes shown in Figure 3.

Author Contributions: Conceptualisation, V.S., B.G. and F.L.; methodology, V.S., L.L., D.A.W., A.P. and O.N.; formal analysis, V.S. and A.P.; resources, P.P.; data curation, V.S.; writing—original draft preparation, V.S. and F.L.; writing—review and editing, B.G. and P.P.; supervision, P.P., B.G. and E.L.; project administration, P.P.; funding acquisition, P.P. All authors have read and agreed to the published version of the manuscript.

Funding: This research was funded by the Italian Ministry of University and Research, grant number 20173EWRT9 and Scuola Superiore Sant'Anna. This work was also supported from European Regional Development Fund-Project "Plants as a tool for sustainable global development" (No. CZ.02.1.01/0.0/0.0/16_019/0000827).

Conflicts of Interest: The authors declare no conflict of interest.

References

1. Akgerman, A.; Gainer, J.L. Diffusion of Gases in Liquids. *Ind. Eng. Chem. Fundam.* **1972**. [CrossRef]
2. Gibbs, J.; Greenway, H. Mechanisms of anoxia tolerance in plants. I. Growth, survival and anaerobic catabolism. *Funct. Plant Biol.* **2003**, *30*, 1–47. [CrossRef] [PubMed]
3. Blokhina, O.; Fagerstedt, K.V. Oxidative metabolism, ROS and NO under oxygen deprivation. *Plant Physiol. Biochem.* **2010**, *48*, 359–373. [CrossRef] [PubMed]

4. van Dongen, J.T.; Licausi, F. Oxygen Sensing and Signaling. *Annu. Rev. Plant Biol.* **2014**, *66*, 150112150216002. [CrossRef] [PubMed]

5. Voesenek, L.A.C.J.; Bailey-Serres, J. Flood adaptive traits and processes: an overview. *New Phytol.* **2015**, *206*, 57–73. [CrossRef] [PubMed]

6. Mendiondo, G.M.; Gibbs, D.J.; Szurman-Zubrzycka, M.; Korn, A.; Marquez, J.; Szarejko, I.; Maluszynski, M.; King, J.; Axcell, B.; Smart, K.; et al. Enhanced waterlogging tolerance in barley by manipulation of expression of the N-end rule pathway E3 ligase PROTEOLYSIS6. *Plant Biotechnol. J.* **2016**. [CrossRef] [PubMed]

7. Cukrov, D.; Zermiani, M.; Brizzolara, S.; Cestaro, A.; Licausi, F.; Luchinat, C.; Santucci, C.; Tenori, L.; Van Veen, H.; Zuccolo, A.; et al. Extreme Hypoxic Conditions Induce Selective Molecular Responses and Metabolic Reset in Detached Apple Fruit. *Front. Plant Sci.* **2016**. [CrossRef]

8. Bui, L.T.; Giuntoli, B.; Kosmacz, M.; Parlanti, S.; Licausi, F. Constitutively expressed ERF-VII transcription factors redundantly activate the core anaerobic response in Arabidopsis thaliana. *Plant Sci.* **2015**, *236*, 37–43. [CrossRef]

9. Gasch, P.; Fundinger, M.; Müller, J.T.; Lee, T.; Bailey-Serres, J.; Mustroph, A. Redundant ERF-VII transcription factors bind an evolutionarily-conserved cis-motif to regulate hypoxia-responsive gene expression in Arabidopsis. *Plant Cell* **2015**, TPC2015-00866-RA. [CrossRef]

10. Kosmacz, M.; Parlanti, S.; Schwarzländer, M.; Kragler, F.; Licausi, F.; Van Dongen, J.T. The stability and nuclear localization of the transcription factor RAP2.12 are dynamically regulated by oxygen concentration. *Plant, Cell Environ.* **2015**, *38*, 1094–1103. [CrossRef]

11. Gibbs, D.J.; Lee, S.C.; Md Isa, N.; Gramuglia, S.; Fukao, T.; Bassel, G.W.; Correia, C.S.; Corbineau, F.; Theodoulou, F.L.; Bailey-Serres, J.; et al. Homeostatic response to hypoxia is regulated by the N-end rule pathway in plants. *Nature* **2011**, *479*, 415–418. [CrossRef] [PubMed]

12. Licausi, F.; Kosmacz, M.; Weits, D.A.; Giuntoli, B.; Giorgi, F.M.; Voesenek, L.A.C.J.; Perata, P.; van Dongen, J.T. Oxygen sensing in plants is mediated by an N-end rule pathway for protein destabilization. *Nature* **2011**, *479*, 419–422. [CrossRef] [PubMed]

13. Weits, D.A.; Giuntoli, B.; Kosmacz, M.; Parlanti, S.; Hubberten, H.-M.; Riegler, H.; Hoefgen, R.; Perata, P.; van Dongen, J.T.; Licausi, F. Plant cysteine oxidases control the oxygen-dependent branch of the N-end-rule pathway. *Nat. Commun.* **2014**, *5*, 3425. [CrossRef] [PubMed]

14. White, M.D.; Klecker, M.; Hopkinson, R.J.; Weits, D.A.; Mueller, C.; Naumann, C.; O'Neill, R.; Wickens, J.; Yang, J.; Brooks-Bartlett, J.C.; et al. Plant cysteine oxidases are dioxygenases that directly enable arginyl transferase-catalysed arginylation of N-end rule targets. *Nat. Commun.* **2017**, *8*, 1–9. [CrossRef] [PubMed]

15. White, M.D.; Kamps, J.J.A.G.; East, S.; Taylor Kearney, L.J.; Flashman, E. The plant cysteine oxidases from Arabidopsis thaliana are kinetically tailored to act as oxygen sensors. *J. Biol. Chem.* **2018**, *293*, 11786–11795. [CrossRef]

16. Garzón, M.; Eifler, K.; Faust, A.; Scheel, H.; Hofmann, K.; Koncz, C.; Yephremov, A.; Bachmair, A. *PRT6/At5g02310* encodes an *Arabidopsis* ubiquitin ligase of the N-end rule pathway with arginine specificity and is not the *CER3* locus. *FEBS Lett.* **2007**, *581*, 3189–3196. [CrossRef]

17. Paul, M.V.; Iyer, S.; Amerhauser, C.; Lehmann, M.; van Dongen, J.T.; Geigenberger, P. RAP2.12 oxygen sensing regulates plant metabolism and performance under both normoxia and hypoxia. *Plant Physiol.* **2016**, *172*, 00460. [CrossRef]

18. Giuntoli, B.; Shukla, V.; Maggiorelli, F.; Giorgi, F.M.; Lombardi, L.; Perata, P.; Licausi, F. Age-dependent regulation of ERF-VII transcription factor activity in *Arabidopsis thaliana*. *Plant. Cell Environ.* **2017**. [CrossRef]

19. Wright, A.J.; de Kroon, H.; Visser, E.J.W.; Buchmann, T.; Ebeling, A.; Eisenhauer, N.; Fischer, C.; Hildebrandt, A.; Ravenek, J.; Roscher, C.; et al. Plants are less negatively affected by flooding when growing in species-rich plant communities. *New Phytol.* **2017**. [CrossRef]

20. Jitsuyama, Y. Morphological root responses of soybean to rhizosphere hypoxia reflect waterlogging tolerance. *Can. J. Plant Sci.* **2015**, *95*, 999–1005. [CrossRef]

21. Cardoso, J.A.; Jiménez, J.D.L.C.; Rao, I.M. Waterlogging-induced changes in root architecture of germplasm accessions of the tropical forage grass Brachiaria humidicola. *AoB Plants* **2014**. [CrossRef] [PubMed]

22. Grzesiak, M.T.; Ostrowska, A.; Hura, K.; Rut, G.; Janowiak, F.; Rzepka, A.; Hura, T.; Grzesiak, S. Interspecific differences in root architecture among maize and triticale genotypes grown under drought, waterlogging and soil compaction. *Acta Physiol. Plant.* **2014**. [CrossRef]

23. Cornelious, B.; Chen, P.; Chen, Y.; De Leon, N.; Shannon, J.G.; Wang, D. Identification of QTLs underlying water-logging tolerance in soybean. *Mol. Breed.* **2005**. [CrossRef]

24. Hodge, A.; Berta, G.; Doussan, C.; Merchan, F.; Crespi, M. Plant root growth, architecture and function. *Plant Soil* **2009**, *321*, 153–187. [CrossRef]

25. López-Bucio, J.; Cruz-Ramírez, A.; Herrera-Estrella, L. The role of nutrient availability in regulating root architecture. *Curr. Opin. Plant Biol.* **2003**, *6*, 280–287. [CrossRef]

26. Lynch, J. Root Architecture and Plant Productivity. *Plant Physiol.* **1995**, *109*, 7–13. [CrossRef]

27. Vidoz, M.L.; Loreti, E.; Mensuali, A.; Alpi, A.; Perata, P. Hormonal interplay during adventitious root formation in flooded tomato plants. *Plant J.* **2010**, *63*, 551–562. [CrossRef]

28. Visser, E.; Cohen, J.D.; Barendse, G.; Blom, C.; Voesenek, L. An Ethylene-Mediated Increase in Sensitivity to Auxin Induces Adventitious Root Formation in Flooded Rumex palustris Sm. *Plant Physiol.* **1996**. [CrossRef]

29. Joshi, R.; Kumar, P. Lysigenous aerenchyma formation involves non-apoptotic programmed cell death in rice (Oryza sativa L.) roots. *Physiol. Mol. Biol. Plants* **2012**, *18*, 1. [CrossRef]

30. THOMAS, A.L.; GUERREIRO, S.M.C.; SODEK, L. Aerenchyma Formation and Recovery from Hypoxia of the Flooded Root System of Nodulated Soybean. *Ann. Bot.* **2005**, *96*, 1191. [CrossRef]

31. van Dongen, J.T.; Fröhlich, A.; Ramírez-Aguilar, S.J.; Schauer, N.; Fernie, A.R.; Erban, A.; Kopka, J.; Clark, J.; Langer, A.; Geigenberger, P. Transcript and metabolite profiling of the adaptive response to mild decreases in oxygen concentration in the roots of arabidopsis plants. *Ann. Bot.* **2009**, *103*, 269–280. [CrossRef] [PubMed]

32. Mustroph, A.; Zanetti, M.E.; Jang, C.J.H.; Holtan, H.E.; Repetti, P.P.; Galbraith, D.W.; Girke, T.; Bailey-Serres, J. Profiling translatomes of discrete cell populations resolves altered cellular priorities during hypoxia in Arabidopsis. *Proc. Natl. Acad. Sci.* **2009**, *106*, 18843–18848. [CrossRef] [PubMed]

33. Jackson, M.B.; Fenning, T.M.; Jenkins, W. Aerenchyma (gas-space) formation in adventitious roots of rice (*Oryza sativa* L.) is not controlled by ethylene or small partial pressures of oxygen. *J. Exp. Bot.* **1985**, *36*, 1566–1572. [CrossRef]

34. Voesenek, L.A.C.J.; Rijnders, J.H.G.M.; Peeters, A.J.M.; Van De Steeg, H.M.; De Kroon, H. Plant hormones regulate fast shoot elongation under water: From genes to communities. *Ecology* **2004**, *85*, 16–27. [CrossRef]

35. Steffens, B.; Wang, J.; Sauter, M. Interactions between ethylene, gibberellin and abscisic acid regulate emergence and growth rate of adventitious roots in deepwater rice. *Planta* **2006**, *223*, 604–612. [CrossRef]

36. Nishiuchi, S.; Yamauchi, T.; Takahashi, H.; Kotula, L.; Nakazono, M. Mechanisms for coping with submergence and waterlogging in rice. *Rice* **2012**, *5*, 2. [CrossRef]

37. Creelman, R.A.; Mullet, J.E. Biosynthesis and Action of Jasmonates in Plants. *Annu. Rev. Plant Physiol. Plant Mol. Biol.* **1997**, *48*, 355–381. [CrossRef]

38. Staswick, P.E. The tryptophan conjugates of jasmonic and indole-3-acetic acids are endogenous auxin inhibitors. *Plant Physiol.* **2009**. [CrossRef]

39. McConn, M.; Browse, J. The Critical Requirement for Linolenic Acid Is Pollen Development, Not Photosynthesis, in an Arabidopsis Mutant. *Plant Cell* **1996**, *8*, 403–416. [CrossRef]

40. Schommer, C.; Palatnik, J.F.; Aggarwal, P.; Chetelat, A.; Cubas, P.; Farmer, E.E.; Nath, U.; Weigel, D. Control of jasmonate biosynthesis and senescence by miR319 targets. *PLoS Biol.* **2008**, *6*, 1991–2001. [CrossRef]

41. Xie, D. COI1: An Arabidopsis Gene Required for Jasmonate-Regulated Defense and Fertility. *Science* **1998**, *280*, 1091–1094. [CrossRef] [PubMed]

42. Montiel, G.; Zarei, A.; Körbes, A.P.; Memelink, J. The jasmonate-responsive element from the ORCA3 promoter from catharanthus roseus is active in arabidopsis and is controlled by the transcription factor AtMYC2. *Plant Cell Physiol.* **2011**, *52*, 578–587. [CrossRef] [PubMed]

43. Chen, Q.; Sun, J.; Zhai, Q.; Zhou, W.; Qi, L.; Xu, L.; Wang, B.; Chen, R.; Jiang, H.; Qi, J.; et al. The Basic Helix-Loop-Helix Transcription Factor MYC2 Directly Represses PLETHORA Expression during Jasmonate-Mediated Modulation of the Root Stem Cell Niche in Arabidopsis. *Plant Cell Online* **2011**, *23*, 3335–3352. [CrossRef]

44. Sun, J.; Xu, Y.; Ye, S.; Jiang, H.; Chen, Q.; Liu, F.; Zhou, W.; Chen, R.; Li, X.; Tietz, O.; et al. Arabidopsis ASA1 is important for jasmonate-mediated regulation of auxin biosynthesis and transport during lateral root formation. *Plant Cell* **2009**, *21*, 1495–1511. [CrossRef] [PubMed]

45. Petricka, J.J.; Winter, C.M.; Benfey, P.N. Control of Arabidopsis root development. *Annu. Rev. Plant Biol.* **2012**. [CrossRef] [PubMed]

46. Dathe, W.; Rönsch, H.; Preiss, A.; Schade, W.; Sembdner, G.; Schreiber, K. Endogenous plant hormones of the broad bean, Vicia faba L. (-)-jasmonic acid, a plant growth inhibitor in pericarp. *Planta* **1981**, *153*, 530–535. [CrossRef]

47. Fonseca, S.; Chini, A.; Hamberg, M.; Adie, B.; Porzel, A.; Kramell, R.; Miersch, O.; Wasternack, C.; Solano, R. (+)-7-iso-Jasmonoyl-L-isoleucine is the endogenous bioactive jasmonate. *Nat. Chem. Biol.* **2009**, *5*, 344–350. [CrossRef]

48. Wasternack, C.; Hause, B. Jasmonates: Biosynthesis, perception, signal transduction and action in plant stress response, growth and development. An update to the 2007 review in Annals of Botany. *Ann. Bot.* **2013**, *111*, 1021–1058. [CrossRef]

49. Poirier, Y.; Antonenkov, V.D.; Glumoff, T.; Hiltunen, J.K. Peroxisomal β-oxidation—A metabolic pathway with multiple functions. *Biochim. Biophys. Acta - Mol. Cell Res.* **2006**, *1763*, 1413–1426. [CrossRef]

50. Abbas, M.; Berckhan, S.; Rooney, D.J.; Gibbs, D.J.; Vicente Conde, J.; Sousa Correia, C.; Bassel, G.W.; Marín-De La Rosa, N.; León, J.; Alabadí, D.; et al. Oxygen sensing coordinates photomorphogenesis to facilitate seedling survival. *Curr. Biol.* **2015**, *25*, 1483–1488. [CrossRef]

51. Pauwels, L.; Morreel, K.; De Witte, E.; Lammertyn, F.; Van Montagu, M.; Boerjan, W.; Inzé, D.; Goossens, A. Mapping methyl jasmonate-mediated transcriptional reprogramming of metabolism and cell cycle progression in cultured Arabidopsis cells. *Proc. Natl. Acad. Sci. USA* **2008**, *105*, 1380–1385. [CrossRef] [PubMed]

52. Hruz, T.; Laule, O.; Szabo, G.; Wessendorp, F.; Bleuler, S.; Oertle, L.; Widmayer, P.; Gruissem, W.; Zimmermann, P. Genevestigator V3: A Reference Expression Database for the Meta-Analysis of Transcriptomes. *Adv. Bioinformatics* **2008**, *2008*, 1–5. [CrossRef] [PubMed]

53. Pauwels, L.; Goossens, A. The JAZ proteins: a crucial interface in the jasmonate signaling cascade. *Plant Cell* **2011**, *23*, 3089–3100. [CrossRef] [PubMed]

54. Eysholdt-Derzso, E.; Sauter, M. Root bending is antagonistically affected by hypoxia and ERF-mediated transcription via auxin signaling. *Plant Physiol.* **2017**, *175*, 412–423. [CrossRef] [PubMed]

55. Yazdanbakhsh, N.; Fisahn, J. Analysis of Arabidopsis thaliana root growth kinetics with high temporal and spatial resolution. *Ann. Bot.* **2010**. [CrossRef] [PubMed]

56. Bannenberg, G.; Martínez, M.; Hamberg, M.; Castresana, C. Diversity of the enzymatic activity in the lipoxygenase gene family of arabidopsis thaliana. *Lipids* **2009**, *44*, 85–95. [CrossRef]

57. Nguyen, C.T.; Martinoia, E.; Farmer, E.E. Emerging Jasmonate Transporters. *Mol. Plant* **2017**. [CrossRef]

58. Li, Q.; Zheng, J.; Li, S.; Huang, G.; Skilling, S.J.; Wang, L.; Li, L.; Li, M.; Yuan, L.; Liu, P. Transporter-Mediated Nuclear Entry of Jasmonoyl-Isoleucine Is Essential for Jasmonate Signaling. *Mol. Plant* **2017**. [CrossRef]

59. de Marchi, R.; Sorel, M.; Mooney, B.; Fudal, I.; Goslin, K.; Kwaśniewska, K.; Ryan, P.T.; Pfalz, M.; Kroymann, J.; Pollmann, S.; et al. The N-end rule pathway regulates pathogen responses in plants. *Sci. Rep.* **2016**, *6*, 26020. [CrossRef]

60. Ahmad, P.; Rasool, S.; Gul, A.; Sheikh, S.A.; Akram, N.A.; Ashraf, M.; Kazi, A.M.; Gucel, S. Jasmonates: Multifunctional Roles in Stress Tolerance. *Front. Plant Sci.* **2016**. [CrossRef]

61. Huang, H.; Liu, B.; Liu, L.; Song, S. Jasmonate action in plant growth and development. *J. Exp. Bot.* **2017**. [CrossRef] [PubMed]

62. Giuntoli, B.; Lee, S.C.; Licausi, F.; Kosmacz, M.; Oosumi, T.; van Dongen, J.T.; Bailey-Serres, J.; Perata, P. A Trihelix DNA Binding Protein Counterbalances Hypoxia-Responsive Transcriptional Activation in Arabidopsis. *PLoS Biol.* **2014**. [CrossRef] [PubMed]

63. Schweizer, F.; Fernandez-Calvo, P.; Zander, M.; Diez-Diaz, M.; Fonseca, S.; Glauser, G.; Lewsey, M.G.; Ecker, J.R.; Solano, R.; Reymond, P. Arabidopsis Basic Helix-Loop-Helix Transcription Factors MYC2, MYC3, and MYC4 Regulate Glucosinolate Biosynthesis, Insect Performance, and Feeding Behavior. *Plant Cell* **2013**, *25*, 3117–3132. [CrossRef] [PubMed]

64. Mähönen, A.P.; Ten Tusscher, K.; Siligato, R.; Smetana, O.; Díaz-Triviño, S.; Salojärvi, J.; Wachsman, G.; Prasad, K.; Heidstra, R.; Scheres, B. PLETHORA gradient formation mechanism separates auxin responses. *Nature* **2014**, *515*, 125–129. [CrossRef] [PubMed]

65. Larrieu, A.; Champion, A.; Legrand, J.; Lavenus, J.; Mast, D.; Brunoud, G.; Oh, J.; Guyomarc'h, S.; Pizot, M.; Farmer, E.E.; et al. A fluorescent hormone biosensor reveals the dynamics of jasmonate signalling in plants. *Nat. Commun.* **2015**, *6*, 6043. [CrossRef]

66. Armengaud, P. EZ-Rhizo software. *Plant Signal. Behav.* **2009**, *4*, 139–141. [CrossRef]

67. Pauwels, L.; Inzé, D.; Goossens, A. Jasmonate-inducible gene: what does it mean? *Trends Plant Sci.* **2009**, *14*, 87–91. [CrossRef]
68. Shukla, V.; Lombardi, L.; Iacopino, S.; Pencik, A.; Novak, O.; Perata, P.; Giuntoli, B.; Licausi, F. Endogenous hypoxia in lateral root primordia controls root architecture by antagonizing auxin signaling in Arabidopsis. *Mol. Plant* **2019**. [CrossRef]
69. Livak, K.J.; Schmittgen, T.D. Analysis of Relative Gene Expression Data Using Real-Time Quantitative PCR and the 2−ΔΔCT Method. *Methods* **2001**, *25*, 402–408. [CrossRef]
70. Floková, K.; Tarkowská, D.; Miersch, O.; Strnad, M.; Wasternack, C.; Novák, O. UHPLC–MS/MS based target profiling of stress-induced phytohormones. *Phytochemistry* **2014**, *105*, 147–157. [CrossRef]

Article

Ethylene Differentially Modulates Hypoxia Responses and Tolerance across *Solanum* Species

Sjon Hartman *, Nienke van Dongen, Dominique M.H.J. Renneberg, Rob A.M. Welschen-Evertman, Johanna Kociemba, Rashmi Sasidharan * and Laurentius A.C.J. Voesenek *

Plant Ecophysiology, Institute of Environmental Biology, Utrecht University, Padualaan 8, 3584 CH Utrecht, The Netherlands; n.s.vandongen@students.uu.nl (N.v.D.); d.m.h.j.renneberg@uu.nl (D.M.H.J.R.); R.A.M.Welschen@uu.nl (R.A.M.W.-E.); johanna.kociemba@crick.ac.uk (J.K.)
* Correspondence: j.g.w.hartman@uu.nl (S.H.); r.sasidharan@uu.nl (R.S.); l.a.c.j.voesenek@uu.nl (L.A.C.J.V.)

Received: 24 July 2020; Accepted: 10 August 2020; Published: 13 August 2020

Abstract: The increasing occurrence of floods hinders agricultural crop production and threatens global food security. The majority of vegetable crops are highly sensitive to flooding and it is unclear how these plants use flooding signals to acclimate to impending oxygen deprivation (hypoxia). Previous research has shown that the early flooding signal ethylene augments hypoxia responses and improves survival in Arabidopsis. To unravel how cultivated and wild *Solanum* species integrate ethylene signaling to control subsequent hypoxia acclimation, we studied the transcript levels of a selection of marker genes, whose upregulation is indicative of ethylene-mediated hypoxia acclimation in Arabidopsis. Our results suggest that ethylene-mediated hypoxia acclimation is conserved in both shoots and roots of the wild *Solanum* species bittersweet (*Solanum dulcamara*) and a waterlogging-tolerant potato (*Solanum tuberosum*) cultivar. However, ethylene did not enhance the transcriptional hypoxia response in roots of a waterlogging-sensitive potato cultivar, suggesting that waterlogging tolerance in potato could depend on ethylene-controlled hypoxia responses in the roots. Finally, we show that ethylene rarely enhances hypoxia-adaptive genes and does not improve hypoxia survival in tomato (*Solanum lycopersicum*). We conclude that analyzing genes indicative of ethylene-mediated hypoxia acclimation is a promising approach to identifying key signaling cascades that confer flooding tolerance in crops.

Keywords: ethylene; flooding; hypoxia; phytoglobin; VII Ethylene Response Factor; PRT6 N-degron pathway of proteolysis; *Solanum tuberosum*; *Solanum lycopersicum*; *Solanum dulcamara*

1. Introduction

Global food demand is expected to double by 2050, as a result of the rising world population and shifting human diets [1]. However, future food security is highly challenged by a decline in arable land and an increase in crop losses, due to the consequences of a changing climate [1,2]. Indeed, elevated temperatures and extreme precipitation patterns due to climate change have led to an increase in the frequency and severity of flooding events [3]. Floods strongly impair agricultural crop production and exacerbate the food security crisis [3,4]. In order to minimize crop losses and safeguard global food security, it is therefore paramount to develop flood-tolerant crops in the near future [5]. Flooding tolerance in plants is regulated by variety of signals that mediate responses to combat oxygen (O_2) deprivation (hypoxia) and ameliorate toxic reactive oxygen species (ROS) [4,6]. Submerged terrestrial plants ultimately encounter hypoxia, due to impaired gas diffusion under water, and flooding survival therefore strongly depends on traits that enhance hypoxia tolerance [4,7]. In addition, passive ethylene entrapment acts as a rapid signal for submergence, and regulates many flood adaptive responses that include morphological and anatomical modifications to prevent hypoxia [8]. Ethylene also mediates metabolic hypoxia acclimation in the wetland species *Rumex palustris* and the model species Arabidopsis (*Arabidopsis thaliana*) [9–11]. In Arabidopsis, ethylene-mediated hypoxia tolerance is conserved in root and shoot meristems and is associated with enhanced expression of a conserved set of hypoxia-responsive genes when O_2 levels decline [10,12]. These hypoxia adaptive genes are predominantly controlled by the group VII Ethylene Response Factor (ERFVII) transcription factors RELATED TO APETALA2.2 (RAP2.2), RAP2.12 and RAP2.3 [13,14]. ERFVIIs are typically known to control a variety of developmental growth [15–17], biotic [18,19] and abiotic stress responses in plants [20–23]. Arabidopsis ERFVIIs possess a conserved N-terminal sequence that regulates their turnover through the PROTEOLYSIS6 (PRT6) N-degron pathway under O_2 and nitric oxide (NO) replete conditions [24–26]. When the cellular levels of either O_2 or NO decline, ERFVII proteolysis is restricted and these proteins stabilize, ultimately activating target hypoxia adaptive gene expression [24–26]. Ethylene can already stabilize ERFVIIs prior to hypoxia through active NO reduction, by enhancing levels of the NO-scavenging protein PHYTOGLOBIN1 (PGB1) [10]. In addition, ethylene can promote ERFVII production through transcriptional activation. Collectively, these results showed that the cascade initiated by ethylene upon submergence is crucial to prevent N-degron targeted ERFVII proteolysis by the increased production of the NO-scavenger PGB1. This ethylene-induced ERFVII accumulation, in turn, augments the transcription of downstream target genes to improve energy homeostasis and oxidative stress tolerance during hypoxia and re-oxygenation in Arabidopsis [10,27–29].

Interestingly, this work in Arabidopsis revealed that if ethylene was not able to enhance PGB1 or ERFVII levels (through mutant analysis), ethylene was also unable to augment the transcriptional hypoxia response and promote hypoxia tolerance [10]. Similarly, the inability to enhance hypoxia tolerance through early ethylene signaling correlated with the absence of ethylene-induced expression of *PGB1*, *RAP2.12* and hypoxia adaptive gene orthologues in the terrestrial wild plant species *Rumex acetosa* [9]. Contrastingly, but similar to Arabidopsis, *R. palustris* uses the ethylene signal to promote hypoxia survival and this corresponded with *PGB1*, *RAP2.12* and hypoxia adaptive transcript induction [9]. Finally, natural variation for higher flooding tolerance is frequently found to be correlated with enhanced ethylene signaling [30,31], PGB1 levels and NO removal [32,33], ERFVII levels [31,34] and hypoxia adaptive gene transcripts [30,35–38] in crop and ornamental plant species. Taken together, the mechanism of ethylene-mediated hypoxia tolerance could be instrumental in enhancing flooding tolerance of intolerant crops through the manipulation of ethylene responsiveness of *PGB1* and *ERFVII* genes and its downstream signaling targets.

Potato (*Solanum tuberosum*) and tomato (*Solanum lycopersicum*) are the most economically important cultivated vegetable crops in the world (e.g., the global tomato export value exceeded USD 9 billion in 2019 [39]), but are both considered to be highly intolerant to abiotic stresses, including flooding [40,41]. Conversely, the closely related wild plant species bittersweet (*Solanum dulcamara*) is highly tolerant to a variety of abiotic stresses [42]. All three species display multiple (ethylene-mediated) flood adaptive responses [41,43,44], but still show strong differences in flooding tolerance. However, whether ethylene also contributes to metabolic hypoxia acclimation and if this correlates with flooding tolerance in these *Solanum* species is currently unknown. Several studies showed that genes involved in ethylene signaling and ethanolic fermentation are induced in response to flooding and hypoxia, but not consistently across all plant tissues in these *Solanum* species [37,40,41,45,46].

Here, we investigated whether these cultivated and wild *Solanum* species integrate the early flooding signal ethylene to control subsequent hypoxia acclimation. To do so, we studied transcript levels of a selection of representative marker gene orthologues that control ethylene-mediated hypoxia responses in Arabidopsis. In addition, we tested whether an ethylene pre-treatment was able to enhance survival during subsequent hypoxia in roots and shoots of several *Solanum* species. Interestingly, our results revealed that ethylene contributes to hypoxia acclimation responses in bittersweet and a waterlogging-tolerant potato cultivar, but does not in tomato and the roots of waterlogging-sensitive potato cultivar. We propose that the mechanism of ethylene-mediated hypoxia acclimation could be instrumental to uncover signaling cascades that confer flooding tolerance in crops.

2. Results

2.1. Elite Potato Cultivars Show Variation in Waterlogging Tolerance

To unravel whether ethylene-mediated hypoxia acclimation contributes to potential flooding tolerance differences in potato, we first explored whether variation in flooding tolerance exists across six elite potato cultivars. Potato plants grown from tubers were waterlogged for up to 12 days and shoot growth parameters were scored every 3 days. Overall, the cultivars Festien and Avarna retained the highest shoot biomass and leaf area, whereas the cultivars Seresta and Ambition showed the highest loss in shoot performance during waterlogging (Figure 1). The differences between the tolerant cultivar Festien and sensitive Seresta were the most consistent for both biomass and leaf area at every recorded time point during the full duration of the waterlogging experiment. Since ethylene typically accumulates in flooded plant tissues [8], we explored whether differences in ethylene-mediated hypoxia responses could contribute to this contrasting waterlogging tolerance among potato cultivars. For this, we selected the tolerant Festien and sensitive Seresta for further transcriptional analysis.

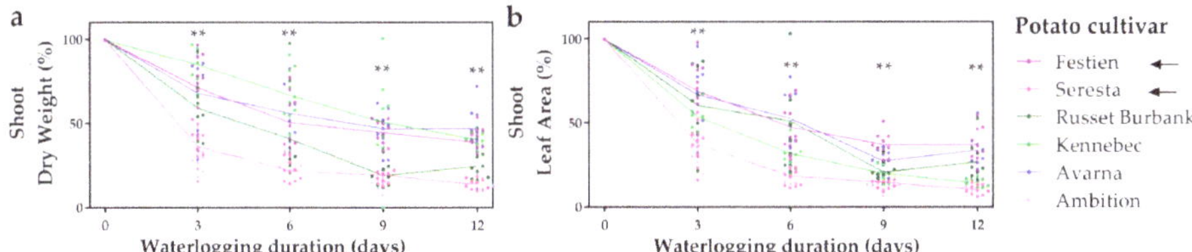

Figure 1. Potato cultivars show variation in waterlogging tolerance. Relative shoot dry weight (**a**), relative leaf area (**b**) of 6 elite potato (*S. tuberosum*) cultivars during 12 days of waterlogging. Values are relative to non-waterlogged plants who were set at 100% per time-point (**a**,**b**). Asterisks indicate significant differences between the Festien (purple) and Seresta (pink) cultivars per time-point and these were the cultivars selected for further analysis (indicated by arrows, ** $p < 0.01$, Generalized linear model, Tukey's honestly significant difference (HSD), $n = 10$ plants).

2.2. Ethylene Differentially Mediates the Transcriptional Hypoxia Response in Potato Roots

In Arabidopsis, augmentation of the transcriptional hypoxia response by ethylene depends on increased levels of ERFVIIs and PGB1 prior to hypoxia. Here, we selected several potato orthologues of Arabidopsis genes associated with this ethylene-controlled hypoxia tolerance as markers for the integration of ethylene signaling with the transcriptional hypoxia response. These included: *ETHYLENE RESPONSE 2 (ETR2)* [47], to test whether ethylene indeed led to ethylene-dependent signaling; *PGB1* as a marker gene for enhanced NO-scavenging capacity [48]; *RAP2.12* to test whether ethylene induces potato ERFVIIs at the transcriptional level and *PYRUVATE DECARBOXYLASE 1 (PDC1)*, encoding the rate limiting protein for hypoxia-induced fermentation [49], as a representative marker gene for the transcriptional hypoxia response. Since *ETR2* and *PGB1* are also among the ERFVII-controlled hypoxia-responsive genes in Arabidopsis [13], they could additionally be interpreted as markers for the hypoxia response and function as an indication for enhanced ERFVII levels, prior to hypoxia.

To unravel how ethylene modulates hypoxia responses in potato, plants were pre-treated with air or ethylene followed by 4 h of hypoxia. Ethylene enhanced the transcript abundance of *ETR2* and *RAP2.12* orthologues in shoot and root tissues of both potato cultivars (Figure 2), showing that the ethylene treatment quickly leads to ethylene-dependent signaling. Interestingly, *PGB1* transcripts were enhanced in response to ethylene in both tissues of the more tolerant cultivar Festien, but were not enriched in the roots of the sensitive cultivar Seresta (Figure 2). Moreover, ethylene pre-treatment also enhanced the hypoxia response of *ETR2*, *PGB1* and *PDC1* transcripts in both the shoots and roots of the tolerant cultivar Festien. However, while ethylene pre-treatment did enhance the hypoxia response of *PGB1* and *PDC1* in the shoots of sensitive cultivar Seresta (Figure 2a), ethylene had no beneficial effect on the hypoxia-responsiveness of *ETR2*, *PGB1* and *PDC1* transcripts in the root tissues (Figure 2b). Together, these data suggest that the capacity to enhance root hypoxia response genes through ethylene signaling corresponds to higher waterlogging tolerance in these two potato cultivars, and could depend on *PGB1* induction to modulate enhanced metabolic hypoxia acclimation.

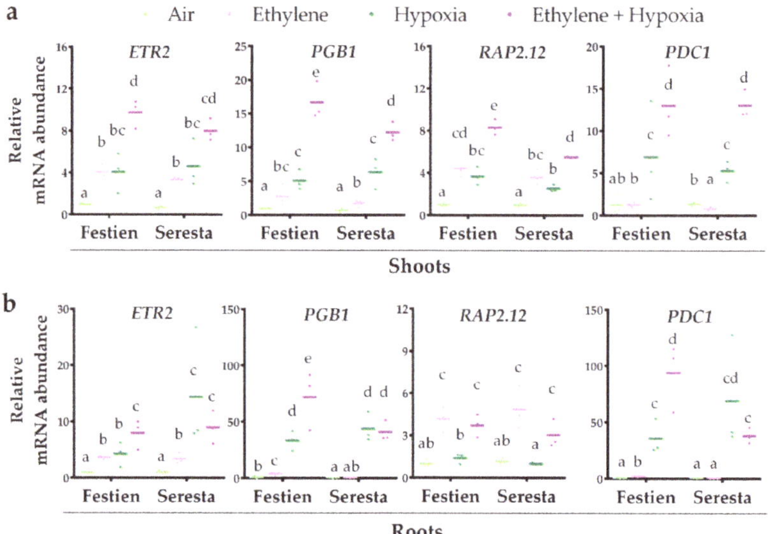

Figure 2. Ethylene differentially modulates hypoxia-responsive transcripts in the roots of elite potato cultivars. Relative mRNA transcript abundance of four marker genes for ethylene-mediated hypoxia tolerance in shoot (**a**) and root tissues (**b**) of the potato cultivars Festien (waterlogging tolerant) and Seresta (waterlogging sensitive) after 4 h of pre-treatment with air (light green) or ~5 μL L^{-1} ethylene (pink), followed by (4 h) hypoxia (green and purple). Marker genes are orthologues of the Arabidopsis genes: ethylene signaling gene *ETR2*, NO-scavenging phytoglobin *PGB1*, ERFVII transcription factor *RAP2.12* and hypoxia adaptive gene *PDC1*. Values are relative to air treated samples of Festien. Different letters indicate significant differences ($p < 0.05$, 2-way ANOVA, Tukey's HSD, $n = 3$ containing 2 shoot meristems (**a**) or ~10 root tips (**b**)).

2.3. Ethylene Differentially Modulates Hypoxia Survival in Solanum Seedlings

While an ethylene-enhanced transcriptional hypoxia response coincides with increased waterlogging tolerance in the previously described potato cultivars, it is unclear whether it also corresponds to ethylene-mediated hypoxia tolerance in *Solanum* species. Previous work has shown that early ethylene signaling is essential to improve the hypoxia tolerance of Arabidopsis and bittersweet roots, but not tomato roots [10]. Here, we tested the root hypoxia tolerance of additional *Solanum* species and the shoot hypoxia tolerance of both bittersweet and tomato. Ethylene pre-treatment strongly enhanced root tip survival in both bittersweet and eggplant (*Solanum melongena*) seedlings, but had no effect on tomato (Moneymaker) and *Solanum pennellii* root tip survival (Figure 3a,b). Furthermore, the ethylene pre-treatment improved the survival and biomass retention (fresh weight) of bittersweet shoot tissues during hypoxia (Figure 3c–e). Similar to root tips, ethylene had no effect on hypoxia tolerance of tomato shoots (Figure 3f,g). Collectively, these results indicate that the ethylene-mediated hypoxia tolerance is conserved in both the shoots and roots of the flood tolerant plant bittersweet, but not in the tomato cultivar Moneymaker.

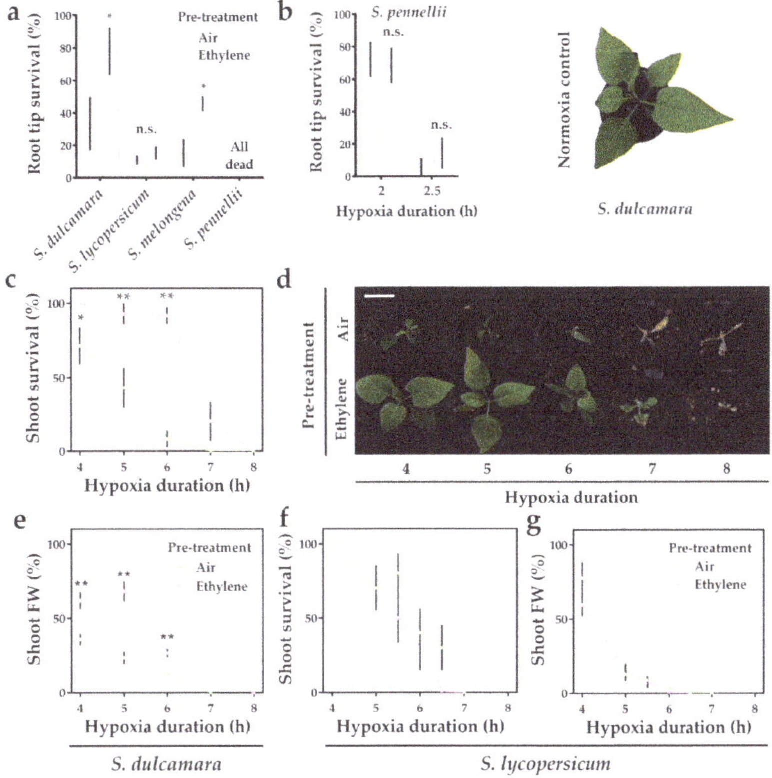

Figure 3. Ethylene differentially mediates hypoxia tolerance in *Solanum* species. Root tip survival of *S. dulcamara*, *S. lycopersicum* (Moneymaker), *S. melongena* and *S. pennellii* seedlings after 4 h of pre-treatment with air (green) or ~5 μL L^{-1} ethylene (purple) followed by 4 h (**a**) or 2 and 2.5 h (**b**) (only *S. pennellii*) of hypoxia and 2 days of recovery. Asterisks indicate significant differences between air and ethylene (error bars are SEM, * $p < 0.05$, Student's *t* test, n = 4–10 rows containing 8–10 seedlings). Shoot survival (**c,f**), phenotypes (**d**) (only *S. dulcamara*) and fresh weight (FW) (**e,g**) of *S. dulcamara* (**c–e**) and *S. lycopersicum* (**f,g**) plants after 4 h of air (green) or ~5 μL L^{-1} ethylene (purple) pre-treatment followed by hypoxia and 7 days recovery. Values are relative to control (normoxia) plants. Scale bar = 3 cm. Asterisks indicate significant differences between air and ethylene (error bars are SEM, ** $p < 0.01$, Student's *t* test, n = 10–13 plants).

2.4. Ethylene Differentially Mediates the Transcriptional Hypoxia Response between Bittersweet and Tomato

Next, we assessed whether the contrasting tolerance between tomato and bittersweet corresponds to how these species use an early ethylene signal to control its subsequent transcriptional hypoxia response. Similar to the analysis in potato, we quantified transcript levels of the representative marker genes studied earlier (Figure 2), but now also included the orthologues of ethylene response marker *ACC OXIDASE 1* (*ACO1*), NO-scavenging genes *PGB2* and *PGB3* [50], the ERFVII *RAP2.3* and the hypoxia markers *SIMILAR TO RCD ONE 5* (*SRO5*) and *ALCOHOL DEHYDROGENESE1* (*ADH1*) to give a better understanding of the transcriptional response in both tomato and bittersweet.

In bittersweet, ethylene typically led to ethylene signalling and subsequent up-regulation of *PGB1*, *PGB3*, *RAP2.3* and *RAP2.12* transcripts in both root and shoot tissues (Figure 4). Moreover, the results revealed that ethylene strongly augmented the transcriptional hypoxia response. In bittersweet shoots, seven out of the eight hypoxia-responsive genes were enhanced by ethylene upon hypoxia. In the roots, all 10 genes tested were hypoxia responsive and an ethylene pre-treatment enhanced eight of these 10 transcripts (Figure 4). In tomato root and shoot tissues, ethylene also switched on ethylene signalling and the subsequent up-regulation of *PGB1*, *RAP2.3* and *RAP2.12* transcripts (Figure 5). However, ethylene rarely further increased mRNA levels of hypoxia-responsive genes. In tomato shoot tissues, ethylene only enhanced one (*ACO1*) out of the eight hypoxia-responsive genes during hypoxia. Out of the seven hypoxia-responsive genes in tomato roots, ethylene promoted transcript levels of *ACO1*, *PGB1* and *PDC1* upon hypoxia. Collectively, these results reveal that ethylene enhances hypoxia responses and tolerance in bittersweet, but not in tomato.

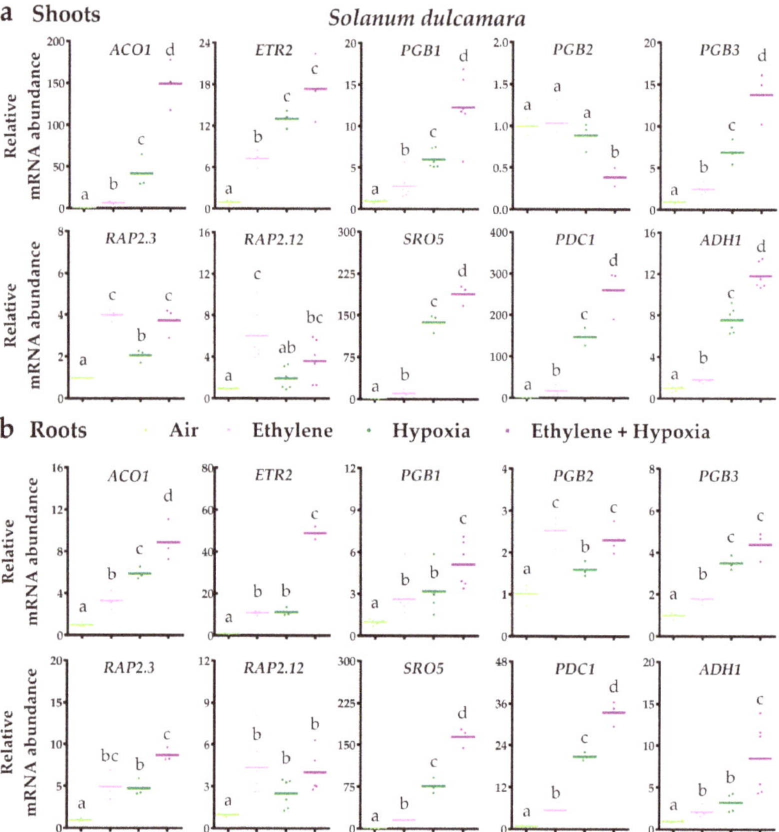

Figure 4. Ethylene augments hypoxia-responsive transcripts in bittersweet. Relative mRNA transcript abundance of 10 marker genes for ethylene-mediated hypoxia tolerance in bittersweet (*S. dulcamara*) shoot (**a**) and root tissues (**b**) after 4 h of pre-treatment with air (light green) or ~5 µL L^{-1} ethylene (pink), followed by (4 h) hypoxia (green and purple). Marker genes are orthologues of the Arabidopsis genes: ethylene signaling genes *ACO1* and *ETR2*, NO-scavenging phytoglobins *PGB1*, *PGB2* and *PGB3*, ERFVII transcription factors *RAP2.3* and *RAP2.12* and hypoxia adaptive genes *SRO5*, *PDC1* and *ADH1*. Values are relative to air treated samples. Different letters indicate significant differences ($p < 0.05$, 1-way ANOVA, Tukey's HSD, n = three or six biological replicates containing 2–4 shoot meristems (**a**) or ~10 root tips (**b**)).

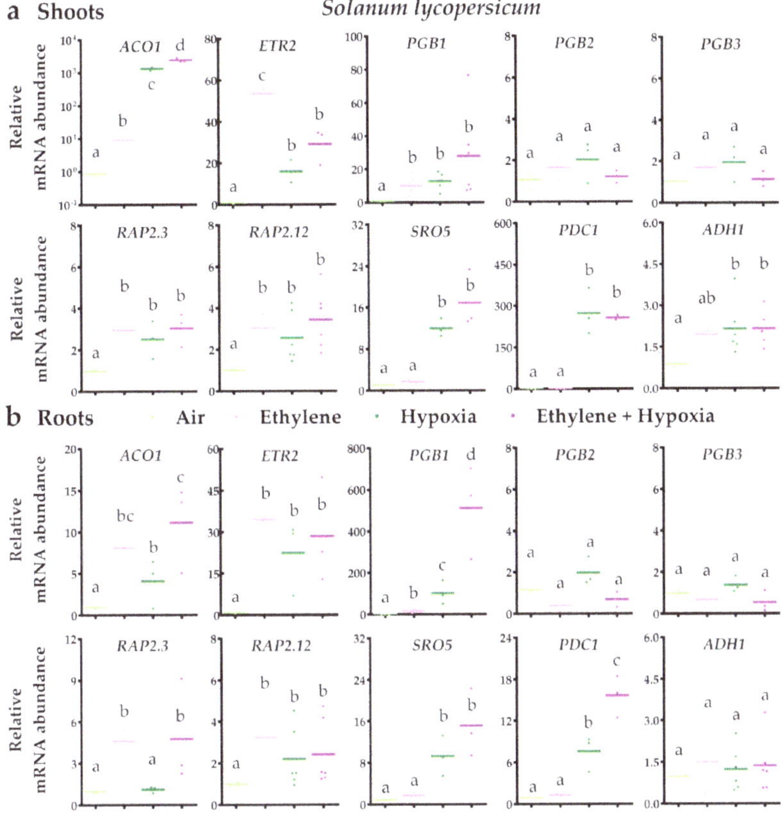

Figure 5. Ethylene rarely enhances hypoxia-responsive transcripts in tomato. Relative mRNA transcript abundance of 10 marker genes for ethylene-mediated hypoxia tolerance in tomato (*S. lycopersicum* cv MoneyMaker) shoot (**a**) and root tissues (**b**) after 4 h of pre-treatment with air (light green) or ~5 µL L^{-1} ethylene (pink), followed by (4 h) hypoxia (green and purple). Marker genes are orthologues of the Arabidopsis genes: ethylene signaling genes *ACO1* and *ETR2*, NO-scavenging phytoglobins *PGB1*, *PGB2* and *PGB3*, ERFVII transcription factors *RAP2.3* and *RAP2.12* and hypoxia adaptive genes *SRO5*, *PDC1* and *ADH1*. Values are relative to air treated samples. Different letters indicate significant differences ($p < 0.05$, 1-way ANOVA, Tukey's HSD, n = three or six biological replicates containing 2–4 shoot meristems (**a**) or ~10 root tips (**b**)).

3. Discussion

In this study, we made a first step in translating the mechanism of ethylene-mediated hypoxia tolerance as established in Arabidopsis [10], to vegetable crops. We show that ethylene can enhance hypoxia tolerance in bittersweet and eggplant, but not in tomato and *S. pennelli*. Moreover, our results reveal that ethylene generally leads to up-regulation of ethylene signaling, *PGB1* and *ERFVII* transcripts in potato, bittersweet and tomato, but that the more downstream hypoxia-responsive genes are not consistently enhanced by early ethylene signaling when O$_2$ levels decline (overview in Figure 6). In addition, the capacity to augment the transcriptional hypoxia response through ethylene signaling correlated with higher waterlogging or ethylene-enhanced hypoxia tolerance in the tested *Solanum* species (Figure 6).

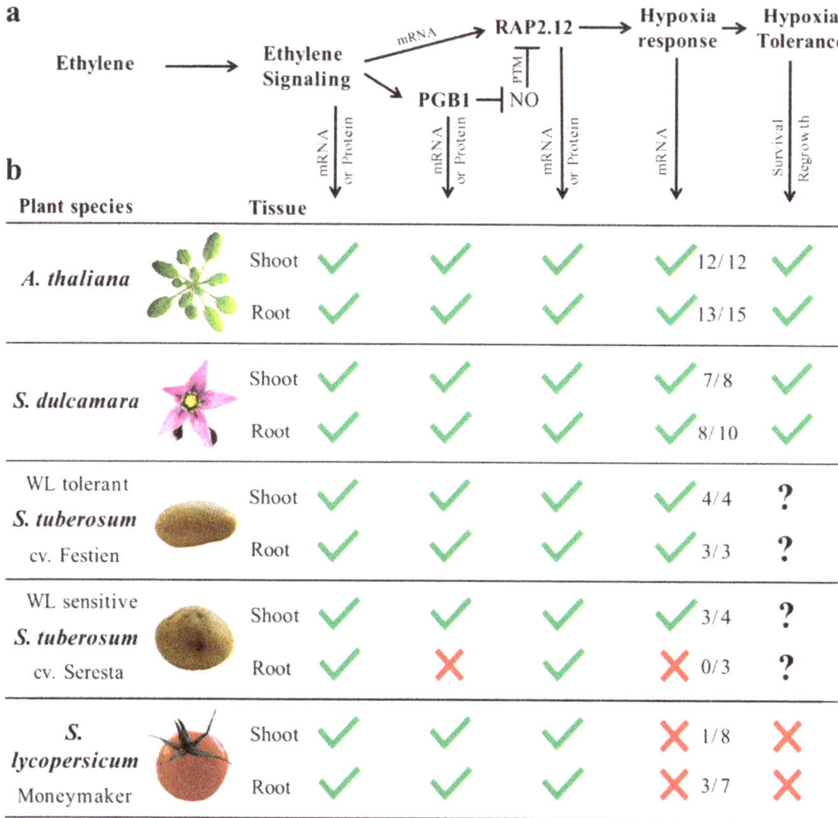

Plant species		Tissue	mRNA or Protein	mRNA or Protein	mRNA or Protein	mRNA	Survival Regrowth
A. thaliana		Shoot	✓	✓	✓	✓ 12/12	✓
		Root	✓	✓	✓	✓ 13/15	✓
S. dulcamara		Shoot	✓	✓	✓	✓ 7/8	✓
		Root	✓	✓	✓	✓ 8/10	✓
WL tolerant *S. tuberosum* cv. Festien		Shoot	✓	✓	✓	✓ 4/4	?
		Root	✓	✓	✓	✓ 3/3	?
WL sensitive *S. tuberosum* cv. Seresta		Shoot	✓	✓	✓	✓ 3/4	?
		Root	✓	✗	✓	✗ 0/3	?
S. lycopersicum Moneymaker		Shoot	✓	✓	✓	✗ 1/8	✗
		Root	✓	✓	✓	✗ 3/7	✗

Figure 6. Overview of ethylene-mediated hypoxia responses in *Solanum* species. (**a**) Schematic showing the proposed mechanism of ethylene-mediated hypoxia tolerance, as discovered in Arabidopsis [10]. Arrows pointing downward indicate key processes that were verified in Arabidopsis at the mRNA and protein level, or tested in the *Solanum* species at mRNA level in both roots and shoots. (**b**) Table showing whether marker genes or hypoxia tolerance (as assessed by regrowth capacity following stress removal) were enhanced by an early ethylene treatment (green check mark = yes, red cross = no, question mark = not tested). The n/n in the hypoxia response column indicates the amount of hypoxia genes that are ethylene-enhanced and hypoxia-responsive upon hypoxia compared to the total amount of hypoxia-responsive genes tested. This table is based on the experimental data shown in Figures 1–5 and [10,27].

While ethylene enhanced transcript levels of *PGB1* and *RAP2.12* orthologues, it is unclear whether this also led to increased corresponding protein levels and biological function. However, our results suggest that ethylene may indeed stimulate ERFVII protein levels in bittersweet and the more waterlogging tolerant potato cultivar Festien, based on the ethylene-enhanced upregulation of typical ERFVII-target hypoxia gene transcripts during subsequent hypoxia treatments (Figures 2 and 4). One explanation could be that, similar to Arabidopsis, ethylene impairs NO-dependent ERFVII proteolysis in these tolerant *Solanum* species through the up-regulation of NO scavenging PGBs, causing enhanced ERFVII stability [10]. Consistently, the lack of *PGB1* up-regulation in the roots of potato cultivar Seresta may explain why ethylene failed to enhance hypoxia-responsive genes in these tissues (Figure 2). Additionally, ethylene may also directly promote protein synthesis of one or more *Solanum* ERFVIIs that escape PRT6 N-degron mediated proteolysis through structural differences, as shown for the ethylene-mediated rice ERFVII SUB1A-1 [51]. Further research is required to determine if ethylene indeed leads to higher *Solanum* ERFVII levels under normoxia, and whether this is mediated through enhanced PGB1-dependent NO scavenging.

Interestingly, while an ethylene pre-treatment increased upstream marker genes *PGB1*, *RAP2.3* and *RAP2.12* transcripts in tomato, it did not enhance mRNA levels of most hypoxia-responsive gene orthologues upon hypoxia (Figures 5 and 6). These results suggest that ethylene-mediated ERFVII levels could in some cases be uncoupled from enhanced target-gene expression upon hypoxia. Strikingly, ethylene did augment three hypoxia-responsive gene transcripts in tomato root tissues, hinting that not all hypoxia-responsive genes are uncoupled from ethylene-enhanced ERFVII regulation in the same fashion. One explanation could be that ethylene only enhances a subset of ERFVII proteins, and that specific ERFVIIs mediate different target genes. Alternatively, it is possible that ethylene does not lead to sufficiently enriched PGB1 protein levels, that PGB1 does not effectively scavenge NO, or that the proteolysis of some ERFVIIs is not NO-dependent in tomato plants and in the roots of the potato cultivar Seresta.

It is currently unclear which hypoxia-responsive genes actually contribute to enhanced hypoxia tolerance. For instance, several studies suggest transcriptional *SUS1* or *ADH1* induction may not be required for anaerobic fermentation, maintenance of glycolysis and subsequent hypoxia tolerance in potato tubers and Arabidopsis, respectively [40,52]. Previous research showed that protein levels of ADH1 in tomato roots increased only 1.7-fold after 24 h of waterlogging, whereas PDC1 levels were even down-regulated [53]. It is likely that the functional implications of specific up-regulated hypoxia genes for hypoxia tolerance are dependent on the plant tissue and determined by the conditions and severity of the stress [6]. In the future, other (ERFVII-dependent) hypoxia-responsive genes should be tested to uncover their role in ethylene-mediated hypoxia acclimation. Especially oxidative stress amelioration is thought to be a major determinant of flooding and hypoxia tolerance [11,27,29,54], and it would therefore be interesting to assess the effect of ethylene on ROS removal during hypoxia and re-oxygenation in *Solanum* species in the future.

Collectively, we show that ethylene-mediated hypoxia responses are conserved in *S. dulcamara* and correspond to enhanced tolerance similar to Arabidopsis and *R. palustris* [9,10]. In potato cultivars, the induction of this mechanism was also conserved in shoots and roots of the waterlogging-tolerant cultivar Festien. Conversely, early ethylene signaling was uncoupled from *PGB1* induction and an enhanced hypoxia response in the roots of waterlogging-intolerant cultivar Seresta, suggesting that waterlogging tolerance in potato could be dependent on ethylene-mediated hypoxia responses in the roots. Similarly, ethylene signaling did not augment most of the transcriptional hypoxia response and hypoxia survival in tomato. While it is unclear whether the (re-)introduction of ethylene-mediated hypoxia genes would enhance flooding tolerance of tomato and more waterlogging sensitive potato cultivars, future work could aim to introduce ethylene-regulated promoter elements into key genes involved in ethylene-mediated hypoxia tolerance (Figure 6). However, it remains to be established why tomato plants and roots of the potato cultivar Seresta do not integrate ethylene signaling to enhance its transcriptional hypoxia response. We conclude that studying genes indicative of ethylene-mediated hypoxia acclimation is a useful approach to explore the universality of this mechanism across species and helps to identify key signaling cascades that confer flooding and hypoxia tolerance in vegetable crops.

4. Materials and Methods

4.1. Plant Material and Growth Conditions

Plant material: potato (*S. tuberosum* L.) tubers of the elite cultivars Russet Burbank, Kennebec, Festien, Seresta, Avarna and Ambition were obtained from Jan De Haas of HZPC Research B.V, Metslawier, The Netherlands and Mariëlle Muskens of Agrico Research B.V., Bant, The Netherlands. Seeds of bittersweet (*S. dulcamara*) and *S. pennellii* were obtained from Dr. Eric Visser and Dr. Ivo Rieu, respectively, Radboud University, Nijmegen, The Netherlands. Tomato (*S. lycopersicum* L.) of the variety Moneymaker and eggplant (*S. melongena*) seeds were obtained from a local garden center (Intratuin).

Growth conditions potted plants: potato tubers were kept in the dark at room temperature for 1 month, until the eye buds started to sprout. Tubers were placed individually in 11 cm × 11 cm sized square pots filled with a sand soil mixture (1:2), were covered by ~5 cm of soil mixture and stratified at 4 °C in the dark for 3 days. Seedlings of tomato and bittersweet were germinated on agar plates as described below and subsequently transplanted in small round pots (diameter = 5 cm) filled with a sand soil mixture (1:2). Pots were then transferred to a growth chamber for germination/growth under short day conditions (9:00–17:00, temperature (T) = 20 °C, photosynthetic photon flux density (PPFD) = ~130 μmol m^{-2} s^{-1}, relative humidity (RH) = 70%). Potato plants were used for waterlogging and ethylene/hypoxia gassing experiments 3 weeks after potting. Tomato and bittersweet plants were 2 weeks old for the shoot survival experiments described below. Per species, homogeneous groups of plants were selected and randomized over treatment groups before phenotypic and molecular analysis. Tomato and bittersweet used for hypoxia survival experiments were transferred back to the same growth room conditions after treatments to recover for 7 days before scoring shoot survival and fresh weight.

Growth conditions seedlings: tomato, bittersweet, eggplant and *S. pennellii* seeds were vapor sterilized through incubation with a beaker containing a mixture of 50 mL bleach (5%) and 3 mL of 37% fuming HCl in a gas tight desiccator jar for 4 h. Seeds were then individually transplanted in 2 rows of 10 seeds on sterile square petri dishes containing 25 mL autoclaved and solidified 1/4 MS, 1% plant agar without sucrose supplement. Petri dishes were sealed with gas-permeable tape (Leukopor, Duchefa) and kept at 4 °C in the dark for 3 days. Seedlings were when grown vertically on the agar plates under short day conditions (9:00–17:00, T = 20 °C, PPFD = ~130 μmol m^{-2} s^{-1}, RH = 70%) for 7 days, before performing root tip survival experiments and shoot and root harvests for RT-qPCR.

4.2. Tolerance Assays, Treatments and Sample Harvests

Waterlogging tolerance of potato cultivars: plants were waterlogged by submerging the pots in large tubs under long day conditions (7:00–23:00, T = 20 °C, PPFD = ~150 μmol m^{-2} s^{-1}) in water that was left stagnant 1 day prior to the experiment. During the experiments, the water surface was maintained 1 cm above the soil. Control plants were placed in similar tubs without water. Of each cultivar, 10 pots were taken out of control and waterlogging treatments after 3, 6, 9 and 12 days. Shoot biomass (dry weight) and total leaf area were measured at the start of the treatment (t = 0) and at each subsequent harvest time-point. Root tip hypoxia survival performance assays of tomato, bittersweet, eggplant and *S. pennellii* 7-day old seedlings grown vertically on agar plates were performed as in [10]. Shoot meristem survival was performed on 2-week-old plants grown in pots, as described for Arabidopsis rosettes in [10]. Samples for mRNA analysis of potato, tomato and bittersweet were harvested after 4 h of air and ethylene (~5 μL L^{-1}) treatments and after 4 h of subsequent hypoxia (<0.00% O_2) treatment, similar to [10]. For shoot tissues, 1–4 primary shoot meristems were harvested including the first young leaf. For root tissues, multiple segments (5–10) of ~1 cm long root tips were harvested per sample.

4.3. RNA Extraction, cDNA Synthesis and RT-qPCR

RNA extraction, cDNA synthesis, RT-qPCR were performed as described in [10]. Gene orthologue sequences were obtained by BLASTing the protein coding sequences of Arabidopsis genes against the bittersweet transcriptome [45], tomato genome (version SL3.0 and Annotation ITAG3.10) and potato genome [55] using the tblastx tool on [56]. For potato, the orthologue of *ACTIN11* was used as a reference gene. For tomato and bittersweet, the orthologues of *TUBULIN6* and *TIP41* were used as reference genes and the mean CT was used to calculate transcript fold changes. Gene annotations and primers used in this study are listed in Table S1.

4.4. Statistics

The data were plotted and the figures were designed using Graphpad Prism software (San Diego, CA, USA). The statistical tests were performed using either Graphpad Prism (San Diego, CA, USA) or R software and the "LSmeans and "multmultcompView" packages. Potato tolerance data were analyzed through a generalized linear modeling (GLM) approach. A negative binomial error structure was used for the GLM. The other data were analyzed with either a Students *t*-test, 1-way or 2-way ANOVA. If necessary, data was log transformed to meet ANOVA prerequisites. Tukey's honestly significant difference (HSD) tests were used to correct for multiple comparisons.

Supplementary Materials: The following are available online at http://www.mdpi.com/2223-7747/9/8/1022/s1, Table S1: RT-qPCR primers used in this study.

Author Contributions: Conceptualization, S.H., R.S. and L.A.C.J.V.; methodology and experiments, S.H., N.v.D., D.M.H.J.R., R.A.M.W.-E. and J.K.; manuscript preparation, S.H., R.S. and L.A.C.J.V.; funding acquisition, S.H., R.S. and L.A.C.J.V. All authors have read and agreed to the published version of the manuscript.

Funding: This research was funded by grants from the Netherlands Organization for Scientific Research (831.15.001 to S.H., 824.14.007 to L.A.C.J.V. and BB.00534.1 to R.S.).

Acknowledgments: We thank Mariëlle Muskens of Agrico Research B.V. and Jan De Haas and Robert Graveland of HZPC Research B.V. for valuable feedback and providing tubers of the potato cultivars used in this study (in kind donations). In addition, we kindly acknowledge Eric Visser & Ivo Rieu (both Radboud University Nijmegen, NL) for providing the seeds of *S. dulcamara* and *S. pennellii*, respectively. Finally, we thank Emilie Reinen, Sarah Courbier and Ankie Ammerlaan of Utrecht University for technical assistance.

Conflicts of Interest: The funders had no role in the design of the study; in the collection, analyses, or interpretation of data; in the writing of the manuscript, or in the decision to publish the results.

References

1. Myers, S.S.; Smith, M.R.; Guth, S.; Golden, C.D.; Vaitla, B.; Mueller, N.D.; Dangour, A.D.; Huybers, P. Climate Change and Global Food Systems: Potential Impacts on Food Security and Undernutrition. *Annu. Rev. Public Health* **2017**, *38*, 259–277. [CrossRef] [PubMed]
2. Lobell, D.B.; Burke, M.B.; Tebaldi, C.; Mastrandrea, M.D.; Falcon, W.P.; Naylor, R.L. Prioritizing climate change adaptation needs for food security in 2030. *Science* **2008**, *319*, 607–610. [CrossRef] [PubMed]
3. Hirabayashi, Y.; Mahendran, R.; Koirala, S.; Konoshima, L.; Yamazaki, D.; Watanabe, S.; Kim, H.; Kanae, S. Global flood risk under climate change. *Nat. Clim. Chang.* **2013**, *3*, 816–821. [CrossRef]
4. Voesenek, L.A.C.J.; Bailey-Serres, J. Flood adaptive traits and processes: An overview. *New Phytol.* **2015**, *206*, 57–73. [CrossRef]
5. Mustroph, A. Improving Flooding Tolerance of Crop Plants. *Agronomy* **2018**, *8*, 160. [CrossRef]
6. Sasidharan, R.; Hartman, S.; Liu, Z.; Martopawiro, S.; Sajeev, N.; van Veen, H.; Yeung, E.; Voesenek, L.A.C.J. Signal Dynamics and Interactions during Flooding Stress. *Plant Physiol.* **2018**, *176*, 1106–1117. [CrossRef]
7. Shiono, K.; Takahashi, H.; Colmer, T.D.; Nakazono, M. Role of ethylene in acclimations to promote oxygen transport in roots of plants in waterlogged soils. *Plant Sci.* **2008**, *175*, 52–58. [CrossRef]
8. Voesenek, L.A.C.J.; Sasidharan, R. Ethylene—and Oxygen Signalling—Drive Plant Survival During Flooding. *Plant Biol.* **2013**, *15*, 426–435. [CrossRef]
9. Van Veen, H.; Mustroph, A.; Barding, G.A.; Eijk, M.V.; Welschen-Evertman, R.A.M.; Pedersen, O.; Visser, E.J.W.; Larive, C.K.; Pierik, R.; Bailey-Serres, J.; et al. Two Rumex Species from Contrasting Hydrological Niches Regulate Flooding Tolerance through Distinct Mechanisms. *Plant Cell* **2013**, *25*, 4691–4707. [CrossRef]
10. Hartman, S.; Liu, Z.; van Veen, H.; Vicente, J.; Reinen, E.; Martopawiro, S.; Zhang, H.; van Dongen, N.; Bosman, F.; Bassel, G.W.G.W.; et al. Ethylene-mediated nitric oxide depletion pre-adapts plants to hypoxia stress. *Nat. Commun.* **2019**, *10*, 4020. [CrossRef]
11. Hartman, S.; Sasidharan, R.; Voesenek, L.A.C.J. The role of ethylene in metabolic acclimations to low oxygen. *New Phytol.* **2020**. [CrossRef] [PubMed]
12. Mustroph, A.; Zanetti, M.E.; Jang, C.J.H.; Holtan, H.E.; Repetti, P.P.; Galbraith, D.W.; Girke, T.; Bailey-Serres, J. Profiling translatomes of discrete cell populations resolves altered cellular priorities during hypoxia in Arabidopsis. *Proc. Natl. Acad. Sci. USA* **2009**, *106*, 18843–18848. [CrossRef] [PubMed]

13. Gasch, P.; Fundinger, M.; Müller, J.T.; Lee, T.; Bailey-Serres, J.; Mustroph, A. Redundant ERF-VII Transcription Factors Bind to an Evolutionarily Conserved cis-Motif to Regulate Hypoxia-Responsive Gene Expression in Arabidopsis. *Plant Cell* **2016**, *28*, 160–180. [CrossRef] [PubMed]

14. Bui, L.T.; Giuntoli, B.; Kosmacz, M.; Parlanti, S.; Licausi, F. Constitutively expressed ERF-VII transcription factors redundantly activate the core anaerobic response in *Arabidopsis thaliana*. *Plant Sci.* **2015**, *236*, 37–43. [CrossRef]

15. Shukla, V.; Lombardi, L.; Iacopino, S.; Pencik, A.; Novak, O.; Perata, P.; Giuntoli, B.; Licausi, F. Endogenous Hypoxia in Lateral Root Primordia Controls Root Architecture by Antagonizing Auxin Signaling in Arabidopsis. *Mol. Plant* **2019**, *12*, 538–551. [CrossRef]

16. Hattori, Y.; Nagai, K.; Furukawa, S.; Song, X.-J.; Kawano, R.; Sakakibara, H.; Wu, J.; Matsumoto, T.; Yoshimura, A.; Kitano, H.; et al. The ethylene response factors SNORKEL1 and SNORKEL2 allow rice to adapt to deep water. *Nature* **2009**, *460*, 1026–1030. [CrossRef]

17. Abbas, M.; Berckhan, S.; Rooney, D.J.; Gibbs, D.J.; Vicente Conde, J.; Sousa Correia, C.; Bassel, G.W.; Marín-De La Rosa, N.; León, J.; Alabadí, D.; et al. Oxygen sensing coordinates photomorphogenesis to facilitate seedling survival. *Curr. Biol.* **2015**, *25*, 1483–1488. [CrossRef]

18. Kim, N.Y.; Jang, Y.J.; Park, O.K. AP2/ERF Family Transcription Factors ORA59 and RAP2.3 Interact in the Nucleus and Function Together in Ethylene Responses. *Front. Plant Sci.* **2018**, *9*, 1675. [CrossRef]

19. Zhao, Y.; Wei, T.; Yin, K.-Q.; Chen, Z.; Gu, H.; Qu, L.-J.; Qin, G. Arabidopsis RAP2.2 plays an important role in plant resistance to Botrytis cinerea and ethylene responses. *New Phytol.* **2012**, *195*, 450–460. [CrossRef]

20. Fukao, T.; Yeung, E.; Bailey-Serres, J. The submergence tolerance regulator SUB1A mediates crosstalk between submergence and drought tolerance in rice. *Plant Cell* **2011**, *23*, 412–427. [CrossRef]

21. Xu, K.; Xu, X.; Fukao, T.; Canlas, P.; Maghirang-Rodriguez, R.; Heuer, S.; Ismail, A.M.; Bailey-Serres, J.; Ronald, P.C.; Mackill, D.J. Sub1A is an ethylene-response-factor-like gene that confers submergence tolerance to rice. *Nature* **2006**, *442*, 705–708. [CrossRef] [PubMed]

22. Vicente, J.; Mendiondo, G.M.; Movahedi, M.; Peirats-Llobet, M.; Juan, Y.-T.; Shen, Y.-Y.; Dambire, C.; Smart, K.; Rodriguez, P.L.; Charng, Y.-Y.; et al. The Cys-Arg/N-End Rule Pathway is a General Sensor of Abiotic Stress in Flowering Plants. *Curr. Biol.* **2017**, *27*, 3183–3190. [CrossRef] [PubMed]

23. Papdi, C.; Pérez-Salamó, I.; Joseph, M.P.; Giuntoli, B.; Bögre, L.; Koncz, C.; Szabados, L. The low oxygen, oxidative and osmotic stress responses synergistically act through the ethylene response factor VII genes RAP2.12, RAP2.2 and RAP2.3. *Plant J.* **2015**, *82*, 772–784. [CrossRef] [PubMed]

24. Gibbs, D.J.; Lee, S.C.; Md Isa, N.; Gramuglia, S.; Fukao, T.; Bassel, G.W.; Correia, C.S.; Corbineau, F.; Theodoulou, F.L.; Bailey-Serres, J.; et al. Homeostatic response to hypoxia is regulated by the N-end rule pathway in plants. *Nature* **2011**, *479*, 415–418. [CrossRef] [PubMed]

25. Gibbs, D.J.; Md Isa, N.; Movahedi, M.; Lozano-Juste, J.; Mendiondo, G.M.; Berckhan, S.; Marín-de la Rosa, N.; Vicente Conde, J.; Sousa Correia, C.; Pearce, S.P.; et al. Nitric Oxide Sensing in Plants Is Mediated by Proteolytic Control of Group VII ERF Transcription Factors. *Mol. Cell* **2014**, *53*, 369–379. [CrossRef]

26. Licausi, F.; Kosmacz, M.; Weits, D.A.; Giuntoli, B.; Giorgi, F.M.; Voesenek, L.A.C.J.; Perata, P.; van Dongen, J.T. Oxygen sensing in plants is mediated by an N-end rule pathway for protein destabilization. *Nature* **2011**, *479*, 419–422. [CrossRef]

27. Hartman, J.G.W. The Early Flooding Signal Ethylene Acclimates Plants to Survive Low-Oxygen Stress. Ph.D. Thesis, Utrecht University, Utrecht, The Netherlands, 2020.

28. Perata, P. Ethylene Signaling Controls Fast Oxygen Sensing in Plants. *Trends Plant Sci.* **2019**. [CrossRef]

29. Liu, Z. Ethylene-Mediated Hypoxia Tolerance in *Arabidopsis thaliana*. Ph.D. Thesis, Utrecht University, Utrecht, The Netherlands, 2019.

30. Zhao, N.; Li, C.; Yan, Y.; Cao, W.; Song, A.; Wang, H.; Chen, S.; Jiang, J.; Chen, F. Comparative transcriptome analysis of waterlogging-sensitive and waterlogging-tolerant Chrysanthemum morifolium cultivars under waterlogging stress and reoxygenation conditions. *Int. J. Mol. Sci.* **2018**, *19*, 1455. [CrossRef]

31. Yu, F.; Liang, K.; Fang, T.; Zhao, H.; Han, X.; Cai, M.; Qiu, F. A group VII ethylene response factor gene, ZmEREB180, coordinates waterlogging tolerance in maize seedlings. *Plant Biotechnol. J.* **2019**, *17*, 2286–2298. [CrossRef]

32. Rivera-Contreras, I.K.; Zamora-Hernández, T.; Huerta-Heredia, A.A.; Capataz-Tafur, J.; Barrera-Figueroa, B.E.; Juntawong, P.; Peña-Castro, J.M. Transcriptomic analysis of submergence-tolerant and sensitive Brachypodium distachyon ecotypes reveals oxidative stress as a major tolerance factor. *Sci. Rep.* **2016**, *6*, 27686. [CrossRef]

33. Li, X.; Peng, R.-H.; Fan, H.-Q.; Xiong, A.-S.; Yao, Q.-H.; Cheng, Z.-M.; Li, Y. Vitreoscilla hemoglobin overexpression increases submergence tolerance in cabbage. *Plant Cell Rep.* **2005**, *23*, 710–715. [CrossRef] [PubMed]

34. Wei, X.; Xu, H.; Rong, W.; Ye, X.; Zhang, Z. Constitutive expression of a stabilized transcription factor group VII ethylene response factor enhances waterlogging tolerance in wheat without penalizing grain yield. *Plant Cell Environ.* **2019**, *42*, 1471–1485. [CrossRef] [PubMed]

35. Zhang, Y.; Kong, X.; Dai, J.; Luo, Z.; Li, Z.; Lu, H.; Xu, S.; Tang, W.; Zhang, D.; Li, W.; et al. Global gene expression in cotton (*Gossypium hirsutum* L.) leaves to waterlogging stress. *PLoS ONE* **2017**, *12*, e0185075. [CrossRef] [PubMed]

36. Pan, R.; He, D.; Xu, L.; Zhou, M.; Li, C.; Wu, C.; Xu, Y.; Zhang, W. Proteomic analysis reveals response of differential wheat (*Triticum aestivum* L.) genotypes to oxygen deficiency stress. *BMC Genom.* **2019**, *20*, 60. [CrossRef]

37. Reynoso, M.A.; Kajala, K.; Bajic, M.; West, D.A.; Pauluzzi, G.; Yao, A.I.; Hatch, K.; Zumstein, K.; Woodhouse, M.; Rodriguez-Medina, J.; et al. Evolutionary flexibility in flooding response circuitry in angiosperms. *Science* **2019**, *365*, 1291–1295. [CrossRef]

38. Van Veen, H.; Vashisht, D.; Akman, M.; Girke, T.; Mustroph, A.; Reinen, E.; Hartman, S.; Kooiker, M.; van Tienderen, P.; Eric Schranz, M.; et al. Transcriptomes of Eight *Arabidopsis thaliana* Accessions Reveal Core Conserved, Genotype- and Organ-Specific Responses to Flooding Stress. *Plant Physiol.* **2016**, *172*, 668–689. [CrossRef]

39. Workman, D. Tomatoes Exports by Country. Available online: http://www.worldstopexports.com/tomatoes-exports-country/ (accessed on 6 August 2020).

40. Biemelt, S.; Hajirezaei, M.R.; Melzer, M.; Albrecht, G.; Sonnewald, U. Sucrose synthase activity does not restrict glycolysis in roots of transgenic potato plants under hypoxic conditions. *Planta* **1999**, *210*, 41–49. [CrossRef]

41. Vidoz, M.L.; Loreti, E.; Mensuali, A.; Alpi, A.; Perata, P. Hormonal interplay during adventitious root formation in flooded tomato plants. *Plant J.* **2010**, *63*, 551–562. [CrossRef]

42. Zhang, Q.; Peters, J.L.; Visser, E.J.W.; de Kroon, H.; Huber, H. Hydrologically contrasting environments induce genetic but not phenotypic differentiation in *Solanum dulcamara*. *J. Ecol.* **2016**, *104*, 1649–1661. [CrossRef]

43. Dawood, T.; Rieu, I.; Wolters-Arts, M.; Derksen, E.B.; Mariani, C.; Visser, E.J.W. Rapid flooding-induced adventitious root development from preformed primordia in *Solanum dulcamara*. *AoB Plants* **2014**, *6*, plt058. [CrossRef]

44. Gururani, M.A.; Upadhyaya, C.P.; Baskar, V.; Venkatesh, J.; Nookaraju, A.; Park, S.W. Plant Growth-Promoting Rhizobacteria Enhance Abiotic Stress Tolerance in *Solanum tuberosum* through Inducing Changes in the Expression of ROS-Scavenging Enzymes and Improved Photosynthetic Performance. *J. Plant Growth Regul.* **2013**, *32*, 245–258. [CrossRef]

45. Dawood, T.; Yang, X.; Visser, E.J.W.; Te Beek, T.A.H.; Kensche, P.R.; Cristescu, S.M.; Lee, S.; Floková, K.; Nguyen, D.; Mariani, C.; et al. A Co-Opted Hormonal Cascade Activates Dormant Adventitious Root Primordia upon Flooding in *Solanum dulcamara*. *Plant Physiol.* **2016**, *170*, 2351–2364. [CrossRef] [PubMed]

46. Safavi-Rizi, V.; Herde, M.; Stöhr, C. RNA-Seq reveals novel genes and pathways associated with hypoxia duration and tolerance in tomato root. *Sci. Rep.* **2020**, *10*, 1692. [CrossRef] [PubMed]

47. Chang, K.N.; Zhong, S.; Weirauch, M.T.; Hon, G.; Pelizzola, M.; Li, H.; Huang, S.C.; Schmitz, R.J.; Urich, M.A.; Kuo, D.; et al. Temporal transcriptional response to ethylene gas drives growth hormone cross-regulation in Arabidopsis. *Elife* **2013**, *2*, e00675. [CrossRef] [PubMed]

48. Hebelstrup, K.H.; Hunt, P.; Dennis, E.; Jensen, S.B.; Jensen, E.Ø. Hemoglobin is essential for normal growth of Arabidopsis organs. *Physiol. Plant* **2006**, *127*, 157–166. [CrossRef]

49. Mithran, M.; Paparelli, E.; Novi, G.; Perata, P.; Loreti, E. Analysis of the role of the pyruvate decarboxylase gene family in *Arabidopsis thaliana* under low-oxygen conditions. *Plant Biol.* **2014**, *16*, 28–34. [CrossRef]

50. Hebelstrup, K.H.; van Zanten, M.; Mandon, J.; Voesenek, L.A.C.J.; Harren, F.J.M.; Cristescu, S.M.; Møller, I.M.; Mur, L.A.J. Haemoglobin modulates NO emission and hyponasty under hypoxia-related stress in *Arabidopsis thaliana*. *J. Exp. Bot.* **2012**, *63*, 5581–5591. [CrossRef]

51. Lin, C.-C.; Chao, Y.-T.; Chen, W.-C.; Ho, H.-Y.; Chou, M.-Y.; Li, Y.-R.; Wu, Y.-L.; Yang, H.-A.; Hsieh, H.; Lin, C.-S.; et al. Regulatory cascade involving transcriptional and N-end rule pathways in rice under submergence. *Proc. Natl. Acad. Sci. USA* **2019**, *116*, 3300–3309. [CrossRef]

52. Bui, L.T.; Novi, G.; Lombardi, L.; Iannuzzi, C.; Rossi, J.; Santaniello, A.; Mensuali, A.; Corbineau, F.; Giuntoli, B.; Perata, P.; et al. Conservation of ethanol fermentation and its regulation in land plants. *J. Exp. Bot.* **2019**. [CrossRef]

53. Ahsan, N.; Lee, D.G.; Lee, S.H.; Lee, K.W.; Bahk, J.D.; Lee, B.H. A proteomic screen and identification of waterlogging-regulated proteins in tomato roots. *Plant Soil* **2007**, *295*, 37–51. [CrossRef]

54. Yeung, E.; Bailey-Serres, J.; Sasidharan, R. After the Deluge: Plant Revival Post-Flooding. *Trends Plant Sci.* **2019**, *24*, 443–454. [CrossRef] [PubMed]

55. Potato Genome Sequencing Consortium; Xu, X.; Pan, S.; Cheng, S.; Zhang, B.; Mu, D.; Ni, P.; Zhang, G.; Yang, S.; Li, R.; et al. Genome sequence and analysis of the tuber crop potato. *Nature* **2011**, *475*, 189–195. [CrossRef] [PubMed]

56. Sol Genomics Network. Available online: https://solgenomics.net/tools/blast/ (accessed on 6 August 2020).

Article

Potassium Efflux and Cytosol Acidification as Primary Anoxia-Induced Events in Wheat and Rice Seedlings

Vladislav V. Yemelyanov [1,2,*], Tamara V. Chirkova [2], Maria F. Shishova [2] and Sylvia M. Lindberg [3]

[1] Department of Genetics and Biotechnology, Saint-Petersburg State University, Universitetskaya em., 7/9, 199034 Saint-Petersburg, Russia

[2] Department of Plant Physiology and Biochemistry, Saint-Petersburg State University, Universitetskaya em., 7/9, 199034 Saint-Petersburg, Russia; mim39@mail.ru (T.V.C.); mshishova@mail.ru (M.F.S.)

[3] Department of Ecology, Environment and Plant Sciences, Stockholm University, SE-106 91 Stockholm, Sweden; sylvia.lindberg@su.se

* Correspondence: bootika@mail.ru

Received: 7 August 2020; Accepted: 15 September 2020; Published: 16 September 2020

Abstract: Both ion fluxes and changes of cytosolic pH take an active part in the signal transduction of different environmental stimuli. Here we studied the anoxia-induced alteration of cytosolic K^+ concentration, $[K^+]_{cyt}$, and cytosolic pH, pH_{cyt}, in rice and wheat, plants with different tolerances to hypoxia. The $[K^+]_{cyt}$ and pH_{cyt} were measured by fluorescence microscopy in single leaf mesophyll protoplasts loaded with the fluorescent potassium-binding dye PBFI-AM and the pH-sensitive probe BCECF-AM, respectively. Anoxic treatment caused an efflux of K^+ from protoplasts of both plants after a lag-period of 300–450 s. The $[K^+]_{cyt}$ decrease was blocked by tetraethylammonium (1 mM, 30 min pre-treatment) suggesting the involvement of plasma membrane voltage-gated K^+ channels. The protoplasts of rice (a hypoxia-tolerant plant) reacted upon anoxia with a higher amplitude of the $[K^+]_{cyt}$ drop. There was a simultaneous anoxia-dependent cytosolic acidification of protoplasts of both plants. The decrease of pH_{cyt} was slower in wheat (a hypoxia-sensitive plant) while in rice protoplasts it was rapid and partially reversible. Ion fluxes between the roots of intact seedlings and nutrient solutions were monitored by ion-selective electrodes and revealed significant anoxia-induced acidification and potassium leakage that were inhibited by tetraethylammonium. The K^+ efflux from rice was more distinct and reversible upon reoxygenation when compared with wheat seedlings.

Keywords: anoxic signaling; potassium; pH; acidification; fluorescence microscopy; *Triticum aestivum*; *Oryza sativa*

---

1. Introduction

A wide range of signal perception and transduction systems in plant cells is responsible for distinguishing and triggering a correct adaptive response [1,2]. Signal transduction pathways that are specific for different stressors include several steps that are shown to be rather universal. One of them is a cytosolic Ca^{2+} elevation at internal and external signals application [3–5]. Plant cells carefully maintain a low Ca^{2+} concentration in the cytosol and a significant gradient between the cell wall and a number of organelles. This balance is actively regulated by a variety of membrane transport systems (recently reviewed in [6]). Another well-documented primary signaling event is the accumulation of reactive oxygen species (ROS) [7–10]. The appearance of ROS caused by different stressors triggers a wide spectrum of reactions and is quickly eliminated by the antioxidant machinery. Both the second messengers (Ca^{2+} and ROS) are closely linked to each other and are involved in the transduction of different environmental stressors and internal signals such as phytohormones, regulatory proteins, RNAs and metabolites [6,7,11]. One important mechanism is the Ca^{2+}-induced activation of NADPH-oxidase activity via its integration in sterol-rich lipid rafts [12]. On the other hand, ROS activates Ca^{2+} channels in the plasma membrane [3] and Ca^{2+}-ATPases, which interfere with Ca^{2+} homeostasis [13]. Initially, weak signals of both Ca^{2+} and ROS were hypothesized to be amplified through the so-called ROS-Ca^{2+} hub [6]. The change of Ca^{2+} and ROS to an inactive state is also a necessity and supposed to be highly regulated. The cytosol Ca^{2+} elevation triggered by a variety of internal and external signals coincides with cytosolic acidification (reviewed in [14,15]). It was questionable for some time if protons had a signaling role, probably because of its involvement in metabolism. It is well known that Ca^{2+} elevation in the cytosol affects the pH level. The duration and intensity of Ca^{2+} increases might vary and thus determine the specificity of response to diverse signals (stressors, hormones, light, etc.). Different systems of H^+ transport through the plasma membrane and tonoplast are involved. Proton pumps are supposed to be regulated on transcriptional and posttranslational levels, which make the H^+ signature highly specific.

Electrolyte leakage is another process accompanying transient Ca^{2+}, ROS and proton increases under stress conditions. A number of experimental data reveals that electrolyte leakage is mainly defined as K^+ efflux from plant cells [16,17].

Potassium is essential for plant cells/organisms in many aspects. It is a well-known macro-nutrient. Deficiency of K^+ results in growth arrest especially in seedlings and young organs. This ion is important for plant metabolism due to its facility to activate more than 70 enzymes [17,18]. Besides that, K^+ serves as a charge-balancing ion, which plays an important role in the transport through the plasma membrane under the limitation of ATP and the depolarization of the membrane potential. The K^+ gradient maintains turgor and serves as a source of energy to stimulate sucrose loading into the phloem [19,20]. Accumulated evidence indicates that the efflux depends on the type and the intensity of the stressor as well as on affected plant species and tissue. The cytosolic K^+ concentration in plant cells is about 70–200 mM [21]. Several transporters have been shown to be involved in K^+ accumulation: the High Affinity K^+ transporter (HAK)–K^+ uniporter and the Arabidopsis K^+ Transport system 1 (AKT1)–K^+/H^+ symporter [18,22]. These processes require energy and depend on the external K^+ concentration and the K^+ vacuolar pool. The priority role in stress-induced K^+ leakage is given to another system: Gated Outward Rectifying K^+ efflux (GORK) channels [23]. By activation, it decreases the cytosolic potassium concentration to 10–30 mM [21]. The activity of these channels is sensitive to membrane potential depolarization through a clustering mechanism [24]. A number of experimental data combined with a bioinformatics approach suggest a possible ligand regulation of K^+ flux through GORK channels [25]. Cyclic nucleotides (CNs), gamma-aminobutyric acid (GABA), G proteins, protein phosphatases, inositol, ROS and ATP are on a list of potential GORK regulating ligands. The presented data even stronger introduced both K^+ and GORK in signaling cascades triggered by stress factors and led to the conclusion that potassium fulfills the role of a second messenger [17,25]. A signaling role of potassium is well documented for salt stress [18].

Oxygen deficiency is another stress factor that affects K^+ efflux and causes severe damage to plant organisms [26–30]. Surprisingly, plants known as strict aerobic organisms might be affected not only by an external lack of oxygen. Different tissues or even groups of cells experience hypoxia conditions during normal plant development [31]. Thus, it becomes even more important to discover the steps of early hypoxia signal transduction. The mechanism of oxygen sensing in plants and animals are strictly diverse. However, low oxygen regulates the function of various K^+ channels in mammals. Recently, plant cell ion channels have been identified as potential candidates for low oxygen sensing in flooded roots [32]. However, additional studies are required to estimate a possible modulation of K^+ transport through an ERF-VII-mediated response to a lack of oxygen. Nevertheless, hypoxic/anoxic environment causes K^+ efflux from plant cells and GORK channels are supposed to play a crucial role in this process [25]. Under hypoxic stress, cytosolic K^+ participates in the regulation of several physiological processes possibly including the formation of aerenchyma [33].

Oxygen deprivation triggers Ca^{2+} signaling. The alteration in $[Ca^{2+}]_{cyt}$ is a fast and intensive reaction for limitation in oxygen supply and energy deficiency [34–37]. Ca^{2+} elevation in cytosol coincides with acidification [15]. A probable reason is the hypoxia-induced inhibition of the plasma membrane H^+-ATPase and tonoplast H^+-ATPase activities due to the lack of ATP.

Signaling via K^+ efflux is transient and therefore the timing of this event should be evaluated. It was found that plant species and even plant organs differ in their tolerance to hypoxia. Some data revealed that the intensity of K^+ efflux corresponded to plant sensitivity to oxygen deprivation and depended on metabolic activity [30]. It would be of interest to integrate $[K^+]_{cyt}$ into the schedule of other intracellular signaling events such as Ca^{2+} elevation and acidification under a lack of oxygen. Earlier we provided a comparison of calcium signaling during anoxia in two well-known agricultural plants such as rice and wheat, which differ in tolerance to oxygen limitation [35]. Rice cells were shown to be more reactive to oxygen depletion and depended on both external and internal Ca^{2+} stores. This study focuses on a possible alteration of intracellular potassium concentration, $[K^+]_{cyt}$, and cytosolic pH, pH_{cyt}, during anoxic signal transduction in the protoplasts from wheat and rice.

The mentioned ions have not only a signaling role but also are very important in the regulation of metabolic processes including those at stress conditions. We therefore also estimated the cell-level events at a whole plant level. We investigated the ion changes after reoxygenation as well. It is of special interest how proton and Ca^{2+} accumulation inside cells and active potassium efflux would reflect processes during long-time stress applications and regulate the ion exchange with an external medium.

2. Results

2.1. Influence of Oxygen Deprivation on $[K^+]_{cyt}$ in Wheat and Rice Leaf Protoplasts

We studied the potassium and proton concentration changes in wheat and rice leaf mesophyll protoplasts under normoxia and anoxia. Wheat was used as it is sensitive to oxygen deficiency and rice as it is resistant to hypoxia. The results of normoxic and anoxic treatment effects on cytosolic potassium concentration, $[K^+]_{cyt}$, are presented in Figure 1. It shows representative traces of $[K^+]_{cyt}$ alteration in a single leaf protoplast. The resting level of cytosolic potassium was around 120 ± 5 mM in wheat leaf protoplasts and 125 ± 5 mM in rice mesophyll protoplasts. The resting level of $[K^+]_{cyt}$ at normoxic conditions had a little decrement within 30 min of measurements of about 7–10 mM in wheat protoplasts and 4–5 mM in rice. Oxygen deprivation led to a more significant drop in K^+ concentration in both plants. An anoxia-induced decrease in $[K^+]_{cyt}$ was 1.5 fold in wheat protoplasts (Figure 1a) and twofold in rice protoplasts (Figure 1b). The K^+ efflux did not start immediately after the imposition of oxygen lack in protoplasts of both experimental plants but after a lag-period. It lasted about 300–400 s in wheat protoplasts and 350–450 s in rice. The duration of the lag-periods for single traces presented in Figure 1 were as follows: 330–350 s in Figure 1a (wheat) and 410–430 s in Figure 1b (rice).

To elucidate the involvement of plasma membrane voltage-gated K^+ channels in the $[K^+]_{cyt}$ decrease, the protoplasts were treated with 1 mM tetraethylammonium (TEA). TEA significantly inhibited efflux from the protoplasts of both plants (Figure 2). Inhibition was 20% in wheat protoplasts and 45% in rice ones.

Thus, anoxia led to leakage of K^+ from the protoplasts of both plants after the lag-period of 300–450 s. The K^+ efflux was blocked by tetraethylammonium and was higher in rice protoplasts.

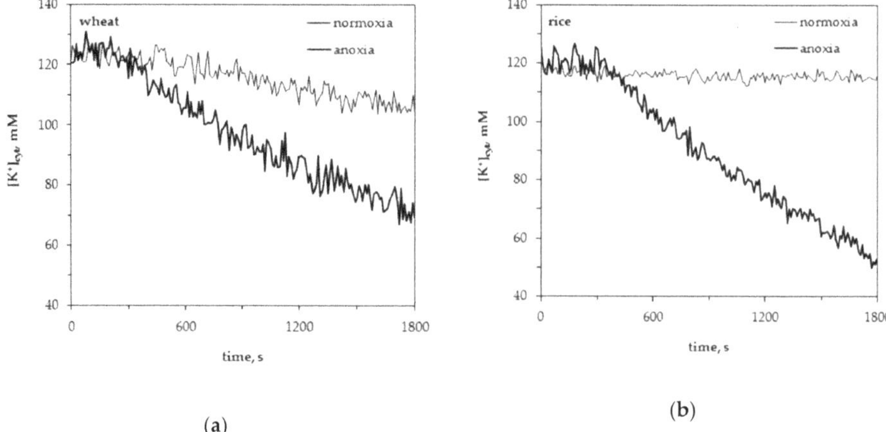

Figure 1. Changes of the free cytosolic K^+ concentration, $[K^+]_{cyt}$, in wheat (**a**) and rice (**b**) leaf protoplasts upon imposition of anoxia. Typical single traces in the presence of 10 mM K^+, 1 mM Ca^{2+} and pH 7.0 in the external medium.

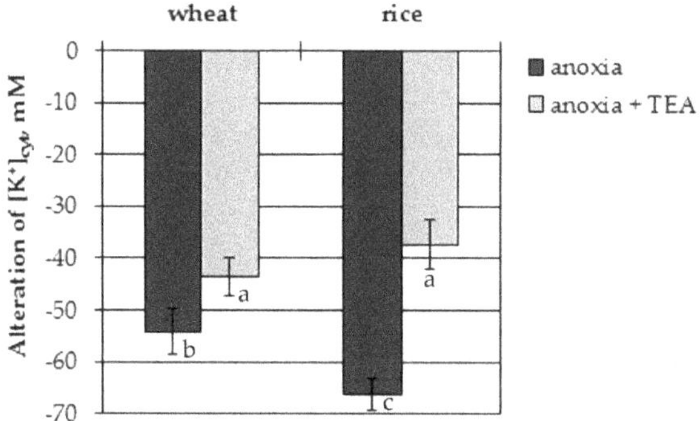

Figure 2. Anoxia-induced alterations of $[K^+]_{cyt}$ in wheat and rice leaf protoplasts with and without tetraethylammonium (TEA, 1 mM, 30 min pre-incubation). Columns represent mean values ± SD. Values with different letters are significantly different at $p < 0.05$, according to the Least Significant Difference LSD test.

2.2. Influence of Oxygen Deprivation on pH_{cyt} in Wheat and Rice Leaf Protoplasts

The representative traces in Figure 3 show anoxia-induced alterations of cytosolic pH in leaf protoplasts of experimental plants. The initial level of pH_{cyt} was somewhat higher in wheat protoplasts (7.3 ± 0.05) than in rice (7.2 ± 0.05). The difference was insignificant. Similar to the measurements of $[K^+]_{cyt}$, the resting level of pH_{cyt} at normoxic conditions had a little decrement within 30 min of measurements. It was about 0.2 pH units in wheat protoplasts and 0.1 pH units in rice. Anoxic treatment led to a decrease of pH_{cyt} starting after a 200–300 s delay in wheat protoplasts. The pH changed from 7.3 to 6.1–6.2 within 900–1000 s and was kept at a low level until the end of the measurement (Figure 3a). A rapid pH_{cyt} drop in rice protoplasts from 7.2 to 6.2–6.3 occurred during the first 400–500 s of oxygen deprivation (Figure 3b). Followed by several domed oscillations, pH_{cyt} then recovered partially and stabilized at 6.4–6.5 after 1200 s from the beginning of the anoxic treatment. The anoxia-induced pH_{cyt} drop in wheat protoplasts was almost twice as high as in rice (−1.23 and −0.75 units, correspondingly) (Figure 4).

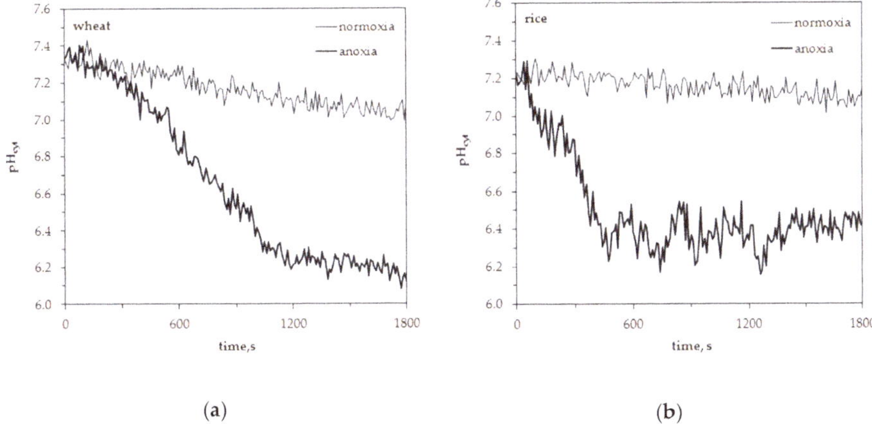

Figure 3. Changes of the free cytosolic H$^+$ concentration, pH$_{cyt}$, in wheat (**a**) and rice (**b**) leaf protoplasts upon imposition of anoxia. Typical single traces.

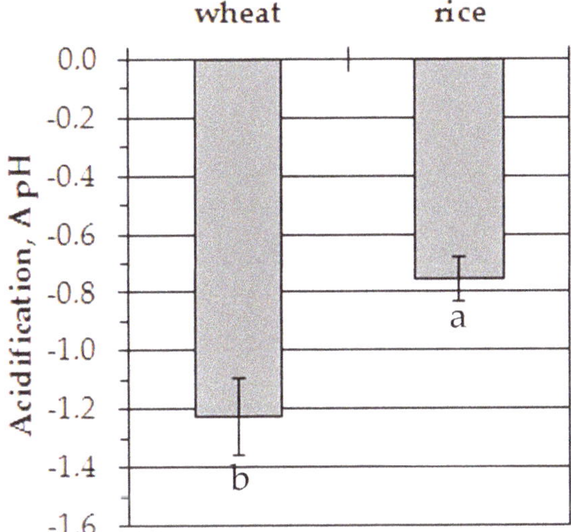

Figure 4. Anoxia-induced alterations of pH$_{cyt}$ in wheat and rice leaf protoplasts. Columns represent mean values ± SD. Values with different letters are significantly different at $p < 0.05$, according to the LSD test.

Thus, anoxia caused the acidification of cytoplasm in the protoplasts of both plants but it was more significant and started after the lag-period of 200–300 s in wheat while in rice protoplasts the anoxia-induced decrease of pH$_{cyt}$ was rapid and partially reversible.

2.3. Influence of Long-Term Anoxia on Potassium Uptake by Intact Wheat and Rice Seedlings

To detect the total fluxes (influx/efflux) of potassium in/from roots of the studied plants upon anoxia treatment we carried out experiments with intact seedlings. When the roots of the seedlings of both tested plants were placed in an aerated Knop solution (normoxic control), the decrease in K$^+$ concentration in the solution, measured by an ion-selective electrode, continued for up to 48 h, reflecting a net uptake of potassium by the roots (Figure 5). The absorption of K$^+$ was somewhat higher in wheat in the first 6 h of normoxia. At the end of measurements (48 h), net K^+ uptake was about the same in both species (18.7 and 20.7 µmol per seedling for wheat and rice, correspondingly). Anoxic treatment arrested the K^+ influx during 6 h in wheat seedlings and during 3 h in rice. Moreover, after 6 h of oxygen deficiency, K^+ started to leak from the seedling roots of both plants. Efflux was significantly different from normoxic control and it was significantly higher in rice at 24 and 48 h of anoxia ($p < 0.05$, according to the LSD test).

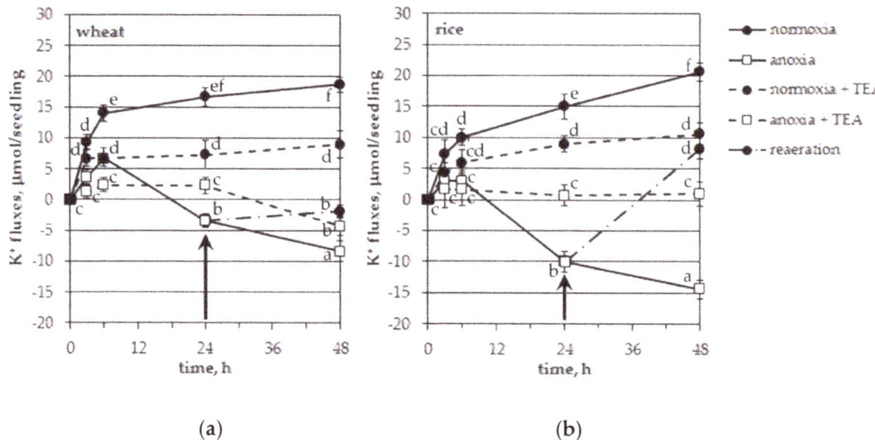

Figure 5. Effects of anoxia and tetraethylammonium (TEA, 0.1 mM) on the net uptake of K^+ by roots of intact wheat (**a**) and rice (**b**) seedlings. Negative values reflect potassium efflux. Arrows indicate the beginning of reoxygenation treatment (24 h). Mean values ± SD. Values with different letters are significantly different at $p < 0.05$, according to the LSD test.

The use of tetraethylammonium (0.1 mM), a blocker of plasma membrane voltage-gated potassium channels, significantly deteriorated both the acquisition and loss of potassium by seedlings under normoxia and anoxia, correspondingly (Figure 5). TEA inhibition of a net uptake of K^+ was about 50% in both plants at normoxic conditions. Anoxia-induced K^+ leakage was completely blocked in rice seedlings while in wheat plants TEA failed to totally inhibit K^+ efflux after 24 h of anoxia. The TEA-treated anoxic variant showed significantly lesser leakage than the untreated one (Figure 5a). Reestablishment of aeration of the incubation solution after 24 h of anoxic treatment blocked the K^+ leakage of both plants (Figure 5). Moreover, it even led to a significant reabsorption of K^+ in rice (Figure 5b).

Thus, anoxia-induced potassium leakage from the roots of intact seedlings was significantly inhibited by TEA. The K^+ efflux was higher and reversible in rice when compared with wheat seedlings.

2.4. Influence of Long-Term Anoxia on the pH of the Incubation Medium of Intact Wheat and Rice Seedlings

Incubation solutions from potassium experiments were also monitored for pH changes (Figure 6). Both plants acidified the incubation medium at normoxic conditions. Wheat seedling roots excreted protons somewhat faster than rice. The anoxic treatment accelerated the acidification of the nutrient solution in both cases (differences were significant at $p < 0.05$, according to the post-hoc LSD test) but in the medium where rice seedlings were growing, acidification was much more rapid (from 6.2 to 4.9 during 6 h of anoxia and from 6.2 to 4.4 during 48 h, Figure 6b).

Thus, anoxia resulted in the acidification of the incubation solution of hydroponically growing plants. Moreover, the acidification of the nutrient solution where rice seedlings were growing was faster and more intensive.

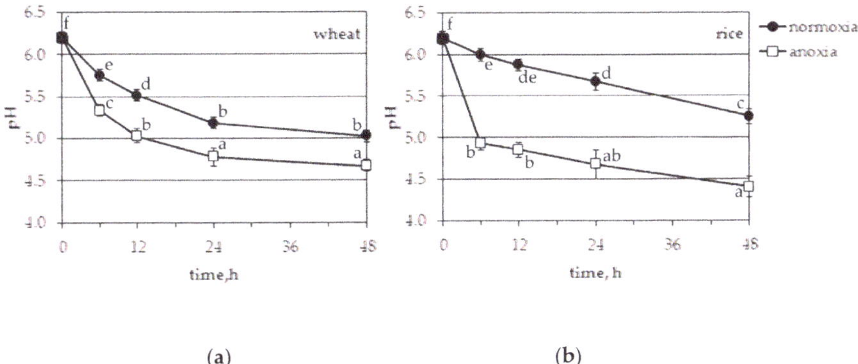

(a)   (b)

Figure 6. Effects of anoxia on the pH of the incubation medium of wheat (**a**) and rice (**b**) seedlings. Mean values ± SD. Values with different letters are significantly different at $p < 0.05$, according to the LSD test.

3. Discussion

3.1. Potassium Changes

The obtained results showed a considerable anoxia-induced decrease of cytosolic potassium concentration, $[K^+]_{cyt}$, in the leaf mesophyll protoplasts of both studied plant species but at different lag-periods and the decrease by efflux was more intensive in hypoxia-tolerant rice seedlings than in hypoxia-sensitive wheat (Figure 1). Moreover, the anoxia-induced K^+ efflux at a cellular level corroborated well with net K^+ fluxes between the roots of intact seedlings and the external nutrient solution (Figure 5). As the efflux from both species was inhibited by tetraethylammonium, it is likely that K^+ mainly was transported by K^+-selective voltage-gated channels. The drop in $[K^+]_{cyt}$ often originates from K^+ leakage into the apoplast, although the sequestration of K^+ ions into the vacuole is also possible during hypoxia [38]. New findings suggest that the potassium efflux might be operated by the activation of GORK channels [25]. The activation depends on depolarization-dependent GORK clustering in the plasma membrane [24]. At normal (unstressed) conditions the activity of GORK is rather limited. It has been suggested for potential dependent channels (including GORK) to form clusters and small-scale rafts that activate channel activity. This process has been shown to be highly regulated by external K^+ concentration [24].

Reaeration after 24 h of anoxia caused quite different reactions in the two species (Figure 5). It stopped the leakage and resulted in a slight K^+ reabsorption in the wheat seedlings (about 1.5 μmol per seedling a day) while K^+ uptake was totally recovered in rice seedlings (18.3 μmol per seedling a day and 15 μmol per seedling a day for the first day of normoxia). This is probably one reason for the anoxia tolerance of rice.

Similar K^+ leakage during oxygen shortage was reported earlier from the roots of intact wheat seedlings [26], barley [30,39], cucumber [40], roots of rooted cuttings of grape [29] and excised rice coleoptiles [27,28]. The intensity of K^+ efflux depends on the severity of the oxygen deficiency (hypoxia or anoxia), tissue specificity and age. For example, 1 h of hypoxia caused a twofold decrease of $[K^+]_{cyt}$ in the cells of the elongation zone in the root epidermis of wild type Arabidopsis but no changes were observed in the mature zone. However, there was a significant increase of $[K^+]_{cyt}$ in the epidermis of both the elongation and the mature zone under long-term hypoxia (24–72 h) [38]. In barley root, cell hypoxia led to less significant K^+ fluxes than anoxia [30]. Moreover, cells of the elongation zone of a hypoxia-tolerant cultivar lost more potassium upon short-term anoxic treatment than cells of a sensitive one while it was vice versa in the mature zone of the root. There were no differences in the long-term responses of K^+ effluxes to anoxia between hypoxia-tolerant and hypoxia-sensitive barley cultivars [30]. On the other hand, hypoxic treatment of the roots of a cultivar of the grape species *Vitis rupestris* sensitive to oxygen shortage resulted in more significant K$^+$ leakage when compared with a tolerant *V. riparia* [29]. Furthermore, as in rice, there was a total recovery of K$^+$ influx into the roots of *V. riparia* upon the reestablishment of aeration after 20 h of hypoxic treatment while the roots were incapable of K$^+$ reabsorption after long-term hypoxia in *V. rupestris*.

Thus, K$^+$ efflux is a common reaction developed at cellular/tissue/whole plant levels under hypoxic/anoxic treatment but further investigations are required to reveal the strict specificity between the tolerance to oxygen deprivation and the degree of potassium leakage.

3.2. Acidification Caused by Anoxia

The acidification of cytosol is another common reaction in oxygen-deprived cells [14,41]. The imposition of oxygen deficiency is accompanied by a severe decrease in pH$_{cyt}$ in pea, maize [42–44], sycamore (*Acer pseudoplatanus*) [45], wheat and rice [46,47]. Our data revealed an anoxia-induced cytosolic acidification in the protoplasts of both plants (Figure 3) but it was more significant in sensitive wheat leaf protoplasts (Figure 4) and the decrease in pH started after the lag-period of 200–300 s (Figure 3a). The rice protoplasts responded to anoxia by a fast decrease of pH$_{cyt}$, which was partially recovered after 500–1200 s of treatment (Figure 3b). Similar dynamics were obtained at a whole plant level; rice seedlings acidified the anoxic solution faster (by 6 h of treatment) than wheat ones (by 12 h). After 12 h the differences between wheat and rice were insignificant (Figure 6). Analogous patterns of pH$_{cyt}$ decrease in wheat and rice root tip cells were reported earlier [46,47].

The acidification of cytosol upon oxygen limitation is due to several reasons. The major reason is low ATP concentration reducing the activity of the plasma membrane proton pumps [43,45,48]. ATP level in the cell is exhausted within 1–2 min after the switch to anaerobic metabolism [49]. Another important H$^+$ source is the hydrolysis of ATP and other NTPs [45]. Possible sources of protons are the leakage from the vacuole [50], passive influx from outside [47] and accumulation of organic acid intermediates and products of anaerobic metabolism, predominantly lactate [44,47]. Different patterns of anoxia-induced cytosol acidification may result from different mechanisms in wheat and rice tissues. A partial pH$_{cyt}$ recovery shown in rice protoplasts (Figure 3b) and in root tips [46,47] may have come from the stimulation of ethanol fermentation rather than a lactic one [51]; disposal of lactic acid and the alteration of metabolic pathways can lead to end products other than ethanol or lactate [46,51]. Glycolytic production of ATP provides activity of the plasma membrane [47] and tonoplast proton pumps. A biochemical pH-stat consisting of a shuttle of carboxylating/decarboxylating enzymes is also involved in pH regulation [47,50].

3.3. Calcium Involvement in Anoxic Signaling

Both cytosolic [K$^+$]$_{cyt}$ decrease and acidification are induced under anoxic treatment. The reactions manifested specificity and appeared due to complicated active and passive ion transport at the plasma membrane. Earlier obtained data highlighted Ca^{2+}$_{cyt}$ elevation as another fast oxygen-deficiency triggered reaction at a cellular level [32,34,36,37]. This reaction was detected in different plant species with a wide spectrum of methods [35,52,53]. Using plant cell protoplasts and sensitive fluorescent dyes we revealed clear differences in the timing and amplitude of Ca^{2+} increase for tolerant rice and sensitive wheat. Cells of sensitive wheat were characterized by a lower amplitude and a longer lag-phase of [Ca^{2+}]$_{cyt}$ increase compared with rice. Leaf and root protoplasts responded in a similar way to anoxia [35].

3.4. A Suggested Model

The obtained results allow us to suggest a general ion signature schedule triggered by anoxia. The events in tolerant rice protoplasts definitely started with cytosol acidification. This reaction was limited to 0.6–0.8 pH units and had very complicated dynamics, probably reflecting the existence of several involved mechanisms. This effect might be linked primarily with the inhibition of

H$^+$-ATPase due to ATP limitation [43,45,48]. The more rapid reaction in rice seedlings might be linked to fast depolarization of the membrane potential. Within 30–40 s, Ca^{2+} efflux started through potential-dependent plasma membrane channels and further on it accompanied Ca^{2+} release from the intracellular stores [35], predominantly from the mitochondrion due to the dampening of the electron transport chain upon oxygen deprivation [32,34]. The drop in [K$^+$]$_{cyt}$ was the last from the tested reactions. It showed about a 10 min lag-period.

Taken together the schedule of primary events of anoxic signal transduction in rice protoplasts could be described as follows. Anoxia-induced inhibition of H$^+$-ATPase fast depolarizes the plasma membrane and thus activates Ca^{2+} channels. Cytosol acidification and [Ca^{2+}]$_{cyt}$ increases regulate the membrane potential and trigger the complicated mechanism of GORK activation and then provide intensive K$^+$ efflux.

In sensitive wheat leaf protoplasts, the reaction was somewhat different. Anoxia application also inhibited H$^+$-ATPases but this process had a more prolonged character (lag-period was 200–300 s). On the other hand, the amplitude of the acidification was more pronounced. Changes in the membrane potential also activated the plasma membrane Ca^{2+} channels but its role in cytosol elevation was smaller in comparison with intracellular stores. A lag-period was needed and the reaction was less intensive [35]. Such a delay in [Ca^{2+}]$_{cyt}$ shifts explained the lower amplitude of K$^+$ efflux in wheat.

The order of primary anoxia-induced reactions still requires additional experimental confirmation. It also desires the insertion of one more significant event such as the elevation of the ROS. In animal cells hypoxia-induced fluxes of Ca^{2+} and K$^+$ are performed by oxygen-, ROS- and CO-dependent inward calcium and outward potassium channels [32]. Plants do not possess such channels. All putative analogs in Arabidopsis belong to either the plasma membrane inward K$^+$ channels (Shaker type inward rectifier K$^+$ channel (AtAKT2) and tandem-pore K$^+$ channel (AtKCO4/AtTPK4)) or the Arabidopsis Two-Pore Channel 1 (AtTPC1) in the vacuolar membrane [32]. Nonetheless, anoxia-induced accumulation of free cytosolic Ca^{2+} and pH$_{cyt}$ decreases might directly or indirectly affect potassium transport via AKT2 and AtKCO4 [32] and possibly result in a prolonged lag-period of K$^+$ efflux, as revealed in this study.

Potassium efflux was significantly more intensive in rice than in wheat but was reversible after reoxygenation. This fact may be one of the reasons for its tolerance to oxygen deprivation. Taking together, we may conclude that the processes started in anoxia treated cells at a cellular level and then continued to the adaptation of whole plants. Moreover, our results suggest that the dynamic of ion exchange was specific for plants differing in sensitivity to oxygen deprivation.

4. Materials and Methods

4.1. Plant Material and Growing Conditions

Caryopses of wheat (*Triticum aestivum* L. cv. Kadett, Svalöf Weibull AB, Hammenhög, Sweden) and japonica rice (*Oryza sativa* L., cv. Liman, Rice Research Institute, Krasnodar, Russia) were surface-sterilized for 10 min with 5% NaClO solution, washed several times in distilled water and soaked for 1 h in hot water at 50–60 °C for rice and at 45–50 °C for wheat. The seedlings were grown in beakers on double layers of miracloth (LIC, Stockholm, Sweden) using a continuously aerated Knop nutrient solution [54] (0.2 strength) at an irradiance of 118 W m^{-2} at the top of the shoots and at a 14/12 h light/dark photoperiod, 70% relative humidity and temperatures 22 °C and 28 °C for wheat and rice, respectively, as previously described [35].

4.2. Protoplast Isolation and Dye Loading

The isolation of mesophyll protoplasts was carried out by an enzymatic method as described earlier [35]. In brief, the leaves (0.5 g) of 6–7-day-old wheat seedlings were sliced into 0.5–1 mm pieces and treated for 2 h with 1% (*w/v*) cellulase from *Tricoderma resei* (EC 3.2.1.4, lyophilized, 10.0 units mg^{-1} solid, Sigma, St. Louis, MO, USA) and 0.3% (*w/v*) macerase, Maceroenzyme R-10 from *Rhizopus* sp.(EC 3.2.1.4, lyophilized, 0.6 units mg^{-1} solid, Serva, Heidelberg, Germany) solution as described by Edwards et al. [55] with modifications [56,57]. For the isolation of rice mesophyll protoplasts, the concentration of both enzymes was twice higher and the digestion time was 3 h [35]. The protoplasts were purified on a sorbitol gradient [58] and loaded in darkness with potassium- or pH-sensitive fluorescent probes in the form of acetoxy methyl-esters (AM). For measurements of cytosolic potassium concentration, $[K^+]_{cyt}$, the acetoxy methyl of potassium-binding benzofuran isophthalate (PBFI-AM) was loaded for 3 h at room temperature (20–23 °C) [56,58]. For the monitoring of cytosolic pH, pH_{cyt}, the tetra (acetoxy methyl) ester of bis-carboxyethylcarboxyfluorescein (BCECF-AM) was loaded for 1 h at 4 °C [57]. Both ion-sensitive fluorescent dyes were purchased from Molecular Probes® (Eugene, OR, USA). Before measurements, protoplasts were kept in darkness at room temperature for 30 min.

4.3. Fluorescence Measurements and In Situ Calibration

The fluorescence intensity ratio was measured with an epifluorescence microscope Axiovert 10 (Zeiss, Oberkochen, Germany) supplied with an electromagnetic filter exchanger (Zeiss), xenon lamp (XBO 75), photometer (Zeiss 01), microprocessor (MSP 201) and a personal computer. Excitation wavelength ratios for the PBFI dye were 340/380 nm and for the BCECF dye 485/436 nm. Emission wavelengths were 500–530 nm for the PBFI and 510–550 nm for the BCECF. All measurements were performed on single protoplasts with a Planneofluar ×40/0.75 objective (Zeiss) for phase contrast. The ratio measurements were performed only with protoplasts of a similar size and properly loaded into the cytosol. Adjustments for signals and noise were made automatically.

For the standard determination of $[K^+]_{cyt}$, measurements of PBFI-fluorescence were undertaken with protoplasts in separate suspension solutions with concentrations of 0, 10, 20, 50, 80, 100 and 150 mM KCl. NaCl was added to the solutions to give a final concentration of 150 mM $[Na^+ + K^+]$ to obtain iso-osmotic conditions [58]. The standard measurements were performed 5–10 min after the addition of 10 μM gramicidin (Sigma, St. Louis, MO, USA) to equilibrate the intracellular and extracellular potassium concentrations. Nigericin (Sigma) was simultaneously added at a final concentration of 5 μM to avoid the pH effect [59].

For pH_{cyt} in situ calibration, the BCECF-AM fluorescence ratio (485/436 nm) corresponding to different pH values was measured on protoplasts in separate standard solutions with pH values ranging from 5.0 to 8.0 with a 0.5 step. The measurements were undertaken 5–10 min after the addition of 5 μM nigericin to equilibrate the intracellular and extracellular pH values [56].

4.4. In Situ Anoxic Treatment

The fluorescence measurements were undertaken in a POC-chamber (Bachhofer Lab Equipment, Fisher Scientific, NJ, USA). Micro slides were covered with poly-L-lysine (MW 150,000–300,000, Sigma) at room temperature. Fifty μL of protoplast suspension was placed onto a micro slide and the POC-chamber was mounted above it [35]. The incubation buffer contained 0.5 M sorbitol, 0.05% polyvinylpyrrolidone, 0.2% bovine serum albumin, 5 mM Hepes (Serva), 1 mM $CaCl_2$ and 10 mM KCl at pH 7.0. It was degassed for 20–30 min under reduced pressure (−70 kPa) and then saturated for at least 30 min with gaseous nitrogen to an oxygen level less than 25–30 nmol mL^{-1}. For the anoxic treatment, protoplasts were submerged by 2 mL of an oxygen-depleted buffer inside the POC-chamber that was immediately closed. For the normoxic treatment, the protoplasts were submersed by the same amount of aerated incubation buffer containing 270–300 nmol mL^{-1} of oxygen. The oxygen concentration in the incubation buffer was detected with an Oxygraph (Hansatech Instruments, Norfolk, UK). To elucidate the effects of tetraethylammonium (TEA, Sigma), which is known to block voltage-dependent K^+ channels, we pre-treated protoplasts with 1 mM TEACl for 30 min prior to the imposition of anoxia. Measurements were carried out for 30 min (1800 s) and were repeated at least 10 times for each treatment. Figures 1 and 3 demonstrate the representative traces of the specific experiments with single protoplasts from independent cultivations.

The viability of protoplasts was tested using Trypan Blue assay [60]. A sample of the protoplast suspension (50 μL) was mixed with the same volume of 0.4% Trypan Blue (Sigma) solution and was observed under a microscope after 3 min. The percentage of non-stained (alive) protoplasts was calculated from at least seven optical fields on each of four separate slides. The viability of wheat leaf protoplasts was 91.7 ± 3.5% under normoxic conditions and 90.4 ± 3.7% after 30 min of anoxia. The viability of rice protoplasts was 90.5 ± 5.2% and 89.8 ± 5.1% at normoxia and anoxia, respectively.

4.5. Measurement of Potassium Uptake by Roots of Intact Seedlings and pH of Incubation Medium

The absorption of K^+ and the alterations of pH in the incubation medium were tested with 7-day-old seedlings grown as discussed above (4.1.). Prior to the experiments, the caryopses were cut off from the seedlings. Each set of twenty seedlings was placed in glasses containing 20 mL of Knop nutrient solution (0.2 strength). The incubation medium contained 0.636 mM total potassium and had a pH value of 6.2. Glasses were placed into the chamber through which gaseous nitrogen was flushed for 45 min for the creation of anaerobic conditions. The chambers were then hermetically closed and put in the dark in order to prevent the formation of oxygen in the light. Half of the anoxic variants were transferred from anaerobic chambers into normoxic conditions after 24 h of oxygen deprivation and left for 24 h of reaeration. Anaerobic conditions were checked by an Anaerotest® indicator (Merck, Darmstadt, Germany). Control variants were kept in the dark at a normal oxygen level. The concentration of K^+ was measured with a membrane K^+-selective electrode EM-101 K (Analit, St. Petersburg, Russia) in the incubation solution after 3, 6, 24 and 48 h of treatment. An EVL-1 M non-polarisable silver chloride electrode was used as a reference electrode. During measurements, the reference electrode was connected to the incubation medium via a U-shaped glass tube 1 mm in diameter and filled with 2% agar in 0.1 M KCl. The electrodes were calibrated with solutions of different concentrations of KCl. The precision of the electrodes was checked by the measurement of the K^+ level with a flame photometer Flapho-4 (Carl Zeiss, Jena, Germany). For the elucidation of the possible role of voltage-dependent K^+ channels in total K^+-fluxes during oxygen depletion, a part of the intact seedlings was treated with 0.1 mM TEA. The blocker was present in the incubation medium throughout the experiment. The shift in K^+ concentration was calculated in μmol per seedling. The pH of the incubation solution was measured using a Seven Easy S20 pH Meter (Mettler Toledo, Columbus, OH, USA) after 6, 12, 24 and 48 h of treatment.

4.6. Statistics

Data in Figures 2 and 4–6 are presented as mean ± SD for 4–10 experiments. Analysis of variance was done with a GraphPad Prism 5 for Windows. Values with different letters were significantly different at $p < 0.05$ according to the Least Significant Difference (LSD) test. Experiments with protoplasts were carried out in at least four independent cultivations. Fluorescence measurements were made on single protoplasts with at least 10 replicates for each treatment. Experiments with intact seedlings were performed five times.

Author Contributions: Conceptualization, V.V.Y. and T.V.C.; methodology, V.V.Y., M.F.S. and S.M.L.; validation, V.V.Y. and M.F.S.; formal analysis, V.V.Y. and M.F.S.; investigation, V.V.Y. and M.F.S.; data curation, S.M.L.; writing—original draft preparation, V.V.Y. and M.F.S.; writing—review and editing, V.V.Y., T.V.C., M.F.S. and S.M.L.; visualization, V.V.Y.; supervision, S.M.L.; project administration, V.V.Y.; funding acquisition, V.V.Y., M.F.S. and S.M.L. All authors have read and agreed to the published version of the manuscript.

Funding: This research was funded by Russian Foundation for Basic Research (grant numbers 18-04-00157a and 19-04-00655a), Russian Ministry of Education and Science (grant number 2006-RI-111.0/002/037) and by the Swedish Institute (grant number 01716/2005).

Acknowledgments: The authors would like to thank G. Wingstrand for the help with the protoplast pH calculations.

Conflicts of Interest: The authors declare no conflict of interest. The funders had no role in the design of the study; in the collection, analyses or interpretation of data; in the writing of the manuscript or in the decision to publish the results.

References

1. Geisler, M.; Venema, K. *Transporters and Pumps in Plant Signaling Editors. Signaling and Communication in Plants (Book 7)*; Springer: Berlin/Heidelberg, Germany, 2011; 388p. [CrossRef]
2. Khan, M.I.R.; Reddy, P.S.; Ferrante, A.; Khan, N.A. *Plant Signaling Molecule: Role and Regulation under Stressful Environments*; Elsevier (Woodhead Publishing): Cambridge, UK, 2019; 596p. [CrossRef]
3. Bose, J.; Pottosin, I.I.; Shabala, S.S.; Palmgren, M.G.; Shabala, S. Calcium efflux systems in stress signaling and adaptation in plants. *Front. Plant Sci.* **2011**, *2*, 85. [CrossRef] [PubMed]
4. Edel, K.H.; Marchadier, E.; Brownlee, C.; Kudla, J.; Hetherington, A.M. The evolution of calcium-based signalling in plants. *Curr. Biol.* **2017**, *27*, 667–679. [CrossRef]
5. Demidchik, V.; Shabala, S. Mechanisms of cytosolic calcium elevation in plants: The role of ion channels, calcium extrusion systems and NADPH oxidase mediated 'ROS-Ca^{2+} Hub'. *Funct. Plant Biol.* **2018**, *45*, 9–27. [CrossRef] [PubMed]
6. Demidchik, V.; Shabala, S.; Isayenkov, S.; Cuin, T.A.; Pottosin, I. Calcium transport across plant membranes: Mechanisms and functions. *New Phytol.* **2018**, *220*, 49–69. [CrossRef] [PubMed]
7. Tripathy, B.C.; Oelmüller, R. Reactive oxygen species generation and signaling in plants. *Plant Signal. Behav.* **2012**, *7*, 1621–1633. [CrossRef] [PubMed]
8. Das, K.; Roychoudhury, A. Reactive oxygen species (ROS) and response of antioxidants as ROS-scavengers during environmental stress in plants. *Front. Environ. Sci.* **2014**, *2*, 53. [CrossRef]
9. Waszczak, C.; Carmody, M.; Kangasjarvi, J. Reactive oxygen species in plant signaling. *Annu. Rev. Plant Biol.* **2018**, *69*, 209–236. [CrossRef]
10. Talaat, N.B. Role of reactive oxygen species signaling in plant growth and development. Chapter 10. In *Reactive Oxygen, Nitrogen and Sulfur Species in Plants: Production, Metabolism, Signaling and Defense Mechanisms*; Hasanuzzaman, M., Fotopoulos, V., Nahar, K., Fujita, M., Eds.; Wiley: New York, NY, USA, 2019; pp. 225–266. [CrossRef]
11. Choi, W.G.; Miller, G.; Wallace, I.; Harper, J.; Mittler, R.; Gilroy, S. Orchestrating rapid long-distance signaling in plants with Ca^{2+}, ROS and electrical signals. *Plant J.* **2017**, *90*, 698–707. [CrossRef]
12. Liu, P.; Li, R.L.; Zhang, L.; Wang, Q.L.; Niehaus, K.; Baluška, F.; Šamaj, J.; Lin, J.X. Lipid microdomain polarization is required for NADPH oxidase dependent ROS signaling in *Picea meyeri* pollen tube tip growth. *Plant J.* **2009**, *60*, 303–313. [CrossRef]
13. Romani, G.; Bonza, M.C.; Filippini, I.; Cerana, M.; Beffagna, N.; De Michelis, M.I. Involvement of the plasma membrane Ca^{2+}-ATPase in the short-term response of *Arabidopsis thaliana* cultured cells to oligogalacturonides. *Plant. Biol.* **2004**, *6*, 192–200. [CrossRef]
14. Felle, H.H. pH: Signal and messenger in plant cells. *Plant Biol.* **2001**, *3*, 577–591. [CrossRef]
15. Behera, S.; Xu, Z.; Luoni, L.; Bonza, M.C.; Doccula, F.G.; De Michelis, M.I.; Morris, R.J.; Schwarzländer, M.; Costa, A. Cellular Ca^{2+} signals generate defined pH signatures in plants. *Plant Cell* **2018**, *30*, 2704–2719. [CrossRef] [PubMed]
16. Demidchik, V.; Straltsova, D.; Medvedev, S.S.; Pozhvanov, G.A.; Sokolik, A.; Yurin, V. Stress-induced electrolyte leakage: The role of K$^+$-permeable channels and involvement in programmed cell death and metabolic adjustment. *J. Exp. Bot.* **2014**, *65*, 1259–1270. [CrossRef]
17. Shabala, S. Signalling by potassium: Another second messenger to add to the list? *J. Exp. Bot.* **2017**, *68*, 4003–4007. [CrossRef] [PubMed]
18. Rubio, F.; Nieves-Cordones, M.; Horie, T.; Shabala, S. Doing 'business as usual' comes with a cost: Evaluating energy cost of maintaining plant intracellular K$^+$ homeostasis under saline conditions. *New Phytol.* **2020**, *225*, 1097–1104. [CrossRef]
19. Gajdanowicz, P.; Michard, E.; Sandmann, M.; Rocha, M.; Corrêa, L.G.G.; Ramírez-Aguilar, S.J.; Gomez-Porras, J.L.; González, W.; Thibaud, J.-B.; van Dongen, J.T.; et al. Potassium (K$^+$) gradients serve as a mobile energy source in plant vascular tissues. *Proc. Natl. Acad. Sci. USA* **2011**, *108*, 864–869. [CrossRef]
20. Dreyer, I.; Gomez-Porras, J.L.; Riedelsberger, J. The potassium battery: A mobile energy source for transport processes in plant vascular tissues. *New Phytol.* **2017**, *216*, 1049–1053. [CrossRef]

21. Shabala, S.; Demidchik, V.; Shabala, L.; Cuin, T.A.; Smith, S.J.; Miller, A.J.; Davies, J.M.; Newman, I.A. Extracellular Ca^{2+} ameliorates NaCl induced K^+ loss from Arabidopsis root and leaf cells by controlling plasma membrane K^+-permeable channels. *Plant Physiol.* **2006**, *141*, 1653–1665. [CrossRef]

22. Rubio, F.; Fon, M.; Rodenas, R.; Nieves-Cordones, M.; Aleman, F.; Rivero, R.M.; Martinez, V. A low K^+ signal is required for functional high-affinity K^+ uptake through HAK5 transporters. *Physiol. Plant.* **2014**, *152*, 558–570. [CrossRef]

23. Shabala, L.; Zhang, J.; Pottosin, I.; Bose, J.; Zhu, M.; Fuglsang, A.T.; Velarde-Buendia, A.; Massart, A.; Hill, C.B.; Roessner, U.; et al. Cell-type-specific H^+-ATPase activity in root tissues enables K^+ retention and mediates acclimation of barley (*Hordeum vulgare*) to salinity stress. *Plant Physiol.* **2016**, *172*, 2445–2458. [CrossRef]

24. Eisenach, C.; Papanatsiou, M.; Hillert, E.-K.; Blatt, M.R. Clustering of the K^+ channel GORK of Arabidopsis parallels its gating by extracellular K^+. *Plant J.* **2014**, *78*, 203–214. [CrossRef]

25. Adem, G.D.; Chen, G.; Shabala, L.; Chen, Z.-H.; Shabala, S. GORK channel: A master switch of plant metabolism? *Trends Plant Sci.* **2020**, *25*, 434–445. [CrossRef] [PubMed]

26. Buwalda, F.; Thomson, C.J.; Steigner, W.; Barrett-Lennard, E.G.; Gibbs, J.; Greenway, H. Hypoxia induces membrane depolarization and potassium-loss from wheat roots but does not increase their permeability to sorbitol. *J. Exp. Bot.* **1988**, *39*, 1169–1183. [CrossRef]

27. Colmer, T.D.; Huang, S.; Greenway, H. Evidence for down-regulation of ethanolic fermentation and K^+ effluxes in the coleoptile of rice seedlings during prolonged anoxia. *J. Exp. Bot.* **2001**, *52*, 1507–1517. [CrossRef]

28. Huang, S.; Ishizawa, K.; Greenway, H.; Colmer, T.D. Manipulation of ethanol production in anoxic rice coleoptiles by exogenous glucose determines rates of ion fluxes and provides estimates of energy requirements for cell maintenance during anoxia. *J. Exp. Bot.* **2005**, *56*, 2453–2463. [CrossRef] [PubMed]

29. Mancuso, S.; Marras, A.M. Adaptative response of *Vitis* root to anoxia. *Plant Cell Physiol.* **2006**, *47*, 401–409. [CrossRef]

30. Zeng, F.; Konnerup, D.; Shabala, L.; Zhou, M.; Colmer, T.D.; Zhang, G.; Shabala, S. Linking oxygen availability with membrane potential maintenance and K^+ retention of barley roots: Implications for waterlogging stress tolerance. *Plant Cell Environ.* **2014**, *37*, 2325–2338. [CrossRef]

31. Loreti, E.; Perata, P. The many facets of hypoxia in plants. *Plants* **2020**, *9*, 745. [CrossRef]

32. Wang, F.; Chen, Z.-H.; Shabala, S. Hypoxia sensing in plants: On a quest for ion channels as putative oxygen sensors. *Plant Cell Physiol.* **2017**, *58*, 1126–1142. [CrossRef]

33. Shabala, S.; Shabala, L.; Barcelo, J.; Poschenrieder, C. Membrane transporters mediating root signalling and adaptive responses to oxygen deprivation and soil flooding. *Plant Cell Environ.* **2014**, *37*, 2216–2233. [CrossRef]

34. Subbaiah, C.C.; Sachs, M.M. Molecular and cellular adaptations of maize to flooding stress. *Ann. Bot.* **2003**, *91*, 119–127. [CrossRef] [PubMed]

35. Yemelyanov, V.V.; Shishova, M.F.; Chirkova, T.V.; Lindberg, S.M. Anoxia-induced elevation of cytosolic Ca^{2+} concentration depends on different Ca^{2+} sources in rice and wheat protoplasts. *Planta* **2011**, *234*, 271–280. [CrossRef] [PubMed]

36. Lindberg, S.; Kader, M.A.; Yemelyanov, V. Calcium signalling in plant cells under environmental stress. In *Plant Adaptations and Stress Tolerance of Plants in the Era of Climate Change*; Ahmad, P., Prasad, M.N.V., Eds.; Springer: New York, NY, USA; Dordrecht, The Netherlands; Heidelberg, Germany; London, UK, 2012; pp. 325–360. [CrossRef]

37. Igamberdiev, A.U.; Hill, R.D. Elevation of cytosolic Ca^{2+} in response to energy deficiency in plants: The general mechanism of adaptation to low oxygen stress. *Biochem. J.* **2018**, *475*, 1411–1425. [CrossRef] [PubMed]

38. Wang, F.; Chen, Z.-H.; Liu, X.; Colmer, T.D.; Shabala, L.; Salih, A.; Zhou, M.; Shabala, S. Revealing the roles of GORK channels and NADPH oxidase in acclimation to hypoxia in Arabidopsis. *J. Exp. Bot.* **2017**, *68*, 3191–3204. [CrossRef]

39. Pang, J.Y.; Newman, I.; Mendham, N.; Zhou, M.; Shabala, S. Microelectrode ion and O_2 fluxes measurements reveal differential sensitivity of barley root tissues to hypoxia. *Plant Cell Environ.* **2006**, *29*, 1107–1121. [CrossRef]

40. He, L.; Li, B.; Lu, X.; Yuan, L.; Yang, Y.; Yuan, Y.; Du, J.; Guo, S. The effect of exogenous calcium on mitochondria, respiratory metabolism enzymes and ion transport in cucumber roots under hypoxia. *Sci. Rep.* **2015**, *5*, 11391. [CrossRef]

41. Felle, H.H. pH regulation in anoxic plants. *Ann. Bot.* **2005**, *96*, 519–532. [CrossRef]

42. Roberts, J.K.M.; Callis, J.; Jardetzky, O.; Walbot, V.; Freeling, M. Cytoplasmic acidosis as a determinant of flooding intolerance in plants. *Proc. Natl. Acad. Sci. USA* **1984**, *81*, 6029–6033. [CrossRef]

43. Saint-Ges, V.; Roby, C.; Bligny, R.; Pradet, A.; Douce, R. Kinetic studies of the variation of cytoplasmic pH, nucleotide triphosphates (^{31}P-NMR) and lactate during normoxic and anoxic transitions in maize root tips. *Eur. J. Biochem.* **1991**, *200*, 477–482. [CrossRef]

44. Fox, G.G.; McCallan, N.R.; Ratcliffe, R.G. Manipulating cytoplasmic pH under anoxia: A critical test of the role of pH in the switch from aerobic to anaerobic metabolism. *Planta* **1995**, *195*, 324–330. [CrossRef]

45. Gout, E.; Boisson, A.-M.; Aubert, S.; Douce, R.; Bligny, R. Origin of the cytoplasmic pH changes during anaerobic stress in higher plant cells. Carbon-13 and Phosphorous-31 nuclear magnetic resonance studies. *Plant Physiol.* **2001**, *125*, 912–925. [CrossRef] [PubMed]

46. Menegus, F.; Cattaruzza, L.; Mattana, M.; Beffagna, N.; Ragg, E. Response to anoxia in rice and wheat seedlings changes in the pH of intracellular compartments, glucose-6-phosphate level, and metabolic rate. *Plant Physiol.* **1991**, *95*, 760–767. [CrossRef] [PubMed]

47. Kulichikhin, K.Y.; Aitio, O.; Chirkova, T.V.; Fagerstedt, K.V. Effect of oxygen concentration on intracellular pH, glucose-6-phosphate and NTP content in rice (*Oryza sativa*) and wheat (*Triticum aestivum*) root tips: In vivo ^{31}P-NMR study. *Physiol. Plant.* **2007**, *129*, 507–518. [CrossRef]

48. Greenway, H.; Gibbs, J. Mechanisms of anoxia tolerance in plants. II. Energy requirement for maintenance and energy distribution to essential processes. *Funct. Plant Biol.* **2003**, *30*, 999–1036. [CrossRef]

49. Drew, M.C. Oxygen deficiency and root metabolism: Injury and acclimation under hypoxia and anoxia. *Annu. Rev. Plant Physiol. Plant Mol. Biol.* **1997**, *48*, 223–250. [CrossRef] [PubMed]

50. Sakano, K.; Kiyota, S.; Yasaki, Y. Acidification and alkalization of culture medium by *Carathanus roseus* cells—Is anoxic production of lactate a cause of cytoplasmic acidification? *Plant Cell Physiol.* **1997**, *38*, 1053–1059. [CrossRef]

51. Chirkova, T.; Yemelyanov, V. The study of plant adaptation to oxygen deficiency in Saint Petersburg University. *Biol. Commun.* **2018**, *63*, 17–31. [CrossRef]

52. Subbaiah, C.; Bush, D.S.; Sachs, M. Elevation of cytosolic calcium precedes anoxic gene expression in maize suspension cultured cells. *Plant Cell* **1994**, *6*, 1747–1762. [CrossRef]

53. Sedbrook, J.C.; Kronebusch, P.J.; Borisy, G.G.; Trewavas, A.J.; Masson, P.H. Transgenic AEQUORIN reveals organ-specific cytosolic Ca^{2+} responses to anoxia in *Arabidopsis thaliana* seedlings. *Plant Physiol.* **1996**, *111*, 243–257. [CrossRef]

54. Biemelt, S.; Keetman, U.; Albrecht, G. Re-aeration following hypoxia or anoxia leads to activation of the antioxidative defense system in roots of wheat seedlings. *Plant Physiol.* **1998**, *116*, 651–658. [CrossRef]

55. Edwards, G.E.S.P.; Robinson, S.P.; Tyler, N.J.E.; Walker, D.A. Photosynthesis by isolated protoplasts, protoplast extracts and chloroplasts of wheat. *Plant Physiol.* **1978**, *62*, 313–319. [CrossRef] [PubMed]

56. Lindberg, S.; Strid, H. Aluminium induces rapid changes in cytosolic pH and free calcium and potassium concentrations in root protoplasts of wheat (*Triticum aestivum*). *Physiol. Plant.* **1997**, *99*, 405–414. [CrossRef]

57. Shishova, M.; Lindberg, S. Auxin-Induced cytosolic acidification in wheat leaf protoplasts depends on external concentration of Ca^{2+}. *J. Plant Physiol.* **1999**, *155*, 190–196. [CrossRef]

58. Lindberg, S. In-situ determination of intracellular concentrations of K$^+$ in barley (*Hordeum vulgare* L. cv. Kara) using the K$^+$-binding fluorescent dye benzofuran isophthalate. *Planta* **1995**, *195*, 525–529. [CrossRef]

59. Morgan, S.H.; Maity, P.J.; Geilfus, C.-M.; Lindberg, S.; Mühling, K.H. Leaf ion homeostasis and plasma membrane H$^+$-ATPase activity in *Vicia faba* change after extra calcium and potassium supply under salinity. *Plant Physiol. Biochem.* **2014**, *82*, 244–253. [CrossRef]

60. Phillips, H.J. Dye expulsion tests for cell viability. In *Tissue Cultures: Methods and Application. Chapter 3, Section VIII—Evaluation of Culture Dynamics*; Kruse, P.F., Patterson, M.K., Eds.; Academic Press: New York, NY, USA, 1973; p. 406.

Article

NADPH Oxidase RbohD and Ethylene Signaling are Involved in Modulating Seedling Growth and Survival Under Submergence Stress

Chen-Pu Hong [1], Mao-Chang Wang [2] and Chin-Ying Yang [1],*

[1] Department of Agronomy, National Chung Hsing University, Taichung 40227, Taiwan; xo7961@hotmail.com
[2] Department of Accounting, Chinese Culture University, Taipei 11114, Taiwan; wmaochang@yahoo.com.tw
* Correspondence: emiyang@dragon.nchu.edu.tw; Tel.: +886-4-22840777 (ext. 608); Fax: +886-4-22877054

Received: 2 March 2020; Accepted: 7 April 2020; Published: 8 April 2020

Abstract: In higher plants under low oxygen or hypoxic conditions, the phytohormone ethylene and hydrogen peroxide (H_2O_2) are involved in complex regulatory mechanisms in hypoxia signaling pathways. The respiratory burst oxidase homolog D (RbohD), an NADPH oxidase, is involved in the primary stages of hypoxia signaling, modulating the expression of downstream hypoxia-inducible genes under hypoxic stress. In this study, our data revealed that under normoxic conditions, seed germination was delayed in the *rbohD/ein2-5* double mutant, whereas postgermination stage root growth was promoted. Under submergence, the *rbohD/ein2-5* double mutant line had an inhibited root growth phenotype. Furthermore, chlorophyll content and leaf survival were reduced in the *rbohD/ein2-5* double mutant compared with wild-type plants under submerged conditions. In quantitative RT-PCR analysis, the induction of *Ethylene-responsive factor 73/hypoxia responsive 1 (AtERF73/HRE1)* and *alcohol dehydrogenase 1 (AtADH1)* transcripts was lower in the *rbohD/ein2-5* double mutant during hypoxic stress than in wild-type plants and in *rbohD* and *ein2-5* mutant lines. Taken together, our results indicate that an interplay of ethylene and RbohD is involved in regulating seed germination and post-germination stages under normoxic conditions. Moreover, ethylene and RbohD are involved in modulating seedling root growth, leaf chlorophyll content, and hypoxia-inducible gene expression under hypoxic conditions.

Keywords: ethylene; Ein2; germination; RbohD; submergence; hypoxia

1. Introduction

Climate change—induced flooding is a major global natural disaster. Flooding and heavy rain can cause soil compaction, reducing soil oxygen concentration and resulting in hypoxic stress—induced plant damage. The plant hormone ethylene participates in regulating stress-inducible genes to help plants to adapt to various environmental stresses, especially in hypoxia signaling caused by flooding [1–3].

Ethylene controls diverse physiological pathways involved in plant growth and developmental processes including the regulation of seed germination and leaf senescence and the promotion of pollen tube growth and fruit ripening [4–6]. Ethylene also contributes to plant responses to different biotic and abiotic stresses such as insect or microbial infections, drought, and salt conditions [7–9]. Several studies have shown that the ER-located membrane protein ETHYLENE INSENSITIVE 2 (EIN2) acts as a key signal transducer that positively signals downstream to members of the ETHYLENE INSENSITIVE 3 (EIN3) family of transcription factors located in the nucleus. EIN3 further binds to the promoters of ethylene-response genes, activating their expression in an ethylene-dependent manner to activate downstream ethylene responses [10,11].

Under oxygen deficient conditions, ethylene plays a major role in the regulation of abscisic acid (ABA) and gibberellic acid (GA) to influence cell elongation [12]. In addition to cell elongation, ethylene-induced programmed cell death through reactive oxygen species (ROS) production leads to the formation of aerenchyma and emergence of adventitious roots in maize and rice [13–15]. Hydrogen peroxide (H_2O_2) acts as a secondary messenger downstream of the ethylene signal to promote aerenchyma formation under oxygen-deficient conditions [16]. H_2O_2 is a type of ROS. ROS include H_2O_2, superoxide anion radicals (O_2^-), hydroxyl radicals (OH), and singlet oxygen ($^1O_2$), all of which are byproducts of aerobic metabolism [17]. NADPH oxidases, also named respiratory burst oxidase homologs (RBOHs), catalyze superoxide radical generation in plant apoplasts. Subsequently, superoxide radicals are converted into H_2O_2 by the activity of cell wall-localized antioxidant enzymes termed superoxide dismutases (SODs). H_2O_2 regulates the induction of *ethylene-responsive factor 73* (*ERF73*) and *alcohol dehydrogenase 1* (*ADH1*) expression under oxygen deprivation [18,19]. In *Arabidopsis thaliana*, RBOHs are members of a multigene family composed of 10 RBOH genes (AtRBOH A-J) that participate in ROS production in response to environmental stresses [20].

Plant RBOHs display different expression patterns during developmental processes and in response to various abiotic stresses and biotic interactions, either pathogenic or symbiotic [21]. AtRBOHC was revealed to be activated in Ca_2^+-dependent signaling triggered by the mechanical stimulation of root hairs [22]. Mild salt stress reportedly increases AtRbohD transcript levels, and AtRbohD and AtRbohF are involved in the ABA signaling network in guard cells [23,24]. AtRbohD mediates rapid systemic signaling in response to wounding, heat, cold, high-intensity light, and salinity stresses [25].

Our previous studies revealed that the accumulation of H_2O_2 was reduced in *ein2-5* and *rbohD*-knockout (*rbohD*-ko) mutants during hypoxic stress. The induction of hypoxia-inducible genes was also reduced in *rbohD*-ko mutants under hypoxic stress. AtRbohD plays a major role in the early stages of the stress response to oxygen deprivation [19,26]. Although AtRbohD is involved in hypoxia signaling, little is known about the relationship between ethylene and RbohD in submergence stress. In this study, to clarify the functional relationship between ethylene and H_2O_2 in the hypoxia signaling pathway, we analyzed *rbohD/ein2-5* double mutant and wild-type plants under hypoxic stress. Our results demonstrate that the interplay between ethylene and H_2O_2 is involved in modulating seed germination, seedling root growth, leaf chlorophyll content, and hypoxia-inducible gene expression under hypoxic conditions.

2. Results

2.1. The rbohd/ein2-5 Double Mutant Exhibited Ethylene Insensitive, Delayed Seed Germination and Increased Postgermination Root Growth

The induction of H_2O_2 and levels of *AtERF73/HRE1* and *ADH1* are reduced in *rbohD* and *ein2-5* single mutants under hypoxic stress [19,26]. To further investigate the interplay of ethylene and H_2O_2 signaling during hypoxia, we obtained *rbohD/ein2-5* double mutant lines by crossing single *rbohD* and *ein2-5* homozygous mutants. After seeds were grown on 1/2 strength Murashige and Skoog (MS) medium with or without 1-aminocyclopropane-1-carboxylic acid (ACC) treatment for 4 d in dark conditions, the etiolated seedlings of the *rbohD/ein2-5* double mutant line did not exhibit the triple response (Figure 1a). Increased *RbohD* transcript levels were not induced in the *rbohD/ein2-5* double mutant line after hypoxic treatment (Figure 1b).

The *rbohD/ein2-5* double mutant seeds germinated more slowly than wild-type, *rbohD*, and *ein2-5* single mutants under normaxic conditions; 90%, 85%, and 91% of wild-type, *rbohD*, and *ein2-5* single mutant seeds germinated after 2 d, whereas only 56% of *rbohD/ein2-5* double mutant seeds germinated. Germination was largely completed after 3 d incubation (Figure 2a,b). The mean germination time of seeds from wild-type, *rbohD* and *ein2-5* single mutants, and that from the *rbohD/ein2-5* double mutant, was 1.98, 2.10, 2.03, and 2.35 d, respectively (Figure 2c).

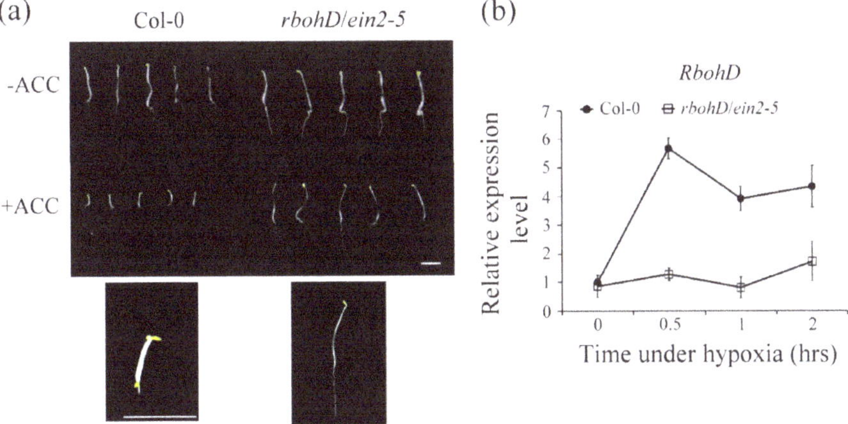

Figure 1. Phenotypes of wild-type Col-0 and *rbohD/ein2-5* double mutants in *Arabidopsis*. (**a**) The wild-type (Col-0) and double mutant (*rbohD/ein2-5*) seeds were grown on 1/2 MS medium with 5 µM ACC or without ACC for 4 d under dark conditions. Bar = 0.5 cm. (**b**) Quantitative RT-PCR analyses of transcript levels of *RbohD* in *Arabidopsis* Col-0 and *rbohD/ein2-5* double mutants in response to hypoxic stress for 0, 0.5, 1, and 2 h. Total RNAs were isolated from roots of 14-day-old seedlings after hypoxia treatment at indicated times and levels of *RbohD* mRNA were determined. Relative transcript amounts were calculated and normalized to *Actin* mRNA levels. Values represent means ± SD from five biologically independent experiments.

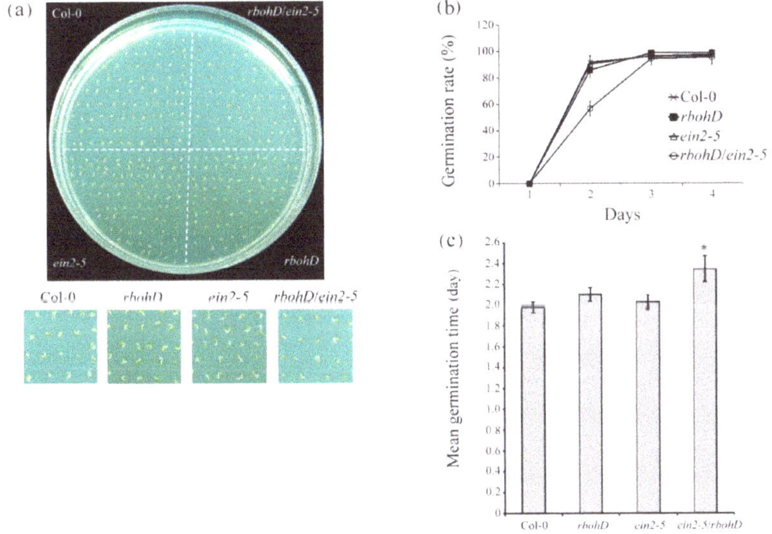

Figure 2. Germination assay of wild-type, *rbohD*, *ein2-5* and *rbohD/ein2-5* mutants. (**a**) Germination of wild-type, *rbohD* and *ein2-5* single mutants, and double mutant *rbohD/ein2-5* seeds. Seeds were sown in 1/2 MS medium and grown for 3 d at 4 °C in dark conditions; photographs were taken after 2 d. (**b,c**) Germination rates and mean germination times of wild-type, *rbohD* and *ein2-5* single mutants, and double mutant *rbohD/ein2-5* seeds. Seeds were scored for 1, 2, 3, and 4 d after sowing from at least 100 seeds of each genotype with three replicates each. Error bars represent SD. * $p < 0.05$, versus wild-type (Student's *t* test).

To determine whether ethylene and H_2O_2 signaling affected postgermination seedling development, the *rbohD/ein2-5* double mutant seedlings were grown for 4, 6, 8, 10, and 12 d. The root lengths of the *rbohD/ein2-5* double mutant lines were significantly longer than those of the wild-type seedlings (Figure 3a). When grown for 4, 6, 8, 10, and 12 d, wild-type seedlings developed roots with average lengths of 0.42, 1.00, 1.57, 2.09, and 2.32 cm, respectively, whereas the average root length of *rbohD/ein2-5* double mutants was 0.36, 1.45, 2.52, 2.95, and 3.09 cm, respectively (Figure 3b). These results indicate that in the *rbohD/ein2-5* double mutant, seed germination was delayed and the postgermination root growth rate increased.

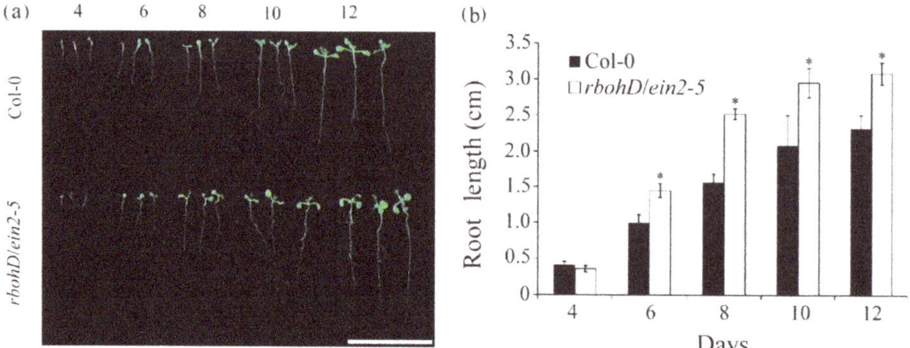

Figure 3. Comparison of root growth in wild-type and *rbohD/ein2-5* double mutant seedlings. (**a**) Phenotypes of seedlings grown on 1/2 MS medium for 4, 6, 8, 10, and 12 d. Bar = 2 cm. (**b**) Average root length of seedlings grown on 1/2 MS. Error bars represent SD of 100 seedlings of each genotype, obtained from three biologically independent experiments. * $p < 0.05$, versus wild-type (Student's *t* test).

2.2. The rbohD/ein2-5 Double Mutant Line Exhibited no Inhibition of Root Growth Following Submergence

Ethylene plays a role in inhibiting root length during plant developmental processes. Hydrogen peroxide also functions in inhibiting root length [27,28]. To determine whether ethylene signaling and H_2O_2 affect root growth in response to submergence, 7-day-old wild-type and *rbohD/ein2-5* double mutant seedlings were treated with normoxic or submerged conditions for 7 d. Root lengths were significantly reduced in wild-type plants compared with the *rbohD/ein2-5* double mutant lines (Figure 4a). The average root lengths of wild-type and the *rbohD/ein2-5* double mutant under the normaxic condition were 1.94 and 2.83 cm, respectively, whereas the average root lengths of wild-type and the *rbohD/ein2-5* double mutant under submergence were 1.34 and 3.19 cm, respectively (Figure 4b). The root lengths of the *rbohD/ein2-5* double mutant increased by 113% under submerged conditions compared with under the normaxic condition, whereas those of wild-type plants decreased by 69%. These results indicate that root growth was not inhibited in submerged *rbohD/ein2-5* double mutant lines.

2.3. Reduced Chlorophyll Content and Survival of rbohD/ein2-5 Double Mutant Plants under Submerged Conditions

To further assess the role of ethylene and H_2O_2 signaling under submerged conditions, chlorophyll content was determined during submergence. The 7-day-old wild-type and *rbohD/ein2-5* double mutant seedlings were submerged for 12 d. The *rbohD/ein2-5* double mutant exhibited more severe leaf etiolation than the wild-type plants did under submergence (Figure 5a). Contents of chlorophyll a and total chlorophyll were significantly decreased in the *rbohD/ein2-5* double mutant line after submergence treatment (Figure 5b). This effect was also observed in 7-day-old seedlings submerged for 12 d then left to recover for 5 d. The degree of leaf damage following submergence of the etiolated seedlings was analyzed and classified according to an index scale ranging from 1 to 4, where index scores of 1 and 2 represent 100% and over 50% green leaf area, respectively, and scores 3 and 4 represent less than 50%

and 0% green leaf area, respectively. These data indicate that more than 95% of the *rbohD/ein2-5* double mutant seedlings exhibited a phenotype with leaf damage (Figure 6a,b), suggesting that ethylene and H_2O_2 signaling positively regulate aboveground leaf survival under submergence stress.

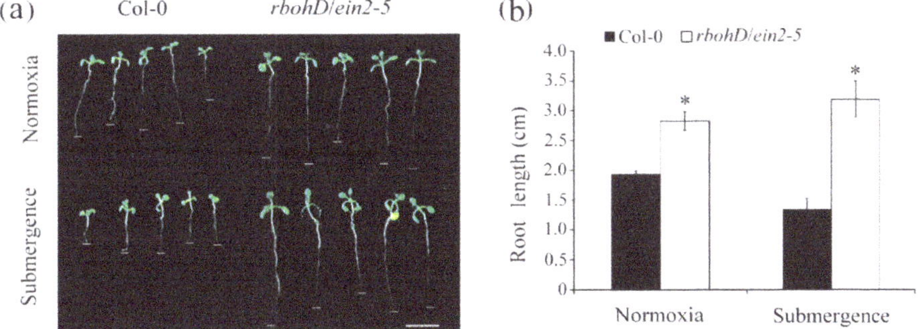

Figure 4. Comparison of root growth in wild-type and *rbohD/ein2-5* double mutant seedlings after submergence. (**a**) Phenotypes of seedlings grown on 1/2 MS medium for 14 d and then treated with submerged conditions for an additional 7 d. Bar = 1 cm. (**b**) Average root length of seedlings grown under normaxic condition and submerged conditions. Error bars represent SD for at least 100 seedlings of each genotype, obtained from three biologically independent experiments. * $p < 0.05$, versus wild-type plants (Student's *t* test).

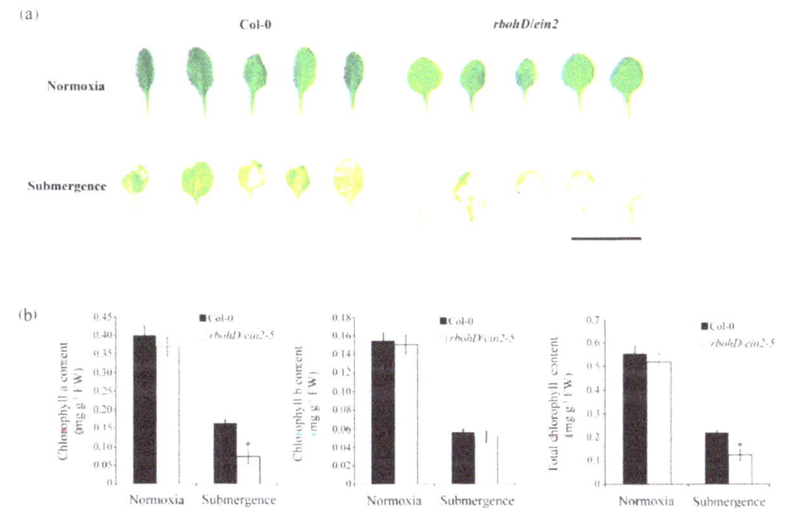

Figure 5. Chlorophyll content of wild-type plants and *rbohD/ein2-5* double mutants after submergence stress. (**a**) Phenotype of 7-day-old wild-type plants (Col-0) and double mutant *rbohD/ein2-5* leaves after treatment under normoxic conditions and submerged conditions for 12 d. Bar = 1 cm. (**b**) Contents of chlorophyll a, b, and total chlorophyll of 7-day-old seedlings after treatment under normoxic conditions and submerged conditions for 12 d. Values represent means ± SD from three biologically independent experiments. * $p < 0.05$, versus wild-type (Student's *t* test).

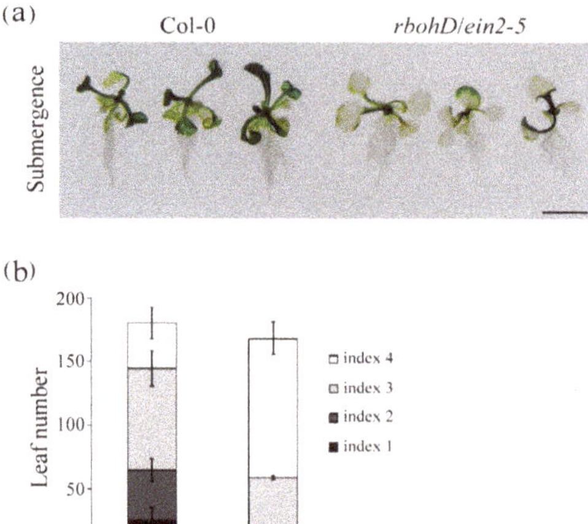

Figure 6. Sensitivity of the *rbohD/ein2-5* double mutant to submerged conditions. (**a**) Phenotypes of 7-day-old wild-type plants and *rbohD/ein2-5* double mutant seedlings after 12 d of submergence and 5 d of subsequent recovery. Bar = 1 cm. The photograph provides the results for three independent seedlings. (**b**) Quantification of the phenotype. Levels of damage were defined using an index based on the percentage of chlorotic leaves. Index 1 = 100% green leaves, Index 2 = over 50% green leaves, Index 3 = less than 50% green leaves, and Index 4 = 0% green leaves. The same results were obtained from three independent experiments ($n \geq 25$).

2.4. Effects of the rbohD/ein2-5 Double Mutant Line on Hypoxia-Inducible Genes under Hypoxic Stress

To determine how ethylene and H_2O_2 signaling affects hypoxia responses, we examined the expression of *AtERF73/HRE1* and *ADH1* in wild-type, *rbohD* and *ein2-5* single mutants, and the *rbohD/ein2-5* double mutant using qRT-PCR. The results indicate that under hypoxic conditions, the induction of *AtERF73/HRE1* and *ADH1* mRNA was significantly reduced in *rbohD* and *ein2-5* single mutants and in the *rbohD/ein2-5* double mutant lines compared with wild-type plants (Figure 7a,b). The *rbohD/ein2-5* double mutant had the most pronounced reduction in hypoxia-induced transcription. Taken together, these results suggest that ethylene and H_2O_2 signaling synergistically regulated hypoxia-inducible gene responses.

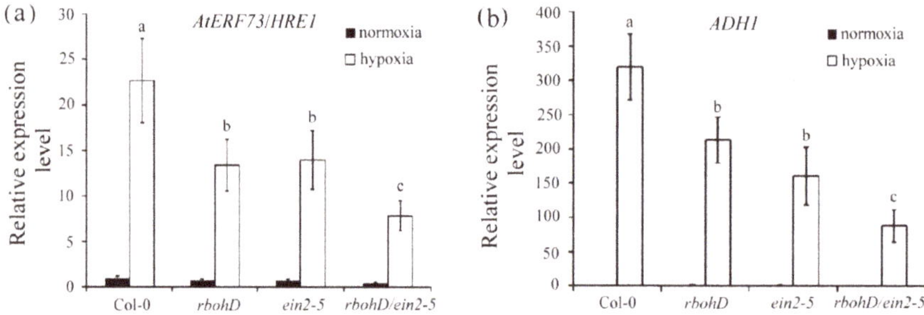

Figure 7. Transcript levels of hypoxia-inducible genes encoding AtERF73/HRE1 and ADH1 in *rbohD/ein2-5* lines after hypoxia treatment. Total RNAs were isolated from roots of 14-day-old Col-0, *rbohD*, *ein2-5*, and *rbohD/ein2-5* seedlings after 3 h of hypoxia treatment. Quantitative RT-PCR analyses of transcript levels of *AtERF73/HRE1* (**a**) and *ADH1* (**b**) genes. Values are means ± SD from three biologically independent experiments ($n = 3$). Values with different letters are significantly different at $p < 0.05$, according to a post-hoc LSD test.

3. Discussion

Plants are sessile organisms that have evolved integrated complex regulation systems to respond to stress. Oxygen deprivation is a common abiotic stress often associated with flooding. It affects plant growth, development, and survival. The role of ethylene in oxygen sensing and signaling is complex [29]. Recent studies have demonstrated that when submergence is established, gas diffusion of ethylene is restricted, and it is trapped in the plant, triggering its signaling pathway at early submergence. When submergence is prolonged, ethylene synthesis is reduced due to hypoxia; group VII ethylene response factor (ERFVII) activity is activated by hypoxia signaling [30–32]. In our previous reports, we demonstrated that the NADPH oxidase RbohD plays a major role in the early stages of hypoxic stress responses. Ethylene signaling modulates H_2O_2 signaling by regulating the expression of Rboh genes [19,26]. RbohD triggers cell death following fungal infection, and crosstalk between the salicylic acid and ethylene signaling pathways inhibits death in neighboring cells [33]. Furthermore, the double null mutant *atrbohD/F* is more sensitive to oxygen deficiency than wild-type plants and single mutants *atrbohD* and *atrbohF* are [34]. Little is known about the ethylene-mediated regulation of RbohD in the hypoxia signaling pathway.

The present study revealed that *rbohD/ein2-5* double mutants had delayed seed germination but increased root growth compared with wild-type plants (Figures 2 and 3). During seed germination, endogenous ethylene promotes seed germination by decreasing sensitivity to endogenous ABA [35]. H_2O_2, acting as a signaling molecule, mediates seed dormancy breaking and germination through the upregulation of ABA catabolism and GA biosynthesis [36]. Our data demonstrate that the mean seed germination time was significantly reduced in wild-type plants and the *rbohD* and *ein2-5* single mutants compared with the *rbohD/ein2-5* double mutant (Figure 2c), indicating that the interplay between ethylene signaling and H_2O_2 signaling modulates seed germination. Previously, we determined that under normoxic conditions, postgermination seedling root lengths of wild-type plants and *rbohD* single mutants do not differ [26]. Notably, the current study revealed that the *rbohD/ein2-5* double mutant had faster root growth 6–12 d after germination under both normoxic (Figure 3b) and submerged (Figure 4) conditions. Ethylene has been demonstrated to inhibit root cell elongation by upregulating auxin biosynthesis in Arabidopsis seedlings; accumulated H_2O_2 also inhibits root cell elongation and root growth [37–39]. Under submerged conditions, root length is reduced by oxygen deficiency [40]. However, the *rbohD/ein2-5* double mutant did not display inhibited root growth under submergence; instead, it had longer roots than wild-type plants did. These results suggest that crosstalk between ethylene signaling and H_2O_2 signaling, in addition to delaying seed germination, affects the root growth rate of postgermination seedlings under both normoxic and hypoxic conditions.

Chlorophyll absorbs the wavelengths associated with violet-blue and orange-red light within the visible light spectrum that are used in photosynthesis during plant growth and development. When plants are exposed to abiotic and biotic stresses such as drought, extreme temperature, cold, or heavy metals, chlorophyll concentration is decreased and protein degradation is increased [41–43]. Many studies have reported that ethylene is involved in the regulation of chlorophyll degradation during development or under stress conditions. When plants are submerged, chloroplasts disintegrate and photosynthetic capacity is lost [44–46]. In the present study, chlorophyll a and total chlorophyll concentration decreased significantly in *rbohD/ein2-5* double mutants after submergence treatment (Figure 5). Furthermore, the seedlings of the *rbohD/ein2-5* double mutant line had more leaf damage compared with wild-type plants after submergence stress (Figure 6). Some APETALA2/ethylene response factors function in an ethylene-controlled signal transduction pathway to regulate tolerance to hypoxia stress [47,48]. The secondary messenger, H_2O_2 production, has also been implicated in regulating tolerance to oxygen deprivation [18,49]. Our results indicate that under hypoxic stress, the induction of transcript levels of the hypoxia-inducible genes encoding ethylene-responsive factor 73 (ERF73) and alcohol dehydrogenase 1 (ADH1) was increased less in the *rbohD/ein2-5* double mutant line than in the *rbohD* and *ein2-5* single mutants (Figure 7). Notably, increased transcript levels of *AtERF73/HRE1* and *ADH1* were induced in all the lines under hypoxic conditions. This suggests that there are signaling elements other than ethylene and RbohD that also modulate the expression of hypoxia-responsive genes, such as *AtERF73/HRE1* and *ADH1*. Our previous results reveal that the accumulation of H_2O_2 was reduced in *rbohD*-ko and *ein2-5* during hypoxic stress [19,26]. In this study, both RbohD and EIN2 had a synergistic effect on seed germination and root growth at the postgermination stage under normoxic conditions and influenced submergence tolerance under oxygen deprivation conditions. Taken together, the results reported here indicate the presence of a synergistic interaction between ethylene and H_2O_2 signaling in plant development under normoxic conditions and in response to oxygen deprivation (Figure 8).

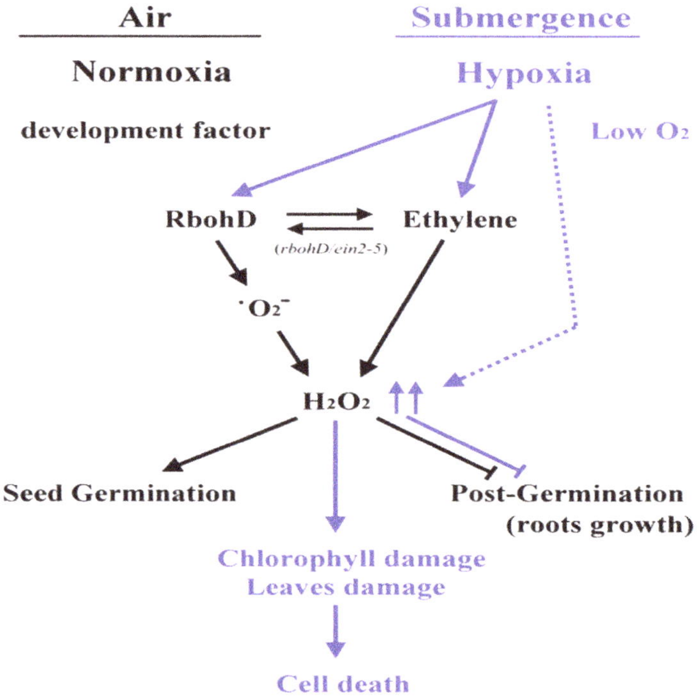

Figure 8. Schematic of the effects of RbohD-derived H_2O_2 and ethylene signaling under normoxic and hypoxic conditions. Both RbohD and EIN2 had a synergistic effect on seed germination and root growth at the postgermination stage under normoxic conditions (black line). Hypoxia triggers complex signaling pathways to increase the production of H_2O_2, leading to chlorophyll and leaves damages influencing submergence tolerance in *Arabidopsis* (blue line).

4. Materials and Methods

4.1. Plant Materials and Growth Conditions

Wild-type *Arabidopsis thaliana* (ecotype Col-0) was used in this study. The single *ein2-5* and *rbohD* (*At5g47910*) T-DNA insertion mutants (salk_005253C) were obtained from Dr. L.-C. Wang and the Arabidopsis Biological Resource Center (Ohio State University, Columbus, OH, USA), respectively, with the homozygous T-DNA insertion sites confirmed by RT-PCR before use. Double homozygous mutant lines, *rbohD/ein2-5*, were obtained by crossing the two single mutants and then screening the F2 and F3 generation as previously described [26]. Seeds were surface sterilized in 70% ethanol for 2 min and then in a 1% bleach solution for 15 min before being washed with sterilized water three times and incubated at 4 °C in the dark for 3 d. Seeds were sown on plates with 0.7% (*w/v*) agar in 1/2-strength Murashige and Skoog (MS) medium containing 0.5% sucrose at pH 5.7. The plates were transferred to a growth chamber at 22 °C with a long photoperiod regimen of 16 h of light (236 μmoLm^{-2}s^{-1})/ and 8 h of darkness.

For the seed germination assay, data were collected from 100 seeds of each genotype over three independent experiments. Germination was considered to have occurred when the radicles were 1 mm long. Germination percentage was recorded every 24 h for 4 d. The number of germinated seeds was expressed as a percentage of the total number of seeds plated for the indicated times. The mean germination time was calculated to assess the time spent to germinate or emerge [50].

4.2. Seedling Hypoxia, Submergence, and ACC Treatments

For the hypoxia treatment, 14-day-old seedlings were placed on damp filter paper for a 10 min pretreatment and then onto a floating platform with their roots immersed in 1/2 MS solution. The solution was constantly supplied with 3% O_2/97% N_2 gas in the dark to replace the oxygen in the solution. The seedling roots were collected for analysis. Other 14-day-old seedlings were placed in the dark without immersion in 1/2 MS solution as controls. For the submergence treatment, seeds were grown on 1/2-strength MS medium in a tissue cultivation box (L:W:H, 12 cm × 9 cm × 7 cm) for 7 d and then treated with or without 250 mL of sterile deionized water for 12 d. The water was drained out for the the subsequent 5-day recovery period. The root lengths of the seedlings (under normoxic conditions or submerged conditions) were measured. Data were collected from at least 100 seedlings of each genotype in three independent experiments. Photographs were taken at the end of the treatment period. For the ACC treatment, seeds were grown on 1/2-strength MS medium with or without 5 μM ACC in the dark for 4 d. The phenotypes were observed, and photographs were taken at the end of the treatment period.

4.3. Plant Chlorophyll Content and Leaf Damage Index Measurements

For chlorophyll content assays, the content was determined according to Wintermans and De Mots (1965) after extraction in ethanol [51]. The 7-day-old seedlings were treated under normoxic conditions or submerged conditions for 12 d. Aboveground tissue (50 mg) was collected and ground in 2 mL of sodium phosphate buffer (50 mM pH 6.8), 40 μL of which was added to 960 mL of 99% ethanol and incubated for 30 min at room temperature in the dark with gentle shaking. After centrifugation at 4 °C for 15 min at 1000× g, the absorbance values of the supernatant were measured at 665 and 649 nm with a spectrophotometer (Metertec SP8001) to determine to concentrations of chlorophyll a, b, and total chlorophyll. Data were collected from three independent experiments. For leaf damage index measurement, the 7-day-old seedlings were treated under submerged conditions for 12 d followed by 5 d of recovery. The levels of damage were defined as follows: Index 1 = 100% green leaves, Index 2 = over 50% green leaves, Index 3 = less than 50% green leaves, and Index 4 = 0% green leaves. Data were collected from three independent experiments. Each experiment involved the use of at least 25 seedlings. Photographs were taken at the end of experiment. For the hypoxia treatment, 14-day-old seedlings were placed on a floating platform with their roots immersed in 1/2 MS solution supplemented with 0.5% (*w/v*) Suc at pH 5.7. The solution was constantly supplied with 3% (*v/v*) O_2 and 97% (*v/v*) N_2 in the dark for 3 h. The root samples were collected and frozen for analysis.

4.4. RNA Extraction and Quantitative RT-PCR (qRT-PCR) Analyses

The 14-day-old seedlings were hypoxia-treated for 3 h. Then, root samples were collected in liquid nitrogen and stored at −80 °C until use. Total RNA was extracted using TRIzol (Invitrogen, Carlsbad, CA, USA) and then subjected to DNase treatment using the TURBO DNA-free kit (Ambion, Austin, TX, USA). RNA concentrations were determined, and samples were then reverse-transcribed into cDNAs by using Moloney murine leukemia virus reverse transcriptase (Invitrogen). qRT-PCR was conducted as previously described [52] and performed using a Rotor-Gene 3000 instrument (Corbett Research, Sydney, Australia) with Power SYBR Green PCR master mix (GeneMark, Taipei, Taiwan) in accordance with the manufacturers' recommendations. The Actin gene was used as an internal control for normalization. Relative expression levels were analyzed with Rotor-Gene 6 software (Corbett). Experiments were repeated five times independently with duplicate samples. The sequences of primers used for qRT-PCR are presented in Table 1.

Table 1. Primers used for quantitative RT-PCR experiments.

Gene Name	Primer Sequence
rAtERF73/HRE1-forward	5′-atcatgggcgatgcgaataa-3′
rAtERF73/HRE1-reverse	5′-gcgagaaatattcggtctggtt-3′
rADH1-forward	5′-catgaacaaggagctggagcttg-3′
rADH1-reverse	5′-ctctcccttcagcatgtaatcaaagg-3′
rRbohD-forward	5′-ccgagcagacggaggagat-3′
rRbohD-reverse	5′-tggaccgtcgataaggacctt-3′

5. Conclusions

In our previous studies, the data indicated that an increase in the transcript levels of *RbohD* was induced at a relatively early stage during hypoxic stress; for other hypoxia-inducible RBOHs, such as *RbohA, B, F, G,* and *I*, transcript expressions were also affected by oxygen deprivation but not at a relatively early stage [24]. The involvement of other signaling elements in the complex signaling network associated with oxygen deprivation requires further evaluation.

Author Contributions: C.-P.H. conducted experiments and analyzed the data. M.-C.W. helps the statistical analysis and manuscript preparation. C.-Y.Y. conceived, designed research and wrote the manuscript. All authors have read and agreed to the published version of the manuscript.

Funding: This work was supported by the National Science Council, Taiwan (NSC101-2311-B-005-001) to Chin-Ying Yang.

Acknowledgments: We would like to thanks Ching-Hsiu Tsai for his kind help in the instrument assistance of quantitative RT-PCR.

Conflicts of Interest: The authors declare that they have no conflict of interest.

References

1. Wen, X.; Wang, J.; Zhang, D.; Wang, Y. A gene regulatory network controlled by bperf2 and bpmyb102 in birch under drought conditions. *Int. J. Mol. Sci.* **2019**, *20*, 3071. [CrossRef] [PubMed]

2. Savada, R.P.; Ozga, J.A.; Jayasinghege, C.P.A.; Kosala, D.; Waduthanthri, K.D.; Reinecke, D.M. Heat stress differentially modifies ethylene biosynthesis and signaling in pea floral and fruit tissues. *Plant Mol. Biol.* **2017**, *95*, 313–331. [CrossRef] [PubMed]

3. Huang, Y.C.; Yeh, T.H.; Yang, C.Y. Ethylene signaling involves in seeds germination upon submergence and antioxidant response elicited confers submergence tolerance to rice seedlings. *Rice* **2019**, *12*, 23. [CrossRef] [PubMed]

4. Abts, W.; Van de Poel, B.; Vandenbussche, B.; De Proft, M.P. Ethylene is differentially regulated during sugar beet germination and affects early root growth in a dose-dependent manner. *Planta* **2014**, *240*, 679–686. [CrossRef] [PubMed]

5. Choudhury, S.R.; Roy, S.; Das, R.; Sengupta, D.N. Differential transcriptional regulation of banana sucrose phosphate synthase gene in response to ethylene, auxin, wounding, low temperature and different photoperiods during fruit ripening and functional analysis of banana SPS gene promoter. *Planta* **2008**, *229*, 207–223. [CrossRef] [PubMed]

6. Eccher, G.; Begheldo, M.; Boschetti, A.; Ruperti, B.; Botton, A. Roles of ethylene production and ethylene receptor expression in regulating apple fruitlet abscission. *Plant Physiol.* **2015**, *169*, 125–137. [CrossRef] [PubMed]

7. Kepczynska, E.; Zielinska, S.; Kepczynski, J. Ethylene production by *Agrobacterium rhizogenes* strains in vitro and in vivo. *Plant Growth Regul.* **2003**, *39*, 13–17. [CrossRef]

8. Wang, H.H.; Liang, X.L.; Wan, Q.; Wang, X.M.; Bi, Y.R. Ethylene and nitric oxide are involved in maintaining ion homeostasis in *Arabidopsis* callus under salt stress. *Planta* **2009**, *230*, 293–307. [CrossRef]

9. Yu, Y.W.; Yang, D.X.; Zhou, S.R.; Gu, J.T.; Wang, F.R.; Dong, J.G.; Huang, R.F. The ethylene response factor OsERF109 negatively affects ethylene biosynthesis and drought tolerance in rice. *Protoplasma* **2017**, *254*, 401–408. [CrossRef]

10. Li, W.Y.; Ma, M.D.; Feng, Y.; Li, H.J.; Wang, Y.C.; Ma, Y.T.; Li, M.Z.; An, F.Y.; Guo, H.W. EIN2-directed translational regulation of ethylene signaling in Arabidopsis. *Cell* **2015**, *163*, 670–683. [CrossRef]

11. Wang, K.L.C.; Li, H.; Ecker, J.R. Ethylene biosynthesis and signaling networks. *Plant. Cell* **2002**, *14*, S131–S151. [CrossRef]

12. Jackson, M.B. Ethylene-promoted elongation: An adaptation to submergence stress. *Ann. Bot.* **2008**, *101*, 229–248. [CrossRef]

13. Lenochova, Z.; Soukup, A.; Votrubova, O. Aerenchyma formation in maize roots. *Biol. Plant.* **2009**, *53*, 263–270. [CrossRef]

14. Yamauchi, T.; Yoshioka, M.; Fukazawa, A.; Mori, H.; Nishizawa, N.K.; Tsutsumi, N.; Yoshioka, H.; Nakazono, M. An NADPH oxidase RBOH functions in rice roots during lysigenous aerenchyma formation under oxygen-deficient conditions. *Plant Cell* **2017**, *29*, 775–790. [CrossRef] [PubMed]

15. Yamauchi, T.; Colmer, T.D.; Pedersen, O.; Nakazono, M. Regulation of root traits for internal aeration and tolerance to soil waterlogging-flooding stress. *Plant Physiol.* **2018**, *175*, 1118–1130. [CrossRef] [PubMed]

16. Steffens, B.; Geske, T.; Sauter, M. Aerenchyma formation in the rice stem and its promotion by H_2O_2. *New Phytol.* **2011**, *190*, 369–378. [CrossRef] [PubMed]

17. Choudhury, F.K.; Rivero, R.M.; Blumwald, E.; Mittler, R. Reactive oxygen species, abiotic stress and stress combination. *Plant J.* **2017**, *90*, 856–867. [CrossRef]

18. Baxter-Burrell, A.; Yang, Z.B.; Springer, P.S.; Bailey-Serres, J. RopGAP4-dependent Rop GTPase rheostat control of Arabidopsis oxygen deprivation tolerance. *Science* **2002**, *296*, 2026–2028. [CrossRef]

19. Yang, C.Y. Hydrogen peroxide controls transcriptional responses of ERF73/HRE1 and ADH1 via modulation of ethylene signaling during hypoxic stress. *Planta* **2014**, *239*, 877–885. [CrossRef]

20. Oda, T.; Hashimoto, H.; Kuwabara, N.; Akashi, S.; Hayashi, K.; Kojima, C.; Wong, H.L.; Kawasaki, T.; Shimamoto, K.; Sato, M.; et al. Structure of the N-terminal regulatory domain of a plant NADPH oxidase and its functional implications. *J. Biol. Chem.* **2010**, *285*, 1435–1445. [CrossRef]

21. Marino, D.; Dunand, C.; Puppo, A.; Pauly, N. A burst of plant NADPH oxidases. *Trends Plant Sci.* **2012**, *17*, 9–15. [CrossRef] [PubMed]

22. Monshausen, G.B.; Bibikova, T.N.; Weisenseel, M.H.; Gilroy, S. Ca^{2+} regulates reactive oxygen species production and pH during mechanosensing in *Arabidopsis* roots. *Plant Cell* **2009**, *21*, 2341–2356. [CrossRef] [PubMed]

23. Xie, Y.J.; Xu, S.; Han, B.; Wu, M.Z.; Yuan, X.X.; Han, Y.; Gu, Q.A.; Xu, D.K.; Yang, Q.; Shen, W.B. Evidence of Arabidopsis salt acclimation induced by up-regulation of HY1 and the regulatory role of RbohD-derived reactive oxygen species synthesis. *Plant J.* **2011**, *66*, 280–292. [CrossRef] [PubMed]

24. Zhang, Y.Y.; Zhu, H.Y.; Zhang, Q.; Li, M.Y.; Yan, M.; Wang, R.; Wang, L.L.; Welti, R.; Zhang, W.H.; Wang, X.M. Phospholipase D alpha 1 and phosphatidic acid regulate NADPH oxidase activity and production of reactive oxygen species in ABA-mediated stomatal closure in *Arabidopsis*. *Plant. Cell* **2009**, *2*, 2357–2377. [CrossRef] [PubMed]

25. Miller, G.; Schlauch, K.; Tam, R.; Cortes, D.; Torres, M.A.; Shulaev, V.; Dangl, J.L.; Mittler, R. The plant NADPH oxidase RbohD mediates rapid systemic signaling in response to diverse stimuli. *Sci. Signal.* **2009**, *2*, ra45. [CrossRef] [PubMed]

26. Yang, C.Y.; Hong, C.P. The NADPH oxidase Rboh D is involved in primary hypoxia signalling and modulates expression of hypoxia-inducible genes under hypoxic stress. *Environ. Exp. Bot.* **2015**, *115*, 63–72. [CrossRef]

27. Dunand, C.; Crevecoeur, M.; Penel, C. Distribution of superoxide and hydrogen peroxide in *Arabidopsis* root and their influence on root development: Possible interaction with peroxidases. *New Phytol.* **2007**, *174*, 332–341. [CrossRef]

28. Ruzicka, K.; Ljung, K.; Vanneste, S.; Podhorska, R.; Beeckman, T.; Friml, J.; Benkova, E. Ethylene regulates root growth through effects on auxin biosynthesis and transport-dependent auxin distribution. *Plant Cell* **2007**, *19*, 2197–2212. [CrossRef]

29. Loreti, E.; van Veen, H.; Perata, P. Plant responses to flooding stress. *Curr. Opin. Plant Biol.* **2016**, *33*, 64–71. [CrossRef]

30. White, M.D.; Kamps, J.; East, S.; Taylor Kearney, L.J.; Flashman, E. The plant cysteine oxidases from Arabidopsis thaliana are kinetically tailored to act as oxygen sensors. *J. Biol. Chem.* **2018**, *293*, 11786–11795. [CrossRef]

31. Hartman, S.; Liu, Z.; van Veen, H.; Vicente, J.; Reinen, E.; Martopawiro, S.; Zhang, H.; van Dongen, N.; Bosman, F.; Bassel, G.W.; et al. Ethylene-mediated nitric oxide depletion pre-adapts plants to hypoxia stress. *Nat. Commun.* **2019**, *10*, 4020. [CrossRef] [PubMed]

32. Perata, P. Ethylene signaling controls fast oxygen sensing in plants. *Trends Plant Sci.* **2020**, *25*, 3–6. [CrossRef] [PubMed]

33. Pogany, M.; von Rad, U.; Grun, S.; Dongo, A.; Pintye, A.; Simoneau, P.; Bahnweg, G.; Kiss, L.; Barna, B.; Durner, J. Dual roles of reactive oxygen species and NADPH oxidase RbohD in an *Arabidopsis*-Alternaria pathosystem. *Plant Physiol.* **2009**, *151*, 1459–1475. [CrossRef] [PubMed]

34. Liu, B.; Sun, L.; Ma, L.; Hao, F.S. Both AtrbohD and AtrbohF are essential for mediating responses to oxygen deficiency in *Arabidopsis. Plant Cell Rep.* **2017**, *36*, 947–957. [CrossRef]

35. Beaudoin, N.; Serizet, C.; Gosti, F.; Giraudat, J. Interactions between abscisic acid and ethylene signaling cascades. *Plant Cell* **2000**, *12*, 1103–1115. [CrossRef]

36. Liu, Y.G.; Ye, N.H.; Liu, R.; Chen, M.X.; Zhang, J.H. H_2O_2 mediates the regulation of ABA catabolism and GA biosynthesis in Arabidopsis seed dormancy and germination. *J. Exp. Bot.* **2010**, *61*, 2979–2990. [CrossRef]

37. Bai, L.; Zhou, Y.; Zhang, X.R.; Song, C.P.; Cao, M.Q. Hydrogen peroxide modulates abscisic acid signaling in root growth and development in *Arabidopsis. Chin. Sci. Bull.* **2007**, *52*, 1142–1145. [CrossRef]

38. Ivanchenko, M.G.; den Os, D.; Monshausen, G.B.; Dubrovsky, J.G.; Bednarova, A.; Krishnan, N. Auxin increases the hydrogen peroxide (H_2O_2) concentration in tomato (*Solanum lycopersicum*) root tips while inhibiting root growth. *Ann. Bot.* **2013**, *112*, 1107–1116. [CrossRef]

39. Swarup, R.; Perry, P.; Hagenbeek, D.; Van Der Straeten, D.; Beemster, G.T.S.; Sandberg, G.; Bhalerao, R.; Ljung, K.; Bennett, M.J. Ethylene upregulates auxin biosynthesis in Arabidopsis seedlings to enhance inhibition of root cell elongation. *Plant Cell* **2007**, *19*, 2186–2196. [CrossRef]

40. Sauter, M. Root responses to flooding. *Curr. Opin. Plant Biol.* **2013**, *16*, 282–286. [CrossRef]

41. Munne-Bosch, S.; Alegre, L. Changes in carotenoids, tocopherols and diterpenes during drought and recovery, and the biological significance of chlorophyll loss in *Rosmarinus officinalis* plants. *Planta* **2000**, *210*, 925–931. [CrossRef] [PubMed]

42. Shi, Q.; Bao, Z.; Zhu, Z.; Ying, Q.; Qian, Q. Effects of different treatments of salicylic acid on heat tolerance, chlorophyll fluorescence, and antioxidant enzyme activity in seedlings of *Cucumis sativa* L. *Plant. Growth Regul.* **2006**, *48*, 127–135. [CrossRef]

43. Song, W.Y.; Park, J.; Mendoza-Cozatl, D.G.; Suter-Grotemeyer, M.; Shim, D.; Hortensteiner, S.; Geisler, M.; Weder, B.; Rea, P.A.; Rentsch, D.; et al. Arsenic tolerance in Arabidopsis is mediated by two ABCC-type phytochelatin transporters. *Proc. Natl. Acad. Sci. USA* **2010**, *107*, 21187–21192. [CrossRef] [PubMed]

44. Herrera, A. Responses to flooding of plant water relations and leaf gas exchange in tropical tolerant trees of a black-water wetland. *Front. Plant Sci.* **2013**, *4*, 150. [CrossRef] [PubMed]

45. Jacob-Wilk, D.; Holland, D.; Goldschmidt, E.E.; Riov, J.; Eyal, Y. Chlorophyll breakdown by chlorophyllase: Isolation and functional expression of the Chlase1 gene from ethylene-treated Citrus fruit and its regulation during development. *Plant J.* **1999**, *20*, 653–661. [CrossRef] [PubMed]

46. Qiu, K.; Li, Z.P.; Yang, Z.; Chen, J.Y.; Wu, S.X.; Zhu, X.Y.; Gao, S.; Gao, J.; Ren, G.D.; Kuai, B.K.; et al. EIN3 and ORE1 accelerate degreening during ethylene-mediated leaf senescence by directly activating chlorophyll catabolic genes in Arabidopsis. *PLoS Genet.* **2015**, *11*, e1005399. [CrossRef]

47. Gibbs, D.J.; Lee, S.C.; Isa, N.M.; Gramuglia, S.; Fukao, T.; Bassel, G.W.; Correia, C.S.; Corbineau, F.; Theodoulou, F.L.; Bailey-Serres, J.; et al. Homeostatic response to hypoxia is regulated by the N-end rule pathway in plants. *Nature* **2011**, *479*, 415–418. [CrossRef]

48. Hinz, M.; Wilson, I.W.; Yang, J.; Buerstenbinder, K.; Llewellyn, D.; Dennis, E.S.; Sauter, M.; Dolferus, R. *Arabidopsis* RAP2.2: An ethylene response transcription factor that is important for hypoxia survival. *Plant Physiol.* **2010**, *153*, 757–772. [CrossRef]

49. Ying, G.; Wang, Z.H.; Yang, Z.B. ROP/RAC GTPase: An old new master regulator for plant signaling. *Curr. Opin. Plant Biol.* **2004**, *7*, 527–536.

50. Matthews, S.; Khajeh-Hosseini, M. Length of the lag period of germination and metabolic repair explain vigour differences in seed lots of maize (*Zea mays*). *Seed Sci. Technol.* **2007**, *35*, 200–212. [CrossRef]

51. Wintermans, J.F.G.M.; De Mots, A. Spectrophotometric characteristics of chlorophyll a and b and their pheophytins in ethanol. *Biochim. Biophys. Acta* **1965**, *109*, 448–453. [CrossRef]
52. Lin, I.S.; Wu, Y.S.; Chen, C.T.; Chen, G.H.; Hwang, S.G.; Jauh, G.Y.; Tzen, J.T.C.; Yang, C.Y. AtRBOH I confers submergence tolerance and is involved in auxin-mediated signaling pathways under hypoxic stress. *Plant Growth Regul.* **2017**, *83*, 277–285. [CrossRef]

Article

The Seedlings of Different *Japonica* Rice Varieties Exhibit Differ Physiological Properties to Modulate Plant Survival Rates under Submergence Stress

Yu-Syuan Li, Shang-Ling Ou and Chin-Ying Yang *

Department of Agronomy, National Chung Hsing University, Taichung 40227, Taiwan;
shaneimayday@gmail.com (Y.-S.L.); slou@dragon.nchu.edu.tw (S.-L.O.)
* Correspondence: emiyang@dragon.nchu.edu.tw; Tel.: +886-4-22840777 (ext. 608); Fax: +886-4-22877054

Received: 17 July 2020; Accepted: 1 August 2020; Published: 3 August 2020

Abstract: *Oryza sativa* is a major food crop in Asia. In recent years, typhoons and sudden downpours have caused field flooding, which has resulted in serious harm to the production of rice. In this study, our data revealed that the plant heights of the five *Japonica* varieties increased during submergence. The elongation rates of TN14, KH139, and TK9 increased significantly during submergence. Chlorophyll contents of the five varieties significantly decreased after submergence and increased after recovery. Moreover, the chlorophyll content of KH139 was significantly higher than those of the other four varieties after recovery. The plant survival rates of the five varieties were higher than 50% after four-day submergence. After eight-day submergence, the survival rate of KH139 remained at 90%, which was the highest among the different varieties. The KH139 presented lower accumulation of hydrogen peroxide and the catalase activity than those of the other four varieties under submergence. The *sucrose synthase 1* and *alcohol dehydrogenase 1* were induced in KH139 under submergence. The results presented that different varieties of *japonica* rice have different flood tolerances, especially KH139 under submergence was superior to that of the other four varieties. These results can provide crucial information for future research on *japonica* rice under flooding stress.

Keywords: *Oryza sativa*; Submergence; Activity of antioxidant enzymes; Chlorophyll content

---

1. Background

Global climate change has led to extreme climate in recent years. Global warming has caused sea-level rise, affected crop cultivation, and led to decreased crop production. The frequency of flooding disasters worldwide caused by biological stresses, such as high temperature, low temperature, droughts, and salt-damaged soils caused by extreme climates, has increased by approximately 65% in the past 25 years [1]. *Oryza sativa* is a major food crop worldwide and it is widely cultivated in Asia. Flooding disasters have considerably affected agricultural production in Asia, particularly in several Southeast Asian countries, such as the Philippines, Myanmar, and Indonesia, and have led to a considerable decrease in crop yield. Thus, flooding disasters have become an issue that must be urgently confronted [2].

In a normal growth environment, higher plants perform aerobic respiration and transport the products of photosynthesis from the source to the sink through the glycolysis pathway, so that the related reactions of carbon metabolism can occur [3]. Sucrose can be decomposed into glucose and fructose in two ways. Sucrose can be catalyzed into fructose and glucose through sucrose invertase, or it can be catalyzed into fructose and uracil-diphosphate glucose through sucrose synthase (SUS). When hypoxia is caused by flooding stress, a plant limits the pathway of sucrose invertase, reduces energy consumption, and increases SUS activity for preserving the source of reactants that are essential for glycolysis [4]. Research has indicated that during flooding treatment, the expression of the *SUS* genes of rice increases with the flooding period. Moreover, ethylene induces submergence1 (*SUB1*), which regulates the expression of *SUS* genes [5]. The expression of *SUS* genes increases when corn is subjected to hypoxia treatment. Corn under hypoxic stress is speculated to induce sucrose for accelerated decomposition in order to maintain the glycolysis reaction and reduce the ATP consumption [6]. In addition, research have shown that plants lacking genes that required for fermentation, such as *alcohol dehydrogenase* (*ADH*) and *pyruvate decarboxylase* (*PDC*), or mutants defective in sucrose metabolism, such as *SUS*, revealed lower tolerance in hypoxia condition [7]. When crops are fully submerged, the gas exchange between plants and air significantly decreases. Gases, such as O_2 (molecular oxygen), CO_2 (carbon dioxide), and ethylene, spread very slowly in water, preventing effective exchange and causing hypoxia in the plant's root, stem and leaves, which causes hypoxic stress [8]. Because the aerobic respiration of the rice plant is restricted under flooding stress, the plant performs anaerobic respiration and enters the fermentation route, which leads to the accumulation of toxic substances, such as ethanol and lactic acid, in the cells. The accumulation of ethanol damages the cell membrane and the membrane structure of the organelle. Lactic acid acidifies the cytoplasm, affects enzyme action in crucial metabolic pathways, and influences crop viability in a hypoxic environment [9].

Reactive oxygen species (ROS) also play a major role when the plants are under hypoxic stress. ROS are produced through induction and cause oxidative stress inside the plant cell [10,11]. Some ROS include the superoxide anion (O_2^-), H_2O_2, and the hydroxyl radical (OH). The presence of excessive free radicals damages the DNA or RNA in the cells or causes protein or lipid peroxidation. Under hypoxic stress, the content of H_2O_2 as well as the gene expressions of *ascorbate peroxidase* (*APX*) and *superoxide dismutase* (*SOD*) in grape sprouts increase significantly, which indicates that the activated oxygen group can participate in the regulation of antioxidant enzyme transcription [12]. Studies have indicated that the H_2O_2 in ROS can act as a signaling molecule to participate in the regulation of the reaction, growth, and development of plants under environmental stress [13].

Under flooding stress, FR13A lowland rice and deepwater rice exhibit two adaptation strategies. The quiescence strategy is one strategy. The accumulation of ethylene in the plant cells of FR13A lowland rice increases under flooding stress. The accumulation of ethylene induced the expression of the ethylene transcription factor submergence 1A (*SUB1A*) for downstream signal regulation. SUB1A influence gibberellin signal transduction and the metabolic pathway of carbohydrate are affected, which leads to a two-week stillness of FR13A lowland rice under flooding stress. The rice growth becomes normal again after the water levels recede [14]. The other strategy is the escape strategy. The accumulation of ethylene inside the plant cells of deepwater rice under flooding stress causes the ethylene transcription factors Snorkel1 (SK1) and Snorkel2 (SK2) to directly or indirectly regulate the synthetic pathway and signal transduction process of gibberellin. Thus, the above-ground parts of rice are elongated and further protrude from the water to perform gas exchange [15]. Ethylene accumulation is also found within the plant of the rice varieties without *SK1*, *SK2*, and *SUB1A* under flooding stress. This accumulation regulates gibberellin signal transduction and the metabolic pathway of carbohydrate, such that the above-ground parts may protrude from the water to perform gas exchange. However, because most of its energy is consumed, the plant often cannot protrude from the water in time. Thus, it displays the physiognomic traits of withering, yellowing, and even death. These traits indicate the occurrence of low-oxygen-escape syndrome (LOES) [4].

Numerous studies have examined the physiological and molecular reaction mechanisms of flood-resistant rice with *SUB1* locus in a fully submerged environment. However, the physiological and molecular performance of the commonly cultivated varieties of *japonica* rice under flooding stress has rarely been investigated. Therefore, five varieties of good-quality *japonica* rice that are cultivated in Taiwan were subjected to submergence experiments and plant survival rate tests. Moreover, their total chlorophyll contents and antioxidant enzyme activities were determined. The physiological and molecular reactions of the seedlings of the *japonica* rice varieties were investigated under submergence conditions. Thus, crucial information was obtained on the properties of the aforementioned rice varieties under flooding stress.

2. Results

2.1. Plant Heights, Dry and Fresh Weights of the Five Varieties of Japonica Rice Under Diverse Submergence Periods

Under flooding stress, varieties of rice that are not flood-resistant exhibit LOES. Their petioles, stems, and leaves elongate as the flooding treatment period increase [4]. Thus, five varieties of good-quality *japonica* rice commonly grown in Taiwan were subjected to different tests to understand the physiological traits of *japonica* rice varieties under diverse flooding periods. The five varieties used in this study were Tainan 11 (TN11), Tainan 14 (TN14), Kaohsiung 139 (KH139), Taiken 9 (TK9), and Tainung 71 (TNG71). The 9-day-old seedlings were subjected to full-submergence treatments for 0, two, four, six, and eight days and compared with the control group without treatment. The changes in the heights of the plants as well as the dry and fresh weights of their above-ground parts were measured after the aforementioned periods. Following two-day submergence treatment, TNG71 exhibited complete lodging and TN11, TN14, KH139, and TK9 exhibited half-lodging. On the eighth day of the submergence treatment, the five rice varieties exhibited the physiognomic traits of significant withering and yellowing, slender leaves, and lodging (Figure 1). The results indicated that the plant heights of the five varieties increased as the flooding treatment period increased (Figure 1b,c). The plant elongation rates of TN14, KH139, and TK9 increased significantly (by 82.09%, 76.18%, and 109.51%, respectively) on the sixth day of the submergence treatment (Figure 1d).

To explore the effects of submergence stress on the dry matter losses of the five *japonica* rice varieties, nine-day-old seedlings of TN11, TN14, KH139, TK9, and TNG71 were subjected to full-submergence treatments for 0, two, four, six, and eight days. After the treatment, the fresh weight, dry weight, and ratio of dry weight to fresh weight of the above-ground part were determined for the plants of the five rice varieties (Table 1). The results indicated that the fresh weight of the above-ground part of TN14 following two-day full-submergence treatment was 36% higher than that of the control group. This increase in the fresh weight was significantly higher than those of the other four varieties. No significant differences were observed in the fresh weights following four- and six-day full-submergence treatments. Following 8-day full-submergence treatment, the fresh weight of TNG71 increased by 57.5%, which was significantly higher than the fresh weight increased for the other four varieties (3.7%, 7.9%, 23.7%, and 12.8%) (Figure 2a). Following eight-day full-submergence treatment, the dry weight of KH139 was 22.6% lower than that of the control group. However, the dry weight of KH139 was higher than those of the other four varieties (65.6%, 48.7%, 41.4%, and 68.9% lower than that of the control group) (Figure 2b). Following eight-day full-submergence treatment, the dry mass change of TNG71 was 47% lower than that of the control group. The dry mass change of TNG71 was significantly lower than those of the other four varieties (whose decreases in percentage of dry mass change were 29.6%, 27.4%, 17.7%, and 21.1%) (Figure 2c). Thus, the dry matter loss of TNG71 following eight-day full-submergence treatment was higher than those of the other four varieties.

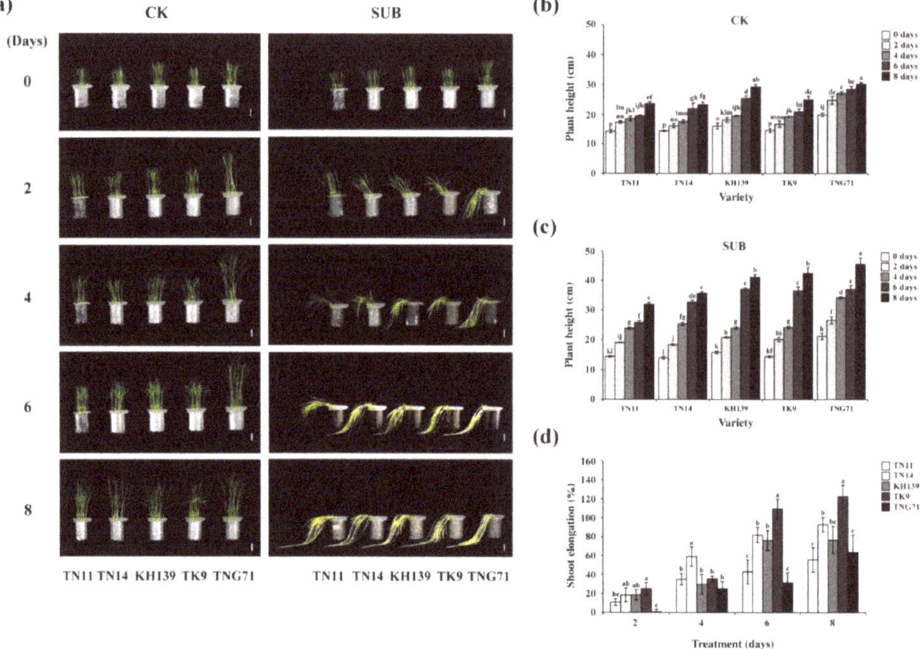

Figure 1. Characterization of the *japonica* rice varieties seedlings grown during submergence stress. (**a**) Photographs of nine-day-old rice seedlings after full-submergence treatment for 0, two, four, six, and eight days; (**b**) plant heights of the nine-day-old seedlings following normal growth for 0, two, four, six, and eight days; (**c**) plant heights of the nine-day-old seedlings subjected to full-submergence treatment for 0, two, four, six, and eight days; and, (**d**) elongation rates of the above-ground parts of the nine-day-old seedlings subjected to full-submergence treatment for two, four, six, and eight days. CK (control check) was the control group; SUB (submergence) was the full-submergence treatment group; and scale bar = 5 cm. Statistical Analysis System (SAS) 9.4 was used to conduct Duncan's analysis. The different alphabets in (**a**,**b**) indicate the significant differencesin the number of treatment days and performance among the five rice varieties ($p < 0.05$). The different alphabets in (**c**) indicate the significant difference between the performance of the rice varieties ($p < 0.05$). Each treatment involved at least 30 plants, and independent experiments were repeated more than three times.

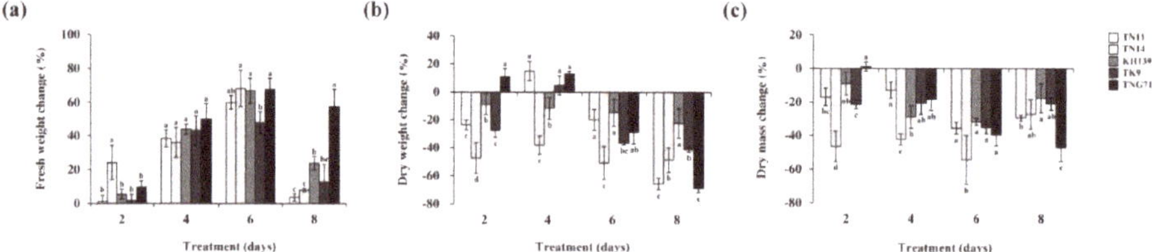

Figure 2. Comparison of fresh, dry weight, and ratio of dry weight to fresh weight in *japonica* rice varieties for different submergence periods. (**a**) The fresh weights, (**b**) dry weights and (**c**) ratio of dry weight to fresh weight of nine-day-old seedlings subjected to full-submergence treatment for two, four, six, and eight days compared with those of the control group. SAS 9.4 was used to conduct Duncan's analysis. The different alphabets inidacte significant difference in the performance of different varieties for the same number of treatment days ($p < 0.05$). Each treatment involved at least 30 plants, and independent experiments were repeated more than three times.

Table 1. The fresh weight, dry weight and ratio of dry mass in five japonica varieties during submergence stress.

Treatment [*]	Days / Variety [**]	Fresh Weight (mg·plant [1])					Dry Weight (mg·plant [1])					Ratio of Dry Mass (%)				
		0	2	4	6	8	0	2	4	6	8	0	2	4	6	8
CK	TN11	42.2 ± 0.1c [***]	62.6 ± 1.3b	65.0 ± 2.0b	77.8 ± 0.2b	97.2 ± 0.5c	5.5 ± 0.2b	8.7 ± 0.2a	8.7 ± 0.4b	11.2 ± 0.2b	13.8 ± 0.2a	13.1 ± 0.1b	13.9 ± 0.1a	13.4 ± 0.2a	14.4 ± 0.3a	14.2 ± 0.1a
	TN14	41.4 ± 1.1c	58.5 ± 0.4c	63.0 ± 2.2b	73.0 ± 1.5c	82.8 ± 2.4d	4.7 ± 0.2c	8.2 ± 0.3b	8.5 ± 0.5b	10.1 ± 0.5c	10.3 ± 0.6c	11.4 ± 0.4c	14.0 ± 0.5a	13.4 ± 0.4a	13.9 ± 0.5ab	12.4 ± 0.4c
	KH139	47.6 ± 0.7a	69.0 ± 1.2a	72.2 ± 1.9a	86.7 ± 1.9a	106.7 ± 2.1a	7.0 ± 0.3a	8.8 ± 0.2a	10.7 ± 0.7a	12.5 ± 0.3a	13.7 ± 0.3a	14.7 ± 0.4a	12.7 ± 0.1b	14.9 ± 0.8b	14.3 ± 0.6a	12.8 ± 0.5c
	TK9	42.4 ± 1.2b	58.0 ± 2.6c	65.4 ± 4.1b	83.5 ± 3.2a	101.5 ± 1.1b	5.6 ± 0.2b	7.8 ± 0.2bc	9.2 ± 0.2b	11.4 ± 0.4b	12.6 ± 0.2b	13.3 ± 0.5b	13.4 ± 0.2ab	14.0 ± 1.1ab	13.7 ± 0.9ab	12.4 ± 0.3c
	TNG71	44.8 ± 0.9c	67.2 ± 1.4a	72.7 ± 3.1a	86.4 ± 2.0a	103.4 ± 1.4b	5.3 ± 0.3b	7.5 ± 0.5c	8.9 ± 0.1b	11.4 ± 0.2b	14.0 ± 0.3a	11.9 ± 0.8c	11.2 ± 0.9c	12.2 ± 0.6c	13.2 ± 0.3b	13.5 ± 0.4b
SUB	TN11	42.9 ± 1.0b	64.0 ± 1.4b	82.4 ± 2.4b	104.7 ± 2.1b	100.3 ± 1.1d	5.6 ± 0.3bc	7.5 ± 0.2b	9.7 ± 0.6a	10.2 ± 0.4ab	10.3 ± 0.4b	13.0 ± 0.3ab	11.7 ± 0.4a	11.7 ± 0.5a	9.8 ± 0.5a	10.3 ± 0.3ab
	TN14	38.4 ± 1.4c	63.4 ± 0.9b	72.1 ± 1.3c	93.9 ± 3.6c	79.8 ± 2.5e	5.2 ± 0.1c	6.5 ± 0.5c	7.3 ± 0.3b	8.4 ± 0.8c	8.7 ± 0.6c	13.4 ± 0.5ab	10.2 ± 0.8b	10.1 ± 0.5bc	9.0 ± 0.9a	10.9 ± 0.7a
	KH139	47.9 ± 1.0a	71.9 ± 1.2a	93.5 ± 3.0a	119.1 ± 1.9a	118.6 ± 1.4b	6.5 ± 0.4a	7.6 ± 0.5ab	9.1 ± 0.2a	10.6 ± 0.7a	11.2 ± 0.6a	13.5 ± 0.4a	10.5 ± 0.9b	9.8 ± 0.1c	8.9 ± 0.5a	9.4 ± 0.5b
	TK9	42.5 ± 1.4b	58.9 ± 1.8c	83.8 ± 3.6b	103.9 ± 3.9b	107.2 ± 5.7c	5.6 ± 0.2b	6.2 ± 0.3c	9.4 ± 0.5a	9.4 ± 0.4bc	10.2 ± 0.3b	13.3 ± 0.7ab	10.5 ± 0.3b	11.3 ± 0.3a	9.0 ± 0.6a	9.6 ± 0.8b
	TNG71	43.8 ± 2.1b	69.8 ± 1.2a	92.8 ± 0.7a	113.9 ± 3.2a	126.1 ± 3.2a	5.4 ± 0.2bc	8.2 ± 0.3a	9.8 ± 0.1a	10.1 ± 0.2ab	10.5 ± 0.2ab	12.4 ± 0.7b	11.8 ± 0.4a	10.5 ± 0.1b	8.9 ± 0.3a	8.3 ± 0.3c

[*] CK (control check) was the control group; SUB (submergence) was the full-submergence treatment group. [**] Variety: TN11; Tainan 11, TN14; Tainan 14, KH139; Kaohsiung 139, TK9; Taiken 9, TNG71; Tainung 71. [***] Statistical Analysis System (SAS) 9.4 was used to conduct Duncan's analysis. The different alphabets indicate the significant differences in the number of treatment days and performance among the five rice varieties ($p < 0.05$). Each treatment involved at least 30 plants, and independent experiments were repeated least three times.

2.2. Changes in the Chlorophyll Contents and Survival Rates in the Five Varieties of Japonica Rice Under Submergence Stress

The chlorophyll a, chlorophyll b, and total chlorophyll contents of the above-ground parts of TN11, TN14, KH139, TK9, and TNG71 were measured following eight-day full-submergence treatment to explore the changes in the chlorophyll contents of the five *japonica* rice varieties under submergence stress. The experimental results indicated that the chlorophyll a, b, and total chlorophyll contents of the five rice varieties significantly decreased under flooding stress and increased following seven-day recovery after the flood. TK9 exhibited the highest rates of decline in the chlorophyll a, b, and total chlorophyll contents under submergence stress (78%, 76%, and 77%, respectively). TN14 exhibited the lowest rates of decline in the chlorophyll a, b, and total chlorophyll contents under flooding stress (70%, 68%, and 70%, respectively). KH139 exhibited the highest rates of increase in the chlorophyll a and total chlorophyll contents (126% and 121%, respectively) after recovery. TNG71 displayed the highest rate of increase in the chlorophyll b content (121%) after recovery. It is worth noting that TN14 also exhibited the lowest rates of increase in the chlorophyll a, b, and total chlorophyll contents (75%, 66%, and 73%, respectively) after recovery (Figure 3).

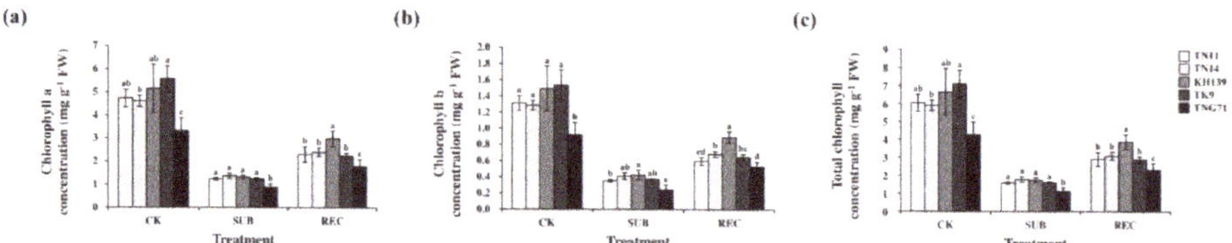

Figure 3. The chlorophyll contents of five *japonica* rice varieties for different submergence periods. The measured chlorophyll a, chlorophyll b, and total chlorophyll contents of the above-ground parts of nine-day-old seedlings subjected to full-submergence treatment for 8 days followed by 7-day recovery. (**a**) Determination of the chlorophyll a content. (**b**) Determination of the chlorophyll b content. (**c**) Determination of the total chlorophyll content. CK (control check) was the control group, SUB (submergence) was the full-submergence treatment group, and REC (recovery) was the recovery group. SAS 9.4 was used to conduct Duncan's analysis. The different alphabets indicatea significant difference in the performances of the various rice varieties for the same number of treatment days ($p < 0.05$). Independent experiments were repeated more than three times.

Studies indicate that submergence-tolerant lowland rice varieties contain *SUB1A-1* genes conferring flooding stress tolerance [5]. The plant survival rates of the rice varieties were examined under flooding stress due to the absence of *SUB1A* genes in the five varieties of *japonica* rice considered in this study. The results indicated that the plant survival rates of TN11, TN14, KH139, TK9, and TNG71 following four-day full-submergence treatment were 94.3%, 87.6%, 98.1%, 81.9%, and 61.0%, respectively. The plant survival rates of the aforementioned varieties following full-submergence treatment for eight-day were 12.4%, 66.7%, 91.4%, 64.8%, and 8.6%, respectively (Figure 4a,b). The results indicate that the plant survival rates of the five varieties following a four-day flood were higher than 50%. Following an eight-day flood, KH139 had the highest plant survival rate, which remained at 90%. The plant survival rates of TN11 and TNG71 were less than 15%.

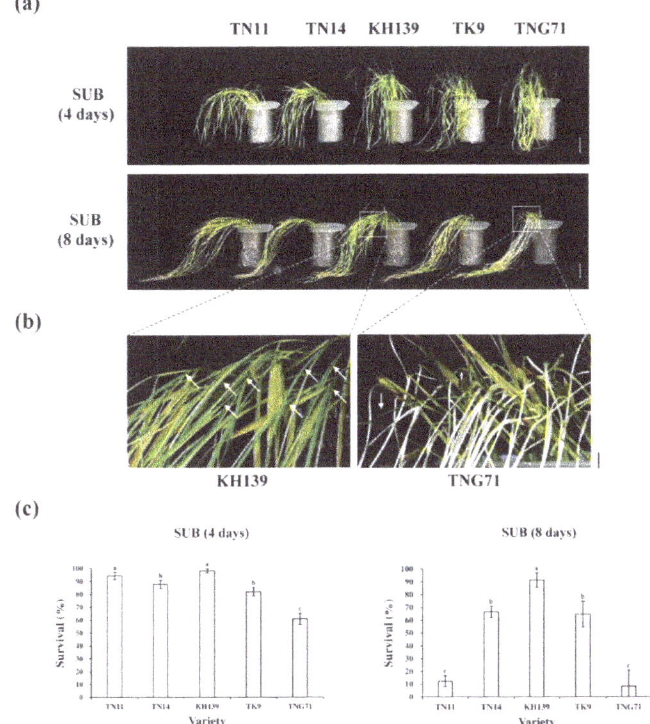

Figure 4. Characterization and survival rates of the five *japonica* rice varieties after recovery following different submergence periods. (**a**) Photographs of 9-day-old seedlings subjected to full-submergence treatment for four and eight days followed by seven-day recovery (bar = 5 cm); (**b**) Photographs of enlarged images of KH139 and TNG71 (bar = 1 cm); and (**c**) survival rates of 9-day-old seedlings subjected to full-submergence treatment for four and eight days followed by seven-day recovery. SUB (submergence) was the full-submergence treatment group. SAS 9.4 was used to conduct Duncan's analysis. The different alphabets indicate significant differences in the performance of the different rice varieties ($p < 0.05$). Each treatment involved at least 35 plants, and independent experiments were repeated more than three times.

2.3. H_2O_2 Accumulation and Antioxidant Enzyme Activities of the Five Varieties of Japonica Rice Under Submergence Stress

To understand the H_2O_2 accumulation in the five varieties under submergence stress, nine-day-old seedlings of TN11, TN14, KH139, TK9, and TNG71 were subjected to eight-day full-submergence treatment, followed by 3, 3′-diaminobenzidine (DAB) staining on the second leaf of each plant. The experiment results indicated that in the control group, the second leaves of TN11, TN14, KH139, TK9, and TNG71 did not appear reddish brown. After full-submergence treatment, all of the second leaves of the aforementioned five varieties appeared reddish brown. A large quantity of H_2O_2 accumulated in the leaf tips of TN11 and leaf margins of TN14. Such accumulation was also found in the leaf margins and bodies of TK9 and TNG71. The H_2O_2 accumulation of KH139 was significantly lower than that of the other four rice varieties (Figure 5a).

Analysis of the antioxidant enzyme activity revealed that the catalase (CAT) activities of the five rice varieties under full-submergence stress decreased to a greater extent than those of the control group, with the CAT activity of KH139 exhibiting the largest decrease. Analysis of the total peroxidase (POD) activity revealed that the POD activities of KH139 and TNG71 in the control group were significantly higher than those of TN11, TN14, and TK9. The POD activities of the five rice varieties under full-submergence stress exhibited a considerably larger increase than those of the control group (Figure 5b).

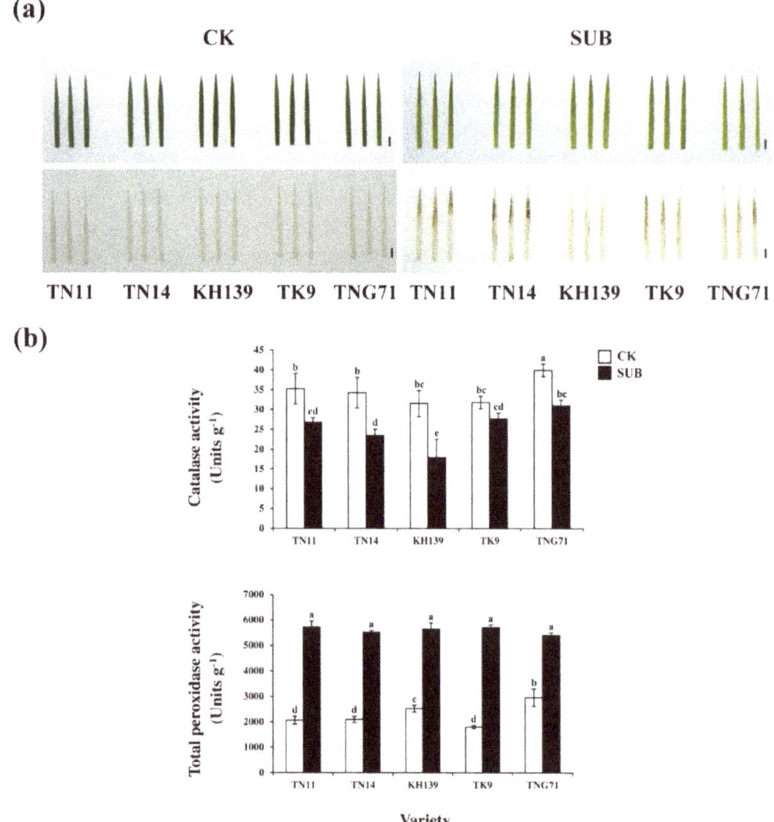

Figure 5. H_2O_2 accumulation and activities of antioxidative enzymes of the *japonica* rice varieties under submergence stress. (**a**) The upper panel of pictures showed the leaves before DAB staining. The down panel of pictures showed the DAB (3,3′-diaminobenzidine) staining of H_2O_2 in nine-day-old seedlings subjected to eight-day full-submergence treatment (bar = 1 cm). (**b**) Determination of the antioxidant enzyme activities of 9-day-old seedlings subjected to eight-day submergence treatment. CK (control check) was the control group, SUB (submergence) was the full-submergence treatment group, SAS 9.4 was used to conduct Duncan's analysis, The different alphabets indicate significant differencesin the performance of the five rice varieties ($p < 0.05$). Independent experiments were repeated more than three times.

2.4. Anaerobic-respiration-related Gene Expressions in the Five Varieties of Japonica Rice Seedlings Under Submergence Stress

Nine-day-old seedlings of TN11, TN14, KH139, TK9, and TNG71 were subjected to full-submergence treatment for two days in order to examine the anaerobic-respiration-related gene expressions of the five varieties of *japonica* rice under submergence stress. The seedlings were then tested for the gene expressions of sucrose synthase 1 (*SUS1*) and alcohol dehydrogenase 1 (*ADH1*). The results indicated that, in the control group, the *SUS1* expressions of TN14 and KH139 were higher than those of TN11, TK9, and TNG71. The *SUS1* expression of KH139 that underwent submergence treatment increased to a considerably greater extent than those of the control group (Figure 6a). In the control group, the *ADH1* expression of TN14 was higher than those of TN11, KH139, TK9, and TNG71. The *ADH1* expressions of TN14 and KH139 that underwent submergence treatment decreased to a considerably greater extent than those of the control group (Figure 6b). The results indicated that the expressions of *SUS1* and *ADH1* were induced in the KH139 plant under submergence stress.

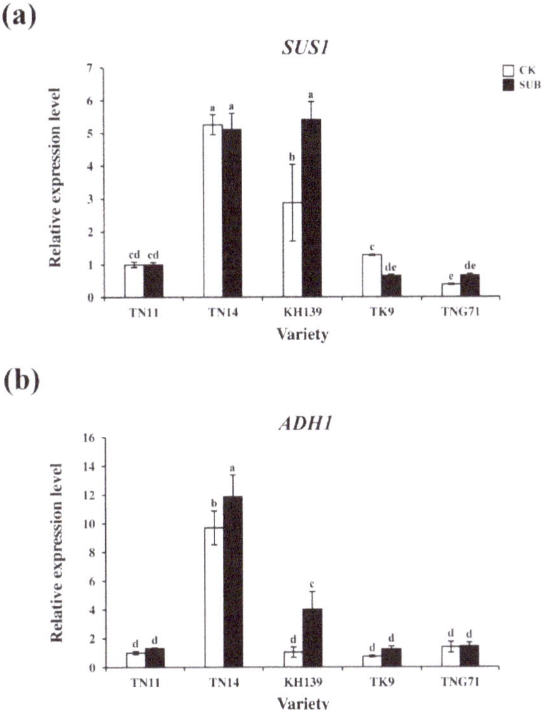

Figure 6. Transcript levels of hypoxia-related genes for the *japonica* rice varieties seedlings under submergence stress. 9-day-old seedlings were subjected to full-submergence treatment for two days, and their above-ground parts were then subjected to qRT-PCR analysis. (**a**) Expressions of *SUS1* of glycolysis-related genes and (**b**) *ADH1* of alcoholic fermentation-related genes. CK (control check) was the control group, and SUB (submergence) was the full-submergence treatment group. SAS 9.4 was used to conduct Duncan's analysis. The different alphabets indicate significant differencesin the performance of the five rice varieties ($p < 0.05$). Independent experiments were repeated four times.

3. Discussion

Rice is the main food crop in Asia. Typhoons and sudden downpours are the major causes of agriculture-related disasters in Taiwan. The aforementioned natural disasters cause field flooding, severe harm to rice cultivation, and considerable losses in rice production. Under flooding stress, the oxygen that is essential for crops cannot be effectively exchanged in water, which causes hypoxic stress. The ability of crops to perform photosynthesis is impeded in a hypoxic environment, which causes problems such as leaf yellowing, impeded root system growth, and decreased mitochondrion metabolism. The cultivation of direct-seeded rice has become common in Asia in recent years. When compared with transplanted cultivation, direct seeding can considerably reduce field water consumption and labor as well as increase productivity. Direct-seeded rice can be divided into dry-, wet-, and water-seeded rice [16]. Water-seeded rice has the best emergence effect among the aforementioned three rice types; however, the cultivation of water-seeded rice involves difficulties that are related to the emergence of its seedlings in flooded fields or the survival of its seedlings after emergence, especially when the seeds fall into deep soil layers. Thus, the rate of successful cultivation of direct-seeded rice can be increased by selecting seedlings of rice varieties with high flood tolerance [17].

Research on varieties of rice that are not flood-resistant indicates that flooding stress causes an increase in the ethylene concentration of the plant tissue due to hypoxia and it enhances the gene expression of *SUB1C*. Moreover, it induces GA downstream gene expression, which causes the elongation of the petioles, stems, and leaves of rice, as well as LOES [4]. Our experimental results indicated that the plant heights of TN11, TN14, KH139, TK9, and TNG71 increased with the submergence treatment period (Figure 1). The plant elongation rates of TN14, KH139, and TK9 significantly increased on the sixth day of the flooding treatment (Figure 1). Thus, the above-ground parts of the five rice varieties were elongated under flooding stress, which indicated that the five rice varieties experienced LOES under submergence stress. Under flooding stress, chloroplasts can cause damage and reduce the photosynthesis efficiency in plants [18]. In addition, the accumulation of ethylene in rice varieties that are not flood-resistant causes the elongation of their above-ground parts, which results in the consumption of carbohydrate in the plant, photo damage to the plant, and degradation of the chlorophyll content in the leaves during flooding stress [19,20]. Stresses lead to the accumulation of excessive activated oxygen groups, such as H_2O_2 and $O_2{}^-$, in the leaves of plants, which causes issues, such as oxidative stress and lipid peroxidation in the plants [21,22]. Our results revealed that the chlorophyll a, chlorophyll b, and total chlorophyll contents of the five varieties of *japonica* rice decreased significantly following eight-day full-submergence treatment. However, the values of the aforementioned parameters increased significantly following seven-day recovery after the full-submergence treatment. The chlorophyll content of KH139 was significantly higher than those of the other four varieties (Figure 3), which indicated that the cells of KH139 were less harmed by flooding stress than those of the other four varieties were and that KH139 returned to normal growth faster than the other four varieties did. Moreover, the results of DAB staining indicated that a low amount of H_2O_2 accumulated in KH139 under submergence stress (Figure 5a). Biotic or abiotic stresses cause the accumulation of activated oxygen groups inside the plant cells. To preserve the oxidation equilibrium inside their cells, plants induce the antioxidant enzyme system to remove the activated oxygen groups for preventing these groups from damaging their cells and for enhancing their stress tolerance [23–25]. Analysis of the antioxidant enzyme activity revealed that the CAT activity of KH139 was significantly lower than those of the other four varieties, probably due to its low H_2O_2 accumulation under submergence stress. Under submergence stress, no significant difference was noted in the total POD activities of KH139 and the other four rice varieties (Figure 5b). The genes *SUS1* and *ADH1* were induced in the plants of KH139 under submergence stress, and KH139 had higher plant survival rates following eight-day submergence treatment (Figures 4 and 6). These results indicated that the survival rate of KH139 might increase under flooding stress due to the enhancement of its hypoxia-related gene expression.

The seedlings of the five varieties exhibited the physiognomic traits of withering, yellowing, slender leaves, and lodging under flooding stress. However, four days after the flooding treatment, the plant survival rates of the five rice varieties were higher than 50%. Excluding TNG71, the plant survival rates of the other four rice varieties were 80% or higher, which indicated that, although *japonica* rice varieties do not contain *SUB1A*, their plant survival rates remain high if the flooding period is short. Moreover, following eight-day submergence treatment, the plant survival rates of three varieties, namely TN14, KH139, and TK9, were 60% or higher (Figure 4). Previous scholars have demonstrated that the *SUB1A* gene is a crucial response gene for rice plants to become flood-resistant through ethylene signaling. The *Indica* rice cultivars, such as FR13A, which contains *SUB1A* revealed extremely submergence tolerant; 100% of 10-day-old seedlings survived 7 days of complete submergence [26]. In addition to *SUB1A*, this study indicated that other crucial reaction mechanisms regulate plant flood tolerance in rice varieties that do not comprise this gene.

4. Conclusions

The changes in the plant heights of five *japonica* rice varieties after full-submergence treatment indicated that the chlorophyll content of the plants decreased significantly and the plants experienced LOES under flooding stress. However, the accumulation of H_2O_2 in the leaves of the five varieties and their peroxidase activities were different. Moreover, molecular-level analysis indicated that the expressions of *SUS1* and *ADH1*, which were induced by hypoxia signaling, were different, which suggested that different varieties of *japonica* rice have different flood tolerances. In this study, we recommend KH139 as a good potential breeding material for flood tolerance when compared with the other four *japonica* varieties of rice seedlings. These results can provide crucial information for future research on *japonica* rice under flooding stress and on direct-seeded rice.

5. Materials and Methods

5.1. Plant Materials and Growth Conditions

Five good-quality *japonica* rice varieties that are commonly planted in Taiwan, namely Tainan 11 (TN11), Kaohsiung 139 (KH139), Taiken 9 (TK9), Tainung 71 (TNG71), and Tainan 14 (TN14) were selected for experiments. These rice plants were irrigated lowland rice. Their seeds were cleaned and disinfected with 3% sodium hypochlorite for 30 min, followed by cleaning with sterile deionized water three times for completely removing the sodium hypochlorite residue and introducing sterile deionized water. The seeds were then placed in a growth box at 28 °C with lighting cycles of 16 h of light and eight hours of darkness (the brightness was 36 μmol m^{-2}s^{-1}). The seeds that germinated after 2-day wet cultivation were placed on iron grids in 500-mL beakers wrapped with aluminum foil and then cultivated with the Kimura B solution [27]. A total of 35 to 40 seedlings were cultivated in each beaker. The hydroponic growing medium was renewed every two days until the seedlings were nine days old. Subsequently, experiments and treatments were performed. Flooding treatment was conducted by placing nine-day-old seedlings in a water tank (40-cm long, 40-cm wide, and 60-cm high) with a water level of 55 cm (with this water level, plants do not protrude from the water following eight-day flooding treatment) for 0, two, four, six, and eight days. In the growth box, the temperature was 28 °C, the lighting alternated between 16 h of light and eight hours of darkness, and the brightness was 236 μmol m^{-2}s^{-1}. The physiognomic traits of the plants were investigated, and the plant materials were subjected to further analyses.

5.2. Seedlings Plant High, Fresh Weight, Dry Weight Chlorophyll Content and Survival Rate Determination

The nine-day-old seedlings were subjected to full-submergence treatments for 0, two, four, six, and eight days. The above-ground heights of their plants were measured after the aforementioned full-submergence treatment durations. For plant height measurement, the above-ground height of the plant was defined as the distance between the base of a straightened seedling and the tip of its longest leaf. In each independent experiment, the mean heights of at least 30 plants were measured. The experiments were repeated at least three times. The plant elongation rates were calculated using the technique of Zhu et al. [28]. The plant elongation rate denotes the difference in the length of the plant's above-ground part before and after treatment. The equation for calculating the plant elongation rate is as follows (Equation (1)):

$$\text{Plant elongation rate (\%)} = [(PH_2 - PH_1)/PH_1] \times 100\% \tag{1}$$

where PH_1 and PH_2 indicate the above-ground heights before and after treatment, respectively.

The dry weight was the measured weight of the above-ground part of each plant following full-submergence treatment for 0, two, four, six, and eight days. Tissues (whose weight was equal to the fresh weight) were dried in an oven at 60 °C for two days. The weight of each plant, which represented the dry weight, was then measured. In each independent experiment, the mean dry weights of at least 30 plants were measured. The experiments were repeated more than three times.

To measure the chlorophyll content, nine-day-old seedlings were subjected to full-submergence treatment for eight days and a recovery period of seven days. The above-ground parts (30 mg) of the rice seedlings were ground with liquid nitrogen. A total of 2 mL of sodium phosphate buffer (50 mM, pH 6.8) was added to the ground seedlings. A total of 960 μL of 99% ethanol was blended into 40 μL of the extract. The mixture was evenly blended in a 1.5-mL microcentrifuge tube (Eppendorf tube) and placed at 4 °C in darkness for 30 min, followed by 1000× g centrifugation at 4 °C for 15 min. The absorbance of the supernatant at OD.665 and 649 was determined with a spectrophotometer (Metertec SP8001). The blank was 99% ethanol. More than three independent trials were conducted. The equations for calculating the chlorophyll content are as follows (Equations (2)–(4)):

$$\text{Chlorophyll a} = (13.7 \times A_{665}) - (5.76 \times A_{649}) \ [\mu gChL \ (40 \ \mu L)^{-1}] \tag{2}$$

$$\text{Chlorophyll b} = (25.8 \times A_{649}) - (7.6 \times A_{665}) \ [\mu gChL \ (40 \ \mu L)^{-1}] \tag{3}$$

$$\text{Total chlorophyll} = (6.1 \times A_{665}) + (20.04 \times A_{649}) \ [\mu gChL \ (40 \ \mu L)^{-1}] \tag{4}$$

5.3. Histochemical Staining and Antioxidative Enzyme Activity Assay

The accumulation of H_2O_2 in cells was visualized by 3, 3′-diaminobenzidine staining as previously described [29]. The experiments were repeated three times. Protein quantitation of the samples were performed after their extraction by using the Bradford protein assay in order to determine the antioxidant enzyme activity [30]. The above-ground parts of rice were ground into powder with liquid nitrogen and analyzed according to the method modify from Kar and Mishra to determine the CAT activity [31]. A total of 4 mL of 50 mM sodium phosphate buffer (pH 6.8) was added in the pre-chilled mortar for performing homogeneous grinding. The mixture was then subjected to centrifugation at 12,000× g and 4 °C for 20 min. The obtained supernatant was the enzyme extract. Protein quantitation was performed on the enzyme extract by using the Bradford assay. Enzyme extract with a volume corresponding to 1 mg of protein and 100 mM of sodium phosphate buffer (pH 7.0) (the total volume of the extract and buffer was 2.9 mL) were evenly blended with 100 μL of 1 M H_2O_2. A spectrophotometer (Metertec SP8001) was used at a wavelength of 240 nm to continuously detect the change in absorbance during a five-minute period. The detection interval was 10 s, and 99% ethanol was used as the blank. The unit of enzyme activity was μmol of H_2O_2 consumed per minute. The above-ground parts of rice were ground into powder with liquid nitrogen in a pre-chilled mortar and analyzed according to the method of MacAdam et al. to determine the total POD activity [32]. Lin and Kao [33] proposed that water-soluble and ionic-bonded peroxidase can be extracted by adding potassium chloride (KCl) in buffer solution. They defined the sum of the aforementioned two peroxidase types as the total activity. In the experiment, 4 mL of 50 mM potassium phosphate buffer (pH 5.8), which included 0.8 M KCl, was added in a pre-chilled mortar for obtaining homogeneous powder. The mixture was then subjected to centrifugation at 12,000× g and 4 °C for 20 min. The obtained supernatant was the enzyme extract. Protein quantitation was performed on the extract by using the Bradford assay. Enzyme extract with a volume corresponding to 1 mg of protein and 50 mM of potassium phosphate buffer (pH 5.8) were homogeneously blended with 1 mL of 21.6 mM guaiacol and 0.9 mL of 39 mM H_2O_2. A spectrophotometer (MetertecSP8001) was used at a wavelength of 470 nm to continuously detect the changes in the absorbance during a 30-s period. The detection interval was 10 s and 99% ethanol was used as the blank. The unit of enzyme activity was μmol of tetra guaiacol produced per minute.

5.4. Quantitative RT-PCR Analyses

The rice samples were placed in a mortar and ground into powder using liquid nitrogen. A total of 2 mL of TRIZOL reagent (Roche Applied Science, Penzberg, Upper Bavaria, Germany) was added to the mortar for extracting the total RNA. The total RNA extracted was then treated using TURBO DNase (Ambion, Austin, TX, USA). A Nano Drop Lite ultra-micro spectrophotometer (Thermo Scientific, Waltham, MA, USA) was used in order to measure the optical density of 2 μL of RNA and examine its concentration and quality. An MMLV First-Strand Synthesis Kit (Gene Direx, Grand Island, NY, USA) was used to synthesize the first strand of cDNA [reverse transcription (RT)]. qRT-PCR was performed, as previously described [34] using a Bio-Rad CFX instrument (CFX Connect™, Bio-Rad, Hercules, CA, USA) with Power SYBR Green PCR Master Mix (Gene-Mark, Taipei, Taiwan), according to the manufacturer's recommendations. The ubiquitin gene was used as an internal control for normalization. The relative expression levels were analyzed using Bio-Rad CFX Manager (version 3.1). The experiments were repeated three times independently with duplicate samples. Table 2 presents the primer sequences for qRT-PCR.

Table 2. Primers used for quantitative RT-PCR experiments.

Gene Name	Primer Sequence
OsUbiquitin—forward	5′-aaccagctgaggcccaaga-3′
OsUbiquitin—reverse	5′-acgattgatttaaccagtccatga-3′
OsSUS1—forward	5′-catctcaggctgagactctga-3′
OsSUS1—reverse	5′-caaattcaatcgaccttactt-3′
OsADH1—forward	5′-gcaaatttctggctttgtcaatcagta-3′
OsADH1—reverse	5′-cgccaaaagatcactgattcttaacaa-3′

5.5. Statistical Analysis

In this experiment, Statistical Analysis System (SAS) software version 9.4 was used for statistical analysis. We used analysis of variance (ANOVA) for pre-comparison, and then used Duncan's Multiple Range Test for multiple comparisons. The different alphabets indicate significant differences in the performance of the five rice varieties ($p < 0.05$). All of the tests were conducted in more than three independent tests.

Author Contributions: Y.-S.L. conducted experiments and analyzed the data. S.-L.O. helps the statistical analysis. C.-Y.Y. conceived, designed research, and wrote the manuscript. All authors read and approved the final manuscript.

Funding: This work was partially supported by the National Science Council, Taiwan, R.O.C. (NSC109-2313-B-005-023) to C.-Y.Y.

Acknowledgments: Not applicable.

References

1. Food and Agriculture Organization of the UN; International Fund for Agricultural Development; UNICEF; World Food Programme; WHO. *The State of Food Security and Nutrition in the World*; Food and Agriculture Organization of the UN: Rome, Italy, 2018.
2. Boyland, M. In pursuit of effective flood risk management in the Mekong region. *Focus* **2015**, 2.
3. Hirose, T.; Scofield, G.N.; Terao, T. An expression analysis profile for the entire sucrose synthase gene family in rice. *Plant Sci.* **2008**, *174*, 534–543. [CrossRef]
4. Bailey-Serres, J.; Voesenek, L.A.C.J. Flooding stress: Acclimations and genetic diversity. *Annu. Rev. Plant Biol.* **2008**, *59*, 313–339. [CrossRef] [PubMed]
5. Fukao, T.; Xu, K.; Ronald, P.C.; Bailey-Serres, J. A variable cluster of ethylene response factor-like genes regulates metabolic and developmental acclimation responses to submergence in rice. *Plant Cell* **2006**, *18*, 2021–2034. [CrossRef] [PubMed]
6. Subbaiah, C.C.; Palaniappan, A.; Duncan, K.; Rhoads, D.M.; Huber, S.C.; Sachs, M.M. Mitochondrial localization and putative signaling function of sucrose synthase in maize. *J. Biol. Chem.* **2006**, *281*, 15625–15635. [CrossRef]
7. Loreti, E.; Valeri, M.C.; Novi, G.; Perata, P. Gene regulation and survival under hypoxia requires starch availability and metabolism. *Plant Physiol.* **2018**, *176*, 1286–1298. [CrossRef]

8. Shabala, S.; Munns, R. Salinity Stress: Physiological Constraints and Adaptive Mechanisms. In *Plant Stress Physiology*; Shabala, S., Ed.; CAB International: Oxford, UK; Oxford, MS, USA, 2012; pp. 59–93.

9. Loreti, E.; van Veen, H.; Perata, P. Plant responses to flooding stress. *Curr. Opin. Plant Biol.* **2016**, *33*, 64–71. [CrossRef]

10. Garnczarska, M.; Bednarski, W. Effect of a short-term hypoxic treatment followed by re-aeration on free radicals level and antioxidative enzymes in lupine roots. *Plant Physiol. Biochem.* **2004**, *42*, 233–240. [CrossRef]

11. Geigenberger, P. Response of plant metabolism to too little oxygen. *Curr. Opin. Plant Biol.* **2003**, *6*, 247–256. [CrossRef]

12. Vergara, R.; Parada, F.; Rubio, S.; Pérez, F.J. Hypoxia induces H_2O_2 production and activates antioxidant defence system in grapevine buds through mediation of H_2O_2 and ethylene. *J. Exp. Botany* **2012**, *63*, 4123–4131. [CrossRef]

13. Chapman, J.M.; Muhlemann, J.K.; Gayomba, S.R.; Muday, G.K. RBOH-dependent ROS synthesis and ROS scavenging by plant specialized metabolites to modulate plant development and stress responses. *Chem. Res. Toxicol.* **2019**, *32*, 370–396. [CrossRef] [PubMed]

14. Fukao, T.; Bailey-Serres, J. Submergence tolerance conferred by Sub1A is mediated by SLR1 and SLRL1 restriction of gibberellin responses in rice. *Proc. Natl. Acad. Sci. USA* **2008**, *105*, 16814–16819. [CrossRef] [PubMed]

15. Hattori, Y.; Nagai, K.; Ashikari, M. Rice growth adapting to deep water. *Curr. Opin. Plant Biol.* **2011**, *14*, 100–105. [CrossRef] [PubMed]

16. Farooq, M.; Siddique, K.H.M.; Rehman, H.; Aziz, T.; Lee, D.J.; Wahid, A. Rice direct seeding: Experiences, challenges and opportunities. *Soil Tillage Res.* **2011**, *111*, 87–98. [CrossRef]

17. Kashiwagi, J.; Hamada, K.; Jitsuyama, Y. Rice (*Oryza sativa* L.) germplasm with better seedling emergence under direct sowing in flooded paddy field. *Plant Genet. Resour. Charact. Util.* **2018**, *16*, 352–358. [CrossRef]

18. Panda, D.; Sarkar, R.K. Leaf photosynthetic activity and antioxidant defense associated with Sub1 QTL in rice subjected to submergence and subsequent re-aeration. *Rice Sci.* **2012**, *19*, 108–116. [CrossRef]

19. Matile, P.; Hortensteiner, S.; Thomas, H.; Krautler, B. Chlorophyll breakdown in senescent leaves. *Plant Physiol.* **1996**, *112*, 1403–1409. [CrossRef]

20. Sone, C.; Sakagami, J.I. Physiological mechanism of chlorophyll breakdown for leaves under complete submergence in rice. *Crop Sci.* **2017**, *57*, 2729–2738. [CrossRef]

21. Blokhina, O.; Virolainen, E.; Fagerstedt, K.V. Antioxidants, oxidative damage and oxygen deprivation stress: A review. *Ann. Bot.* **2003**, *91*, 179–194. [CrossRef]

22. Anjum, S.A.; Xie, X.; Wang, L.; Saleem, M.; Man, C.; Lei, W. Morphological, physiological and biochemical responses of plants to drought stress. *Afr. J. Agric. Res.* **2011**, *6*, 2026–2032.

23. Gill, S.S.; Tuteja, N. Reactive oxygen species and antioxidant machinery in abiotic stress tolerance in crop plants. *Plant Physiol. Biochem.* **2010**, *48*, 909–930. [CrossRef] [PubMed]

24. Ali, S.; Huang, Z.; Li, H.X.; Bashir, M.H.; Ren, S.X. Antioxidant enzyme influences germination, stress tolerance, and virulence of *Isaria Fumosorosea*. *J. Basic Microbiol.* **2013**, *53*, 489–497. [CrossRef] [PubMed]

25. Liu, J.; Sun, X.; Xu, F.; Zhang, Y.; Zhang, Q.; Miao, R.; Zhang, J.; Liang, J.; Xu, W. Suppression of *OsMDHAR4* enhances heat tolerance by mediating H_2O_2-induced stomatal closure in rice plants. *Rice* **2018**, *11*, 38. [CrossRef] [PubMed]

26. Bailey-Serres, J.; Fukao, T.; Ronald, P.; Ismail, A.; Heuer, S.; Mackill, D. Submergence tolerant rice: SUB1's journey from landrace to modern cultivar. *Rice* **2010**, *3*, 138–147. [CrossRef]

27. Yoshida, S.; Forno, D.A.; Cock, J.H.; Gomez, K.A. Routine Procedure for Growing Rice Plants in Culture solution. In *Laboratory Manual for Physiological Studies of Rice*, 3rd ed.; The International Rice Research Institute: Los Baños, Philippines, 1976; pp. 61–66.

28. Zhu, G.L.; Chen, Y.T.; Ella, E.S.; Ismail, A.M. Mechanisms associated with tiller suppression under stagnant flooding in rice. *J. Agron. Crop. Sci.* **2019**, *205*, 235–247. [CrossRef]

29. Huang, Y.; Yeh, T.; Yang, C. Ethylene signaling involves in seeds germination upon submergence and antioxidant response elicited confers submergence tolerance to rice seedlings. *Rice* **2019**, *12*, 23. [CrossRef]

30. Bradford, M.M. A rapid and sensitive method for the quantitation of microgram quantities of protein utilizing the principle of protein-dye binding. *Anal. Biochem.* **1976**, *72*, 248–254. [CrossRef]

31. Kar, M.; Mishra, D. Catalase, peroxidase and polyphenol oxidase activities during rice leaf senescence. *Plant Physiol.* **1976**, *57*, 315–319. [CrossRef]

32. MacAdam, J.W.; Nelson, C.J.; Sharp, R.E. Peroxidase activity in the leaf elongation zone of tall fescue1. *Plant Physiol.* **1992**, *99*, 872–878. [CrossRef]

33. Lin, C.C.; Kao, C.H. NaCl induced changes in ionically bound peroxidase activity in roots of rice seedlings. *Plant Soil* **1999**, *216*, 147–153. [CrossRef]

34. Yang, S.Y.; Wu, Y.S.; Chen, C.T.; Lai, M.H.; Yen, H.M.; Yang, C.Y. Physiological and molecular responses of seedlings of an upland rice ('Tung Lu 3') to total submergence compared to those of a submergence-tolerant lowland rice ('FR13A'). *Rice* **2017**, *10*, 42. [CrossRef] [PubMed]

Article

Some Accessions of Amazonian Wild Rice (*Oryza glumaepatula*) Constitutively Form a Barrier to Radial Oxygen Loss along Adventitious Roots under Aerated Conditions

Masato Ejiri, Yuto Sawazaki and Katsuhiro Shiono *

Graduate School of Bioscience and Biotechnology, Fukui Prefectural University, 4-1-1 Matsuoka-Kenjojima, Eiheiji, Fukui 910-1195, Japan; s2093001@g.fpu.ac.jp (M.E.); s2073004@g.fpu.ac.jp (Y.S.)
* Correspondence: shionok@fpu.ac.jp; Tel.: +81-776-61-6000

Received: 29 May 2020; Accepted: 10 July 2020; Published: 13 July 2020

Abstract: A barrier to radial oxygen loss (ROL), which reduces the loss of oxygen transported via the aerenchyma to the root tips, enables the roots of wetland plants to grow into anoxic/hypoxic waterlogged soil. However, little is known about its genetic regulation. Quantitative trait loci (QTLs) mapping can help to understand the factors that regulate barrier formation. Rice (*Oryza sativa*) *inducibly* forms an ROL barrier under stagnant conditions, while a few wetland plants *constitutively* form one under aerated conditions. Here, we evaluated the formation of a constitutive ROL barrier in a total of four accessions from two wild rice species. Three of the accessions were wetland accessions of *O. glumaepatula*, and the fourth was a non-wetland species of *O. rufipogon*. These species have an AA type genome, which allows them to be crossed with cultivated rice. The three *O. glumaepatula* accessions (W2165, W2149, and W1183) formed an ROL barrier under aerated conditions. The *O. rufipogon* accession (W1962) did not form a constitutive ROL barrier, but it formed an inducible ROL barrier under stagnant conditions. The three *O. glumaepatula* accessions should be useful for QTL mapping to understand how a constitutive ROL barrier forms. The constitutive barrier of W2165 was closely associated with suberization and resistance to penetration by an apoplastic tracer (periodic acid) at the exodermis but did not include lignin at the sclerenchyma.

Keywords: apoplastic barrier; barrier to radial oxygen loss (ROL); lignin; *Oryza glumaepatula*; *O. rufipogon*; rice (*O. sativa*); suberin; wild rice

1. Introduction

Under waterlogged conditions, plants can suffer from hypoxia or anoxia because the ability of oxygen to diffuse through the water to the soil is extremely low [1]. Other problems associated with waterlogging are the accumulation of phytotoxic compounds in the soil and a decline in the availability of some nutrients [2]. The roots of wetland plants contain a large volume of aerenchyma, which provides a low-resistance pathway for the diffusion of oxygen from the shoot to the root [3–5]. Some wetland species also form a barrier to radial oxygen loss (ROL) [2,6–8]. The ROL barrier forms at the basal part of roots and reduces the loss of oxygen transported via the aerenchyma to the root tips. In roots with an ROL barrier, oxygen in the root tips and short lateral roots can be maintained at a higher level to allow root elongation into hypoxic/anoxic soil [5,6]. An ROL barrier in roots is a key feature contributing to long-distance oxygen transport and waterlogging tolerance in wetland species.

Lignin and suberin act as an apoplastic diffusion barrier in the root. Lignin is a complex of polyphenolic polymers [9], which are the main components of Casparian strips [10]. Suberin is a hydrophobic macromolecule built from long-chain fatty acids, glycerol, and aromatic polymers [11], and is the main component of suberin lamellae [12]. An ROL barrier is thought to be formed by deposits of suberin in the hypodermis/exodermis but not by deposits of lignin [6,13]. Suberin deposits in the apoplast (the outer cellular space) prevent movement of ions and mycorrhizal fungi through the apoplast and thus act as an apoplastic barrier [14–16]. At the basal part, the ROL barrier, in addition to restricting oxygen loss, also could reduce the entry of soil phytotoxins (e.g., Fe^{2+}) into roots in waterlogged soils [6,17]. Another possible benefit of the ROL barrier is that it would allow oxygen to reach the root tip. Because the root tip does not have an ROL barrier, the ROL at the root tip could detoxify toxic reduced substances in the waterlogged soil [2].

Some wetland plants, including cultivated rice (*Oryza sativa*), form an inducible ROL barrier under waterlogged soil or stagnant conditions, while they remain leaky to oxygen under well-drained or aerated conditions [2,4,18]. Some soil phytotoxins produced by anaerobic bacteria in waterlogged soils (e.g., Fe^{2+}, NH_4, sulfide, and/or carboxylic acids) seem to act as environmental triggers to induce ROL barrier formation [5,19–21], although the signaling pathway to induce the ROL barrier is not known [6]. On the other hand, a few wetland plants (e.g., wild *Echinochloa* species) form a constitutive ROL barrier, i.e., even in the absence of waterlogging [13,22,23].

The molecular mechanisms that control whether a plant forms an ROL barrier inducibly or constitutively are not understood. One approach to this problem is to map the quantitative trait loci (QTLs) that control constitutive ROL barrier formation. To do this in rice, it is necessary to have a rice accession that constitutively forms an ROL barrier. However, none of 14 rice cultivars examined for a constitutive ROL barrier were found to have one [24–26]. So far, 24 wild rice species have been identified in tropical regions [27]. Seven of them, including *O. glumaepatula* and *O. rufipogon*, have an AA-genome [27], the same as the genome of cultivated rice (*O. sativa*), which would allow cross-breeding. We thus examined three accessions of *O. glumaepatula* and one accession of *O. rufipogon* to see if they constitutively form an ROL barrier and aerenchyma. In one of the accessions that formed a constitutive ROL barrier, we evaluated its chemical composition and ability to act as an apoplastic barrier to better characterize its properties.

2. Results

2.1. Discovery of Wild Rice Accessions That Constitutively Form a Radial Oxygen Loss (ROL) Barrier

To determine whether the four wild rice accessions formed an ROL barrier under aerated conditions, oxygen leakage was measured along adventitious roots (about 120 mm long) in three wetland accessions of *O. glumaepatula* (W2165, W2149, and W1183) and one non-wetland accession of *O. rufipogon* (W1962) (Figure 1). Under aerated conditions, all three *O. glumaepatula* accessions (W2165, W2149 and W1183) had an oxygen loss of about 30 ng cm^{-2} min^{-1} in the apical 5 mm and much less oxygen loss ($\leq$20 ng cm^{-2} min^{-1}) in the basal region (Figure 1a–c, open circles), which are the characteristics of an ROL barrier. In the accessions of *O. rufipogon* (W1962) under aerated conditions, the oxygen flux from the basal part remained high ($\geq$50 ng cm^{-2} min^{-1} at 100 mm from the apex) (Figure 1d, open circles), indicating that they did not form a constitutive ROL barrier.

2.2. Assessment of Inducible ROL Barrier Formation

When W2165, W2149, and W1183 of *O. glumaepatula* were grown under stagnant conditions, the barrier became stronger from the basal part to the middle part of the roots (Figure 1a–c, closed circles). The rates of ROL at the basal parts were close to zero, and oxygen leakage at 40 mm from the root apex was even lower (40 mm: under 20 ng cm^{-2} min^{-1}). When W1962 of *O. rufipogon* was grown under stagnant conditions, the oxygen flux at the basal regions (40–100 mm from the root apex) was low (1–12 ng cm^{-2} min^{-1}) (Figure 1d, closed circles), indicating it had a tight ROL barrier. When cv. Nipponbare was grown under stagnant conditions, the oxygen flux at the basal to middle regions (20–100 mm from the root apex) was low (1–19 ng cm^{-2} min^{-1}) (Figure 1e, closed circles), indicating it had a tight barrier to ROL. In W2165 under aerated conditions, the rates of ROL at the basal part were comparable to those of cv. Nipponbare roots under stagnant conditions, although the region with a strong barrier in W2165 was smaller than the region with an inducible ROL barrier in cv. Nipponbare under stagnant conditions.

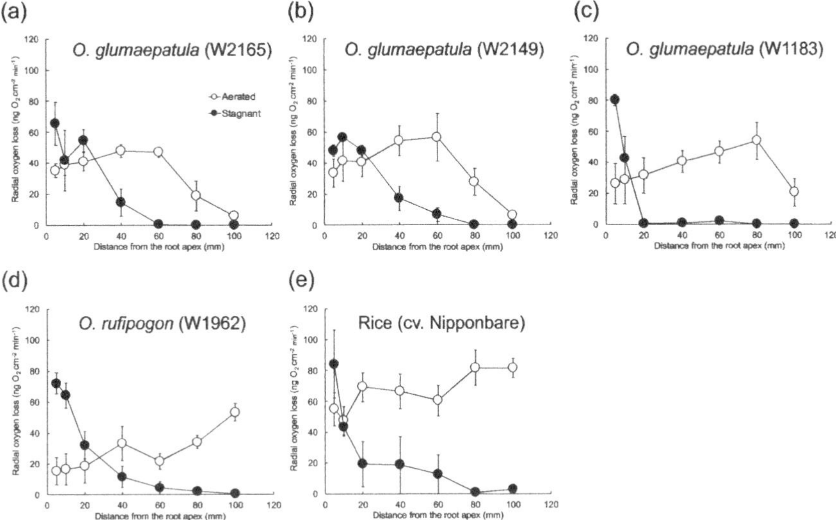

Figure 1. Rates of radial oxygen loss (ROL) along adventitious roots in accessions of *O. glumaepatula*, *O. rufipogon*, and rice (cv. Nipponbare) under aerated or stagnant conditions. W2165 (**a**), W2149 (**b**), and W1183 (**c**) of *O. glumaepatula* came from wetlands. W1962 (**d**) of *O. rufipogon* came from habitats other than wetlands. For a control that forms an inducible ROL barrier, cv. Nipponbare (**e**) was used. ROL along adventitious roots (115–120 mm length) was measured by Pt cylindrical electrode. Plants were grown in aerated nutrient solution for eight days and then transferred to deoxygenated stagnant 0.1% agar solution or continued aerated solution for 13–15 days. Mean ± standard error (SE). *n* = 3 or 4.

2.3. Assessment of Aerenchyma Formation

Under aerated conditions, all wild rice accessions of *O. glumaepatula* and *O. rufipogon* had well-developed aerenchyma [21 to 48% of the cortex cross-section at the basal part (100 mm from the root apex) (Table 1)]. These values were as high as the percentage in cv. Nipponbare, which indicates that these accessions form aerenchyma constitutively. All wild rice accessions formed aerenchyma constitutively. However, stagnant conditions induced additional increases in aerenchyma of almost 50% in three of the four wild accessions (W2149, W1183, and W1962) (Table 1). All wild rice accessions formed aerenchyma constitutively and induced aerenchyma by stagnant treatment, but not superior to cultivated rice cv. Nipponbare.

2.4. Suberin and Lignin Accumulation in W2165

The basal parts (100 mm from the root apex) of the adventitious roots of *O. glumaepatula* (W2165) were surrounded by a well-suberized exodermis, as shown by yellowish-green fluorescence of Fluorol Yellow 088 (Figure 2a). The yellow-green fluorescence at the exodermis was patchy in the middle part of the roots (40 mm from the root apex) (Figure 2b), but it was not observed at all in the root tip (5 mm from the root apex) (Figure 2c). In W2165 under stagnant conditions, both the basal and middle parts of the roots were surrounded by well-suberized exodermis (Figure 2d–e). The exodermal fluorescence intensity at 100 mm from the root apex of W2165 under aerated conditions (third box from the left in Figure 2m) was as high as the intensities in the middle and basal parts of W2165 and cv. Nipponbare (where oxygen leakage was reduced) under stagnant conditions (Figure 2m). The exodermal fluorescence intensity at 100 mm from the root apex was significantly higher in W2165 under aerated conditions than in cv. Nipponbare under aerated conditions ($p < 0.05$, *t*-test; Figure 2m).

Table 1. Aerenchyma formation (% aerenchyma/cortex) in accessions of *O. glumaepatula*, *O. rufipogon*, and rice (cv. Nipponbare) under aerated or stagnant conditions.

Species	Accession	Aerenchyma Formation (% aerenchyma/cotex)		*t*-Test
		Aerated	Stagnant	
O. glumaepatula	W2165	48 ± 3 [c]	46 ± 5 [A]	n.s.
	W2149	32 ± 5 [a,b]	52 ± 4 [A]	*
	W1183	21 ± 1 [a]	49 ± 6 [A]	*
O. rufipogon	W1962	38 ± 2 [b,c]	52 ± 2 [A]	*
O. sativa	Nipponbare	48 ± 2 [c]	64 ± 4 [A]	*

Aerenchyma formation (% aerenchyma/cortex) at 100 mm from apex of 115–120 mm-length roots. Asterisks denote significant differences between means of aerated and stagnant conditions (two-sample *t*-test, *: $p < 0.05$). n.s.: not significant. Different lower-case and upper-case letters denote significant differences among accessions under aerated and stagnant conditions, respectively ($p < 0.05$, one-way analysis of variance (ANOVA) and Tukey's honest significant difference (HSD) for multiple comparisons). Plants were grown in aerated nutrient solution for eight days and then transferred to deoxygenated stagnant 0.1% agar solution or continued aerated solution for 13–15 days. Mean ± SE. $n = 3$ or 4.

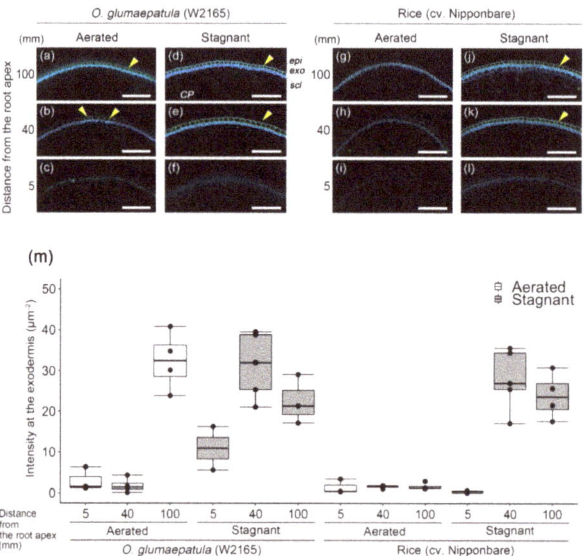

Figure 2. Suberization in the outer part of roots in accession W2165 of *O. glumaepatula* and rice (cv. Nipponbare) under aerated or stagnant conditions. Suberin lamellae were observed in the basal parts (95–105 mm from the root apex; (**a,d,g,j**), middle parts (35–45 mm from the root apex; (**b,e,h,k**) and root tips (2.5–7.5 mm from the root apex; (**c,f,i,l**) of adventitious roots of 115–120 mm length. Suberin lamellae are indicated as yellow-green fluorescence with Fluorol Yellow 088 (yellow arrowhead, mean of intensity value >20.0 μm^{-2}). Blue fluorescence indicates autofluorescence. Plants were grown in aerated nutrient solution for eight days and then transferred to deoxygenated stagnant 0.1% agar solution or continued aerated solution for 13–15 days. *CP*, cortical parenchyma; *epi*, epidermis; *exo*, exodermis; *scl*, sclerenchyma. Scale bars: 100 μm. (**m**) Fluorescence intensity of Fluorol Yellow 088 at the exodermis under aerated (white box) or stagnant (grey box) conditions. Black dots indicate each raw value. $n = 2$–5.

In W2165 under both aerated or stagnant conditions, Basic Fuchsin staining showed that lignin was more developed in the basal regions than in the more apical regions (Figure 3a–f,m) and that it was located mainly in the sclerenchyma (Figure 3a,b,d,e) and partially in the exodermis (Figure 3d). At the basal part of roots (100 mm from the root apex), the intensities of Basic Fuchsin fluorescence at the sclerenchyma under stagnant conditions were relatively higher than those under aerated conditions (Figure 3m). In cv. Nipponbare, as in W2165, lignin was well developed at the basal part of roots under stagnant conditions (Figure 3j,k). The fluorescence intensity at 100 mm from the root apex in cv. Nipponbare (9th box from the left in Figure 3m) was nearly equal in that of W2165 (third box from the left in Figure 3m). The roots of cv. Nipponbare were still leaky to oxygen at 100 mm from the root apex (Figure 1e), but the roots of W2165 were less leaky (Figure 1a). Lignification of the basal part of roots in W2165 under aerated conditions was not associated with ROL barrier formation (Figures 1a and 3a).

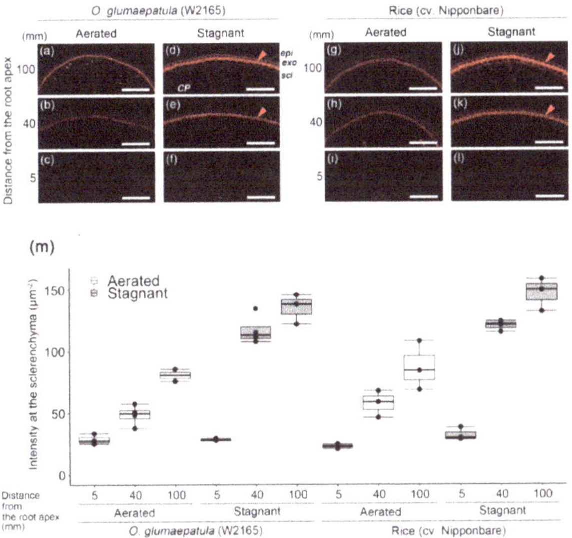

Figure 3. Lignification in the outer part of roots in accession W2165 of *O. glumaepatula* and rice (cv. Nipponbare) under aerated or stagnant conditions. Lignin deposits were observed in the basal parts (95–105 mm from the root apex; (**a,d,g,j**), middle parts (35–45 mm from the root apex; (**b,e,h,k**) and root tips (2.5–7.5 mm from the root apex; (**c,f,i,l**) of adventitious roots of 115–120 mm length. Lignin is indicated as red fluorescence with Basic Fuchsin (red arrowhead, mean of intensity value > 110.0 µm^{-2}). Plants were grown in aerated nutrient solution for eight days and then transferred to deoxygenated stagnant 0.1% agar solution or continued aerated solution for 13–15 days. *CP*, cortical parenchyma; *epi*, epidermis; *exo*, exodermis; *scl*, sclerenchyma. Scale bars: 100 µm. (**m**) Fluorescence intensity of Basic Fuchsin at the sclerenchyma under aerated (white box) or stagnant (grey box) conditions. Black dots indicate each raw value. *n* = 2–4.

2.5. Apoplastic Barrier Assay in W2165

The ability of the ROL barrier to act as an apoplastic barrier was tested with an apoplastic tracer (periodic acid). At the basal parts of roots in W2165 under aerated and stagnant conditions, the purple color of periodic acid was detected only in the epidermal cells (Figure 4a,d,e). The penetration of the tracer was blocked at the outside of the exodermis. In cv. Nipponbare, penetration of the tracer was also blocked at the outside of the exodermis under stagnant conditions (Figure 4j,k), but not under aerated conditions (Figure 4g). Thus, the constitutive ROL barrier in W2165 also acts as an apoplastic barrier at the exodermis.

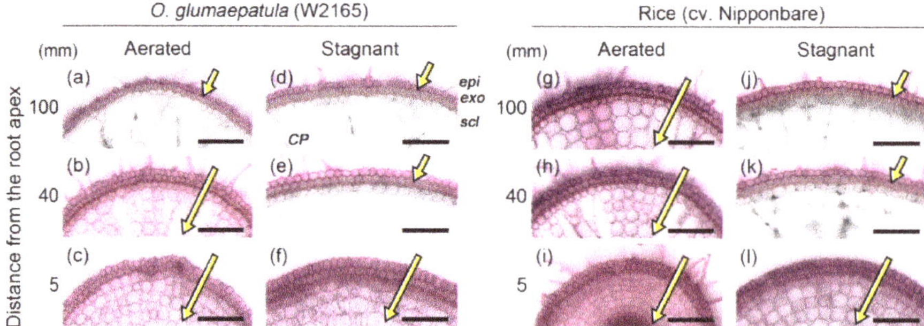

Figure 4. Permeability of the outer part of roots to an apoplastic tracer (periodic acid) in accession W2165 of *O. glumaepatula* and rice (cv. Nipponbare) under aerated or stagnant conditions. The permeability of the exodermis was evaluated at the basal parts (95–105 mm from the root apex; (**a,d,g,j**), middle parts (35–45 mm from the root apex; (**b,e,h,k**) and root tips (2.5–7.5 mm from the root apex; (**c,f,i,l**) of adventitious roots of 115–120 mm length. Purple color indicates that periodic acid penetrated into root tissues. The length of yellow arrows indicates the extent of penetration. Plants were grown in aerated nutrient solution for eight days and then transferred to deoxygenated stagnant 0.1% agar solution or continued aerated solution for 13–15 days. *CP*, cortical parenchyma; *epi*, epidermis; *exo*, exodermis; *scl*, sclerenchyma. Scale bars: 100 μm.

3. Discussion

Until this study, a constitutively formed ROL barrier had not been reported in any rice or wild rice accessions. Here, by screening four accessions of two wild rice species from wetland and non-wetland habitats, we found three accessions of *O. glumaepatula* (W2165, W2149, and W1183) that form a constitutive ROL barrier (Figure 1a–c). Because they have an AA-genome, the three accessions are good candidates for crossing with cultivated rice to identify the QTLs that regulate constitutive ROL barrier formation. Similar hybridization approaches have been used to transfer wild QTLs associated with an inducible ROL barrier formation to wheat [28,29] and maize [30], as well as to understand the genetic regulation of ROL barrier formation [30]. *Hordeum marinum*, a waterlogging-tolerant wild relative of wheat, inducibly forms an ROL barrier under stagnant conditions [31]. To obtain wheat varieties with inducible ROL barriers in the roots, wheat was hybridized with *H. marinum*, producing amphiploids [28] and disomic chromosome addition lines [29]. Under stagnant conditions, two of the amphiploids had tight ROL barriers [28], while no ROL barrier was detected in any of six disomic chromosome addition lines tested [29]. The wild maize *Zea nicaraguensis* inducibly forms an ROL barrier under stagnant conditions, but maize does not [30]. Analyses of *Z. nicaraguensis* introgression lines in the genetic background of maize identified a major locus in a segment of the short arm of chromosome 3 of *Z. nicaraguensis* as this segment conferred inducible ROL barrier formation in maize [32]. So far, little is known about the genetic regulation of constitutive ROL barrier formation. Introgression lines of another accession of *O. glumaepatula* (IRGC105668) in the genetic background of *O. sativa* (cv. Taichung 65) have also been developed to investigate other QTLs [33]. Thus, mapping ROL-related QTLs in *O. glumaepatula* may help to understand how constitutive ROL barrier formation is regulated.

Aerenchyma formation, as well as an ROL barrier formation, are key features contributing to long-distance oxygen transport and waterlogging tolerance [2,5]. Like cultivated rice, all four of the wild accessions formed aerenchyma constitutively (Table 1). Constitutive aerenchyma formation has also been observed in other wild grasses [34–37]. *Z. nicaraguensis* accessions with a higher degree of constitutive aerenchyma formation had better waterlogging tolerance than maize and the other *Z. nicaraguensis* accessions with a lower degree of constitutive aerenchyma formation [38]. Additionally, accessions of *O. glumaepatula* formed a constitutive ROL barrier (Figure 1a–c). Even in cultivated rice, forming an ROL barrier takes 24–48 h after the start of stagnant treatment [25,39]. Additionally, shorter adventitious roots (65–90 mm length roots at the commencement of treatment) need 72–120 h to complete ROL barrier formation [25]. Constitutive ROL barrier is likely to have an advantage for rapid adaptation to waterlogging or flash-flooding. Having a constitutive ROL barrier is likely to be an advantage for plants living in areas prone to flash-flooding, such as accessions of *O. glumaepatula* that grow in the floodplain of the Amazon river [40]. Thus, constitutive aerenchyma and constitutive ROL barrier of *O. glumaepatula* accessions would clearly be an advantage.

Like cv. Nipponbare (*O. sativa* vg. *japonica*), accession W1962 of *O. rufipogon* formed an inducible ROL barrier (Figure 1d). Cultivated rice (*O. sativa* vg. *japonica*) was domesticated from a specific *O. rufipogon* population in southern China thousands of years ago [41]. This raises the possibility that the inducible ROL barrier in cultivated rice is inherited from its ancestor (i.e., *O. rufipogon*). But, there are many morphological and physiological differences between *O. sativa* and *O. rufipogon* [27]. Additionally, we checked only one *O. rufipogon* accession (W1962) (Figure 1d) that came from a non-wetland habitat. *O. rufipogon* grows in deep permanent water [42]. There are over 600 accessions of *O. rufipogon* in the Oryzabase database. Thus, accessions that grow in flooding sites might be more likely to form a constitutive ROL barrier. Further investigations of ROL barrier formation in more accessions of *O. rufipogon* and other wild rice species are needed to reveal how wild species of *Oryza* acquired traits of waterlogging tolerance during the course of their evolution.

Suberin is considered a major component of ROL barriers [6]. A transcriptome analysis using laser-microdissected tissues of the outer part of roots in rice showed that many genes involved in suberin biosynthesis (but not in lignin biosynthesis) were up-regulated during ROL barrier formation in rice [43]. Moreover, a metabolomic analysis in rice clearly showed that rice roots that form an ROL barrier accumulate malic acid and very long-chain fatty acids, which are substrates for suberin biosynthesis [44]. In the present study, the constitutive ROL barrier of W2165 was closely associated with suberization (Figure 2) but not associated with lignification (Figure 3). This is in agreement with observations that suberized exodermis was associated with constitutive ROL barrier formation in wild *Echinochloa* accessions [13] and a tropical forage grass (*Urochloa humidicola* [45]).

Moreover, suberin lamellae inhibited the infiltration of an apoplastic tracer (periodic acid) (Figure 4), suggesting that the ROL barrier can prevent the entry of phytotoxic compounds from the soil. Similar findings were reported for the constitutive ROL barrier in a wild *Echinochloa* species [13] and inducible ROL barrier in rice [46] and *Z. nicaraguensis* [32]. The constitutive ROL barrier of W2165 would have the advantage of detoxifying and preventing the infiltration of toxic ion. The present results open the door to further QTL studies to identify the genes involved in constitutive ROL barrier formation and thus the development of more flood-tolerant varieties of rice.

4. Materials and Methods

4.1. Plant Materials

The *Oryza glumaepatula* accessions (W2165, W2149, and W1183) and *O. rufipogon* accession (W1962) used in this study were kindly provided by the National Institute of Genetics, Japan. For a control that forms an inducible ROL barrier, we used rice (*O. sativa* L. cv. Nipponbare).

4.2. Growth Conditions

To break dormancy, *O. glumaepatula* and *O. rufipogon* seeds were incubated at 35 °C for five days and rice cv. Nipponbare seeds were incubated at 50 °C for three days. Seeds were sterilized for 30 min in 0.6% (*w/v*) sodium hypochlorite, washed thoroughly with deionized water, and for imbibition, placed in Petri dishes (8.5 cm diameter) containing about 6 mL of deionized water (about 1 mm water-depth) at 28 °C under darkness. The plants were grown in a controlled-environment chamber under constant light to avoid effects of circadian rhythm on gene expression (24 h light, 28 °C, relative humidity over 50%, photosynthetic photon flux density at 248.8 μmol m^{-2} s^{-1}). One day after imbibition, seeds were placed on a stainless mesh with attached floats floating on an aerated quarter-strength nutrient solution [24,25] and exposed to light. For the next phase, a soft sponge was floated on a container (380 mm × 260 mm × 160 mm high) containing aerated full-strength nutrient solution. Vertical slits were cut into the edges of the sponge. Six days after imbibition, each plant was inserted into the slit so that the roots were submerged, and the shoot protruded through the sponge into the light. To evaluate constitutive ROL barrier formation under aerated conditions, eight days after imbibition, plants were transplanted into aerated nutrient solution in 5-L pots (120 mm × 180 mm × 250 mm high, three plants per pot) for an additional 13–15 days. In each pot, a rectangular 2 cm-thick piece of foam was placed on the solution, and aluminum foil was placed on the top of the foam to keep the solution dark. Vertical cuts were made on three sides of the foam to accommodate the stems. Then three plants were transferred to each pot, sliding the stems into the cuts. In this way, the roots were kept dark. The nutrient solution was renewed every seven days.

To evaluate inducible ROL barrier formation under stagnant conditions, eight days after imbibition, plants were transplanted into stagnant deoxygenated nutrient solution in 5 L pots for 13–15 days. The stagnant solution was prepared by adding 0.1% (*w/v*) agar (not enough to cause solidification) to the nutrient solution and boiling the solution to dissolve the agar. The low agar concentration produced a viscous liquid rather than a gel. By preventing convective movements, the solution mimics the changes in gas composition found in waterlogged soils (i.e., decreased oxygen, increased ethylene) [47]. The solution was poured into the pots and deoxygenated by bubbling N_2 gas gently through two air stones at a flow rate of about 2.2 L min^{-1} for 15 min per pot. The dissolved oxygen (DO) level was confirmed to be less than 1.0 mg L^{-1} by DO meter (SG6-ELK, Mettler Toledo, Greifensee, Switzerland).

4.3. ROL Barrier Formation

Radial oxygen loss from adventitious roots was measured with Pt cylindrical root-sleeving O_2 electrodes [48,49]. The roots were relatively young (115 to 120 mm long) and had few or no lateral roots. The plants were placed in clear plastic boxes (55 mm × 55 mm × 300 mm high) fitted with rubber lids. The boxes were filled with an O_2-free medium containing 0.1% (*w/v*) agar, 0.5 mM $CaSO_4$, and 5 mM KCl. The plant was inserted through the hole, fixing the shoot base to the rubber lid. Thus, the shoot was in the air, and the root was in the medium. An adventitious root was inserted through the cylindrical root-sleeving O_2 electrode (internal diameter, 2.25 mm; height, 5.0 mm). Root diameters at the point of oxygen measurement were measured with a micrometer caliper. Following Fick's law, the rate of ROL was calculated with the following equation [48]:

$$ROL = -\frac{4.974I}{A},$$

where *ROL* is radial oxygen loss (O_2 ng cm^{-2} root surface min^{-1}), *I* is the diffusion current (μA) with the root in the electrode minus the diffusion current (μA) without the root (current of background), and *A* is the surface area of the root within the electrode (cm^2). The oxygen electrode was 5 mm long, and it was placed at 5, 10 or 20 mm intervals along the root. So, for example, when the electrode was centered at 10 mm, it measured oxygen loss from 7.5 to 12.5 mm. The experiment was conducted in a lighted room kept at a constant 23 °C.

4.4. Aerenchyma Formation

The roots used for aerenchyma measurements were the same ones that were previously used for ROL measurements with the O_2 electrode. Root segments at the basal parts (95–105 mm from the root apex) were prepared from the adventitious roots. Root cross-sections were prepared by hand sectioning with a razor blade. Root cross-sections were photographed using a microscope (Axio Imager.A2, Carl Zeiss, Oberkochen, Germany) under white light with a CCD (charge-coupled device) camera (AxioCam MRc CCD, Carl Zeiss). Areas of cortex and aerenchyma in the cross-section were measured using Fiji (version 2.0.0-rc-69/1.52p), and the percentage of the cortex occupied by aerenchyma was calculated from the cross-sectional areas.

4.5. Histochemical Staining

The adventitious roots (115–120 mm length) in which ROL was measured with the O_2 electrode were cut at the root-shoot junction. Their basal parts (95–105 mm from the root apex), middle parts (35–45 mm from the root apex), and root tips (2.5–7.5 mm from the root apex) were each embedded in 5% (*w/v*) agar, respectively. Root cross-sections of 100 μm thickness were made using a vibrating microtome (Leica VT1200S, Leica Biosystems, Wetzlar, Germany). The cross-sections were made transparent by incubating them in lactic acid saturated with chloral hydrate at 70 °C for 60 min [50]. To detect suberin lamellae, we used 0.01% (*w/v*) Fluorol Yellow 088 in polyethylene glycol 400 as described previously [51]. Suberin lamellae were visualized as a yellowish-green fluorescence excited by ultraviolet (UV) light. For quantification, all cross-sections were photographed with a fluorescence microscope with the following settings [Exposure time: 0.8438 sec; an 02 UV filter set (Excitation G 365 nm, Beamsplitter FT 395, Emission LP 420), an Axio Imager.A2 and an AxioCam MRc CCD camera (all Carl Zeiss)]. Lignin was stained with Basic Fuchsin [52], which causes it to fluoresce red. Before lignin staining, cross-sections were made transparent by incubating them overnight in ClearSee solution at room temperature [53]. Cross-sections were incubated overnight in 0.2% (*w/v*) Basic Fuchsin (Sigma-Aldrich) dissolved in ClearSee solution at room temperature as described previously [54]. Before the observation, the cross-sections were gently washed at least five times with ClearSee solution. Lignin was imaged on a confocal microscope (LSM510 META, Carl Zeiss; Excitation: 543 nm, Detection: 565–651 nm).

4.6. Quantification of Fluorescence Intensity

All images were obtained with the same section thickness, same exposure time, and the same laser power for excitation. In the images obtained by the fluorescence and confocal microscopes (both Carl Zeiss), the fluorescence intensities at each pixel were recorded as 12 bit and 16 bit values, respectively (Supplemental Figures S1a and S2a, respectively). Fluorescence intensities were quantified using Fiji.

To quantify Fluorol Yellow 088 fluorescence (Supplemental Figure S1a), impulse noise was removed from each image with a median filter (Fiji command: Median, radius: 0.5 pixels). Then, the 12 bit color root images were split into red, green, and blue images by Fiji (Fiji command: Split Channels) (Supplemental Figure S1b). To quantify yellow fluorescence, the sum of the red and green intensities at each pixel was subtracted from the blue intensity (Fiji command: Image Calculator). The images were then converted to 8 bit values (0–255) (Fiji command: 8 bit) (Supplemental Figure S1c). The background intensities (i.e., area of the solvent without root cross-section as a red rectangle in Supplemental Figure S1c) in the images were measured (Fiji Command: Measure), and the mean value was 14.733 ± 1.869 (mean ± standard deviation). Background noise was removed by subtracting the mean intensity of blank images (Fiji command: Subtract, value: 14.733) (Supplemental Figure S1d).

To quantify Basic Fuchsin fluorescence (Supplemental Figure S2a), impulse noise was removed from each image with a median filter (Fiji command: Median, radius: 0.5 pixels). Then, intensities of the 16 bit monochrome images were converted to 8 bit values (Fiji command: 8 bit) (Supplemental Figure S2b). The background intensities (i.e., area of the solvent without root cross-section as a red rectangle in Supplemental Figure S2b) in the images were measured (Fiji Command: Measure), and the mean value was 10.934 ± 2.822 (mean ± standard deviation). Background noise was removed by subtracting the mean intensity of blank images (Fiji command: Subtract, value: 10.934).

Fluorescence intensities were calculated for regions of interest (ROIs) that consisted of 20 selected cells. Fluorescence intensity (μm^{-2}) was calculated as F/A_{ROI}, where F is the dimensionless sum of fluorescence intensities at each pixel in the ROI, and A_{ROI} is the area of the ROI (μm^2). For Fluorol Yellow 088, the ROI was in the exodermis (Supplemental Figure S1d) and for Basic Fuchsin, the ROI was in the sclerenchyma (Supplemental Figure S2c).

4.7. Permeability Test

Adventitious roots (115–120 mm length) were cut at the root–shoot junction. The permeabilities of the exodermal layers at the basal parts (95–105 mm from the root apex), middle parts (35–45 mm from the root apex), and root tips (2.5–7.5 mm from the root apex) were assessed with an apoplastic tracer, periodic acid, as described previously [13]. Periodic acid that penetrated the root tissue was visualized as a purple color under white light with the above microscope and camera.

4.8. Statistical Analysis

Means of aerenchyma formation (% aerenchyma/cortex) were compared with one-way analysis of variance (ANOVA) and Tukey's honest significant difference (HSD) for multiple comparisons at the 5% probability level or with a two-sample t-test at the 5% probability level. Means of fluorescence intensities of Fluorol Yellow 088 or Basic Fuchsin between accession W2165 of *O. glumaepatula* and cv. Nipponbare was compared with a two-sample t-test at the 5% probability level. The data were analyzed with R version 3.5.1 [55].

Supplementary Materials: The following are available online at http://www.mdpi.com/2223-7747/9/7/880/s1, Figure S1: Procedure for quantifying Fluorol Yellow 088 fluorescence intensity; Figure S2: Procedure for quantifying Basic Fuchsin fluorescence intensity.

Author Contributions: M.E. and K.S. designed the experiments. M.E. and Y.S. performed the experiments and analyses. K.S. supervised the experiments. M.E. and K.S. wrote the article. All authors have read and agreed to the published version of the manuscript.

Funding: This work was supported by the Japan Society for the Promotion of Science (Grant No. JP16KK0173, JP17K15211, JP19K05978).

Acknowledgments: We thank Takeshi Fukao for stimulating discussions. The wild rice accessions used in the present study were distributed from the National Institute of Genetics supported by the National Bioresource Project (NBRP), AMED, Japan.

Conflicts of Interest: The authors declare no conflict of interest.

References

1. Jackson, M.B.; Fenning, T.M.; Drew, M.C.; Saker, L.R. Stimulation of ethylene production and gas-space (aerenchyma) formation in adventitious roots of *Zea mays* L. by small partial pressures of oxygen. *Planta* **1985**, *165*, 486–492. [CrossRef]
2. Colmer, T.D. Long-distance transport of gases in plants: A perspective on internal aeration and radial oxygen loss from roots. *Plant Cell Environ.* **2003**, *26*, 17–36. [CrossRef]
3. Shiono, K.; Takahashi, H.; Colmer, T.D.; Nakazono, M. Role of ethylene in acclimations to promote oxygen transport in roots of plants in waterlogged soils. *Plant Sci.* **2008**, *175*, 52–58. [CrossRef]
4. Nishiuchi, S.; Yamauchi, T.; Takahashi, H.; Kotula, L.; Nakazono, M. Mechanisms for coping with submergence and waterlogging in rice. *Rice* **2012**, *5*, 2. [CrossRef] [PubMed]
5. Pedersen, O.; Sauter, M.; Colmer, T.D.; Nakazono, M. Regulation of root adaptive anatomical and morphological traits during low soil oxygen. *New Phytol.* **2020**. In Press. [CrossRef]
6. Yamauchi, T.; Colmer, T.D.; Pedersen, O.; Nakazono, M. Regulation of root traits for internal aeration and tolerance to soil waterlogging-flooding stress. *Plant Physiol.* **2018**, *176*, 1118–1130. [CrossRef]
7. Abiko, T.; Miyasaka, S.C. Aerenchyma and barrier to radial oxygen loss are formed in roots of Taro (*Colocasia esculenta*) propagules under flooded conditions. *J. Plant Res.* **2020**, *133*, 49–56. [CrossRef]
8. Manzur, M.E.; Grimoldi, A.A.; Insausti, P.; Striker, G.G. Radial oxygen loss and physical barriers in relation to root tissue age in species with different types of aerenchyma. *Funct. Plant Biol.* **2015**, *42*, 9–17. [CrossRef]
9. Barros, J.; Serk, H.; Granlund, I.; Pesquet, E. The cell biology of lignification in higher plants. *Ann. Bot.* **2015**, *115*, 1053–1074. [CrossRef]
10. Naseer, S.; Lee, Y.; Lapierre, C.; Franke, R.; Nawrath, C.; Geldner, N. Casparian strip diffusion barrier in *Arabidopsis* is made of a lignin polymer without suberin. *Proc. Natl. Acad. Sci. USA* **2012**, *109*, 10101–10106. [CrossRef]
11. Graça, J. Suberin: The biopolyester at the frontier of plants. *Front. Chem.* **2015**, *3*. [CrossRef]

12. Schreiber, L.; Franke, R.B. Endodermis and exodermis in roots. In *eLS*; John Wiley & Sons Ltd.: Chichester, UK, 2011; pp. 1–7. [CrossRef]

13. Ejiri, M.; Shiono, K. Prevention of radial oxygen loss is associated with exodermal suberin along adventitious roots of annual wild species of *Echinochloa*. *Front. Plant Sci.* **2019**, *10*. [CrossRef] [PubMed]

14. Enstone, D.E.; Peterson, C.A.; Ma, F. Root endodermis and exodermis: Structure, function, and responses to the environment. *J. Plant Growth Regul.* **2003**, *21*, 335–351. [CrossRef]

15. Ranathunge, K.; Lin, J.; Steudle, E.; Schreiber, L. Stagnant deoxygenated growth enhances root suberization and lignifications, but differentially affects water and NaCl permeabilities in rice (*Oryza sativa* L.) roots. *Plant Cell Environ.* **2011**, *34*, 1223–1240. [CrossRef]

16. Ranathunge, K.; Schreiber, L.; Franke, R. Suberin research in the genomics era—New interest for an old polymer. *Plant Sci.* **2011**, *180*, 399–413. [CrossRef]

17. Armstrong, J.; Armstrong, W. Rice: Sulfide-induced barriers to root radial oxygen loss, Fe^{2+} and water uptake, and lateral root emergence. *Ann. Bot.* **2005**, *96*, 625–638. [CrossRef]

18. Colmer, T.D.; Gibberd, M.R.; Wiengweera, A.; Tinh, T.K. The barrier to radial oxygen loss from roots of rice (*Oryza sativa* L.) is induced by growth in stagnant solution. *J. Exp. Bot.* **1998**, *49*, 1431–1436. [CrossRef]

19. Mongon, J.; Konnerup, D.; Colmer, T.D.; Rerkasem, B. Responses of rice to Fe^{2+} in aerated and stagnant conditions: Growth, root porosity and radial oxygen loss barrier. *Funct. Plant Biol.* **2014**, *41*, 922–929. [CrossRef]

20. Ranathunge, K.; Schreiber, L.; Bi, Y.M.; Rothstein, S.J. Ammonium-induced architectural and anatomical changes with altered suberin and lignin levels significantly change water and solute permeabilities of rice (*Oryza sativa* L.) roots. *Planta* **2016**, *243*, 231–249. [CrossRef]

21. Colmer, T.D.; Kotula, L.; Malik, A.I.; Takahashi, H.; Konnerup, D.; Nakazono, M.; Pedersen, O. Rice acclimation to soil flooding: Low concentrations of organic acids can trigger a barrier to radial oxygen loss in roots. *Plant Cell Environ.* **2019**, *42*, 2183–2197. [CrossRef] [PubMed]

22. McDonald, M.P.; Galwey, N.W.; Colmer, T.D. Waterlogging tolerance in the tribe Triticeae: The adventitious roots of *Critesion marinum* have a relatively high porosity and a barrier to radial oxygen loss. *Plant Cell Environ.* **2001**, *24*, 585–596. [CrossRef]

23. McDonald, M.P.; Galwey, N.W.; Colmer, T.D. Similarity and diversity in adventitious root anatomy as related to root aeration among a range of wetland and dryland grass species. *Plant Cell Environ.* **2002**, *25*, 441–451. [CrossRef]

24. Colmer, T.D. Aerenchyma and an inducible barrier to radial oxygen loss facilitate root aeration in upland, paddy and deep-water rice (*Oryza sativa* L.). *Ann. Bot.* **2003**, *91*, 301–309. [CrossRef] [PubMed]

25. Shiono, K.; Ogawa, S.; Yamazaki, S.; Isoda, H.; Fujimura, T.; Nakazono, M.; Colmer, T.D. Contrasting dynamics of radial O_2-loss barrier induction and aerenchyma formation in rice roots of two lengths. *Ann. Bot.* **2011**, *107*, 89–99. [CrossRef]

26. Kotula, L.; Ranathunge, K.; Schreiber, L.; Steudle, E. Functional and chemical comparison of apoplastic barriers to radial oxygen loss in roots of rice (*Oryza sativa* L.) grown in aerated or deoxygenated solution. *J. Exp. Bot.* **2009**, *60*, 2155–2167. [CrossRef]

27. Atwell, B.J.; Wang, H.; Scafaro, A.P. Could abiotic stress tolerance in wild relatives of rice be used to improve *Oryza sativa*? *Plant Sci.* **2014**, *215*, 48–58. [CrossRef]

28. Malik, A.I.; Islam, A.K.M.R.; Colmer, T.D. Transfer of the barrier to radial oxygen loss in roots of *Hordeum marinum* to wheat (*Triticum aestivum*): Evaluation of four *H. marinum*–wheat amphiploids. *New Phytol.* **2011**, *190*, 499–508. [CrossRef]

29. Konnerup, D.; Malik, A.I.; Islam, A.K.M.R.; Colmer, T.D. Evaluation of root porosity and radial oxygen loss of disomic addition lines of *Hordeum marinum* in wheat. *Funct. Plant Biol.* **2017**, *44*, 400–409. [CrossRef]

30. Abiko, T.; Kotula, L.; Shiono, K.; Malik, A.I.; Colmer, T.D.; Nakazono, M. Enhanced formation of aerenchyma and induction of a barrier to radial oxygen loss in adventitious roots of *Zea nicaraguensis* contribute to its waterlogging tolerance as compared with maize (*Zea mays* ssp. *mays*). *Plant Cell Environ.* **2012**, *35*, 1618–1630. [CrossRef]

31. Garthwaite, A.J.; von Bothmer, R.; Colmer, T.D. Diversity in root aeration traits associated with waterlogging tolerance in the genus *Hordeum*. *Funct. Plant Biol.* **2003**, *30*, 875–889. [CrossRef]

32. Watanabe, K.; Takahashi, H.; Sato, S.; Nishiuchi, S.; Omori, F.; Malik, A.I.; Colmer, T.D.; Mano, Y.; Nakazono, M. A major locus involved in the formation of the radial oxygen loss barrier in adventitious roots of teosinte *Zea nicaraguensis* is located on the short-arm of chromosome 3. *Plant Cell Environ.* **2017**, *40*, 304–316. [CrossRef] [PubMed]

33. Yoshimura, A.; Nagayama, H.; Kurakazu, T.; Sanchez, P.L.; Doi, K.; Yamagata, Y.; Yasui, H. Introgression lines of rice (*Oryza sativa* L.) carrying a donor genome from the wild species, *O. glumaepatula* Steud. and *O. meridionalis* Ng. *Breed. Sci.* **2010**, *60*, 597–603. [CrossRef]

34. Laan, P.; Berrevoets, M.J.; Lythe, S.; Armstrong, W.; Blom, C.W.P.M. Root morphology and aerenchyma formation as indicators of the flood-tolerance of *Rumex* species. *J. Ecol.* **1989**, *77*, 693–703. [CrossRef]

35. Visser, E.J.W.; Bögemann, G.M. Aerenchyma formation in the wetland plant *Juncus effusus* is independent of ethylene. *New Phytol.* **2006**, *171*, 305–314. [CrossRef] [PubMed]

36. Mano, Y.; Omori, F.; Takamizo, T.; Kindiger, B.; Bird, R.M.; Loaisiga, C.H.; Takahashi, H. QTL mapping of root aerenchyma formation in seedlings of a maize x rare teosinte "*Zea nicaraguensis*" cross. *Plant Soil* **2007**, *295*, 103–113. [CrossRef]

37. Malik, A.I.; English, J.P.; Colmer, T.D. Tolerance of *Hordeum marinum* accessions to O_2 deficiency, salinity and these stresses combined. *Ann. Bot.* **2009**, *103*, 237–248. [CrossRef] [PubMed]

38. Mano, Y.; Omori, F. Relationship between constitutive root aerenchyma formation and flooding tolerance in *Zea nicaraguensis*. *Plant Soil* **2013**, *370*, 447–460. [CrossRef]

39. Insalud, N.; Bell, R.W.; Colmer, T.D.; Rerkasem, B. Morphological and physiological responses of rice (*Oryza sativa*) to limited phosphorus supply in aerated and stagnant solution culture. *Ann. Bot.* **2006**, *98*, 995–1004. [CrossRef] [PubMed]

40. Akimoto, M.; Shimamoto, Y.; Morishima, H. Population genetic structure of wild rice *Oryza glumaepatula* distributed in the Amazon flood area influenced by its life-history traits. *Mol. Ecol.* **1998**, *7*, 1371–1381. [CrossRef]

41. Huang, X.; Kurata, N.; Wei, X.; Wang, Z.X.; Wang, A.; Zhao, Q.; Zhao, Y.; Liu, K.; Lu, H.; Li, W.; et al. A map of rice genome variation reveals the origin of cultivated rice. *Nature* **2012**, *490*, 497–501. [CrossRef] [PubMed]

42. Waters, D.L.E.; Nock, C.J.; Ishikawa, R.; Rice, N.; Henry, R.J. Chloroplast genome sequence confirms distinctness of Australian and Asian wild rice. *Ecol. Evol.* **2012**, *2*, 211–217. [CrossRef] [PubMed]

43. Shiono, K.; Yamauchi, T.; Yamazaki, S.; Mohanty, B.; Malik, A.I.; Nagamura, Y.; Nishizawa, N.K.; Tsutsumi, N.; Colmer, T.D.; Nakazono, M. Microarray analysis of laser-microdissected tissues indicates the biosynthesis of suberin in the outer part of roots during formation of a barrier to radial oxygen loss in rice (*Oryza sativa*). *J. Exp. Bot.* **2014**, *65*, 4795–4806. [CrossRef] [PubMed]

44. Kulichikhin, K.; Yamauchi, T.; Watanabe, K.; Nakazono, M. Biochemical and molecular characterization of rice (*Oryza sativa* L.) roots forming a barrier to radial oxygen loss. *Plant Cell Environ.* **2014**, *37*, 2406–2420. [CrossRef] [PubMed]

45. Jiménez, J.D.L.C.; Kotula, L.; Veneklaas, E.J.; Colmer, T.D. Root-zone hypoxia reduces growth of the tropical forage grass *Urochloa humidicola* in high-nutrient but not low-nutrient conditions. *Ann. Bot.* **2019**, *124*, 1019–1032. [CrossRef]

46. Shiono, K.; Ando, M.; Nishiuchi, S.; Takahashi, H.; Watanabe, K.; Nakamura, M.; Matsuo, Y.; Yasuno, N.; Yamanouchi, U.; Fujimoto, M.; et al. RCN1/OsABCG5, an ATP-binding cassette (ABC) transporter, is required for hypodermal suberization of roots in rice (*Oryza sativa*). *Plant J.* **2014**, *80*, 40–51. [CrossRef]

47. Wiengweera, A.; Greenway, H.; Thomson, C.J. The use of agar nutrient solution to simulate lack of convection in waterlogged soils. *Ann. Bot.* **1997**, *80*, 115–123. [CrossRef]

48. Armstrong, W.; Wright, E.J. Radial oxygen loss from roots: the theoretical basis for the manipulation of flux data obtained by the cylindrical platinum electrode technique. *Physiol. Plant.* **1975**, *35*, 21–26. [CrossRef]

49. Armstrong, W. Polarographic oxygen electrodes and their use in plant aeration studies. *Proc. R. Soc. B Biol. Sci.* **1994**, *102*, 511–527. [CrossRef]

50. Lux, A.; Morita, S.; Abe, J.; Ito, K. An improved method for clearing and staining free-hand sections and whole-mount samples. *Ann. Bot.* **2005**, *96*, 989–996. [CrossRef]

51. Brundrett, M.C.; Kendrick, B.; Peterson, C.A. Efficient lipid staining in plant material with sudan red 7B or fluoral yellow 088 in polyethylene glycol-glycerol. *Biotech. Histochem.* **1991**, *66*, 111–116. [CrossRef]

52. Kapp, N.; Barnes, W.J.; Richard, T.L.; Anderson, C.T. Imaging with the fluorogenic dye Basic Fuchsin reveals subcellular patterning and ecotype variation of lignification in *Brachypodium distachyon*. *J. Exp. Bot.* **2015**, *66*, 4295–4304. [CrossRef] [PubMed]
53. Kurihara, D.; Mizuta, Y.; Sato, Y.; Higashiyama, T. ClearSee: A rapid optical clearing reagent for whole-plant fluorescence imaging. *Development* **2015**, *142*, 4168–4179. [CrossRef] [PubMed]
54. Ursache, R.; Andersen, T.G.; Marhavý, P.; Geldner, N. A protocol for combining fluorescent proteins with histological stains for diverse cell wall components. *Plant J.* **2018**, *93*, 399–412. [CrossRef] [PubMed]
55. Team, R.C. *R: A Language and Environment for Statistical Computing*; R Foundation for Statistical Computing: Vienna, Austria, 2018.

Article

A Role for Auxin in Ethylene-Dependent Inducible Aerenchyma Formation in Rice Roots

Takaki Yamauchi [1,2,*], **Akihiro Tanaka** [3], **Nobuhiro Tsutsumi** [2], **Yoshiaki Inukai** [4] **and Mikio Nakazono** [3,5,*]

[1] Japan Science and Technology Agency, PRESTO, Kawaguchi, Saitama 332-0012, Japan
[2] Graduate School of Agricultural and Life Sciences, The University of Tokyo, Bunkyo, Tokyo 113-8657, Japan; atsutsu@mail.ecc.u-tokyo.ac.jp
[3] Graduate School of Bioagricultural Sciences, Nagoya University, Nagoya, Aichi 464–8601, Japan; tanaka.akihiro.j@outlook.com
[4] International Center for Research and Education in Agriculture, Nagoya University, Nagoya, Aichi 464–8601, Japan; inukaiy@agr.nagoya-u.ac.jp
[5] The UWA School of Agriculture and Environment, Faculty of Science, The University of Western Australia, Crawley, WA 6009, Australia
* Correspondence: atkyama@mail.ecc.u-tokyo.ac.jp (T.Y.); nakazono@agr.nagoya-u.ac.jp; (M.N.)

Received: 24 April 2020; Accepted: 8 May 2020; Published: 11 May 2020

Abstract: Internal oxygen diffusion from shoot to root tips is enhanced by the formation of aerenchyma (gas space) in waterlogged soils. Lysigenous aerenchyma is created by programmed cell death and subsequent lysis of the root cortical cells. Rice (*Oryza sativa*) forms aerenchyma constitutively under aerobic conditions and increases its formation under oxygen-deficient conditions. Recently, we have demonstrated that constitutive aerenchyma formation is regulated by auxin signaling mediated by Auxin/indole-3-acetic acid protein (AUX/IAA; IAA). While ethylene is involved in inducible aerenchyma formation, the relationship of auxin and ethylene during aerenchyma formation remains unclear. Here, we examined the effects of oxygen deficiency and ethylene on aerenchyma formation in the roots of a rice mutant (*iaa13*) in which auxin signaling is suppressed by a mutation in the degradation domain of IAA13 protein. The results showed that AUX/IAA-mediated auxin signaling contributes to ethylene-dependent inducible aerenchyma formation in rice roots. An auxin transport inhibitor abolished aerenchyma formation under oxygen-deficient conditions and reduced the expression of genes encoding ethylene biosynthesis enzymes, further supporting the idea that auxin is involved in ethylene-dependent inducible aerenchyma formation. Based on these studies, we propose a mechanism that underlies the relationship between auxin and ethylene during inducible aerenchyma formation in rice roots.

Keywords: aerenchyma; auxin; ethylene; rice (*Oryza sativa*); root; waterlogging

---

1. Introduction

Internal oxygen movement from shoot to roots through aerenchyma is essential for plants to adapt to waterlogged soils [1]. Lysigenous aerenchyma in roots is created by programmed cell death (PCD) and subsequent lysis of the cortical cells [2,3]. In roots of upland plants, such as maize (*Zea mays* ssp. *mays*) and wheat (*Triticum aestivum*), lysigenous aerenchyma is not generally formed under aerobic conditions, but its formation is induced under oxygen-deficient conditions [4–7]. In roots of the wetland plant rice (*Oryza sativa*), lysigenous aerenchyma is formed constitutively even under aerobic conditions (constitutive aerenchyma formation), and its formation is further increased under oxygen-deficient conditions (inducible aerenchyma formation) [4].

Ethylene is involved in inducible aerenchyma formation in rice roots [4–7]. Under waterlogging, lower diffusion rates of gases to the rhizosphere enhance the ethylene accumulation in roots [8,9]. Ethylene is biosynthesized by the conversion of S-adenosylmethionine to 1-amino-cyclopropane-1-carboxylic acid (ACC) by ACC synthase (ACS) and that of ACC to ethylene by ACC oxidase (ACO) [10]. During inducible aerenchyma formation in rice roots, the expression levels of *ACS1* and *ACO5* are increased, and they contribute to increased ethylene content in the roots [11]. Moreover, ethylene-induced production of reactive oxygen species (ROS) by respiratory burst oxidase homolog H (RBOHH) is involved in inducible aerenchyma formation in rice roots [12]. Ethylene and ROS signaling is also involved in lysigenous aerenchyma formation in rice shoots [13].

Auxin signaling is mediated by a family of transcription factors called auxin response factors (ARFs) [14]. ARF-dependent transcriptional regulation is repressed by the binding of auxin/indole-3-acetic acid proteins (AUX/IAAs; IAAs) to ARFs [15]. The rice genome has 25 *ARF* genes and 31 *IAA* genes [16,17]. IAA proteins have a conserved amino acid sequence motif (AUX/IAA domain II), which is required for auxin-dependent proteolysis of the IAA proteins [18]. The auxin signaling is suppressed in roots of the gain of function (dominant-negative) *iaa13* mutant having a single amino acid substitution in the AUX/IAA domain II of IAA13 protein [19]. Although until recently it remained unclear what triggers constitutive aerenchyma formation in rice roots [6], we demonstrated that constitutive aerenchyma formation in rice roots is regulated by the auxin signaling through the functional analysis of the *iaa13* mutant [20].

Exogenous treatment with ethylene stimulates aerenchyma formation in rice roots even under aerobic conditions [21,22]. While inhibitors of ethylene perception or ethylene action reduce aerenchyma formation in rice roots under oxygen-deficient conditions [12,21], they cannot abolish aerenchyma formation under either oxygen-deficient or aerobic conditions [12,21,23]. On the other hand, an auxin transport inhibitor completely blocks constitutive aerenchyma formation in rice roots under aerobic conditions [20], implying that auxin signaling is required for the ethylene-dependent aerenchyma formation. However, the relationship between auxin and ethylene during inducible aerenchyma formation remains unclear.

The objective of this study was to test the possibility that auxin is involved in ethylene-dependent aerenchyma formation. To this end, we used the *iaa13* mutant in which the dominant negative IAA13 suppressed auxin signaling in the roots [19]. We examined the effect of enhancing ethylene signaling on aerenchyma formation in roots of *iaa13* and its wild type (WT; cv. Taichung 65; T65). We also examined the effects of an auxin transport inhibitor on ethylene-dependent aerenchyma formation and the expression levels of genes encoding ethylene biosynthesis enzymes in the roots of the WT. Finally, we examined the effects of an ethylene precursor on aerenchyma formation in the presence of the auxin transport inhibitor. Our results strongly suggest that auxin is involved in the regulation of ethylene-dependent inducible aerenchyma formation in rice roots.

2. Results

2.1. Effect of Oxygen Deficiency on Aerenchyma Formation

During inducible aerenchyma formation under oxygen-deficient conditions, ethylene accumulation increases in rice roots [11,12]. To test the effect of oxygen deficiency on aerenchyma formation in *iaa13*, 20-d-old aerobically grown WT and *iaa13* seedlings were transferred to aerated or stagnant (deoxygenated) conditions, which mimic the changes in gas composition in waterlogged soils [24], for 48 h. After 48 h, root elongation of the WT was 12.9% less under stagnant conditions than under aerated conditions, while root elongation of *iaa13* was 18.2% less under stagnant conditions (Supplemental Figure S1a). Subsequently, transverse sections along the adventitious roots were prepared (Figure 1a), and the percentage of each cross-section occupied by aerenchyma was determined (Figure 1b,c). Aerenchyma formation in the WT roots was significantly higher at 10, 20, 30, and 40 mm under stagnant conditions than under aerated conditions (Figure 1a,b), whereas aerenchyma formation in the *iaa13* roots was significantly higher at all positions under stagnant conditions (Figure 1a,c). Aerenchyma formation in the WT at 20, 30, 40, and 50 mm was significantly higher than that in *iaa13* both under aerated and stagnant conditions (Supplemental Figure S2a,b), suggesting that difference in aerenchyma formation between the WT and *iaa13* under stagnant conditions is largely affected by the reduced constitutive aerenchyma formation in the *iaa13* roots.

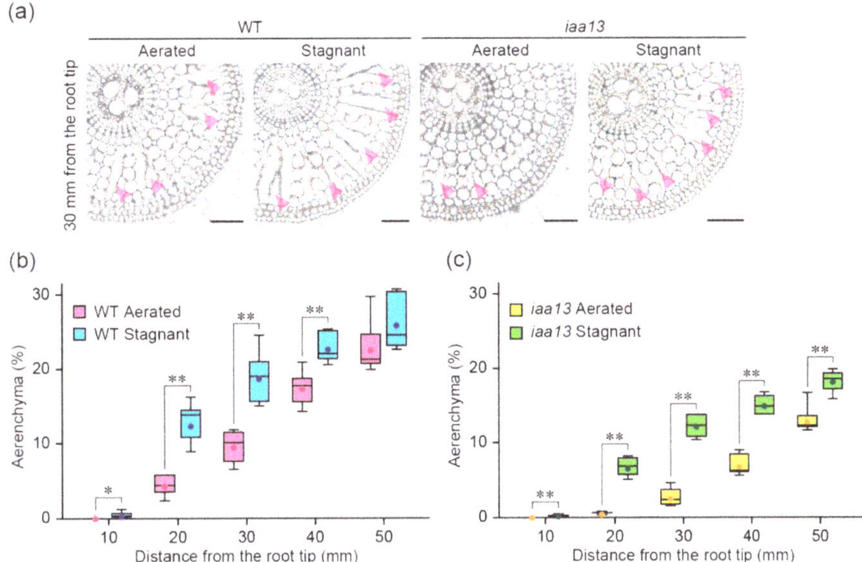

Figure 1. Aerenchyma formation under aerated or stagnant conditions. (**a**) Cross-sections at 30 mm from the tips of adventitious roots of the wild-type (WT) and *iaa13* mutant. Aerenchyma is indicated by magenta arrowheads. Bars = 100 μm. (**b**,**c**) Percentages of aerenchyma in root cross-sectional area at 10, 20, 30, 40, and 50 mm. Twenty-day-old aerobically grown WT (**b**) and *iaa13* (**c**) seedlings were further grown under aerated or stagnant conditions for 48 h. (**b**,**c**) Significant differences between the conditions at $p < 0.01$ are denoted by ** (two-sample *t*-test). Boxplots show the median (horizontal lines), 25th to 75th percentiles (edges of the boxes), minimum to maximum (edges of the whiskers), and mean values (dots in the boxes) ($n = 6$).

2.2. Effect of an Ethylene Precursor on Aerenchyma Formation

Exogenously supplied ethylene stimulates aerenchyma formation in rice roots even under aerobic conditions [21,22], and the treatment with an ethylene precursor ACC also induces its formation [11,23]. To further investigate the effect of ethylene on aerenchyma formation in the *iaa13* roots, 20-d-old aerobically grown WT and *iaa13* seedlings were transferred to aerated conditions with or without 10 μM ACC. After 48 h, root elongation of the WT was 15.1% less under aerated conditions with ACC than without ACC, while root elongation of *iaa13* was 18.7% less under aerated conditions with ACC (Supplemental Figure S1b). The suppression of root elongation in both the WT and *iaa13* by ACC treatment (Supplemental Figure S1b) was similar to the suppression of root elongation in both the WT and *iaa13* by stagnant conditions (Supplemental Figure S1a). This suggests that ethylene accumulation, which is stimulated by ACC or oxygen deficiency, has similar effects on root elongation in the WT and *iaa13*. As is the case with stagnant conditions, aerenchyma formation in the WT roots was significantly higher at 10, 20, 30, and 40 mm from the root tips under aerated conditions with ACC than without ACC (Figure 2a,b), whereas aerenchyma formation in the *iaa13* roots was significantly higher at all positions under aerated conditions with ACC (Figure 2a,c). Aerenchyma formation in the WT at 20, 30, 40 and 50 mm was significantly higher than that in *iaa13* under aerated conditions without ACC, and aerenchyma formation in the WT at 20, 30, and 40 mm was significantly higher than that in *iaa13* under aerated conditions with ACC (Supplemental Figure S2c,d).

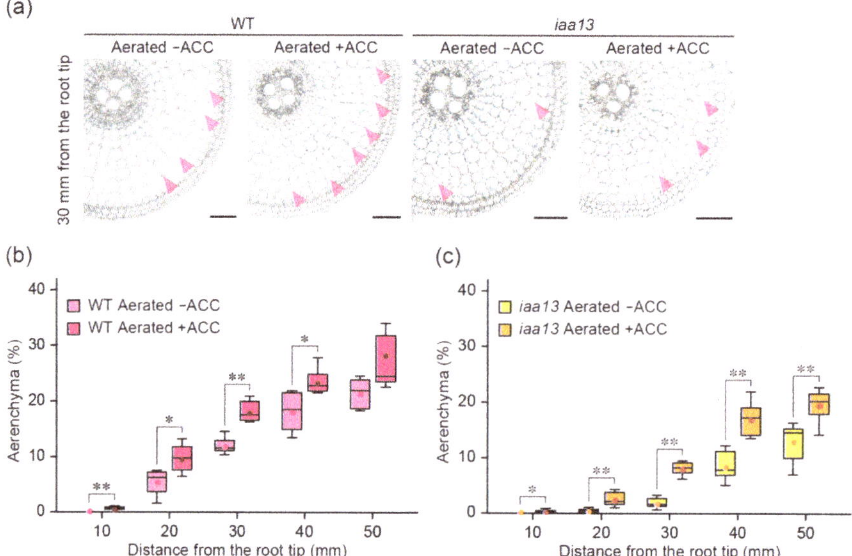

Figure 2. Aerenchyma formation under aerated conditions with or without 1-aminocyclopropane-1-carboxylic acid (ACC) treatment. (**a**) Cross-sections at 30 mm from the tips of adventitious roots of the wild type (WT) and *iaa13* mutant. Aerenchyma is indicated by magenta arrowheads. Bars = 100 μm. (**b,c**) Percentages of aerenchyma in root cross-sectional area at 10, 20, 30, 40, and 50 mm. Twenty-day-old aerobically grown WT (**b**) and *iaa13* (**c**) seedlings were further grown under aerated conditions with or without 10 μM ACC for 48 h. (**b,c**) Significant differences between the conditions at $p < 0.05$ and $p < 0.01$ are denoted by * and **, respectively (two-sample *t*-test). Boxplots show the median (horizontal lines), 25th to 75th percentiles (edges of the boxes), minimum to maximum (edges of the whiskers) and mean values (dots in the boxes) ($n = 6$).

2.3. Differences in Response to Oxygen Deficiency and ACC between the Wild Type and iaa13

Both oxygen deficiency (Figure 1c) and ACC (Figure 2c) increased aerenchyma formation in *iaa13* roots, suggesting that ethylene-dependent inducible aerenchyma formation is not reduced in *iaa13*. Interestingly, longitudinal patterns of the differences in aerenchyma formation between aerated and stagnant conditions (Figure 3a), and between aerated conditions with and without ACC (Figure 3b), were similar to each other. At 10 to 20 mm from the root tips, the differences were larger in the WT roots than in *iaa13* (Figure 3a,b). The differences were comparable at 30 mm and then became smaller in the WT than those in *iaa13* at 40 to 50 mm (Figure 3a,b). These results suggest that ethylene-dependent aerenchyma formation in *iaa13* is reduced at the apical part of the roots.

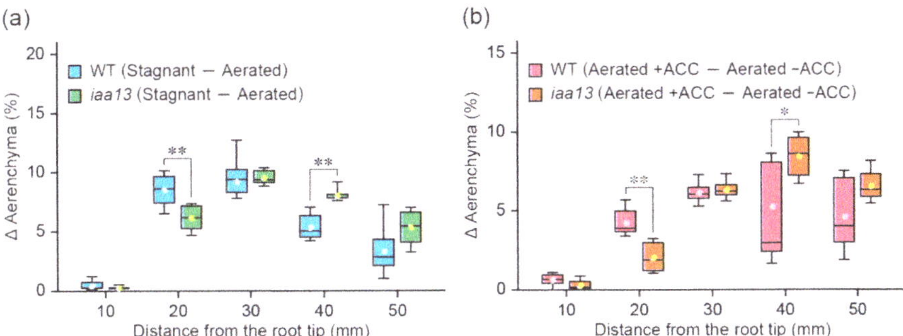

Figure 3. Differences in aerenchyma formation between under aerated and stagnant conditions and between aerated conditions with and without 1-aminocyclopropane-1-carboxylic acid (ACC) treatment. (**a**,**b**) Differences in the percentages of aerenchyma (D Aerenchyma (%)) in root cross-sectional area at 10, 20, 30, 40, and 50 mm from the tips of adventitious roots of the wild type (WT) and *iaa13* mutant. The differences in the percentages of aerenchyma were calculated by the data obtained in Figures 1a and 2b, respectively. (**a**,**b**) Significant differences between the genotypes at $p < 0.05$ and $p < 0.01$ are denoted by * and **, respectively (two-sample *t*-test). Boxplots show the median (horizontal lines), 25th to 75th percentiles (edges of the boxes), minimum to maximum (edges of the whiskers) and mean values (dots in the boxes) (*n* = 6).

2.4. Effect of an Auxin Transport Inhibitor on Aerenchyma Formation

To confirm the effect of auxin on the ethylene-dependent aerenchyma formation, we examined the effect of the auxin transport inhibitor *N*-1-naphthylphthalamic acid (NPA) on aerenchyma formation. Twenty-day-old aerobically grown WT seedlings were transferred to aerated or stagnant conditions with or without 0.5 µM NPA. After 48 h, root elongation of the WT was 19.0% less under aerated conditions with NPA than without NPA, while root elongation of it was 17.4% less under stagnant conditions with NPA (Supplemental Figure S1c). Root elongation in 48 h NPA treatment was 31.3 ± 3.4 mm under aerated conditions and 26.8 ± 4.7 mm under stagnant conditions (Supplemental Figure S1c). These results indicate that root cortical cells at 10 to 30 mm under aerated conditions and at 10 to 20 mm under stagnant conditions are generated in the presence of NPA. Under aerated conditions, NPA completely blocked aerenchyma formation at 10 to 30 mm from the root tips (Figure 4a,b). Interestingly, NPA also completely blocked aerenchyma formation at 10 to 20 mm under stagnant conditions (Figure 4a,c). These results strongly suggest that auxin is also involved in ethylene-dependent aerenchyma formation in rice roots.

2.5. Effect of NPA on the Expression of Ethylene Biosynthesis Genes

To test the effect of the auxin transport inhibitor on the ethylene biosynthesis in rice roots, the transcript levels of ethylene biosynthesis genes were analyzed by quantitative reverse transcription (qRT)-PCR analysis. Previously, we showed that, among six *ACS* and seven *ACO* genes in the rice genome, *ACS1* and *ACO5* had the highest transcript levels during inducible aerenchyma formation in rice roots [23]. The transcript levels of *ACS1* and *ACO5* were significantly increased under stagnant conditions and peaked at 15–25 mm from the root tips (Figure 5a,b), where aerenchyma formation is highly induced (at 20 mm; Figure 4c). NPA significantly reduced the transcript level of *ACS1* at 5–15 mm and 25–35 mm (Figure 5a), and it also reduced that of *ACO5* at 15–25 mm (Figure 5b). By contrast, under aerated conditions, the transcript levels of *ACS1* and *ACO5* with NPA treatment were comparable to those without NPA treatment (Figure 5a,b). These results suggest that auxin contributes to the transcriptional induction of the *ACS1* and *ACO5* genes under stagnant conditions.

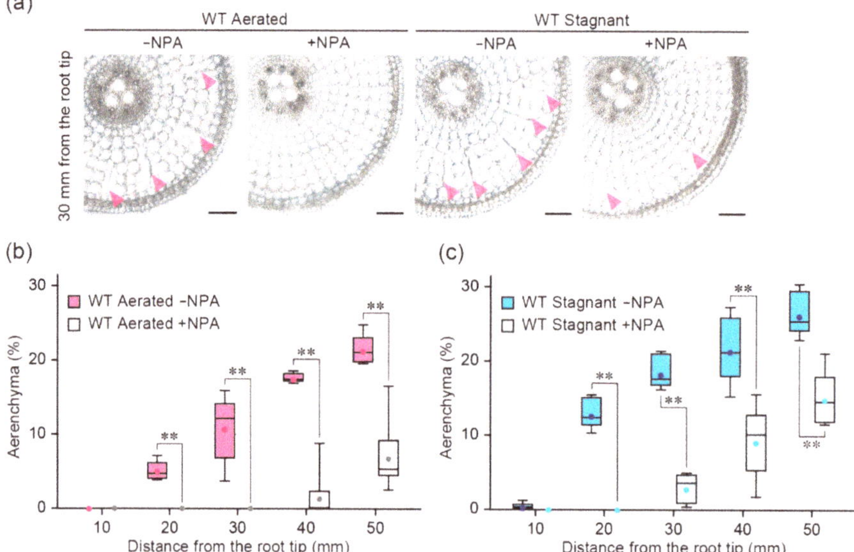

Figure 4. Aerenchyma formation under aerated or stagnant conditions with or without *N*-1-naphthylphthalamic acid (NPA) treatment. (**a**) Cross-sections at 30 mm from the tips of adventitious roots of the wild type (WT). Aerenchyma is indicated by magenta arrowheads. Bars = 100 μm. (**b**,**c**) Percentages of aerenchyma in root cross-sectional area at 10, 20, 30, 40, and 50 mm. Twenty-day-old aerobically grown WT seedlings were further grown under aerated (**b**) or stagnant (**c**) conditions with or without 0.5 μM NPA for 48 h. (**b**,**c**) Significant differences between the conditions at $p < 0.01$ are denoted by ** (two-sample *t*-test). Boxplots show the median (horizontal lines), 25th to 75th percentiles (edges of the boxes), minimum to maximum (edges of the whiskers) and mean values (dots in the boxes) ($n = 6$).

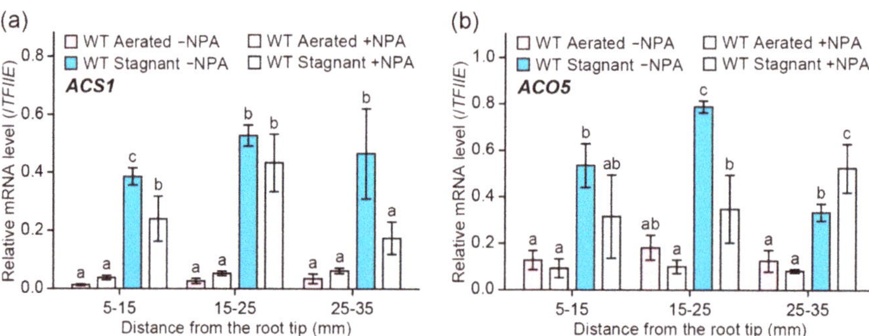

Figure 5. Expression of ethylene biosynthesis genes in adventitious roots under aerated or stagnant conditions with or without *N*-1-naphthylphthalamic acid (NPA) treatment. Twenty-day-old aerobically grown wild type (WT) seedlings were further grown under aerated or stagnant conditions with or without 0.5 μM NPA for 48 h. Relative transcription levels of *ACS1* (**a**) and *ACO5* (**b**) at 5–15, 15–25, and 25–35 mm from the tips of adventitious roots. The gene encoding transcription initiation factor IIE (TFIIE) was used as a control. Different lowercase letters denote significant differences among the conditions ($p < 0.05$, one-way ANOVA followed by Tukey's test for multiple comparisons). Values are means ± SD ($n = 3$).

2.6. Effect of ACC on Aerenchyma Formation in the Presence of NPA

The reduction of ACS1 and ACO5 genes in the WT roots by the NPA treatment suggested that auxin stimulates ethylene biosynthesis through the transcriptional regulation of ethylene biosynthesis genes. To test this hypothesis, the WT roots were treated with the ethylene precursor ACC in the presence of NPA. Twenty-day-old aerobically grown WT seedlings were transferred to aerated conditions with or without 0.5 µM NPA and/or 10 µM ACC. After 48 h, root elongation of NPA treated seedlings (31.3 ± 3.4 mm) was 25.4% less than that of untreated seedlings (42.0 ± 7.0 mm), while root elongation of NPA- and ACC-treated seedlings (14.8 ± 4.1 mm) was 64.9% less than that of untreated seedlings (Supplemental Figure S1d). NPA almost completely blocked aerenchyma formation at 10 to 30 mm from the root tips (Figure 6a,b), whereas ACC restored its formation at 30, 40, and 50 mm (Figure 6a,b). These results indicate that the prevention of ethylene-dependent aerenchyma formation by NPA is at least partly canceled by adding the ethylene precursor ACC.

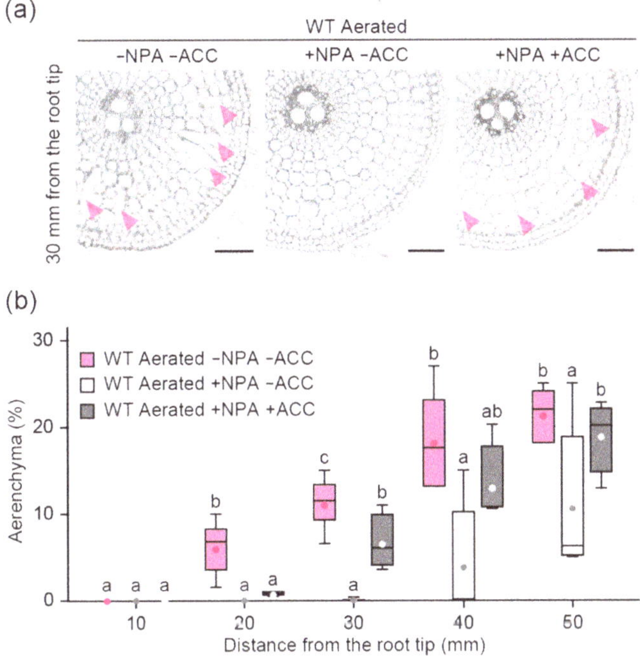

Figure 6. Aerenchyma formation under aerated conditions with or without *N*-1-naphthylphthalamic acid (NPA) or 1-aminocyclopropane-1-carboxylic acid (ACC) treatments. (**a**) Cross-sections at 30 mm from the tips of adventitious roots of the wild type (WT). Aerenchyma is indicated by magenta arrowheads. Bars = 100 µm. (**b**) Percentages of aerenchyma in root cross-sectional area at 10, 20, 30, 40, and 50 mm. Twenty-day-old aerobically grown WT seedlings were further grown under aerated conditions without NPA and ACC, with 0.5 µM NPA and without ACC or with 0.5 µM NPA and 10 µM ACC for 48 h. Different lowercase letters denote significant differences among the conditions ($p < 0.05$, one-way ANOVA followed by Tukey's test for multiple comparisons). Boxplots show the median (horizontal lines), 25th to 75th percentiles (edges of the boxes), minimum to maximum (edges of the whiskers), and mean values (dots in the boxes) ($n = 4$–6).

3. Discussion

The present results demonstrate that auxin is required for inducible aerenchyma formation in rice roots. Aerenchyma formation in the WT and *iaa13* roots was induced by the growth under stagnant conditions (Figure 1b,c), and under aerated conditions with an ethylene precursor ACC treatment

(Figure 2b,c). These results suggest that oxygen deficiency and ethylene can stimulate inducible aerenchyma formation in *iaa13*. On the other hand, the mutated *IAA13* (*iaa13*) gene is expressed predominantly at the apical part of the *iaa13* roots, and the dominant negative effect of the mutated IAA13 protein on aerenchyma formation is restricted to the apical part of the roots [20]. Moreover, the curly root phenotype in the rice *pin2* mutant, which is caused by the asymmetric auxin distribution at the apical part of the roots, is rescued in roots of the *pin2 iaa13* double mutant [25]. These observations suggest that the mutated IAA13 protein affects ethylene-dependent aerenchyma formation at the apical part of the roots. Indeed, the difference in the levels of aerenchyma formation at 20 mm from the root tips between aerated and stagnant conditions, and between aerated conditions with and without ACC, was significantly lower in *iaa13* than that in the WT (Figure 3a,b). These results suggest that the AUX/IAA-mediated auxin signaling is involved in ethylene-dependent inducible aerenchyma formation under oxygen-deficient conditions.

During constitutive aerenchyma formation in rice roots, the auxin transport inhibitor NPA completely blocks the death of the cortical cells [20]. NPA also abolished aerenchyma formation at 10 to 20 mm from the root tips under stagnant conditions (Figure 4c). As the elongation of the WT roots with NPA was ~25 mm under stagnant conditions (Supplemental Figure S1c), aerenchyma formation in the cortex was completely blocked by NPA even under stagnant conditions. These results are interesting because they showed that auxin is required not only for constitutive aerenchyma formation but also for inducible aerenchyma formation in rice roots. In Arabidopsis roots, exogenous ACC treatment was found to reduce lateral root formation by enhancing auxin level at the apical part of the roots, whereas a knockout mutant of the ethylene-signaling gene *ETHYLENE INSENSITIVE2* (*EIN2*) is defective in these responses [26]. Similar observations were reported in the apical hook formation [27] and the control of root gravitropism [28], all of which are directly regulated by the auxin signaling [29–31]. If this is also the case for aerenchyma formation, it is reasonable that inhibition of auxin transport from the shoots to root tips by NPA abolished aerenchyma formation under oxygen-deficient conditions (Figure 4c).

On the other hand, another possibility is that auxin affects ethylene biosynthesis during inducible aerenchyma formation in rice roots, as the NPA treatment reduced the expression levels of *ACS1* and *ACO5* (which have highest expression levels among the *ACS* and *ACO* homologs in rice roots during inducible aerenchyma formation [23]) under stagnant conditions (Figure 5a,b). In Arabidopsis, some *ACS* and *ACO* genes are transcriptionally activated by auxin [32–34]. In roots of maize, exogenous natural auxin treatment increases aerenchyma formation, possibly by stimulating ethylene biosynthesis [35]. In maize roots, auxin-dependent constitutive aerenchyma formation does not generally occur under aerobic conditions [36–38], which suggests that exogenous auxin stimulates the ethylene-dependent pathway [35]. These results further support the idea that auxin is involved in ethylene-dependent aerenchyma formation through the control of ethylene biosynthesis. Indeed, the exogenous treatment of ACC partly restored aerenchyma formation in the WT roots in the presence of NPA (Figure 6b). Although the expression level of *ACO5* was decreased by the NPA treatment (Figure 5b), the exogenously supplied ACC could still enhance the ethylene production by the remaining activity of ACOs, thereby stimulating ethylene-dependent inducible aerenchyma formation in the WT roots (Figure 6b). Similar observations were previously obtained for a rice mutant, which has lower expression levels of *ACS1* and *ACO5* in the roots [11,23]. So far, we cannot rule out the possibility that the reduced root elongation rate by NPA and ACC affects the amounts of aerenchyma formation, as the root elongation of the WT is severely reduced by the NPA and ACC treatments (Supplemental Figure S1d). Further studies using the auxin and ethylene biosynthesis and/or signaling mutants are needed to understand the molecular mechanisms underlying the relationship between auxin and ethylene during inducible aerenchyma formation and how this relationship contributes to the fine-tuning of lysigenous aerenchyma formation in rice roots under oxygen-deficient conditions.

4. Materials and Methods

4.1. Plant Materials and Growth Conditions

Seeds of the rice *iaa13* mutant [19] and its background wild type (cv. Taichung 65; T65) were sterilized in 0.5% (*v/v*) sodium hypochlorite for 30 min and rinsed with deionized water. The seeds were germinated on petri dishes filled up with deionized water and put in a growth chamber at 28 °C under dark conditions. After 2 days, seeds were placed on a mesh floating on top of an aerated quarter strength nutrient solution at 28 °C under 24 h light conditions (photosynthetically active radiation, 200–250 μmol m^{-2} s^{-1}) for 4 d. The composition of the nutrient solution is described by Colmer et al. [22]. Seedlings (6-d-old) were then transferred to 5-L pots (8–12 plants per pot, 250 mm height × 120 mm length × 180 mm width) containing aerated full-strength nutrient solution. After 7 days, 13-day-old rice plants were transferred newly prepared full-strength nutrient solution and further grown for 7 days. Twenty-day-old seedlings were transferred to 5-L pots containing an aerated full-strength nutrient solution or stagnant solution. Stagnant solution, which mimics waterlogged soils [24], contained 0.1% (*w/v*) dissolved agar and was deoxygenated (dissolved O_2, <0.5 mg L^{-1}) prior to use by flushing with N_2 gas.

4.2. Chemical Treatments

For each treatment, 20-day-old rice seedlings were transferred to 2-L pots (4 plants per pot, 250 mm height × 80 mm length × 120 mm width) containing nutrient solution. For the 1-aminocyclopropane-1-carboxylic acid (ACC) and *N*-1-naphthylphthalamic acid (NPA) treatments, 20-d-old aerobically grown rice seedlings were further grown in aerated or stagnant nutrient solutions with or without 10 μM ACC and/or with or without 0.5 μM NPA for 48 h. The stock solutions of ACC and NPA (both Sigma-Aldrich) were prepared to 100 mM in sterilized water and dimethylformamide, respectively.

4.3. Anatomical Observations

Root cross-sections were prepared from 4-mm-long segments of adventitious roots. For analysis of aerenchyma formation, root segments were cut at the indicated distances (±2 mm) from the tips of adventitious roots. Cross-sections were prepared by hand sectioning with a razor blade. The root cross-sections were photographed using an optical microscope (BX60; OLYMPUS) with a charge-coupled device (CCD) camera (DP70; OLYMPUS). The percentages of each cross-section occupied by aerenchyma were determined with ImageJ software (Ver. 1.43u, US National Institutes of Health).

4.4. qRT-PCR Analysis

Root segments at the indicated distances from the tips of adventitious roots were ground in liquid nitrogen. Total RNA was extracted from the frozen fixed tissues using a RNeasy Plant Mini Kit (QIAGEN) according to the instructions of the manufacturer. Transcript levels were measured using a StepOnePlus Real-Time PCR System (Applied Biosystems) and One Step SYBR PrimeScript RT-PCR Kit II (Takara Bio) as described by Yamauchi et al. [12]. Transcript levels were normalized to the transcript level of *transcription initiation factor IIE* (*TFIIE*). Primer sequences used for the qRT-PCR are shown in Supplemental Table S1.

4.5. Statistical Analyses

Statistical differences between means were calculated using two-sample *t*-tests. For multiple comparisons, data were analyzed by one-way ANOVA and post hoc Tukey's test using SPSS Statistics Version 25 (IBM Software).

Supplementary Materials: The following are available online at http://www.mdpi.com/2223-7747/9/5/610/s1, Figure S1: Root elongation under aerated or stagnant conditions with or without the inhibitor treatments; Figure S2: Aerenchyma formation under aerated or stagnant conditions or aerated conditions with or without 1-aminocyclopropane-1-carboxylic acid (ACC) treatment; Table S1: List of primers used for qRT–PCR analysis.

Author Contributions: Investigation, T.Y. and A.T.; project administration, T.Y. and M.N.; validation, A.T., N.T., and Y.I.; writing—original draft, T.Y.; writing—review and editing, Y.I. and M.N. All authors have read and agreed to the published version of the manuscript.

Funding: This work was partly supported by the Japan Society for the Promotion of Science (KAKENHI Grant 18H02175) (to M.N.) and the Japan Science and Technology Agency PRESTO Grants JPMJPR17Q8 (to T.Y.) and JPMJPR15Q3 (to Y.I.).

Acknowledgments: We thank Hiroki Inahashi for his help in phenotyping the *iaa13* mutant. We also thank T. D. Colmer, O. Pedersen, A.I. Marik, H. Takahashi, S. Nishiuchi, and K. Watanabe for stimulating discussions.

Conflicts of Interest: The authors declare no conflict of interest.

References

1. Colmer, T.D. Long-distance transport of gases in plants: A perspective on internal aeration and radial oxygen loss from roots. *Plant Cell Environ.* **2003**, *26*, 17–36. [CrossRef]

2. Jackson, M.B.; Armstrong, W. Formation of aerenchyma and the processes of plant ventilation in relation to soil flooding and submergence. *Plant Biol.* **1999**, *1*, 274–287. [CrossRef]

3. Evans, D.E. Aerenchyma formation. *New Phytol.* **2003**, *161*, 35–49. [CrossRef]

4. Colmer, T.D.; Voesenek, L.A.C.J. Flooding tolerance: Suites of plant traits in variable environments. *Funct. Plant Biol.* **2009**, *36*, 665–681. [CrossRef]

5. Voesenek, L.A.C.J.; Bailey-Serres, J. Flood adaptive traits and processes: An overview. *New Phytol.* **2015**, *206*, 57–73. [CrossRef]

6. Yamauchi, T.; Colmer, T.D.; Pedersen, O.; Nakazono, M. Regulation of root traits for internal aeration and tolerance to soil waterlogging-flooding stress. *Plant Physiol.* **2018**, *176*, 1118–1130. [CrossRef]

7. Pedersen, O.; Sauter, M.; Colmer, T.D.; Nakazono, M. Regulation of root adaptive anatomical and morphological traits during low soil oxygen. *New Phytol.* **2020**, in press. [CrossRef]

8. Sasidharan, R.; Hartman, S.; Liu, Z.; Martopawiro, S.; Sajeev, N.; van Veen, H.; Yeung, E.; Voesenek, L.A.C.J. Signal dynamics and interactions during flooding stress. *Plant Physiol.* **2018**, *176*, 1106–1117. [CrossRef]

9. Mustroph, A.; Steffens, B.; Sasidharan, R. Signalling interactions in flooding tolerance. *Annu. Plant Rev.* **2018**, *1*, 417–458.

10. Yang, S.F.; Hoffman, N.E. Ethylene biosynthesis and its regulation in higher plants. *Annu. Rev. Plant Physiol.* **1984**, *35*, 155–189. [CrossRef]

11. Yamauchi, T.; Shiono, K.; Nagano, M.; Fukazawa, A.; Ando, M.; Takamure, I.; Mori, H.; Nishizawa, N.K.; Kawai-Yamada, M.; Tsutsumi, N.; et al. Ethylene biosynthesis is promoted by very-long-chain fatty acids during lysigenous aerenchyma formation in rice roots. *Plant Physiol.* **2015**, *169*, 180–193. [CrossRef] [PubMed]

12. Yamauchi, T.; Yoshioka, M.; Fukazawa, A.; Mori, H.; Nishizawa, N.K.; Tsutsumi, N.; Yoshioka, H.; Nakazono, M. An NADPH oxidase RBOH functions in rice roots during lysigenous aerenchyma formation under oxygen-deficient conditions. *Plant Cell.* **2017**, *29*, 775–790. [CrossRef] [PubMed]

13. Steffens, B.; Geske, T.; Sauter, M. Aerenchyma formation in the rice stem and its promotion by H_2O_2. *New Phytol.* **2011**, *190*, 369–378. [CrossRef] [PubMed]

14. Ulmasov, T.; Hagen, G.; Guilfoyle, T.J. ARF1, a transcription factor that binds to auxin-response elements. *Science* **1997**, *276*, 1865–1868. [CrossRef]

15. Rouse, D.; Mackay, P.; Stirnberg, P.; Estelle, M.; Leyser, O. Changes in auxin response from mutations in an *AUX/IAA* gene. *Science* **1998**, *279*, 1371–1373. [CrossRef] [PubMed]

16. Gray, W.M.; Kepinski, S.; Rouse, D.; Leyser, O.; Estelle, M. Auxin regulates SCF(^{TIR1})-dependent degradation of AUX/IAA proteins. *Nature* **2001**, *414*, 271–276. [CrossRef]

17. Wang, D.; Pei, K.; Fu, Y.; Sun, Z.; Li, S.; Liu, H.; Tang, K.; Han, B.; Tao, Y. Genome-wide analysis of the *auxin response factors* (ARF) gene family in rice (*Oryza sativa*). *Gene* **2007**, *394*, 13–24. [CrossRef]

18. Jain, M.; Kaur, N.; Garg, R.; Thakur, J.K.; Tyagi, A.K.; Khurana, J.P. Structure and expression analysis of early auxin-responsive *Aux/IAA* gene family in rice (*Oryza sativa*). *Funct. Integr. Genom.* **2006**, *6*, 47–59. [CrossRef]

19. Kitomi, Y.; Inahashi, H.; Takehisa, H.; Sato, Y.; Inukai, Y. OsIAA13-mediated auxin signaling is involved in lateral root initiation in rice. *Plant Sci.* **2012**, *190*, 116–122. [CrossRef]

20. Yamauchi, T.; Tanaka, A.; Inahashi, H.; Nishizawa, N.K.; Tsutsumi, N.; Inukai, Y.; Nakazono, M. Fine control of aerenchyma and lateral root development through AUX/IAA- and ARF-dependent auxin signaling. *Proc. Natl. Acad. Sci. USA* **2019**, *116*, 20770–20775. [CrossRef]

21. Justin, S.H.F.W.; Armstrong, W.R. Evidence for the involvement of ethene in aerenchyma formation in adventitious roots of rice (*Oryza sativa* L.). *New Phytol.* **1991**, *118*, 49–62. [CrossRef]

22. Colmer, T.D.; Cox, M.C.H.; Voesenek, L.A.C.J. Root aeration in rice (*Oryza sativa*): Evaluation of oxygen, carbon dioxide, and ethylene as possible regulators of root acclimatizations. *New Phytol.* **2006**, *170*, 767–778. [CrossRef] [PubMed]

23. Yamauchi, T.; Tanaka, A.; Mori, H.; Takamure, I.; Kato, K.; Nakazono, M. Ethylene-dependent aerenchyma formation in adventitious roots is regulated differently in rice and maize. *Plant Cell Environ.* **2016**, *39*, 2145–2157. [CrossRef] [PubMed]

24. Wiengweera, A.; Greenway, H.; Thomson, C.J. The use of agar nutrient solution to simulate lack of convection in waterlogged soils. *Ann. Bot.* **1997**, *80*, 115–123. [CrossRef]

25. Inahashi, H.; Shelley, I.J.; Yamauchi, T.; Nishiuchi, S.; Takahashi-Nosaka, M.; Matsunami, M.; Ogawa, A.; Noda, Y.; Inukai, Y. *OsPIN2*, which encodes a member of the auxin efflux carrier proteins, is involved in root elongation growth and lateral root formation patterns via the regulation of auxin distribution in rice. *Physiol. Plant.* **2018**, *164*, 216–225. [CrossRef]

26. Negi, S.; Ivanchenko, M.G.; Muday, G.K. Ethylene regulates lateral root formation and auxin transport in *Arabidopsis thaliana*. *Plant J.* **2008**, *55*, 175–187. [CrossRef]

27. Vandenbussche, F.; Petrasek, J.; Zadnikova, P.; Hoyerova, K.; Pesek, B.; Raz, V.; Swarup, R.; Bennett, M.; Zazimalova, E.; Benkova, E.; et al. The auxin influx carriers AUX1 and LAX3 are involved in auxin-ethylene interactions during apical hook development in *Arabidopsis thaliana* seedlings. *Development* **2010**, *137*, 597–606. [CrossRef]

28. Nziengui, H.; Lasok, H.; Kocherrsperger, P.; Ruperrti, B.; Reveille, F.; Palme, K.; Ditengou, F.A. Root gravitropism is regulated by a crosstalk between para-aminobenzoic acid, ethylene, and auxin. *Plant Physiol.* **2018**, *178*, 1370–1389. [CrossRef]

29. Muday, G.K.; Rahman, A.; Binder, B.M. Auxin and ethylene: Collaborators or competitors? *Trends Plant Sci.* **2012**, *17*, 181–195. [CrossRef]

30. Lewis, D.R.; Negi, S.; Sukumar, P.; Muday, G.K. Ethylene inhibits lateral root development, increases IAA transport and expression of PIN3 and PIN7 auxin efflux carriers. *Development* **2011**, *138*, 3485–3495. [CrossRef]

31. Zadnikova, P.; Petrasek, J.; Marhavy, P.; Raz, V.; Vandenbussche, F.; Ding, Z.; Schwarzerova, K.; Morita, M.T.; Tasaka, M.; Hejatko, J.; et al. Role of PIN-mediated auxin efflux in apical hook development of *Arabidopsis thaliana*. *Development* **2010**, *137*, 607–617. [CrossRef] [PubMed]

32. Abel, S.; Nguyen, M.D.; Chow, W.; Theologis, A. *ACS4*, a primary indole acetic acid-responsive gene encoding 1-aminocyclopropane-1-carboxylate synthase in *Arabidopsis thaliana*. *J. Biol. Chem.* **1995**, *270*, 19093–19099. [CrossRef] [PubMed]

33. Tian, Q.; Uhlir, N.J.; Reed, J.W. Arabidopsis SHY2/IAA3 inhibits auxin-regulated gene expression. *Plant Cell* **2002**, *14*, 301–319. [CrossRef] [PubMed]

34. Vandenbussche, F.; Vriezen, W.H.; Samlle, J.; Laarhoven, L.J.J.; Harrenn, F.J.M.; Van Der Straeten, D. Ethylene and auxin control the Arabidopsis response to decreased light intensity. *Plant Physiol.* **2003**, *133*, 517–527. [CrossRef]

35. Justin, S.H.F.W.; Armstrong, W. A reassessment of the influence of NAA on aerenchyma formation in maize roots. *New Phytol.* **1991**, *117*, 607–618. [CrossRef]

36. Mano, Y.; Omori, F. Relationship between constitutive root aerenchyma formation and flooding tolerance in *Zea nicaraguensis*. *Plant Soil* **2013**, *370*, 447–460. [CrossRef]

37. Yamauchi, T.; Abe, F.; Tsutsumi, N.; Nakazono, M. Root cortex provides a venue for gas-space formation and is essential for plant adaptation to waterlogging. *Front. Plant Sci.* **2019**, *10*, 259. [CrossRef]

38. Gong, F.; Takahashi, H.; Omori, F.; Wang, W.; Mano, Y.; Nakazono, M. QTLs for constitutive aerenchyma from *Zea nicaraguensis* improve tolerance of maize to root-zone oxygen deficiency. *J. Exp. Bot.* **2019**, *70*, 6475–6487. [CrossRef]

Article

Overexpression of *Jatropha curcas ERFVII2* Transcription Factor Confers Low Oxygen Tolerance in Transgenic Arabidopsis by Modulating Expression of Metabolic Enzymes and Multiple Stress-Responsive Genes

Piyada Juntawong [1,2,3,*], **Pimprapai Butsayawarapat** [1], **Pattralak Songserm** [1], **Ratchaneeporn Pimjan** [1] **and Supachai Vuttipongchaikij** [1,2,3]

[1] Department of Genetics, Faculty of Science, Kasetsart University, Bangkok 10900, Thailand; pimprapai.bu@ku.th (P.B.); p.songserm017480@gmail.com (P.S.); ratchaneeporn.pj@gmail.com (R.P.); fsciscv@ku.ac.th (S.V.)

[2] Center for Advanced Studies in Tropical Natural Resources, National Research University-Kasetsart University, Bangkok 10900, Thailand

[3] Omics Center for Agriculture, Bioresources, Food and Health, Kasetsart University (OmiKU), Bangkok 10900, Thailand

* Correspondence: fscipdj@ku.ac.th or pjuntawong@gmail.com; Tel.: +66-02-562-5555

Received: 18 July 2020; Accepted: 18 August 2020; Published: 20 August 2020

Abstract: Enhancing crop tolerance to waterlogging is critical for improving food and biofuel security. In waterlogged soils, roots are exposed to a low oxygen environment. The group VII ethylene response factors (ERFVIIs) were recently identified as key regulators of plant low oxygen response. Oxygen-dependent N-end rule pathways can regulate the stability of ERFVIIs. This study aims to characterize the function of the *Jatropha curcas ERFVIIs* and the impact of N-terminal modification that stabilized the protein toward low oxygen response. This study revealed that all three JcERFVII proteins are substrates of the N-end rule pathway. Overexpression of *JcERFVII2* conferred tolerance to low oxygen stress in Arabidopsis. In contrast, the constitutive overexpression of stabilized *JcERFVII2* reduced low oxygen tolerance. RNA-seq was performed to elucidate the functional roles of *JcERFVII2* and the impact of its N-terminal modification. Overexpression of both wildtype and stabilized *JcERFVII2* constitutively upregulated the plant core hypoxia-responsive genes. Besides, overexpression of the stabilized *JcERFVII2* further upregulated various genes controlling fermentative metabolic processes, oxidative stress, and pathogen responses under aerobic conditions. In summary, JcERFVII2 is an N-end rule regulated waterlogging-responsive transcription factor that modulates the expression of multiple stress-responsive genes; therefore, it is a potential candidate for molecular breeding of multiple stress-tolerant crops.

Keywords: abiotic stress; RNA-seq; transcription factor; waterlogging

---

1. Introduction

Waterlogging can damage most crops, creating one of the most significant problems in agriculture worldwide. During the heavy rainy season in the plain area, soil can quickly become waterlogged due to poor drainage, creating a low oxygen environment in the root area underground. Low oxygen stress leads to the induction of a particular set of genes involved in carbohydrate utilization, energy metabolism, and fermentation to sustain ATP production [1]. Over the long term, low oxygen stress means morphological adaptation is required to keep the level of oxygen under control [2].

Since global climate change could increase the number of flooding events, improved crop varieties with waterlogging tolerance are essential [3,4].

The ethylene response factor (*ERF*) family is one of the largest plant-specific transcription factor families characterized by a single DNA-binding domain, APETALA2 (AP2), with expanded functions in hormonal response, development, and tolerance to biotic and abiotic stresses [5–7]. Group VII ERF (ERFVII) transcription factors is a subgroup of ERFs that has been recognized as a critical factor controlling the expression of numerous genes involved in an adaptive response to low oxygen stress in model plants [8–11]. A characteristic feature of all Arabidopsis *ERFVIIs* (*RAP2.2*, *RAP2.3*, *RAP2.12*, *HRE1*, and *HRE2*) is a conserved N-terminal motif (N-degron; [5,12]), which enables them to be degraded by oxygen and nitric oxide (NO)-dependent N-end rule pathways [13–15]. Overexpression of the Arabidopsis *ERFVIIs* enhances flooding or low oxygen stress tolerance in transgenic Arabidopsis plants [13,14,16–21]. Interestingly, overexpression of the stabilized (N-terminal mutation) *HRE1* and *HRE2* in Arabidopsis further improved low oxygen tolerance compared to Arabidopsis lines overexpressing the wildtype *HRE1* and *HRE2* [14]. However, enhanced stability of *RAP2.12* resulted in reduced biomass under aerobic conditions and did not increase tolerance to low oxygen stress in transgenic Arabidopsis plants [13,22]. Transcription of genes encoding for fermentative and starch degradation enzymes were constitutively activated in transgenic Arabidopsis overexpressing stable *RAP2.12*, which negatively affected growth and development under the aerobic condition and reduced tolerance to low oxygen stress [22]. In rice, previous studies identified *ERFVIIs*, *Snorkels*, and *Sub1A*, as a key player orchestrating the escape and quiescence response needed to survive flash-flood and prolonged submergence, respectively [23,24]. Although Sub1A contained an N-degron, it was not a substrate of the N-end rule pathway [11,14]. It has recently been shown that *Sub1A* transcriptionally activates the other *ERFVIIs*, *ERF66*, and *ERF67*, resulting in transcriptional accumulation of anaerobic survival genes and improved submergence tolerance in rice [11]. Remarkably, constitutive expression of the stabilized wheat *ERFVII*, *TaERFVII.1*, enhanced tolerance to waterlogging in transgenic wheat without negative impacts on development and grain yield under aerobic conditions [25]. Thus, identification, selection, and modification of the *ERFVII* genes could be a valuable approach to improve crop waterlogging tolerance.

Jatropha curcas is a drought-tolerant oilseed crop for biodiesel production. It can be grown on marginal land without competing with other food crops [26]. However, waterlogging caused a significant reduction of growth and biomass yield, suggesting that Jatropha is extremely sensitive to waterlogging [27,28]. Undoubtedly, genetic improvement of waterlogging tolerant Jatropha is needed to increase Jatropha oil production. Previously, we transcriptionally profiled gene expression in Jatropha and found that waterlogging promoted anaerobic respiration, but inhibited carbohydrate synthesis, cell wall biogenesis, and plant growth [29]. Based on our previous study, *ERFVIIs* had been proposed as candidate genes for genetic engineering of waterlogging tolerant Jatropha [29].

In this study, we cloned and evaluated the tissue-specific expression and waterlogging responsive pattern of Jatropha *ERFVIIs* (*JcERFVIIs*). Next, we followed up by examining the N-end rule regulated protein stability of *JcERFVIIs* and overexpressing the *JcERFVII2* genes in Arabidopsis to evaluate the flooding and low oxygen tolerant phenotype. Finally, the molecular function of *JcERFVII2* was further investigated by transcriptome profiling of transgenic Arabidopsis lines.

2. Results

2.1. Cloning and Bioinformatics Analysis of JcERFVII Genes

Previously, we identified three *JcERFVIIs*, namely *JcERVII1* (*Jcr4S00420.40*), *JcERFVII2* (*Jcr4S00982.160*), and *JcERFVII3* (*Jcr4S01651.60*), from the Jatropha genome [29]. All three *JcERFVIIs* possess a conserved N-degron signal [NH$_2$-MCGGAII(A/S)D] [29]. The full-length open reading frames (ORFs) of *JcERFVIIs* were cloned. Sequence analysis reveals that *JcERFVII1*, *JcERFVII2*, and *JcERFVII3* are composed of 1158, 762, and 945 nucleotides, respectively (Supplementary Materials Data S1).

Deduced amino acid sequences of *JcERFVII1*, *JcERFVII2*, and *JcERFVII3* provided encoded proteins with 385, 253, and 314 amino acids with predicted molecular weights of 43, 29, and 36 kD, respectively. The amino acid sequences of the *JcERFVIIs* were aligned with amino acid sequences from all five members of the *Arabidopsis ERFVIIs*, including *RAP2.2*, *RAP2.3*, *RAP2.12*, *HRE1*, and *HRE2*, and the phylogenetic relationship was evaluated. The results revealed that *JcERFVII1* clustered with *RAP2.2* and *RAP2.12*, *JcERFVII2* clustered with *RAP2.3*, and *JcERFVII3* clustered with *HRE2* (Figure 1A). Based on the previously reported Arabidopsis ERFVII protein domain data [5], MEME assisted domain analysis also showed the similarity among each phylogenetic cluster (Figure 1B).

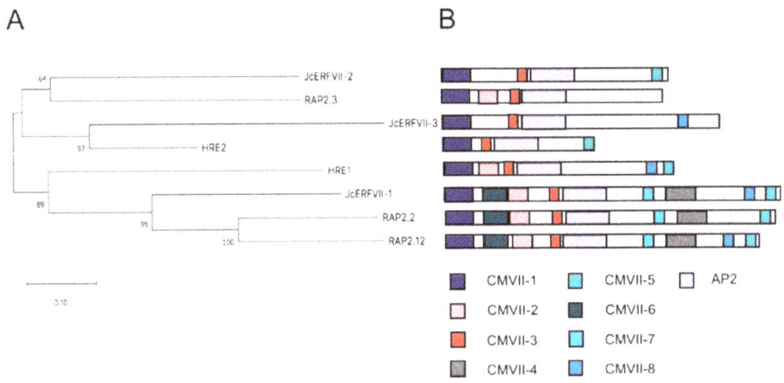

Figure 1. Phylogenetic and domain architecture analysis of Jatropha ERFVIIs. (**A**) The phylogenetic tree based on the amino acid sequence of Arabidopsis and Jatropha ERFVIIs. The numbers are bootstrap values after 1000 replicates. Scale bar represents genetic distance. (**B**) Diagram representing domain architecture of Jatropha ERFVIIs following previously published motifs [5].

2.2. Tissue-Specific and Waterlogging Expression Patterns of JcERFVIIs

We examined the expression pattern of the three *JcERFVIIs* in the tissue of Jatropha seedlings using qRT-PCR (Figure 2). Under aerobic conditions, the expression of all three *JcERFVIIs* can be found in roots, leaves, apical buds, and petioles of Jatropha seedlings. *JcERFVII1* exhibited the highest expression in apical buds and the lowest expression in leaves, while *JcERFVII2* and *JcERVII3* exhibited the highest expression in roots and the lowest expression in leaves. We also compared the expression levels of the three *JcERFVIIs* using the transcriptome data from roots, leaves, stems, and shoot apexes collected in a publically available *J. curcas* database (JCDB) [30]. We found that among the three *JcERFVIIs*, *JcERFVII1*, and *JcERFVII3* displayed the highest and the lowest expression, respectively (Supplementary Materials Figure S1). Moreover, the expression levels of *JcERFVII2* in Jatropha tissues were more uniform than those of the others (Supplementary Materials Figure S1).

To explore whether the *JcERFVIIs* are related to waterlogging response, we examined the expression patterns of *JcERFVIIs* in Jatropha seedlings subjected to 24 h soil waterlogging. In the waterlogged root, the expression of *JcERFVII2* and *JcERFVII3* was significantly increased, while the expression of *JcERFVII1* remained unaffected (Figure 2). Besides, waterlogging resulted in a significant reduction of *JcERFVII1*, *JcERFVII2*, and *JcERFVII3* expression in apical buds (Figure 2).

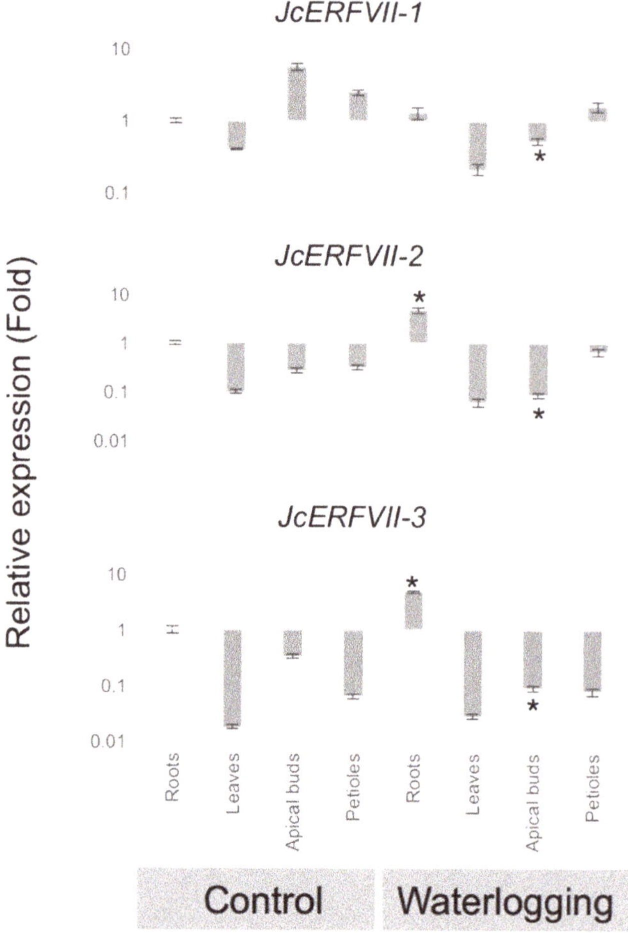

Figure 2. Quantitative analysis of *JcERFVII* expression. Relative expression values of *JcERFVIIs 1–3* from roots, leaves, apical buds, and petioles of Jatropha seedlings under control and 1 d waterlogging. Relative expression was normalized to the abundance of *UBQ10*. Data represent mean ± SE (n = 3). Asterisks indicate $p < 0.05$ (*t*-test).

2.3. Stability of JcERFVIIs In Vitro

Since all three JcERFVII proteins possess a conserved N-degron, we hypothesized that they are targets of the N-end rule pathway. We used a previously established in vitro assay by which proteins are expressed in a rabbit reticulocyte system containing essential components for the N-end rule pathway [14]. Western blot analysis of in vitro translated JcERFVII proteins tagged with a haemagglutinin (3xHA) epitope demonstrated a single band with the migration pattern corresponding to their predicted molecular weight (Figure 3). Our results demonstrated that mutation of cysteine to alanine at amino acid residue position 2 (MA) in all three JcERFVIIs increased protein stability after 60 and 120 min incubation periods (Figure 3). We also showed that supplementation of MG132, a proteasome inhibitor, increased the accumulation of wildtype (MC) JcERFVII proteins in vitro (Figure 3). These data strongly suggest that JcERFVIIs are substrates of the N-end rule pathway.

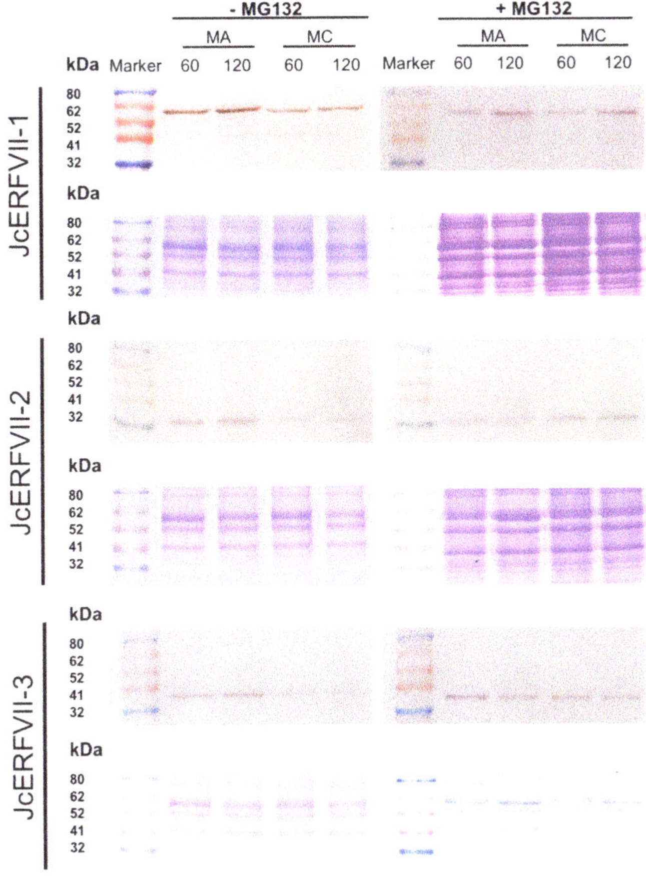

Figure 3. Jatropha ERFVIIs are substrates for the N-end rule pathway in vitro. Western blot analysis of in vitro stability of HA-tagged wildtype (MC) and stable mutation (MA) of JcERFVIIs 1–3 in the absence or presence of proteasome inhibitor (MG132). 60 and 120 indicate incubation time in minutes. Coomassie staining of a similar SDS-PAGE used for western blotting was used as a loading control.

2.4. Overexpression of the JcERFVII2 Enhanced Low Oxygen Tolerance in Arabidopsis

Based on previous studies, among the five members of the Arabidopsis *ERFVIIs*, the role of *RAP2.3* in low oxygen responses has been less explored. Therefore, we aim to characterize *JcERFVII2* function towards low oxygen response. To investigate the function of *JcERFVII2* in providing tolerance to low oxygen stress and whether modulation of its stability could affect the stress tolerance, we generated transgenic Arabidopsis lines overexpressing *MA-* or *MC-JcERFVII2* driven by the CaMV35S promoter. Ten and 5 transgenic lines overexpressing *MA-* and *MC-JcERFVII2* were generated and 4 independent homozygous lines, *35S:MA-JcERFVII2-1* (MA-Line1), *35S:MA-JcERFVII2-7* (MA-Line7), *35S:MC-JcERFVII2-3* (MC-Line3), and *35S:MC-JcERFVII2-5* (MC-Line5), were selected for functional analysis. Semi-quantitative RT-PCR analysis confirmed the expression of *JcERFVII-2* in the transgenic lines (Supplementary Materials Figure S2A).

For submergence stress, the four transgenic lines and the wildtype *A. thaliana* Col-0 were grown until reaching the 10 leaf-stage and subjected to submergence stress for 3 d (Figure 4A). While overexpression of *MC-JcERFVII2* did not show any effect on the phenotype of the transgenic lines (MC-Line3 and MC-Line5) (Figure 4A,B), it considerably improved submergence tolerance with respect to the wildtype, as demonstrated by the increases of dry weight after submergence (Figure 4C). On the other hand, transgenic lines overexpressing *MA-JcERFVII2* (MA-Line1 and MA-Line7) showed reduced plant growth when grown under aerobic conditions (Figure 4A,B) and decreased submergence tolerance when compared with the wildtype (Figure 4A,C).

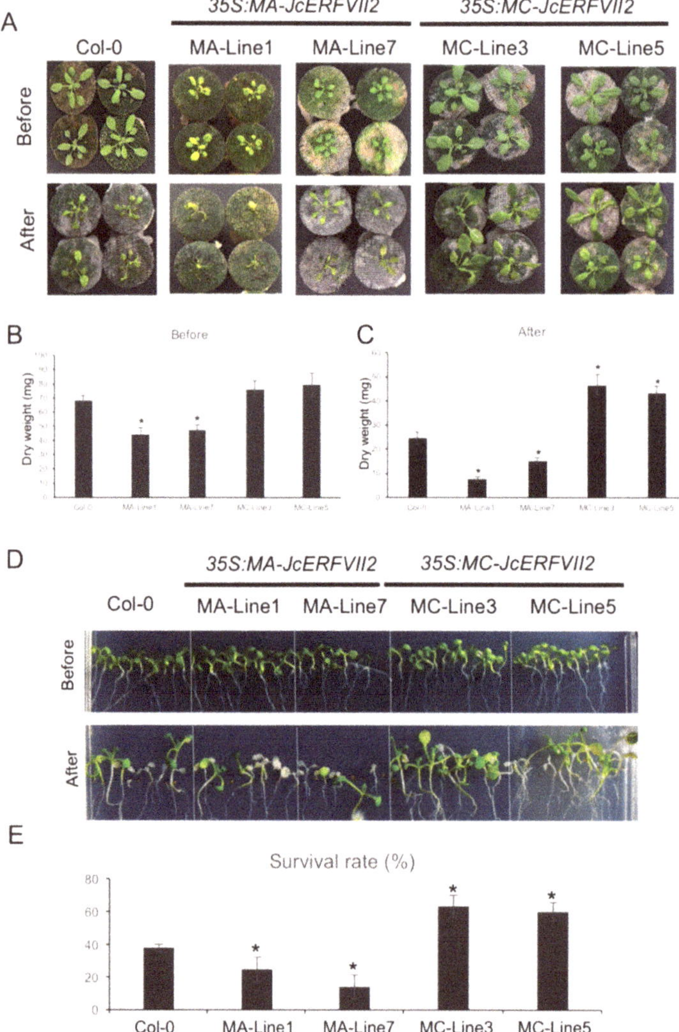

Figure 4. Overexpression of wildtype *JcERFVII2* confer low-oxygen tolerance. (**A**) Phenotype of wildtype Arabidopsis (Col-0) Arabidopsis transgenic lines overexpressing wildtype (MC) and stable (MA) *JcERFVII2* subjected to 3 d submergence stress. (**B**) Dry weight of 3-week-old rosette leaves before submergence (n = 10). (**C**) Dry weight of rosette leaves after 3 d submergence (n = 10). (**D**) Phenotypes of the Arabidopsis transgenic seedlings overexpressing *MA-* and *MC-JcERFVII2* after 3 d hypoxia and 3 d recovery. (**E**) Percentage of seedling survival for wildtype Arabidopsis (Col-0) and the Arabidopsis lines overexpressing *MA* and *MC-JcERFVII2*. Data are means of triplicate experiments. Each experiment contains 6–10 plants/genotype. Error bars represent SD. Asterisks indicate $p < 0.05$ (*t*-test).

For low oxygen survival assay, after 3 d of 2% oxygen and 3 d of recovery under aerobic condition, MC-Line3 and MC-Line5 showed significantly higher survival rate (63% and 60%, respectively) than that of the wildtype (38%) (Figure 4D,E). However, MA-Line1 and MA-Line7 displayed a significantly lower survival rate (24% and 13%, respectively) (Figure 4D,E). Together, these results clearly demonstrated that the constitutive overexpression of *MC-JcERFVII2* could enhance growth and survival under low oxygen in transgenic Arabidopsis, while that of *MA-JcERFVII2* resulted in growth reduction under aerobic conditions and poorly performed under low-oxygen stress.

2.5. Transcriptome Profiling of Transgenic Arabidopsis Overexpressing JcERFVII2

To analysis the impact of the N-terminal modification on the molecular function of the *JcERFVII2* gene, we profiled the transcriptome of transgenic Arabidopsis overexpressing *MA-* and *MC-JcERFVII2* (MA-Line1 and MC-Line3, respectively) using Col-0 as a control genotype. Two biological replicates of total RNAs from 7 d.o. seedlings grown in aerobic conditions were isolated and subjected to RNA-seq. RNA-seq reads were mapped to the *A. thaliana* TAIR10 genome. The number of reads aligned back to each gene was obtained for differential gene expression analysis. Transcriptome analysis identified 344 and 282 differentially expressed genes (DEGs) with significant changes in gene expression as evaluated by false discovery rate (FDR) < 0.05 from *MA-* or *MC-JcERFVII2* overexpressing lines, respectively (Figure 5A; Supplementary Materials Table S1). Of 282 DEGs from the MC-Line3, 29 DEGs (10%) were upregulated, and 253 DEGs (90%) were downregulated (Supplementary Materials Table S1), while, of 344 DEGs in the MA-Line1, 122 DEGs (35%) were upregulated, and 222 DEGs (65%) were downregulated (Supplementary Materials Table S1). Venn's diagram analysis revealed that 112 DEGs were commonly found in both *MA-* and *MC-JcERFVII2* transgenic lines, while 232 DEGs and 170 DEGs were exclusively found in *MA-* and *MC-JcERFVII2* transgenic lines, respectively (Figure 5A). It should be noted that the endogenous ERFVIIs were not differentially expressed in transgenic lines overexpressing both *MA-* and *MC-JcERFVII2* (Supplementary Materials Table S1). To confirm that, we obtained the CPM (count per million) expression values from our RNA-seq data. Mostly, the expression of the endogenous ERFVII genes in transgenic lines is similar to the Col-0 (Supplementary Materials Figure S2B).

Gene ontology (GO) analysis was performed to obtain the overview of *JcERFVII2* regulated genes using an FDR cutoff of $<1.00 \times 10^{-4}$ (Figure 5B). The results demonstrated that *JcERFVII2* regulated genes function in cellular metabolic processes and several aspects of stress responses, as observed in the enriched GO terms derived from both *MA-* and *MC-JcERFVII2* DEGs (Figure 5B). GO terms related to response to stress, stimulus, chemical, and oxygen-containing compounds were enriched in the 112, 170, and 232 DEGs previously described (Figure 5B). Interestingly, 8 out of 49 core hypoxia-responsive genes (*At2g16060: hemoglobin 1 (Hb1), At3g02550: LOB domain-containing protein 41 (LBD41)), At1g43800; Acyl carrier protein (ACP) desaturase 6 (AAD6), At5g15120: Plant cysteine oxidase 1 (PCO1), At5g39890: PCO2, At4g33070: Pyruvate decarboxylase 1 (PDC1), At2g17850, and At5g66985*), which are universally induced under low oxygen [31] can be found in the DEGs of *MA* and *MC-JcERFVII2* overexpressing lines (Supplementary Materials Table S1).

Since we observed more DEGs being upregulated in the *MA-JcERFVII2* overexpressing line than that of the *MC-JcERFVII2*, we carefully examined the expression of the 122 upregulated DEGs from the *MA-JcERFVII2* transgenic line (Figure 5C). Of the 122 upregulated *MA-JcERFVII2* DEGs, 22 of these were also upregulated in the *MC- JcERFVII2* overexpressing line (Supplementary Materials Table S1). The rest of them (100 genes) were not differentially expressed in the *MC-JcERFVII2* overexpressing line (Figure 5C; Supplementary Materials Table S1). A possible explanation for these results is that the increase in *JcERFVII2* protein abundance could elevate the expression of these 100 genes. GO analysis of the upregulated DEGs from the *MA-JcERFVII2* transgenic line revealed their roles in response to multiple stresses, including hypoxia (FDR: 2.60×10^{-7}), oxidative stress (FDR: 6.60×10^{-7}), and other organisms (1.10×10^{-6}) (Figure 5D; Supplementary Materials Table S2).

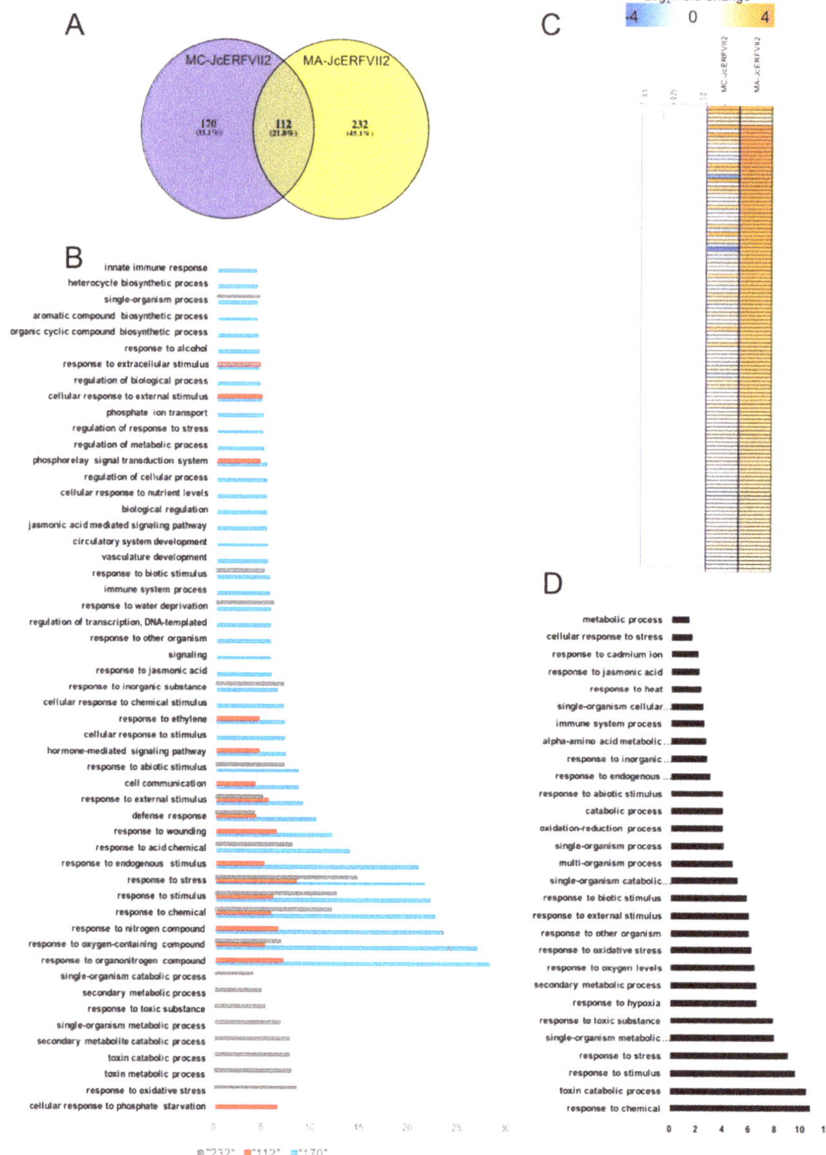

Figure 5. Overexpression of *JcERFVII2* upregulates multiple stress-responsive genes in Arabidopsis. (**A**) Venn diagram of differentially expressed genes (DEGs) from the Arabidopsis lines overexpressing *MA-* and *MC-JcERFVII2*. (**B**) Enrichment of GO terms from DEGs of the Arabidopsis lines overexpressing *MA-* and *MC-JcERFVII2*. Bar chart represents −log10 adjusted *p*-values of enrichments GO terms. (**C**) Heat map represents the expression pattern of the upregulated DEGs derived from the Arabidopsis lines overexpressing *MA-* and *MC-JcERFVII2*. (**D**) Selected enrichment GO terms of the upregulated DEGs found in (**C**). Black bar represents −log10 adjusted *p*-values of enrichments GO terms. Data used to generate this figure can be found in Supplementary Materials Tables S1 and S2.

Based on GO enrichment results, DEGs in some specific classes demonstrated co-expression patterns (Figure 6). Several *Plant defensin* (*PDF*) genes were upregulated in both *MA-* and *MC-JcERFVII2* lines (Figure 6A, Supplementary Materials Table S1). *Glutathione transferase* and *peroxidase* genes were upregulated mainly in the *MA-JcERFVII2* line (Figure 6A, Supplementary Materials Table S1). In contrast, specific genes that function in ABA and JA responses were downregulated in both *MA-* and *MC-JcERFVII2* transgenic lines (Figure 6B; Supplementary Materials Table S1). These results altogether indicate that post-translational modification of JcERFVII2 protein under aerobic conditions can affect its regulative function.

A

	ID	Name	MC-ERFVII2-WT_logFC	MA-ERFVII2-WT_logFC
GST	AT1G02930	glutathione S-transferase 6	0.15	1.73
	AT1G02920	glutathione S-transferase 7	0.02	1.69
	AT5G17220	glutathione S-transferase phi 12	0.41	1.82
	AT4G02520	glutathione S-transferase PHI 2	-0.23	1.61
	AT1G69930	glutathione S-transferase TAU 11	-1.21	2.08
	AT1G69920	glutathione S-transferase TAU 12	0.41	3.53
	AT1G17170	glutathione S-transferase TAU 24	0.47	1.97
	AT1G17180	glutathione S-transferase TAU 25	0.66	1.88
	AT1G17190	glutathione S-transferase tau 26	-0.33	-1.77
	AT2G29470	glutathione S-transferase tau 3	-2.70	3.57
Peroxidase	AT4G08770	Peroxidase superfamily protein	0.60	2.72
	AT1G14540	Peroxidase superfamily protein	0.74	2.97
	AT1G14550	Peroxidase superfamily protein	1.51	3.49
PDF	AT5G44420	plant defensin 1.2	2.76	3.54
	AT2G26020	plant defensin 1.2b	2.60	2.00
	AT5G44430	plant defensin 1.2C	3.46	5.14
	AT2G26010	plant defensin 1.3	2.87	3.00

B

	ID	Name	MCERFVII2-WT_logFC	MAERFVII2-WT_logFC
ABA	AT1G66600	ABA overly sensitive mutant 3	0.3	3.4
	AT3G28580	P-loop containing nucleoside triphosphate hydrolases super	0.0	3.1
	AT2G38340	DREB19	-0.8	3.0
	AT3G28210	Stress associated protein 12	0.2	2.5
	AT3G26830	PHYTOALEXIN DEFICIENT 3	0.0	2.4
	AT1G05680	Uridine diphosphate glycosyltransferase 74E2	0.6	2.2
	AT1G43910	P-loop containing nucleoside triphosphate hydrolases super	0.4	2.2
	AT5G26920	Cam-binding protein 60-like G	-1.4	0.0
	AT5G27420	carbon/nitrogen insensitive 1	-1.8	0.0
	AT2G47190	MYB2	-2.7	-0.4
	AT4G37260	myb domain protein 73	-1.4	-0.5
	AT4G28080	Protein phosphatase 2C family protein	-1.1	-0.6
	AT5G37770	EF hand calcium-binding protein family	-1.4	-0.8
	AT1G23140	Calcium-dependent lipid-binding (CaLB domain) family prot	-1.7	-0.7
	AT1G68670	myb-like transcription factor family protein	-1.2	-0.9
	AT1G61340	F-box family protein	-2.0	-1.0
	AT5G59550	zinc finger (C3HC4-type RING finger) family protein	-2.0	-1.1
	AT4G27410	NAC (No Apical Meristem) domain transcriptional regulator	-1.3	-1.1
	AT3G19580	Zinc finer protein ZF2	-2.2	-1.1
	AT5G40390	Raffinose synthase family protein	-0.8	-1.2
	AT1G20440	cold-regulated 47	-0.3	-1.2
	AT3G15210	ERF4	-1.8	-1.2
	AT1G62660	Glycosyl hydrolases family 32 protein	-0.4	-1.3
	AT1G27730	salt tolerance zinc finger	-2.5	-1.3
	AT2G46510	ABA-INDUCIBLE BHLH-TYPE TRANSCRIPTION FACTOR	-1.8	-1.4
	AT3G46620	zinc finger (C3HC4-type RING finger) family protein	-2.2	-1.4
	AT5G52050	MATE efflux family protein	-2.3	-1.5
	AT2G38940	PHOSPHATE TRANSPORTER 1;4 PHT1;4	-2.4	-1.5
	AT5G59220	highly ABA-induced PP2C gene 1	-2.1	-1.6
	AT1G69260	ABI five binding protein	-0.7	-1.7
	AT1G32640	Basic helix-loop-helix (bHLH) DNA-binding family protein	-2.3	-1.8
	AT5G59320	lipid transfer protein 3	-1.8	-1.9
	AT2G39800	DELTA1-PYRROLINE-5-CARBOXYLATE SYNTHASE 1 P	-1.1	-2.1
	AT5G59310	lipid transfer protein 4	-3.2	-2.6
	AT2G30020	Protein phosphatase 2C family protein	-3.3	-2.5
	AT5G60660	plasma membrane intrinsic protein 2;4	-0.6	-2.9
	AT5G52300	CAP160 protein	-1.6	-2.9
	AT4G19690	iron-regulated transporter 1	-0.3	-3.0
	AT5G66400	Dehydrin family protein	-1.8	-3.0
JA	AT5G44420	plant defensin 1.2	2.8	3.5
	AT3G23250	MYB15	-2.1	0.2
	AT3G15500	NAC3	-1.2	0.0
	AT4G37260	myb domain protein 73	-1.4	-0.5
	AT1G19180	jasmonate-zim-domain protein 1	-1.9	-0.5
	AT5G47220	ethylene responsive element binding factor 2	-1.9	-0.6
	AT1G17380	jasmonate-zim-domain protein 5	-2.2	-0.9
	AT4G11280	1-aminocyclopropane-1-carboxylic acid (acc) synthase 6	-2.7	-0.9
	AT1G30135	jasmonate-zim-domain protein 8	-2.6	-1.0
	AT1G61340	F-box family protein	-2.0	-1.0
	AT1G74950	TIFY domain/Divergent CCT motif family protein	-1.5	-1.0
	AT1G17420	lipoxygenase 3	-1.8	-1.2
	AT3G15210	ERF4	-1.8	-1.2
	AT2G34600	JAZ7	-2.6	-1.7
	AT3G22275	JAZ13	-3.0	-1.7
	AT1G32640	Basic helix-loop-helix (bHLH) DNA-binding family protein	-2.3	-1.8

Figure 6. Gene expression pattern of DEGs related to (**A**) ROS scavenging and pathogen responses and (**B**) ABA and JA responses. Blue and yellow colors indicate upregulation and down regulation, respectively. Data used to generate this figure can be found in Supplementary Materials Table S1.

2.6. Validation of JcERFVII2 Target Genes

For verification of the RNA-seq results, quantitative reverse-transcription polymerase chain reaction (RT-PCR) was used to quantify 6 representative transcripts. The selected mRNAs included three core hypoxia genes (*HB1*, *PDC1*, and *PCO2*), two plant defense responsive genes (*PDF1.2* and *PDF1.3*), and *Alternative oxidase 1D* (*AOX1D*). The analysis confirmed that levels of these mRNAs are more induced in the *MA-JcERFVII2* overexpressing line than those of the *MC-JcERFVII2* and Col-0 grown in aerobic conditions (Figure 7). Furthermore, low oxygen-induced the accumulation of these mRNAs in all genotypes; however, the mRNA accumulation in some of these genes is slightly higher in the *MA* or *MC-JcERFVII2* overexpressing lines (Supplementary Materials Table S2).

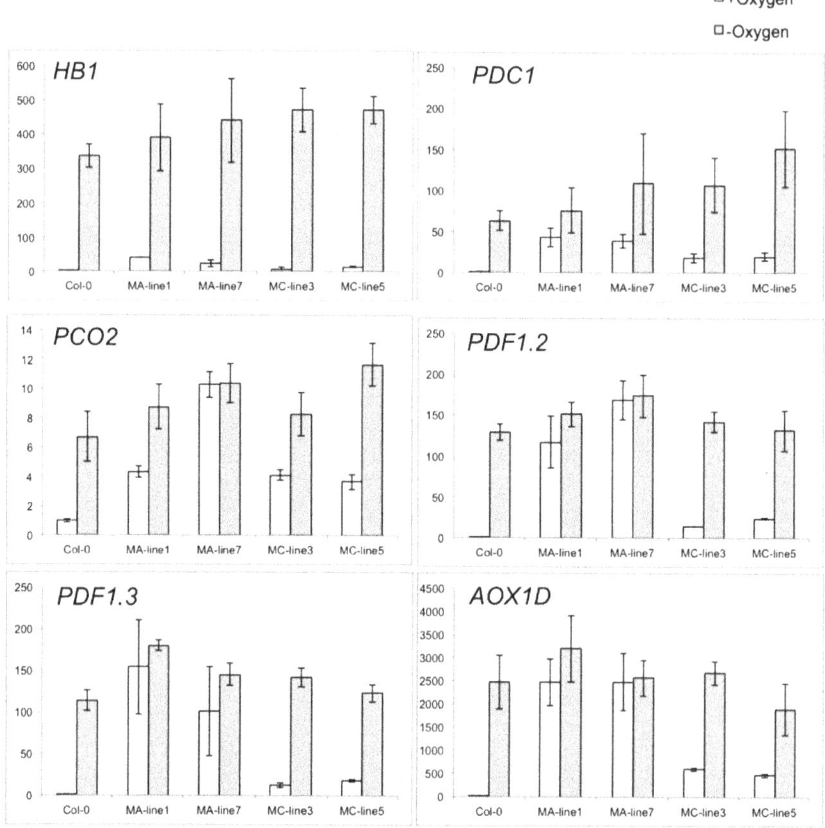

Figure 7. Quantitative real-time PCR validation of transcriptome data for selected genes. Relative expression was normalized to the abundance of *UBQ10*. Data represent mean ± SE (*n* = 3).

3. Discussion

This study focuses on elucidating the roles of *JcERFVIIs* towards waterlogging and low oxygen response. Phylogenetic and domain architecture analyses reveal that *JcERFVII1* and *JcERFVII2* are orthologs of constitutively expressed Arabidopsis *ERFVII* genes, *RAP2.2* and *RAP2.12* and *RAP2.3*, respectively (Figure 1). The last member of this *JcERFVII* family, *JcERFVII3*, is an ortholog of low-oxygen induced Arabidopsis *HRE2* (Figure 1). This study reveals that the expression of *JcERFVII1* is highly constitutive and remains unaffected following waterlogging, while *JcERFVII2* and *JcERFVII3* are upregulated by waterlogging (Figure 2 and Supplementary Materials Figure S1). Analysis of *RAP2.3* in flooding tolerant Brassica species, *Rorippa sylvestris* and *Rorippa amphibia,* demonstrated that under flooding, no induction of *RAP2.3* was observed [12]. Altogether, these data indicate that *JcERFVII2* from waterlogging sensitive Jatropha and *RAP2.3* from *Brassica* plants might undergo divergent evolution in gene expression.

In the dicot model Arabidopsis, all five ERFVIIs possess conserved motif function as N-degron that promotes the degradation of ERFVIIs via oxygen- and nitric oxide (NO) dependent N-end rule pathway of targeted proteolysis [8,10,13,14]. Overexpression of all five Arabidopsis *ERFVIIs* drastically improves low oxygen tolerance by promoting the expression of the genes involved in low oxygen adaptation [13,14,16,20]. Intriguingly, overexpression of stable version of *HRE1* and *HRE2* further improved low oxygen tolerance in Arabidopsis [14], while overexpression of stable version of *RAP2.12* resulted in a reduction of plant growth in air and also decreasing submergence tolerance in Arabidopsis [13,22]. In this study, we demonstrated that the JcERFVIIs 1–3 are targeted at the N-end rule pathway in vitro (Figure 3), leading to a question of whether modulation of the JcERFVII2 stability can further improve low oxygen tolerance. Transgenic Arabidopsis lines overexpressing *MC-JcERFVII2* are highly tolerant of both flooding and low oxygen stress, suggesting that *JcERFVII2* could function as a low-oxygen determinant (Figure 4). In contrast, transgenic Arabidopsis lines overexpressing *MA-JcERFVII2* are highly sensitive to low oxygen stresses (Figure 4). Moreover, overexpression of *MA-JcERFVII2* yields a decrease in rosette size and dry-weight when grown in air (Figure 4A,B), demonstrating that modulation of the JcERFVII2 stability interferes with growth and development.

In this study, transcriptome profiling reveals that modification of JcERFVII2 stability affects transcript accumulation of multiple genes controlling cellular metabolism and stress responses (Figure 5B). Previously, Bui et al. [32] demonstrated transcriptional activity of constitutively expressed *RAP2.2, RAP2.3* or, *RAP2.12* on a set of hypoxia-responsive promoters. Papdi et al. [20] showed that all three *RAP2* genes, when overexpressed, can transactivate *ADH* (*alcohol dehydrogenase*) promoter. In addition, Gasch et al. [33] demonstrated that overexpression of all three *RAP2* genes induced expression of *ADH* in transgenic Arabidopsis. Similarly, our study found that overexpression of *MA-* and *MC-JcERFVII2* upregulated the expression of 8 out of 49 core hypoxia-responsive genes (Supplementary Materials Table S1). Previous studies demonstrated that ectopic expression of *ERFVIIs* in transgenic plants increased tolerance to multiple abiotic stresses [8]. Some evidence suggested that RAP2.3 functions in pathogen response and ROS detoxification. Ogawa et al. [34] showed that tobacco BRIGHT YELLOW-2 cells overexpressing Arabidopsis *RAP2.3* were more tolerant of H_2O_2 and heat stress. Moreover, the expression of *PDF1.2* and *GST6* was enhanced in the transgenic Arabidopsis lines overexpressing *RAP2.3* [34]. Furthermore, overexpression of the *RAP2.3* ortholog, *CaPF1* (*Capsicum annuum pathogen and freezing tolerance-related protein 1*), in Virginia pine upregulated several antioxidant enzymes including ascorbate peroxidase, glutathione reductase and superoxide dismutase [35]. In this study, we found that overexpression of both *MA-* and *MC-JcERFVII2* induced the expression of several *PDF* genes (*PDFs 1.2, 1.2b, 1.2C,* and *1.3*; Figure 6A; Supplementary Materials Table S1). We also observed the upregulation of several *GST* and *peroxidase* genes in the transgenic line overexpressing *MA-JcERFVII2* (Figure 6A; Supplementary Materials Table S1). Altogether, these results demonstrate that *JcERFVII2* may involve in pathogen response and reducing ROS accumulation in plant cells.

Our study demonstrated that overexpression of *MA-JcERFVII2* interferes with growth and development (Figure 4). Paul et al. [22] compared transgenic Arabidopsis lines overexpressing wildtype and stabilized forms of *RAP2.12* under aerobic conditions and found that the stabilized *RAP2.12* affected central metabolic processes by increasing activities of fermentative enzymes and accumulation of fermentative products including ethanol, lactate, alanine and γ-amino butyrate (GABA), therefore resulted in decreased ATP and starch levels. In this study, GO enrichment analysis revealed that the alpha-amino acid metabolic process was enriched in the upregulated DEGs from *MA-ERFVII2* (Figure 5D). This GO category includes genes encoding for several enzymes responsible for glutamate and GABA synthesis (*AT2G02010*: *glutamate decarboxylase 4* (*GAD4*), *AT5G37600*: *glutamine synthase (GSR1)*; *AT5G38200: Class I glutamine amidotransferase-like superfamily protein*, and *AT4G35630: phosphoserine aminotransferase*; Supplementary Materials Table S1), implying the possibility that the transgenic Arabidopsis lines overexpressing *MA-JcERFVII2* could face carbohydrate starvation that leads to reduced growth and development.

In Arabidopsis, transcriptional activation of RAP2.12 can be counterbalanced by a trihelix transcriptional factor, namely HYPOXIA RESPONSE ATTENUATOR1 (HRA1) [36]. Giuntoli et al. [36] demonstrated that the interaction between *RAP2.12* and *HRA1* could enable an adaptive response to low oxygen, required for stress survival. Interestingly, transgenic wheat constitutively expressed the stabilized *TaERFVII.1* showed no growth defect phenotype, which resulted from the upregulation of *TaSAB18.1*, an ortholog of *HRA1*, under aerobic condition [25]. In this study, we did not observe the upregulation of *HRA1* from the transgenic Arabidopsis overexpressing both *MA-* and *MC-JcERFVII2* grown under aerobic conditions (Supplementary Materials Table S1).

Leon et al. [37] recently showed that enhanced RAP2.3 expression reduced NO-triggered transcriptome adjustment, and thus it functions as a brake for NO-triggered responses that included the activation of JA and ABA signaling in Arabidopsis. In addition, Vincente et al. [38] found that *RAP2.3* enhanced abiotic stress responses by interacting with BRM, a chromatin-remodeling ATPase, that repressed ABA responses. Gibb et al. [39] demonstrated that *RAP2.3* regulated the expression of *ABSICISIC ACID INSENSITIVE5* (*ABI5*), a major negative regulator of germination in seed endosperm. Interestingly, our study found that the NO-scavenger gene, *HB1*, was upregulated (Supplementary Materials Table S1). Additionally, genes involved in JA and ABA-activated signaling and responses were mostly down-regulated in transgenic Arabidopsis overexpressing *MA* and *MC-JcERFVII2* (Figure 6B), suggesting *JcERFVII2* could modulate NO accumulation and hormonal response.

In summary, our study demonstrated that JcERFVII2 is an N-end rule regulated waterlogging-responsive transcription factor that functions by modulating gene expression of cellular metabolic and multiple stress-responsive genes, including low-oxygen, oxidative, and pathogen response. Constitutive upregulation of fermentative and stress-responsive genes could compromise growth and development in the transgenic Arabidopsis overexpressing the stabilized *JcERFVII2*. This study highlights several possibilities for future investigation, including genetic manipulation of the *JcERFVII2* gene in Jatropha to determine whether it can improve waterlogging tolerance and elucidation of the *JcERFVII2* roles in controlling physiological responses to multiple abiotic stresses in Jatropha and other crop plants.

4. Materials and Methods

4.1. Multiple Sequence Alignment and Motif Identification

Full-length amino acid coding regions of ERFVIIs were downloaded from the Jatropha genome database (https://www.kazusa.or.jp/jatropha/) and the Arabidopsis information resource (http://www.arabidopsis.org/). Multiple sequence alignment was performed using CLUSTALW, and then a phylogenetic tree was built by the neighbor-joining method (Poisson correction, pairwise deletion of gaps) using the MEGA10 software [40]. Domain analysis was performed using MEME [41] following the models published for Arabidopsis [5].

4.2. Genetic Materials

J. curcas (cv. "Chai Nat"—a local Thai variety) and *A. thaliana* genotypes including the Col-0 accession and *35S:MC-JcERFVII2* and *35S:MA-JcERFVII2* (ectopic expression) transgenic lines were used in this study. The genome of Col-0 has already been sequenced.

4.3. Plant Growth and Stress Condition

J. curcas seedlings were grown and waterlogged, as described in Juntawong et al. [29].

For growth in soil, *A. thaliana* plants were grown in soil containing 50% (*v/v*) peat, 25% (*v/v*) perlite, and 25% (*v/v*) coconut fiber with regular irrigation in a growth room at 120 µmol photon m^{-2} s^{-1} 16 h light/8 h dark, at 23 °C. Submergence stress was performed using 10 leaf-stage plants grown in 5-cm^2 pots by placing them in a plastic container completely filled with water for 3 d.

For growth in sterile culture, *A. thaliana* seeds were surface sterilized, stratified at 4 °C for 48 h and plated on 0.5× solid Murashige and Skoog (MS) medium (0.215% (*w/v*) MS salts containing 1% (*w/v*) agar, pH 5.7) in 20-mm^2 dishes. Growth was in a vertical orientation in a growth room. Hypoxia stress was performed under dim light at the end of the 16-h light cycle in a sealed argon chamber. For hypoxia stress, 98% argon and 2% oxygen mixture was passed through water and into the chamber while ambient air was pushed out by positive pressure. Control was placed in an open chamber side by side.

4.4. Quantitative Reverse Transcription PCR

Total RNA samples were extracted using TRIzol reagents (Thermo Fisher Scientific, Waltham, MA, USA), subjected to DNase treatment, and RNA cleanup using an RNA-mini kit (Qiagen, Hilden, Germany). Three replicates of total RNA samples were used. One microgram of total RNAs was used to construct cDNA using MMuLv reverse transcriptase (Biotechrabbit, Berlin, Germany) in a final volume of 20 µL. The cDNA was diluted five times. Quantitative-realtime PCR (qPCR) reaction was performed according to Butsayawarapat et al. [42] using QPCR Green Master Mix (Biotechrabbit, Berlin, Germany) on a MasterCycler RealPlex4 (Eppendorf, Hamburg, Germany). For each sample, the PCR reaction was performed in triplicate. Each reaction contained 1 µL of diluted cDNA, 0.5 µM of each primer and 10 µL of QPCR Green Master Mix in a final volume of 20 µL. The PCR cycle was 95 °C for 2 min, followed by 45 cycles of 95 °C for 15 s and 60 °C for 30 s. Amplification specificity was validated by melt-curve analysis at the end of each PCR experiment. Relative gene expression was calculated using the $2^{-\Delta\Delta CT}$ method. Primers used to study Jatropha's gene expression were previously reported by Juntawong et al. [29]. The genes and primers used in Arabidopsis are shown in Supplementary Materials Table S3.

4.5. Analysis of Protein Stability

To construct the plasmids used for in vitro protein stability assay, cDNAs were amplified from *J. curcas* total cDNA using gene-specific primers (Supplementary Materials Table S3). The PCR products were ligated into a modified version of the pTNT (Invitrogen, Carlsbad, CA, USA) expression vector (pTNT-3xHA) [14]. N-terminal mutations were incorporated by modifying the forward primer sequences accordingly (Supplementary Materials Table S3).

For in vitro protein expression, TNT T7 Coupled Reticulocyte Lysate System (Promega, Madison, WI, USA) and 2 µg plasmid template was used according to manufacturer's guidelines. Where appropriate, 100 mM MG132 (Sigma, St. Louis, MO, USA) was added. Reactions were incubated at 30 °C. Samples were taken at indicated time points before mixing with protein loading dye to terminate the reaction. Equal amounts of each reaction were subjected to anti-HA immunoblot analysis. All blots were checked for equal loading by Coomassie Brilliant Blue staining.

For immunoblotting, proteins resolved by SDS-PAGE were transferred to PVDF using a MiniTrans-Blot electrophoretic transfer cell (Bio-Rad, Hercules, CA, USA). Membranes were probed with HA-probe (Y-11) HRP (Santa Cruz, CA, USA) at a titer of 1:1000. Immunoblots were detected using TMB (Tetramethyl Benzidine; Thermo Fisher Scientific, Waltham, MA, USA) solution (Invitrogen, Carlsbad, CA, USA).

4.6. Generation of Transgenic Lines

To construct Ti binary plasmids for plant transformation, *JcERFVII2* open reading frame was amplified by RT-PCR from RNA extracted from roots of *J. curcas* using gene-specific primers (Supplementary Materials Table S3). The PCR product was inserted into the pCXSN binary plasmid [43], transformed into *E. coli* DH5α, and selected with 50 µg mL^{-1} kanamycin.

N-terminal mutations were incorporated by changing the forward primer sequences accordingly (Supplementary Materials Table S3). The pCXSN, a plant overexpression vector, provides a CaMV 35S promoter and nopaline synthase terminator sequence in a Ti binary plasmid with a hygromycin-resistant gene. After sequence confirmation, the plasmid was electroporated into *Agrobacterium tumefaciens* GV3101 and colonies selected with 50 μg mL^{-1} kanamycin. Col-0 transformation was performed according to Clough and Bent [44]). T1 seeds were collected, seedlings resistant to 35 μg mL^{-1} hygromycin were propagated, and homozygous single insertion events were established.

4.7. RNA-Seq, Differential Gene Expression Analysis, and Gene Ontology Enrichment

Total RNA samples were extracted using TRIzol reagents (Thermo Fisher Scientific, Waltham, MA, USA), subjected to DNase treatment, and RNA cleanup using an RNA-mini kit (Qiagen, Hilden, Germany). Two replicates of total RNA samples were used for transcriptome analysis according to the ENCODE recommended RNA-seq standards (https://genome.ucsc.edu/ENCODE/protocols/dataStandards/ENCODE_RNAseq_Standards_V1.0.pdf). The integrity of the RNA samples (RIN) was evaluated on an RNA 6000 Nano LapChiprun on Agilent2100 Bioanalyzer (Agilent Technologies, Waldbronn, Germany). Samples with a RIN > 7 were used in RNA-seq library preparation.

For each sample, 3 μg of total RNAs were used to generate a sequencing library using an Illumina$^{®}$ TruSeq^{TM} RNA Sample Preparation Kit v2 (Illumina, San Diego, CA, USA). Paired-end, 100 bp RNA-seq was performed on a NovaSeq6000 platform. FASTQ files were generated with the base caller provided by the instrument. Quality control filtering and 3′ end trimming were analyzed using the FASTX-toolkit (http://hannonlab.cshl.edu/fastx_toolkit/index.html) and Trimmomatic software [45], respectively. The raw read files were deposited in the NCBI GEO database under the accession numbers GSE154601.

Differential gene expression analysis was performed according to Juntawong et al. [29]. The FASTQ files were aligned to the reference transcriptome using TopHat2 software (v2.0.13) [46]. A binary format of sequence alignment files (BAM) was generated and used to create read count tables by the HTseq python library (citation). Differentially-expressed genes were calculated using the edgeR program [47] with an FDR cutoff of <0.05.

Gene ontology (GO) enrichment was analyzed using AgriGO V2.0 [48]. For visualization, REViGO was applied to summarize and removing redundant GO terms [49].

Supplementary Materials: The following are available online at http://www.mdpi.com/2223-7747/9/9/1068/s1, Data S1: Nucleotide and amino acid sequences of JcERFVIIs, Figure S1: Gene expression level of *JcERFVIIs 1, 2,* and *3* retrieved from the JCDB database, Figure S2: Expression analysis of *JcERFVII-2* and five members of the Arabidopsis *ERFVIIs* in transgenic lines. (A) semi quantitative RT-PCR analysis of *JcERFVII-2* in the transgenic Arabidopsis plants overexpressing *MA-* or *MC-JcERFVII-2*. Actin was used as a control. (B) Means of Count Per Million (CPM) gene expression values of the Arabidopsis *ERFVII* genes obtained from RNA-seq experiment, Table S1: Differentially expressed genes, Table S2: Gene ontology enrichment results. Table S3: Primers used in this study.

Author Contributions: Conceived, designed, and supervised research, P.J.; performed research, P.J., P.B., P.S., R.P., and S.V.; performed in-silico analysis, interpreted the results and wrote the manuscript, P.J. All authors have read and agreed to the published version of the manuscript.

Funding: This study was funded by the Faculty of Science, Kasetsart University, Kasetsart University Research and Development Institute, Omics Center for Agriculture, Bioresources, Food and Health, Kasetsart University (OmiKU), and the Thailand Research Fund (RSA6280013).

Acknowledgments: We thank Julia Bailey-Serres for the pTNT-3xHA construct.

Conflicts of Interest: The authors declare no conflict of interest.

References

1. Van Dongen, J.T.; Frohlich, A.; Ramirez-Aguilar, S.J.; Schauer, N.; Fernie, A.R.; Erban, A.; Kopka, J.; Clark, J.; Langer, A.; Geigenberger, P. Transcript and metabolite profiling of the adaptive response to mild decreases in oxygen concentration in the roots of arabidopsis plants. *Ann. Bot.* **2009**, *103*, 269–280. [CrossRef] [PubMed]

2. Bailey-Serres, J.; Voesenek, L.A. Flooding stress: Acclimations and genetic diversity. *Annu. Rev. Plant Biol.* **2008**, *59*, 313–339. [CrossRef] [PubMed]

3. Bailey-Serres, J.; Fukao, T.; Gibbs, D.J.; Holdsworth, M.J.; Lee, S.C.; Licausi, F.; Perata, P.; Voesenek, L.A.; van Dongen, J.T. Making sense of low oxygen sensing. *Trends Plant Sci.* **2012**, *17*, 129–138. [CrossRef] [PubMed]

4. Voesenek, L.A.; Bailey-Serres, J. Flooding tolerance: O$_2$ sensing and survival strategies. *Curr. Opin. Plant Biol.* **2013**, *16*, 647–653. [CrossRef] [PubMed]

5. Nakano, T.; Suzuki, K.; Fujimura, T.; Shinshi, H. Genome-wide analysis of the ERF gene family in Arabidopsis and rice. *Plant Physiol.* **2006**, *140*, 411–432. [CrossRef]

6. Dey, S.; Corina Vlot, A. Ethylene responsive factors in the orchestration of stress responses in monocotyledonous plants. *Front. Plant Sci.* **2015**, *6*, 640. [CrossRef]

7. Licausi, F.; Ohme-Takagi, M.; Perata, P. APETALA2/Ethylene Responsive Factor (AP2/ERF) transcription factors: Mediators of stress responses and developmental programs. *New Phytol.* **2013**, *199*, 639–649. [CrossRef]

8. Gibbs, D.J.; Conde, J.V.; Berckhan, S.; Prasad, G.; Mendiondo, G.M.; Holdsworth, M.J. Group VII Ethylene Response Factors Coordinate Oxygen and Nitric Oxide Signal Transduction and Stress Responses in Plants. *Plant Physiol.* **2015**, *169*, 23–31. [CrossRef]

9. Voesenek, L.A.; Bailey-Serres, J. Flood adaptive traits and processes: An overview. *New Phytol.* **2015**, *206*, 57–73. [CrossRef]

10. Giuntoli, B.; Perata, P. Group VII Ethylene Response Factors in Arabidopsis: Regulation and Physiological Roles. *Plant Physiol.* **2018**, *176*, 1143–1155. [CrossRef]

11. Lin, C.C.; Chao, Y.T.; Chen, W.C.; Ho, H.Y.; Chou, M.Y.; Li, Y.R.; Wu, Y.L.; Yang, H.A.; Hsieh, H.; Lin, C.S.; et al. Regulatory cascade involving transcriptional and N-end rule pathways in rice under submergence. *Proc. Natl. Acad. Sci. USA* **2019**, *116*, 3300–3309. [CrossRef] [PubMed]

12. Van Veen, H.; Akman, M.; Jamar, D.C.; Vreugdenhil, D.; Kooiker, M.; van Tienderen, P.; Voesenek, L.A.; Schranz, M.E.; Sasidharan, R. Group VII ethylene response factor diversification and regulation in four species from flood-prone environments. *Plant Cell Environ.* **2014**, *37*, 2421–2432. [CrossRef] [PubMed]

13. Licausi, F.; Kosmacz, M.; Weits, D.A.; Giuntoli, B.; Giorgi, F.M.; Voesenek, L.A.; Perata, P.; van Dongen, J.T. Oxygen sensing in plants is mediated by an N-end rule pathway for protein destabilization. *Nature* **2011**, *479*, 419–422. [CrossRef] [PubMed]

14. Gibbs, D.J.; Lee, S.C.; Isa, N.M.; Gramuglia, S.; Fukao, T.; Bassel, G.W.; Correia, C.S.; Corbineau, F.; Theodoulou, F.L.; Bailey-Serres, J.; et al. Homeostatic response to hypoxia is regulated by the N-end rule pathway in plants. *Nature* **2011**, *479*, 415–418. [CrossRef] [PubMed]

15. Gibbs, D.J.; Bacardit, J.; Bachmair, A.; Holdsworth, M.J. The eukaryotic N-end rule pathway: Conserved mechanisms and diverse functions. *Trends Cell Biol.* **2014**, *24*, 603–611. [CrossRef]

16. Hinz, M.; Wilson, I.W.; Yang, J.; Buerstenbinder, K.; Llewellyn, D.; Dennis, E.S.; Sauter, M.; Dolferus, R. Arabidopsis RAP2.2: An ethylene response transcription factor that is important for hypoxia survival. *Plant Physiol.* **2010**, *153*, 757–772. [CrossRef]

17. Hess, N.; Klode, M.; Anders, M.; Sauter, M. The hypoxia responsive transcription factor genes ERF71/HRE2 and ERF73/HRE1 of Arabidopsis are differentially regulated by ethylene. *Physiol. Plant* **2011**, *143*, 41–49. [CrossRef]

18. Licausi, F.; van Dongen, J.T.; Giuntoli, B.; Novi, G.; Santaniello, A.; Geigenberger, P.; Perata, P. HRE1 and HRE2, two hypoxia-inducible ethylene response factors, affect anaerobic responses in Arabidopsis thaliana. *Plant J.* **2010**, *62*, 302–315. [CrossRef]

19. Park, H.Y.; Seok, H.Y.; Woo, D.H.; Lee, S.Y.; Tarte, V.N.; Lee, E.H.; Lee, C.H.; Moon, Y.H. AtERF71/HRE2 transcription factor mediates osmotic stress response as well as hypoxia response in Arabidopsis. *Biochem. Biophys. Res. Commun.* **2011**, *414*, 135–141. [CrossRef]

20. Papdi, C.; Perez-Salamo, I.; Joseph, M.P.; Giuntoli, B.; Bogre, L.; Koncz, C.; Szabados, L. The low oxygen, oxidative and osmotic stress responses synergistically act through the ethylene response factor VII genes RAP2.12, RAP2.2 and RAP2.3. *Plant J.* **2015**, *82*, 772–784. [CrossRef]

21. Yang, C.Y.; Hsu, F.C.; Li, J.P.; Wang, N.N.; Shih, M.C. The AP2/ERF transcription factor AtERF73/HRE1 modulates ethylene responses during hypoxia in Arabidopsis. *Plant Physiol.* **2011**, *156*, 202–212. [CrossRef] [PubMed]

22. Paul, M.V.; Iyer, S.; Amerhauser, C.; Lehmann, M.; van Dongen, J.T.; Geigenberger, P. Oxygen Sensing via the Ethylene Response Transcription Factor RAP2.12 Affects Plant Metabolism and Performance under Both Normoxia and Hypoxia. *Plant Physiol.* **2016**, *172*, 141–153. [CrossRef] [PubMed]

23. Xu, K.; Xu, X.; Fukao, T.; Canlas, P.; Maghirang-Rodriguez, R.; Heuer, S.; Ismail, A.M.; Bailey-Serres, J.; Ronald, P.C.; Mackill, D.J. Sub1A is an ethylene-response-factor-like gene that confers submergence tolerance to rice. *Nature* **2006**, *442*, 705–708. [CrossRef] [PubMed]

24. Hattori, Y.; Nagai, K.; Furukawa, S.; Song, X.J.; Kawano, R.; Sakakibara, H.; Wu, J.; Matsumoto, T.; Yoshimura, A.; Kitano, H.; et al. The ethylene response factors SNORKEL1 and SNORKEL2 allow rice to adapt to deep water. *Nature* **2009**, *460*, 1026–1030. [CrossRef] [PubMed]

25. Wei, X.; Xu, H.; Rong, W.; Ye, X.; Zhang, Z. Constitutive expression of a stabilized transcription factor group VII ethylene response factor enhances waterlogging tolerance in wheat without penalizing grain yield. *Plant Cell Environ.* **2019**, *42*, 1471–1485. [CrossRef]

26. Fukao, T.; Barrera-Figueroa, B.E.; Juntawong, P.; Pena-Castro, J.M. Submergence and Waterlogging Stress in Plants: A Review Highlighting Research Opportunities and Understudied Aspects. *Front Plant Sci.* **2019**, *10*, 340. [CrossRef]

27. Gimeno, V.; Syvertsen, J.P.; Simón, I.; Nieves, M.; Díaz-López, L.; Martínez, V.; García-Sánchez, F. Physiological and morphological responses to flooding with fresh or saline water in *Jatropha curcas*. *Environ. Exp. Bot.* **2012**, *78*, 47–55. [CrossRef]

28. Verma, K.C.; Verma, S.K. Biophysicochemical evaluation of wild hilly biotypes of *Jatropha curcas* for biodiesel production and micropropagation study of elite plant parts. *Appl. Biochem. Biotechnol.* **2015**, *175*, 549–559. [CrossRef]

29. Juntawong, P.; Sirikhachornkit, A.; Pimjan, R.; Sonthirod, C.; Sangsrakru, D.; Yoocha, T.; Tangphatsornruang, S.; Srinives, P. Elucidation of the molecular responses to waterlogging in Jatropha roots by transcriptome profiling. *Front Plant Sci.* **2014**, *5*, 658. [CrossRef]

30. Zhang, X.; Pan, B.Z.; Chen, M.; Chen, W.; Li, J.; Xu, Z.F.; Liu, C. JCDB: A comprehensive knowledge base for *Jatropha curcas*, An emerging model for woody energy plants. *BMC Genom.* **2019**, *20*, 958. [CrossRef]

31. Mustroph, A.; Zanetti, M.E.; Jang, C.J.; Holtan, H.E.; Repetti, P.P.; Galbraith, D.W.; Girke, T.; Bailey-Serres, J. Profiling translatomes of discrete cell populations resolves altered cellular priorities during hypoxia in Arabidopsis. *Proc. Natl. Acad. Sci. USA* **2009**, *106*, 18843–18848. [CrossRef] [PubMed]

32. Bui, L.T.; Giuntoli, B.; Kosmacz, M.; Parlanti, S.; Licausi, F. Constitutively expressed ERF-VII transcription factors redundantly activate the core anaerobic response in Arabidopsis thaliana. *Plant Sci.* **2015**, *236*, 37–43. [CrossRef] [PubMed]

33. Gasch, P.; Fundinger, M.; Muller, J.T.; Lee, T.; Bailey-Serres, J.; Mustroph, A. Redundant ERF-VII Transcription Factors Bind to an Evolutionarily Conserved cis-Motif to Regulate Hypoxia-Responsive Gene Expression in Arabidopsis. *Plant Cell* **2016**, *28*, 160–180. [CrossRef] [PubMed]

34. Ogawa, T.; Pan, L.; Kawai-Yamada, M.; Yu, L.H.; Yamamura, S.; Koyama, T.; Kitajima, S.; Ohme-Takagi, M.; Sato, F.; Uchimiya, H. Functional analysis of Arabidopsis ethylene-responsive element binding protein conferring resistance to Bax and abiotic stress-induced plant cell death. *Plant Physiol.* **2005**, *138*, 1436–1445. [CrossRef] [PubMed]

35. Tang, W.; Charles, T.M.; Newton, R.J. Overexpression of the pepper transcription factor CaPF1 in transgenic Virginia pine (Pinus Virginiana Mill.) confers multiple stress tolerance and enhances organ growth. *Plant Mol. Biol.* **2005**, *59*, 603–617. [CrossRef] [PubMed]

36. Giuntoli, B.; Lee, S.C.; Licausi, F.; Kosmacz, M.; Oosumi, T.; van Dongen, J.T.; Bailey-Serres, J.; Perata, P. A trihelix DNA binding protein counterbalances hypoxia-responsive transcriptional activation in Arabidopsis. *PLoS Biol.* **2014**, *12*, e1001950. [CrossRef]

37. Leon, J.; Costa-Broseta, A.; Castillo, M.C. RAP2.3 negatively regulates nitric oxide biosynthesis and related responses through a rheostat-like mechanism in Arabidopsis. *J. Exp. Bot.* **2020**, *71*, 3157–3171. [CrossRef]

38. Vicente, J.; Mendiondo, G.M.; Movahedi, M.; Peirats-Llobet, M.; Juan, Y.T.; Shen, Y.Y.; Dambire, C.; Smart, K.; Rodriguez, P.L.; Charng, Y.Y.; et al. The Cys-Arg/N-End Rule Pathway Is a General Sensor of Abiotic Stress in Flowering Plants. *Curr. Biol.* **2017**, *27*, 3183–3190. [CrossRef]

39. Gibbs, D.J.; Md Isa, N.; Movahedi, M.; Lozano-Juste, J.; Mendiondo, G.M.; Berckhan, S.; Marin-de la Rosa, N.; Vicente Conde, J.; Sousa Correia, C.; Pearce, S.P.; et al. Nitric oxide sensing in plants is mediated by proteolytic control of group VII ERF transcription factors. *Mol. Cell* **2014**, *53*, 369–379. [CrossRef]

40. Kumar, S.; Stecher, G.; Li, M.; Knyaz, C.; Tamura, K. MEGA X: Molecular Evolutionary Genetics Analysis across Computing Platforms. *Mol. Biol. Evol.* **2018**, *35*, 1547–1549. [CrossRef]

41. Bailey, T.L.; Boden, M.; Buske, F.A.; Frith, M.; Grant, C.E.; Clementi, L.; Ren, J.; Li, W.W.; Noble, W.S. MEME SUITE: Tools for motif discovery and searching. *Nucleic Acids Res.* **2009**, *37*, W202–W208. [CrossRef] [PubMed]

42. Butsayawarapat, P.; Juntawong, P.; Khamsuk, O.; Somta, P. Comparative Transcriptome Analysis of Waterlogging-Sensitive and Tolerant Zombi Pea (Vigna Vexillata) Reveals Energy Conservation and Root Plasticity Controlling Waterlogging Tolerance. *Plants* **2019**, *8*, 264. [CrossRef] [PubMed]

43. Chen, S.; Songkumarn, P.; Liu, J.; Wang, G.L. A versatile zero background T-vector system for gene cloning and functional genomics. *Plant Physiol.* **2009**, *150*, 1111–1121. [CrossRef]

44. Clough, S.J.; Bent, A.F. Floral dip: A simplified method for Agrobacterium-mediated transformation of Arabidopsis thaliana. *Plant J.* **1998**, *16*, 735–743. [CrossRef]

45. Bolger, A.M.; Lohse, M.; Usadel, B. Trimmomatic: A flexible trimmer for Illumina sequence data. *Bioinformatics* **2014**, *30*, 2114–2120. [CrossRef]

46. Kim, D.; Pertea, G.; Trapnell, C.; Pimentel, H.; Kelley, R.; Salzberg, S.L. TopHat2: Accurate alignment of transcriptomes in the presence of insertions, deletions and gene fusions. *Genome Biol.* **2013**, *14*, R36. [CrossRef]

47. Robinson, M.D.; McCarthy, D.J.; Smyth, G.K. edgeR: A Bioconductor package for differential expression analysis of digital gene expression data. *Bioinformatics* **2010**, *26*, 139–140. [CrossRef]

48. Tian, T.; Liu, Y.; Yan, H.; You, Q.; Yi, X.; Du, Z.; Xu, W.; Su, Z. agriGO v2.0: A GO analysis toolkit for the agricultural community, 2017 update. *Nucleic Acids Res.* **2017**, *45*, W122–W129. [CrossRef]

49. Supek, F.; Bosnjak, M.; Skunca, N.; Smuc, T. REVIGO summarizes and visualizes long lists of gene ontology terms. *PLoS ONE* **2011**, *6*, e21800. [CrossRef]

Article

Metabolic Responses to Waterlogging Differ between Roots and Shoots and Reflect Phloem Transport Alteration in *Medicago truncatula*

Jérémy Lothier [1,*], Houssein Diab [1], Caroline Cukier [1], Anis M. Limami [1] and Guillaume Tcherkez [1,2,*]

1   Seedling Metabolism and Stress, Université d'Angers, Agrocampus Ouest, INRAE, UMR IRHS, SFR QuaSaV, 49071 Beaucouzé, France; drhd85@hotmail.com (H.D.); caroline.cukier@univ-angers.fr (C.C.); anis.limami@univ-angers.fr (A.M.L.)
2   Research School of Biology, ANU Joint College of Science, Australian National University, Canberra, ACT 2601, Australia
*   Correspondence: jeremy.lothier@univ-angers.fr (J.L.); guillaume.tcherkez@anu.edu.au (G.T.); Tel.: +33-(0)2-49-18-04-76 (J.L.); +33-(0)2-41-22-55-36 (G.T.)

Received: 15 September 2020; Accepted: 2 October 2020; Published: 15 October 2020

Abstract: Root oxygen deficiency that is induced by flooding (waterlogging) is a common situation in many agricultural areas, causing considerable loss in yield and productivity. Physiological and metabolic acclimation to hypoxia has mostly been studied on roots or whole seedlings under full submergence. The metabolic difference between shoots and roots during waterlogging, and how roots and shoots communicate in such a situation is much less known. In particular, the metabolic acclimation in shoots and how this, in turn, impacts on roots metabolism is not well documented. Here, we monitored changes in the metabolome of roots and shoots of barrel clover (*Medicago truncatula*), growth, and gas-exchange, and analyzed phloem sap exudate composition. Roots exhibited a typical response to hypoxia, such as γ-aminobutyrate and alanine accumulation, as well as a strong decline in raffinose, sucrose, hexoses, and pentoses. Leaves exhibited a strong increase in starch, sugars, sugar derivatives, and phenolics (tyrosine, tryptophan, phenylalanine, benzoate, ferulate), suggesting an inhibition of sugar export and their alternative utilization by aromatic compounds production via pentose phosphates and phospho*enol*pyruvate. Accordingly, there was an enrichment in sugars and a decline in organic acids in phloem sap exudates under waterlogging. Mass-balance calculations further suggest an increased imbalance between loading by shoots and unloading by roots under waterlogging. Taken as a whole, our results are consistent with the inhibition of sugar import by waterlogged roots, leading to an increase in phloem sugar pool, which, in turn, exert negative feedback on sugar metabolism and utilization in shoots.

Keywords: waterlogging; hypoxia; metabolomics; phloem

1. Introduction

Flooding is currently one of the most severe factors that reduces crop productivity [1,2]. It is believed that at least 10% of arable fields can be affected by flooding (FAOSTAT, www.fao.org). As such, it concerns many cultivated species, including major crops, like wheat, and alternative agricultural practices have to be found to overcome this problem (reviewed in Manik et al., 2019 [3]). Flooding leads to occasional or prolonged root submergence (waterlogging). O_2 deprivation is the most important biochemical factor during waterlogging [4]: in fact, when air spaces that are normally present in soil are filled with water, root environment becomes hypoxic due to O_2 consumption by respiring roots (and microorganisms). The well-known effects of hypoxia relate to energy limitation (i.e. reduction in respiratory ATP production due to O_2 shortage) and cytoplasm acidification due to a decline in plasma membrane H^+-ATPase activity as well as organic acid generation by metabolism [5]. Although prolonged hypoxia strongly impacts on growth and survival [6], occasional hypoxia can be accommodated via the expression of so-called "core hypoxia-responsive genes" (reviewed in Hsu & Shih, 2013, [7]). Such a genetic response leads to a re-orchestration of root primary metabolism, including (*i*) the induction of ethanolic fermentation (e.g. induction of alcohol dehydrogenase and pyruvate decarboxylase, and in soybean, an increase in enolase and phosphoglycerate kinase [8]), (*ii*) an increase in sucrose synthase to metabolize sugars, and (*iii*) the accumulation of metabolites from alternative pathways e.g. γ-aminobutyrate (GABA) and alanine [9]. Alanine accumulation is a hypoxic biomarker metabolite (reviewed in Planchet et al., 2017, [10]) and, accordingly, a gene encoding alanine aminotransferase has been found amongst core hypoxia-response genes [11].

Up to now, most of the studies regarding hypoxia utilize roots or whole seedlings kept in darkness in a N_2 atmosphere or under submergence [12–17]. Although valuable in understanding root metabolism during waterlogging, these studies provide limited information to delineate metabolic responses at the whole-plant level and potential metabolic differences in other organs. In effect, roots and shoots do not exhibit the same response to hypoxia [18–23]. Recently, Mustroph et al., 2014 [24] addressed this question via a meta-analysis of published data that were obtained in Arabidopsis (*Arabidopsis thaliana*). They found that roots respond more strongly and typically produce GABA, alanine, and lactate, while shoots seem to have less hypoxic symptoms (but accumulate alanine), probably due to O_2 generation by photosynthesis.

In adult plants (rather than seedlings) and in other species, organ-specific metabolic effects of waterlogging are much less known. An accumulation of starch in leaves of waterlogged plants is generally observed (reviewed in Irfan et al., 2010 [25]). Recently, detailed metabolomics analyses have been carried out in sunflower (*Helianthus annuus*) and oil palm (*Elaeis guineensis*) [26,27]. In both species, waterlogging leads to an increase in alanine, GABA, and polyols (glycerol, inositol, ononitol), changes in organic acids (citrate, aconitate, maleate), and a decrease in many amino acids, showing the limitation of catabolism (respiration) and the inhibition of the tricarboxylic acid pathway, which, in turn, impacts on amino acid homeostasis. In leaves, waterlogging leads to a strong increase in sugars (hexoses and polyols derived therefrom) and a decline in several amino acids (sometimes including alanine), suggesting an inhibition of sugar export and N assimilation [26]. In addition, there is a decline in nitrate content in leaves, much less pronounced in roots, suggesting an inhibition of xylem nitrate transport and in fact, computations based on ^{15}N natural abundance have shown a strong effect of waterlogging on nitrate circulation [28]. Additionally, in cotton, waterlogging causes a decline in the expression of nitrite reductase and an increase in nitrite concentration in leaves [29]. In nodulated soybean (*Glycine max*), waterlogging inhibits the transport of ureides and glutamine to the shoot via the xylem [30,31]. Taken as a whole, waterlogging has important consequences on leaf metabolism, including N assimilation, and part of this effect is likely caused by a perturbation of metabolite exchange via phloem (and xylem) sap. In effect, in castor bean (*Ricinus communis*, a species that allows facile phloem sap sampling) under waterlogging, phloem sap flow, and conducting area have been found to decrease while phloem sap sugar concentration slightly increases [32]. Also, in French bean (*Phaseolus vulgaris*) cultivated hydroponically, ^{14}C translocation to roots after the isotopic labelling of leaves with ^{14}C-sucrose is approximately 50% lower when roots are under anoxic conditions [33]. Similarly, when a non-metabolizable radioactive glucose analogue (^{14}C-1-deoxyglucose) is fed to shoots of maize seedlings, the translocation of radioactivity does not reach root tips when the root is kept under anoxic conditions [34].

Nevertheless, in our present knowledge of systemic effects of waterlogging, uncertainty remains as to (*i*) how leaf and root metabolome are coordinated, (*ii*) whether both phloem sap composition and transport rate are affected, and, if so, (*iii*) whether phloem loading or unloading (or both) is primarily affected. Here, we used non-nodulated barrel clover (*Medicago truncatula*) plants and carried out a systematic metabolic analysis using GC-MS profiling in roots, leaves, and also phloem sap exudates at different time points after the onset of waterlogging. We also carried out gas-exchange measurements and elemental content (C, N) determination. We used barrel clover as a model species because of its importance as a cultivated forage crop in the Mediterranean region and Australia, where waterlogging stress periodically occurs due to seasonal heavy rain events. Additionally, we focused on non-nodulated plants, since the establishment of N_2-fixing symbiosis considerably modifies nitrogen metabolism (in favor of ureides, amides, etc.) and the impact of waterlogging on *Rhizobium* would have had to be taken into account. Additionally, this will facilitate the comparison with plants devoid of N_2-fixation in the discussion. We found metabolic changes that are consistent with an inhibition of sugar import (unloading) by waterlogged roots, leading to an increase in the phloem sugar pool, which, in turn, exerts a feedback on sugar export and metabolism in leaves.

2. Results

2.1. Photosynthesis and C:N Composition

Leaf photosynthesis (net CO_2 assimilation) was measured under standard conditions (380 μmol mol^{-1} CO_2, 21% O_2). As expected, there was a significant negative effect of waterlogging on photosynthesis (Figure 1a), due to lower stomatal conductance (Figure 1b). Interestingly, this effect was not more pronounced after 21 d (as compared to 7 d) suggesting that leaf photosynthesis partly acclimated to waterlogging conditions. By contrast, CO_2 evolution in darkness was minimally affected after 7 d and was significantly lower under waterlogging after 21 d (Figure 1c). The overall result of lower carbon assimilation was a lower dry weight (DW) in roots after 7 d and in both roots and shoots after 21 d (Figure S1a,b). Interestingly, the relative water content was higher by 1–2% under waterlogging (Figure S1c,d), which partly compensated for the effect of waterlogging on DW in fresh matter production.

There was no significant change in the carbon elemental content (%C), whereas nitrogen elemental content decreased in shoots and increased (at 21 d) in roots under waterlogging (Figure 2a–d). As a result, after 21 d, the C:N ratio increased a lot in shoots (nearly two-fold increase) and decreased slightly (by about 25%) in roots (Figure 2e,f). Accordingly, there was a four- to 10-fold increase in starch content and 2-fold increase in protein content (expressed in mg g^{-1} DW) in shoots under waterlogging (Figure 3a–c), while, in roots, both starch and proteins increased 2-fold (Figure 3b–d).

2.2. Metabolomics Pattern in Leaves and Roots

Because both C assimilation and N content were affected by waterlogging, we carried out an in-depth analysis with metabolic profiling to identify metabolites of primary C and N metabolism influenced by waterlogging. Metabolomics analyses were carried out on leaves and roots that were collected over four weeks, using GC-MS. This allowed for us to identify and quantify 123 analytes. The effect of waterlogging was analyzed using both univariate (two-way ANOVA) and multivariate (O2PLS) analyses, where two factors were considered (time and waterlogging). In leaves, 33 analytes were found to be significantly affected ($p < 0.01$) by waterlogging, and formed two clusters (labelled 1 and 2 in Figure 4). The first cluster comprised nitrogenous compounds (urea, spermidine, and β-alanine) as well as threitol, which were less abundant under waterlogging. The second cluster was made of metabolites that were more abundant under waterlogging, mostly sugars (fructose, galactose, glucose, mannose, xylose), organic acids (such as succinate, glutarate), aromatics (tyrosine, phenylalanine, tryptophan, ferulate), and some amino acids (glycine, alanine, glutamine, lysine). The multivariate analysis yielded a predictive ($R^2 = 0.88$) and robust ($Q^2 = 0.66$) model that was highly significant ($P_{CV\text{-}ANOVA} = 2.4{\cdot}10^{-9}$) and separated control and waterlogged conditions easily (Figure 4b). When univariate and multivariate analyses are combined into a volcano plot, the best biomarkers of waterlogging appeared to be tyrosine, threonolactone and galactonate (increased), and urea (decreased) (Figure 4c). Interestingly, amongst the ratios of organic acids, the succinate-to-fumarate ratio was significantly higher under waterlogging (Figure 4d), which suggested a specific downregulation of succinate oxidation to fumarate in the Krebs cycle and/or fumarate generation by polyamine synthesis. Additionally, there was a significant increase in photorespiratory intermediates (sum of serine, glycine, glycolate, and glycerate) under waterlogging, which suggested a higher photorespiration rate. This would agree with the lower stomatal conductance that was found by gas exchange (Figure 1b). Three metabolites had a significant interaction (time × waterlogging) effect (Figure 4e): arginine and glutamine (which were significantly higher only at the beginning of waterlogging), and threonolactone (which increased under waterlogging, while it remained constant under control conditions).

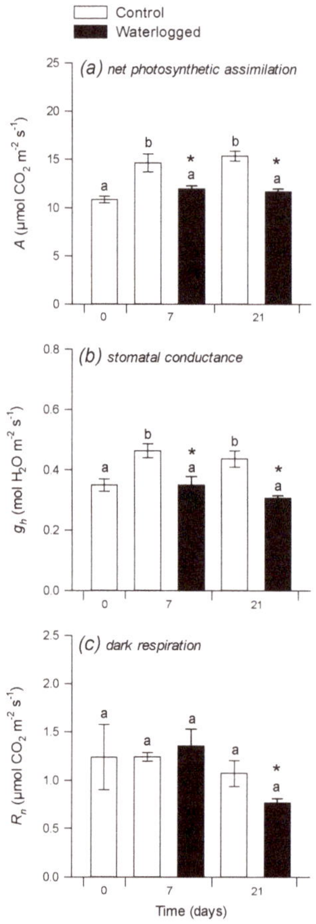

Figure 1. Leaf gas-exchange parameters: net photosynthetic CO_2 assimilation. (**a**), stomatal conductance (**b**) and dark respiration (**c**) 7 and 21 days after the beginning of waterlogging. The zero-time point was just before the onset of waterlogging. Non-waterlogged (control) plants are shown with open bars, waterlogged plants with black bars. Gas exchange was carried out under 380 µmol mol^{-1} CO_2 and 21% O_2 at 400 µmol m^{-2} s^{-1} PPFD. Data are means ± SE (*n* = 4). Letters stand for significantly different statistical classes (one-way ANOVA, $p < 0.05$ with post-hoc Tukey test). In (**c**), the asterisk stands for significance ($p < 0.05$, Welch test) for pair-wise comparison waterlogging vs. control.

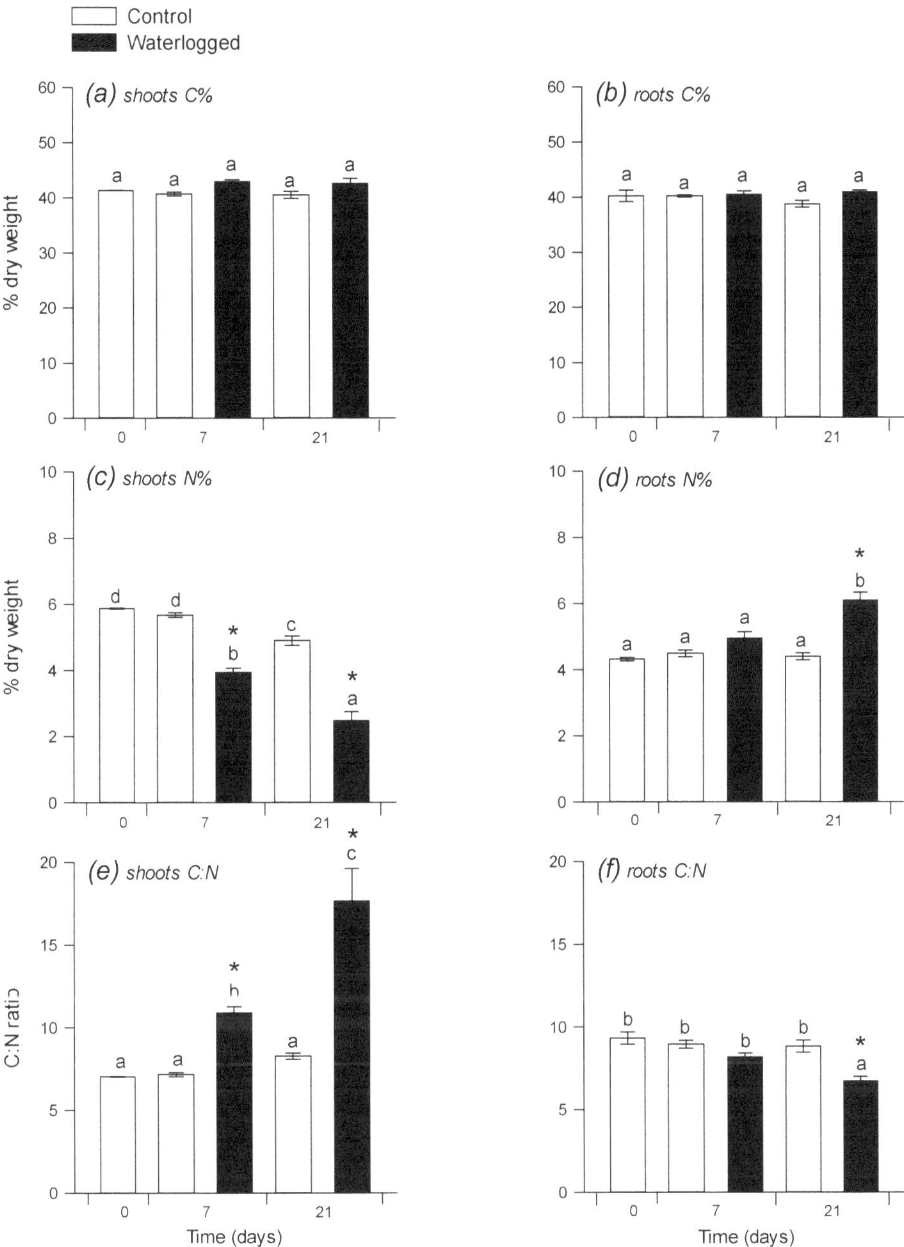

Figure 2. Carbon and nitrogen elemental content. C concentration in shoots (**a**) and roots (**b**) N concentration in shoots (**c**) and roots (**d**) and C:N ratio in shoots (**e**) and roots (**f**) Same legend as in Figure 1. Data are means ± SE (*n* = 3).

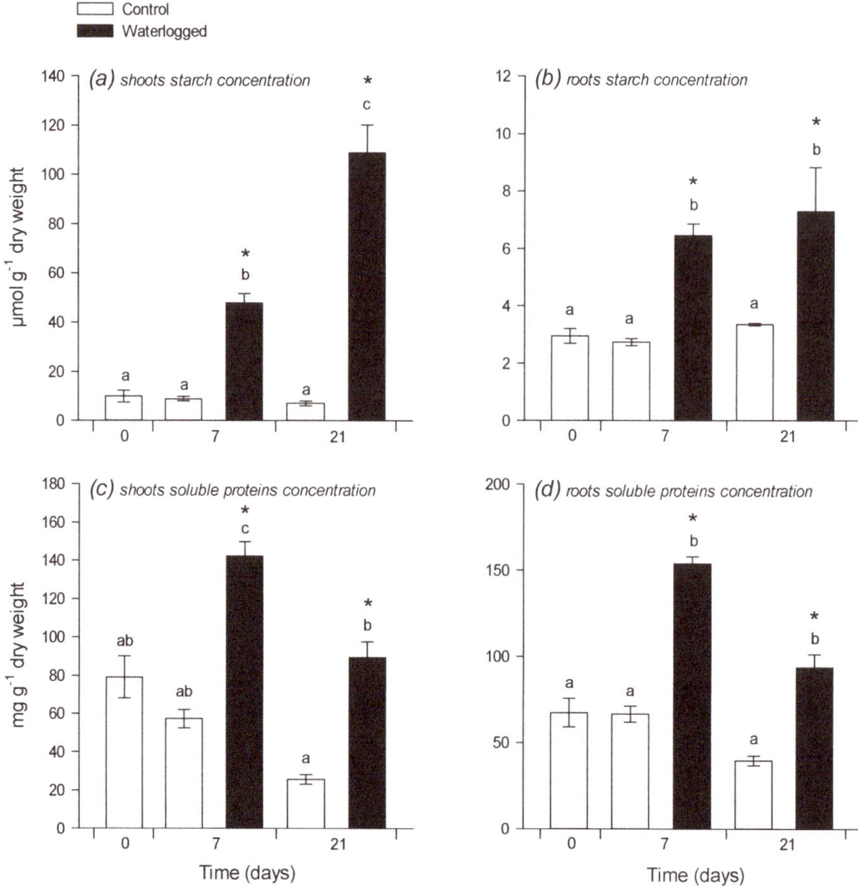

Figure 3. Starch and soluble protein concentrations. Starch content (in μmol hexose equivalents g^{-1} DW) in shoots (**a**) and roots (**b**) and soluble proteins concentration (in mg g^{-1} DW) in shoots (**c**) and roots (**d**) 7 and 21 days after the beginning of waterlogging. Same legend as in Figure 1. Data are means ± SE (*n* = 3).

In roots, metabolites that were significantly more abundant under waterlogging were mostly nitrogenous compounds, including alanine and GABA, as well as glutamine, putrescine, spermidine, and arginine (Figure 5a). Only three metabolites were found to be less abundant under waterlogging: xylose, raffinose, and vanillate. The multivariate analysis yielded a predictive (R^2 = 0.88) and robust (Q^2 = 0.69) model that was highly significant ($P_{CV\text{-}ANOVA}$ = 5.6·10^{-8}) and separated control and waterlogged conditions easily (Figure 5b). When univariate and multivariate analyses were combined into a volcano plot, the best biomarkers of waterlogging appeared to be alanine, GABA and β-alanine (increased) and raffinose (decreased) (Figure 5c). Interestingly, the aminoadipate-to-lysine ratio was significantly lower under waterlogging (Figure 4d), which suggested a specific downregulation of lysine catabolism. By contrast, the putrescine-to-ornithine ratio was significantly higher, showing a higher commitment to polyamine synthesis (Figure 5d).

2.3. Phloem Sap Composition and Movement

Because photosynthesis was altered while leaf sugar content increased, we explore the possibility that sugar transport via phloem sap was altered by waterlogging. Shoot phloem sap exudates were collected in EDTA solution in darkness. We collected phloem sap from whole shoots, since it was not possible to collect enough sap from individual leaves. In addition, collecting phloem sap exuded by whole shoots is useful for performing mass-balance at the plant scale and phloem transfer calculations. The metabolic composition was determined using both HPLC (sugars) and GC-MS (other metabolites). It is important to keep in mind that the exudation technique allows for quantifying the amount of metabolites of interest in the sample and, thus, to convert to a rate in nanomoles shoot^{-1} h^{-1}. However, this cannot be converted into a concentration in phloem sap since the determination of the exudated (very small) volume is not straightforward. Also, the collection of the xylem sap failed under our conditions, including using a pressurized system (Scholander chamber).

Figure 6 shows the composition of phloem sap exudates. Unsurprisingly, sugars prevailed in phloem sap exudates. There was a general increase in sugar exudation under waterlogging (Figure 6a), although it was rather variable (sucrose was more abundant after 21 d under waterlogging, with a *p*-value of 0.07). There was little change in the amino acid composition (Figure 6b) of exudates, with the notable exception of glycine, proline, and alanine, which were more abundant under waterlogging after 21 d. Amongst organic acids, fumarate, isocitrate, and malate were less abundant under waterlogging after 21 d.

The rate of phloem translocation was estimated while using mass-balance. Assuming that carbon transferred from shoots to roots was only made of sucrose, phloem loading and unloading rates were calculated, accounting for sucrose consumption for growth, starch synthesis, and respiration (Figure S2). Calculated average phloem loading rate was lower under waterlogging, in agreement with the lower assimilation rate and lower growth (Figure 7a). The difference between the loading by shoots and unloading by roots was always very small, in the order of 0.6 to 2% of loading; however, there was a clear increase in this difference under waterlogging (Figure 7b,c), showing that waterlogging inhibited phloem unloading by roots.

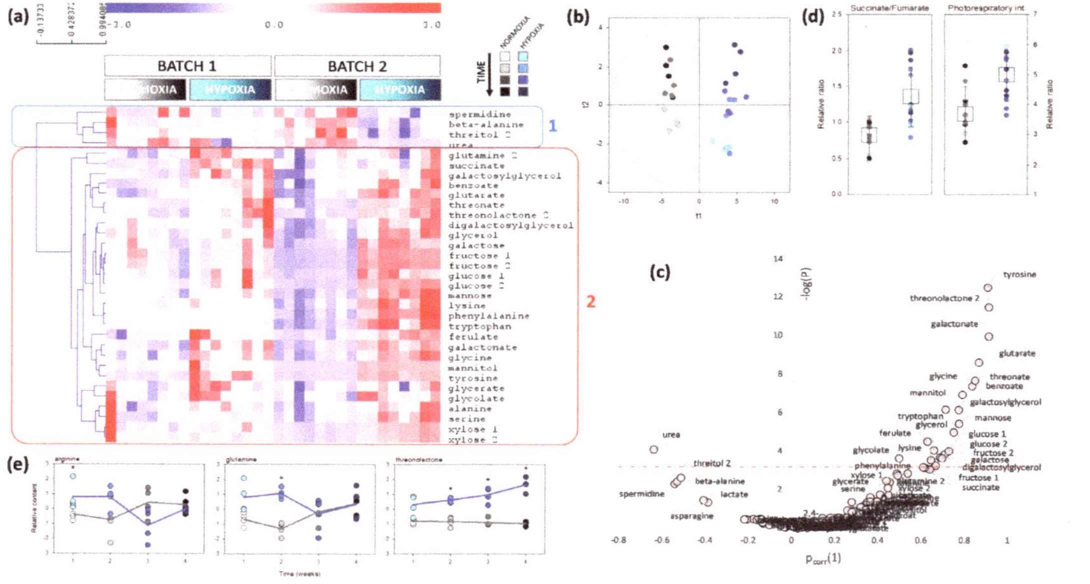

Figure 4. Metabolomics pattern of leaves when roots of the same plant are under control (normoxia) or waterlogging (hypoxia) (grey and blue shades, respectively). (**a**) Heat map showing significant metabolites ($p < 0.01$; two-way ANOVA) with a hierarchical clustering on left (Pearson correlation). The two main groups are framed and numbered (group 1 framed in blue: metabolites decreasing under hypoxia; group 2 framed in red: metabolites increasing under hypoxia). (**b**) Score plot of the multivariate analysis by O2PLS demonstrating the very good sample discrimination. (**c**) Volcano plot showing best discriminating metabolites (waterlogging vs. control) with the *p*-value (*y*-axis) and the loading in the O2PLS (*x*-axis). The horizontal dash-dotted line represents the Bonferroni threshold (0.0005). The two best discriminating features under waterlogging are an increase in tyrosine and a decrease in urea. (**d**) Relative metabolic ratio succinate-to-fumarate (left) and percentage (%) of photorespiratory intermediates (serine + glycine + glycolate + glycerate) in total metabolites (right). For both, the difference between waterlogging and control is significant ($p < 0.01$, Welch). (**e**) Metabolites that have a significant waterlogging × time interaction effect (arginine, glutamine, threonolactone). Conditions under which the difference between control and waterlogging is significant is labelled with an asterisk (*).

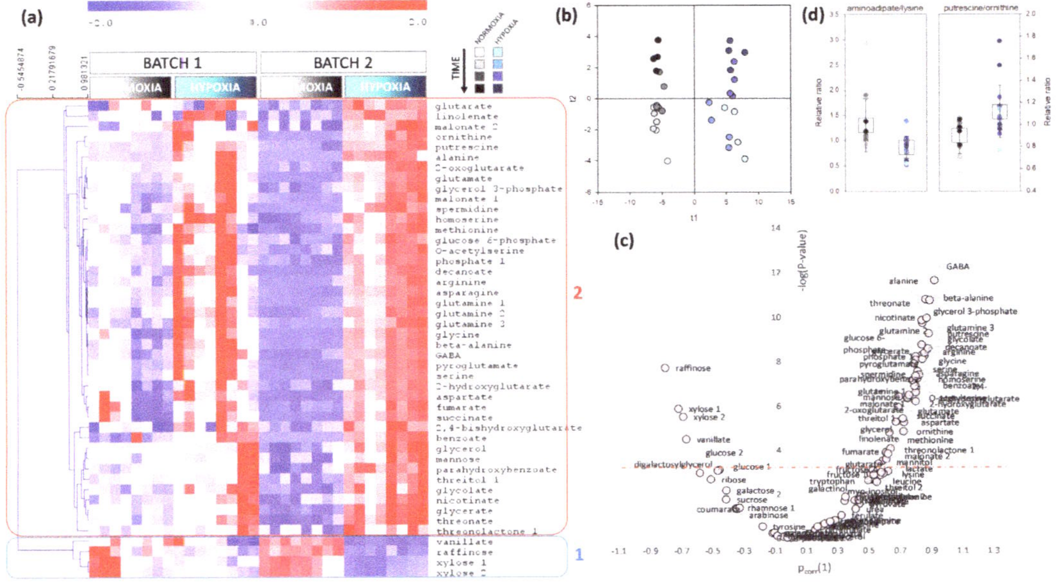

Figure 5. Metabolomics pattern of roots under control (normoxia) or waterlogging (hypoxia) conditions (grey and blue shades, respectively). (**a**) Heat map showing significant metabolites ($p < 0.0005$, i.e. Bonferron. threshold; two-way ANOVA) with a hierarchical clustering on left (Pearson correlation). The two main groups are framed and numbered (group 1 framed in blue: metabolites decreasing under waterlogging; group 2 framed in red: metabolites increasing under waterlogging). (**b**) Score plot of the multivariate analysis by O2PLS demonstrating the very good sample discrimination. (**c**) Volcano plot showing best discriminating metabolites (control vs. waterlogging) with the p-value (y-axis) and the loading in the O2PLS (x-axis). The horizontal dash-dotted line represents the Bonferroni threshold (0.0005). The two best discriminating features under waterlogging are an increase in γ-aminobutyrate (GABA) and a decrease in raffinose. (**d**) Relative metabolic ratios: aminoadipate-to-lysine (left) and putrescine-to-ornithine (right). For both, the difference between waterlogging and control conditions is significant ($p < 0.01$, Welch).

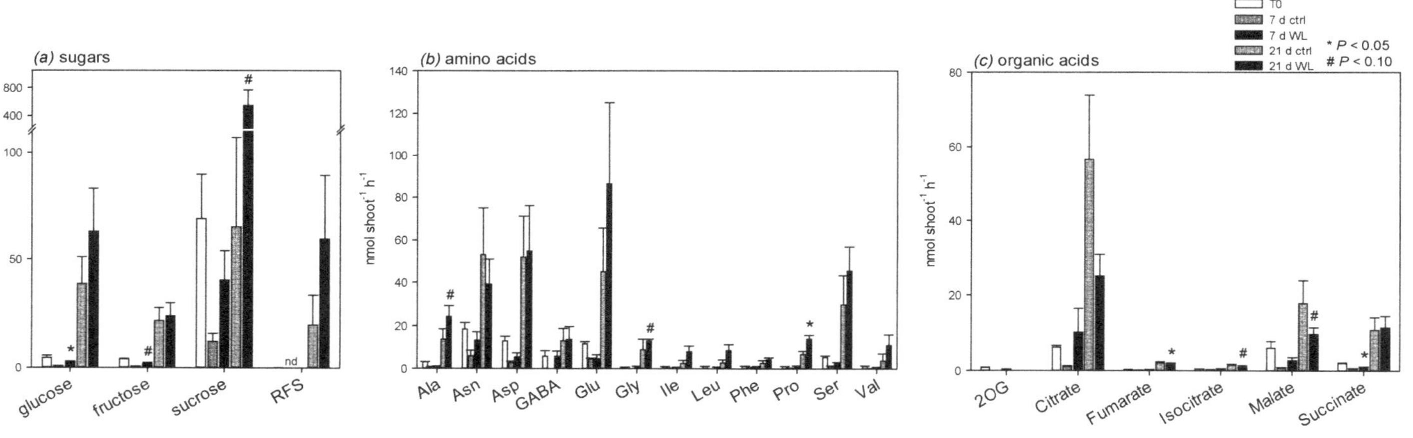

Figure 6. Composition of shoot phloem sap exudate (in EDTA solution) in darkness: sugars (**a**), amino acids (**b**) and organic acids (**c**). Results are expressed in exudation rate in nmol shoot^{-1} h^{-1}. Absolute quantitation in moles was achieved by HLPC-conductimetry (sugars) or GC-MS (other metabolites). Abbreviations: RFS, raffinose-family sugars (raffinose, verbascose, stachyose); nd, not detected. Symbols (*, #) stand for statistical significance (Student-Welsh) between control (ctrl) and waterlogging (WL). Mean ± SE ($n = 4$).

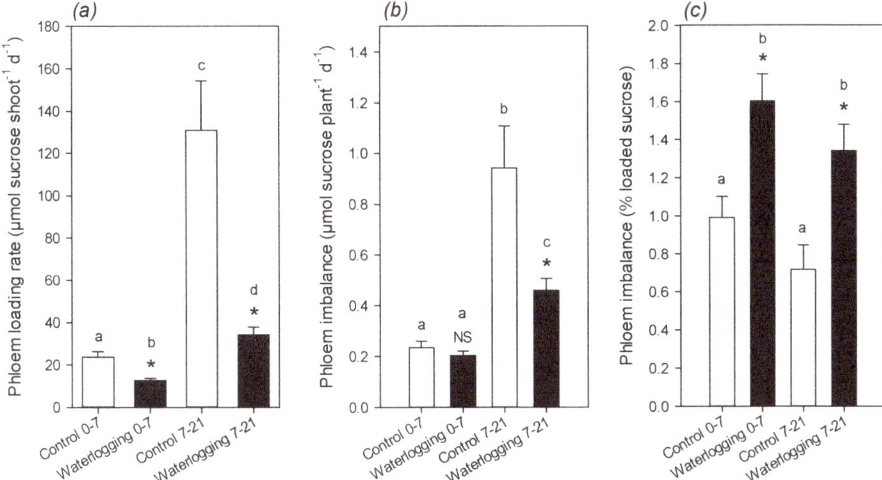

Figure 7. Calculated sucrose phloem loading rate and imbalance using mass-balance between shoots and roots: (**a**) average loading rate during for the two time windows considered (0 to 7 d, and 7 to 21 d) in μmol sucrose shoot^{-1} day^{-1}; (**b**) imbalance (difference between loading by shoots and unloading by roots) in μmol sucrose plant^{-1} day^{-1}; (**c**) imbalance expressed in percentage of loaded sucrose. Letters stand for significantly different classes in one-way ANOVA and asterisks stand for statistical significance (Welch test) between control and waterlogging. NS, non-significant. See Supplementary Material for calculation details.

3. Discussion

3.1. Differential Effect of Waterlogging on Leaf and Root Metabolome

It is generally accepted that there are metabolic effects of waterlogging in leaves, despite the fact that waterlogging (as opposed to full submergence) does not lead to hypoxic conditions in leaves. A common symptom of waterlogging is leaf starch accumulation [25], suggesting a change in sugar export and/or in partitioning between sucrose and starch synthesis during photosynthesis. Here, we also found a strong increase in leaf starch content, which represented up to 1.8% of shoot DW (~1000 μmol hexose equivalents g^{-1} DW) (Figure 2). This effect is not caused by an increase in photosynthesis, since waterlogging rather induces a decline in CO_2 assimilation (Figure 1). Hexoses (including galactose) and polyols (glycerol, mannitol) also both increased under waterlogging (Figure 4) and a similar effect has been found in other species [26,27]. The accumulation of hexoses under waterlogging is likely the result of the higher activity of sucrose synthase [9,10], which can regenerate UDP-glucose and, thus, in turn, feed galactose synthesis and metabolism (galactose, galactonate, threonate, and galactosylglycerol are significantly increased by waterlogging). Additionally, three metabolites of the polyamine pathway (urea, β-alanine, and spermidine) were less abundant, which suggested a downregulation of the entire pathway, which leads to a lower consumption and, in turn, a transient increase in arginine (Figure 4). This strong effect on polyamine metabolism has also been found in sunflower, including under K-deficient conditions, which normally induce putrescine accumulation [26]. The general increase in aromatics implies an increase in the consumption of their precursors, phospho*enol*pyruvate (PEP) and erythrose 4-phosphate, likely at the expense of both organic acid synthesis by PEP carboxylase and synthesis of erythrose derivatives, such as threitol (effectively decreased). The rationale of all of these metabolic changes probably relates to the stimulation of alternative pathways to consume hexoses via glycolysis, galactose metabolism, and the oxidative pentose phosphate pathway.

There was a clear opposite effect of waterlogging on root metabolism, with less sugars (and one aromatic compound, vanillate), and an increase in hypoxic biomarkers (alanine, GABA) reflecting the recycling of pyruvate and 2-oxoglutarate when catabolism is inhibited by the lack of oxygen. In fact, waterlogging is accompanied by an inhibition of the Krebs cycle for two reasons: first, a limitation of NADH reoxidation due to low O_2 concentration; second, hypoxia leads to an increase in nitric oxide (NO) production, which is a potent inhibitor of aconitase [35]. In effect, NO inhibits the second step (i.e., conversion of aconitate to isocitrate) of aconitase catalysis [36–38]. Additionally, there was an increase in many nitrogenous compounds, including β-alanine, putrescine, arginine, and the non-proteinogenic amino acid ornithine (Figure 5). This suggests that there was an increase in polyamines biosynthesis, which was perhaps associated with the urea cycle (urea was not significantly increased in roots under our conditions). Polyamines play a role in tolerance against abiotic stress, including hypoxia [39–41]. In addition, it has been proposed that the involvement of amino acid metabolism and polyamines under waterlogging could reflect an aspartate cycle [26], whereby arginine synthesis would represent a source of fumarate (citrulline + aspartate $\rightarrow\rightarrow$ arginine + fumarate), consumes reducing equivalents, and might be degraded to NO. Aspartate may then be reformed via malate and oxaloacetate (last steps of the Krebs cycle) or via the action of PEP carboxylase.

Interestingly, despite the general depletion in sugars and the high requirement to sustain substrate-level ATP synthesis (anaerobic glycolysis), waterlogged roots contained more starch (Figure 3), which suggested an inhibition of starch breakdown. This phenomenon is probably the result of the limitation in ATP generation and the utilization of glucose 6-phosphate for catabolism at the expense of phosphotransfer to glucose 1-phosphate and then starch non-reducing ends.

3.2. Phloem Composition and Translocation under Waterlogging

The opposite situation that is associated with sugar content in leaves and roots suggests that the inhibition of sugar translocation via the phloem is a major driver of metabolic changes. Our analysis of exudate composition shows that sugar—in particular sucrose—was more abundant under waterlogging. This effect could have come from a higher sugar concentration in sap and/or a higher sap flow. It is rather unlikely that sap flow was larger, since waterlogging is generally associated with lower stem capacitance (hydraulic conductance) and low stomatal conductance (Figure 1), which, in principle, inhibits both xylem and phloem sap flow. In waterlogged castor bean plants, phloem sap flow (bleeding rate) has been found to be lower while sugar concentration was slightly higher [32]. In *Eucalyptus*, waterlogging leads to a decrease in sucrose, but a strong increase in raffinose in phloem sap [42]. Thus, it appears more likely that the sucrose concentration in phloem sap was higher under waterlogging, and this was caused by an inhibition of translocation. Our mass-balance calculation further supports this hypothesis, with a higher relative imbalance (loading rate minus unloading rate) under waterlogging (Figure 7). This also agrees with the known depressing effect of hypoxia or anoxia on sugar import by sink organs, while source leaves and stems show little change in export and transfer rate, respectively (see Introduction and the review in Geiger & Sovonick, 1975, [43]).

In principle, the effect of waterlogging on phloem unloading by roots must also alter metabolites other than sugars. Here, we found that phloem exudates contained amino acids (with glutamate being the most represented) and organic acids (mostly citrate and malate). The general increase in amino acids under waterlogging (significant for alanine, proline, and glycine) not only reflects the lower consumption by roots, but also likely, the higher production by leaves (Figure 4). The cause for the lower content in organic acids is presently unclear, since malate and citrate in particular are not significantly less abundant in leaf tissue (Figure 4). Waterlogging has consequences on inorganic anions and cations, with a well-known inhibition of nitrate absorption and translocation to shoots via the xylem. Because nitrate is a major counter-anion that is associated with potassium, it is possible that, under our conditions, waterlogging caused a decrease in cation (K^+, Ca^{2+}, Na^+) content in phloem sap, and this was compensated for by lower organic acid content. In waterlogged *Eucalyptus*, there is a decrease in both Na^+ and Ca^{2+} in phloem sap [42].

3.3. Conclusions and Perspectives

Taken as a whole, our results show that roots and leaves have distinct metabolic responses, not only driven by the effect of oxygen shortage (in roots), but also by the side effect of root hypoxia on sap homeostasis. In particular, our metabolomics data suggest that the decline in unloading efficiency in roots alters sugar translocation, and this exerts a feedback on leaf metabolism so as to redirect sugars to other metabolic pathways. Of course, our present data are limited by the present technology and further work would be needed to provide a full characterization of phloem and xylem composition and flow rate. This would be useful to better appreciate the impact of waterlogging on metabolite circulation. For example, the enhancement of polyamine metabolism in roots and its repression in shoots could be associated with a change in polyamine circulation in both xylem and phloem saps. The exchange of metabolites between roots and other compartments is important, because it could reflect a strategy to export redox power when oxygen limitation that is caused by waterlogging impedes NADH reoxidation in roots. Solving metabolite circulation would require not only metabolomics analyses, but also multiple isotopic tracing, as well as proteomics analyses to identify key enzymes that are up-regulated by waterlogging in an organ-specific manner. This will be addressed in a subsequent study.

4. Materials and Methods

4.1. Plant Material and Growing Conditions

Barrel clover *Medicago truncatula* (line A17) were grown in pots on neutral substrate and watered each other day with $\frac{1}{2}$ Murashige and Skoog (MS) nutrient solution. The plants were randomly distributed in the growth chamber. Growth conditions were: constant temperature $22 \pm 0.2\ °C$, relative humidity of $70 \pm 2\ \%$, long days (16:8 hours light/dark). Three weeks after germination, young plants were subdivided into a set of control plants and one set of plants subjected to waterlogging. Waterlogging was done by filling pots with the nutrient solution. The nutrient solution was replaced every 72 h (with degassed solution). The establishment of hypoxia was checked by measuring dissolved O_2 content in the pots while using an electrochemical oxygen meter (GMH 3630, Greisinger, Regenstauf, Germany). The control plants were watered with the 1/2 MS nutrient solution. Plants were harvested at the onset of waterlogging treatment (time 0) and after seven and 21 days.

4.2. Biomass and C and N Elemental Content

At each sampling time, six control and waterlogged plants were harvested. The roots and shoots were collected separately, weighted, frozen in liquid nitrogen, and then stored at $-80\ °C$ before further experiments. The plant material was then freeze-dried and grounded in order to obtain a homogeneous fine powder. A sub-sample of 4 mg was used to determine the total nitrogen and carbon content using an elemental analyzer (Thermoquest FlashEA1112®, Thermo Fisher Scientific, Waltham, MA, USA).

4.3. Gas Exchange Measurements

The measurement of net photosynthetic assimilation, stomatal conductance, and respiration rates was carried out on four intact plants while using a gas exchange open system Li-Cor 6400 *xt* with the $2\ cm^2$ fluorescence chamber (Li-Cor, Austin, TX, USA). Net photosynthesis and stomatal conductance were measured under typical conditions (380 $\mu mol\ mol^{-1}$ CO_2, 21% O_2, 22 °C, 400 $\mu mol·m^{-2}s^{-1}$ PAR, 10% blue). Dark respiration was measured under the same gaseous conditions after 30 min. dark acclimation.

4.4. Determination of Starch and Soluble Protein Concentrations

Soluble proteins were extracted from 50 mg frozen plant tissue. Samples were ground to a fine powder, extracted with 2 mL of 100 mM sodium phosphate buffer (pH 7.4) at 4 °C, vortexed three times for 30 s, and then centrifuged at 10,000 g for 20 min. at 4 °C, and the supernatant was collected. Soluble proteins were analyzed in the supernatant with a commercially available kit (Coomassie Protein assay reagent; Bio-Rad, Les Ulis, France) while using bovine serum albumin as a standard. Starch was extracted from 15 mg frozen-dried (lyophilized) plant tissue ground to a fine powder. The samples were depigmented with 1 mL of 80% ethanol and then centrifuged at 10,000 g for 20 min. at 4 °C. Starch was extracted from the pellet by adding 0.2 U of α-amyloglucosidase (Sigma Chemical Co., St. Louis, MO, USA) and 40 U of α-amylase (Sigma Chemical Co., St. Louis, USA) in 20 mM sodium acetate buffer (pH 5.1). Starch was quantified in glucose equivalents determined spectrophotometrically (hexokinase-glucose 6-phosphate dehydrogenase assay).

4.5. Phloem Sap Exudation

Shoot phloem exudates were collected over a period of 2 h. The shoots were sectioned at the collar, recut in a Petri dish filled with exudation buffer (5 mM EDTA, pH 6.0), and then immersed in 500 µL of the same buffer. To avoid transpiration, the shoots were placed in dark under very high hygrometry (100% RH). Three biological replicates were used for each point and each nutrition treatment. The exudates were stored at −80 °C until metabolic analysis.

4.6. Metabolomics Analyses

Gas chromatography coupled to mass spectrometry (GC-MS) analyses were carried out as in Cui et al., 2019 [26,27]. We used 80% methanol extracts that were derivatized with methoxamine and N-methyl-N-(trimethylsilyl)trifluoroacetamide (MSTFA) in pyridine. Ribitol was used as an internal standard. Two batches of samples were used for metabolomics, which corresponded to two sets of plant cultivation. The use of two batches for metabolomics was decided to ensure that statistical analysis of metabolomics data (with many variables) account for intra-group variability and a decrease the likelihood of false discoveries upon multiple univariate analysis.

4.7. Statistics

Physiological variables (photosynthesis, respiration, etc.; Figures 1–3) were examined using univariate analysis (one-way or two-way ANOVA, followed by Tukey post-hoc test, and pair-wise comparison with Welch tests, as specified in figure legends). The number of replicates (3 to 5) is indicated in figure legends.

Metabolomics data were analyzed using both univariate and multivariate statistics. Supervised multivariate analysis was performed with orthogonal projection on latent structure (O2PLS) with Simca 13 (Umetrics, Umea, Sweden), using waterlogging and developmental time as predicted Y variables, and metabolites (UV-scaled) as predicting X variables. We first verified the absence of statistical outliers with principal component analysis (PCA, also done using Simca 13, with UV-scaled metabolite contents as variables), where no data point was outside the 99% confidence Hotelling region. The performance of the multivariate model was assessed while using the determination coefficient R^2 and the predictive power was quantified by the cross-validated determination coefficient, Q^2. The significance of the statistical O2PLS model was tested using a χ^2 comparison with a random model (average ± random error), and the associated p-value ($P_{\text{CV-ANOVA}}$) is reported. The best marker metabolites were visualized with volcano plots, where the logarithm of the p-value that was obtained in univariate analysis (two-way ANOVA) was plotted against the rescaled loading (p_{corr}) obtained in O2PLS. In such a representation, the best biomarkers have both maximal $-\log(P)$ and p_{corr} values.

4.8. Mass-Balance Calculation of Phloem Loading and Imbalance

Carbon exchange between shoots and roots via the phloem is simplified using a three-compartment model (Figure S2a), where only sucrose is considered. By mass-balance, assimilated carbon is partitioned to net starch synthesis (positive when starch is synthesized; negative when it is degraded), respiration, growth, and export (loading). Similarly, root imported carbon (unloading) is partitioned to respiration, growth, and net starch synthesis. Growth was calculated using the biomass increment (Figure S1) and expressed in sucrose equivalents using %C (Figure 2). Respiration was estimated using the respiration rate that was measured in leaves (Figure 1), converted to a dry mass basis, and rescaled to the total biomass of the organ considered. Starch content was measured directly (Figure 2). When rates are expressed in µmol sucrose shoot^{-1} d^{-1}, the mass-balance applied to the shoot sucrose total pool (S_s) is so that:

$$\frac{dS_s}{dt} = A - R_s - G_s - \sigma_s - C \tag{1}$$

Because the total sucrose pool equals the average concentration (ω_s, µmoles per g DW) times biomass (B_s, in g DW), we have $S_s = B_s\omega_s$. Assuming that variations in ω_s are small as compared to biomass increase (i.e. The order of magnitude of sucrose concentration does not change dramatically), we have:

$$\frac{dS_s}{dt} \approx \omega_s\frac{dB_s}{dt} = \omega_s\xi G_s \tag{2}$$

where ξ is the conversion factor of growth (G_s) from dry mass to µmoles sucrose (0.00036 g DW µmol^{-1} sucrose). Combining (1) and (2) gives:

$$C = A - R_s - G_s \cdot (1 + \xi\omega_s) - \sigma_s \tag{3}$$

Similarly, for roots, we obtain:

$$D = R_r + G_r \cdot (1 + \xi\omega_r) + \sigma_r \tag{4}$$

where ω_r is sucrose concentration in roots.

The phloem imbalance (denoted as i) is then given by the difference between loading and unloading, $i = C - D$. By definition, this imbalance represents the incremental change in total phloem sucrose pool (S_p, in µmol phloem sucrose plant^{-1}) with time, which is:

$$\frac{dS_p}{dt} = i = C - D \tag{5}$$

When this difference is positive, S_p increases. This is the general case, since the plant size increases and so must be total phloem volume. Nevertheless, because $S_p = V_p \cdot \omega_p$ (where V_p is total phloem volume and ω_p phloem sucrose concentration) and V_p is not readily accessible, S_p cannot be converted into a concentration. Computations indicate that S_p is within 1 and 10 µmol plant^{-1}. For example, if we assume that V_p is about 2.5% of total plant volume, it means an average concentration ω_p of 40 mM (14 mg mL^{-1}) in phloem sap, which is a realistic value. In Figure 7, the imbalance was also expressed in percentage (denoted as p) of sucrose loading, as follows:

$$p = \frac{i}{C} \tag{6}$$

Supplementary Materials: The following are available online at http://www.mdpi.com/2223-7747/9/10/1373/s1, Figure S1: Plant biomass and relative water content, Figure S2: Model used to estimate the rate of phloem loading.

Author Contributions: Data acquisition and experiments, H.D., C.C. and J.L.; conceptualization, J.L., A.M.L. and G.T.; data analysis, J.L. and G.T.; writing—original draft preparation, G.T. and J.L.; writing—review and editing, A.M.L.; All authors have read and agreed to the published version of the manuscript.

Funding: This research was funded by the Région Pays de la Loire and Angers Loire Métropole, via the grant Connect Talent Isoseed.

Acknowledgments: The authors thank the Plateforme Métabolisme-Métabolome (IPS2) for metabolomics analyses, and also thankPascale Satour for her help to interpret HPLC profiles for sugar analysis.

Conflicts of Interest: The authors declare no conflict of interest.

References

1. Bailey-Serres, J.; Fukao, T.; Gibbs, D.J.; Holdsworth, M.J.; Lee, S.C.; Licausi, F.; Perata, P.; Voesenek, L.A.; van Dongen, J.T. Making sense of low oxygen sensing. *Trends Plant Sci.* **2012**, *17*, 129–138. [CrossRef] [PubMed]

2. Licausi, F. Molecular elements of low-oxygen signaling in plants. *Physiol. Plant.* **2013**, *148*, 1–8. [CrossRef] [PubMed]

3. Manik, S.; Pengilley, G.; Dean, G.; Field, B.; Shabala, S.; Zhou, M. Soil and crop management practices to minimize the impact of waterlogging on crop productivity. *Front. Plant Sci.* **2019**, *10*, 140. [CrossRef] [PubMed]

4. Jackson, M.; Colmer, T. Response and adaptation by plants to flooding stress. *Ann. Bot.* **2005**, *96*, 501–505. [CrossRef] [PubMed]

5. Perata, P.; Alpi, A. Plant responses to anaerobiosis. *Plant Sci.* **1993**, *93*, 1–17. [CrossRef]

6. Bailey-Serres, J.; Voesenek, L. Flooding stress: Acclimations and genetic diversity. *Annu. Rev. Plant Biol.* **2008**, *59*, 313–339. [CrossRef]

7. Hsu, F.-C.; Shih, M.-C. Plant defense after flooding. *Plant Signal. Behav.* **2013**, *8*, 2699–2713. [CrossRef]

8. Alam, I.; Lee, D.-G.; Kim, K.-H.; Park, C.-H.; Sharmin, S.A.; Lee, H.; Oh, K.-W.; Yun, B.-W.; Lee, B.-H. Proteome analysis of soybean roots under waterlogging stress at an early vegetative stage. *J. Biosci.* **2010**, *35*, 49–62. [CrossRef]

9. Drew, M.C. Oxygen deficiency and root metabolism: Injury and acclimation under hypoxia and anoxia. *Annu. Rev. Plant Biol.* **1997**, *48*, 223–250. [CrossRef]

10. Planchet, E.; Lothier, J.; Limami, A.M. Hypoxic respiratory metabolism in plants: Reorchestration of nitrogen and carbon metabolisms. In *Plant Respiration: Metabolic Fluxes and Carbon Balance*; Tcherkez, G., Ghashghaie, J., Eds.; Springer: Berlin, Germany, 2017; pp. 209–226.

11. Mustroph, A.; Zanetti, M.E.; Jang, C.J.; Holtan, H.E.; Repetti, P.P.; Galbraith, D.W.; Girke, T.; Bailey-Serres, J. Profiling translatomes of discrete cell populations resolves altered cellular priorities during hypoxia in Arabidopsis. *Proc. Natl. Acad. Sci. USA* **2009**, *106*, 18843–18848. [CrossRef]

12. Loreti, E.; Poggi, A.; Novi, G.; Alpi, A.; Perata, P. A genome-wide analysis of the effects of sucrose on gene expression in Arabidopsis seedlings under anoxia. *Plant Physiol.* **2005**, *137*, 1130–1138. [CrossRef] [PubMed]

13. Bond, D.M.; Dennis, E.S.; Finnegan, E.J. Hypoxia: A novel function for *VIN3*. *Plant Signal. Behav.* **2009**, *4*, 773–776. [CrossRef]

14. Christianson, J.A.; Wilson, I.W.; Llewellyn, D.J.; Dennis, E.S. The low-oxygen-induced NAC domain transcription factor ANAC102 affects viability of Arabidopsis seeds following low-oxygen treatment. *Plant Physiol.* **2009**, *149*, 1724–1738. [CrossRef]

15. Van Dongen, J.T.; Fröhlich, A.; Ramírez-Aguilar, S.J.; Schauer, N.; Fernie, A.R.; Erban, A.; Kopka, J.; Clark, J.; Langer, A.; Geigenberger, P. Transcript and metabolite profiling of the adaptive response to mild decreases in oxygen concentration in the roots of Arabidopsis plants. *Ann. Bot.* **2009**, *103*, 269–280. [CrossRef] [PubMed]

16. Banti, V.; Mafessoni, F.; Loreti, E.; Alpi, A.; Perata, P. The heat-inducible transcription factor HsfA2 enhances anoxia tolerance in Arabidopsis. *Plant Physiol.* **2010**, *152*, 1471–1483. [CrossRef]

17. Licausi, F.; Van Dongen, J.T.; Giuntoli, B.; Novi, G.; Santaniello, A.; Geigenberger, P.; Perata, P. HRE1 and HRE2, two hypoxia-inducible ethylene response factors, affect anaerobic responses in Arabidopsis thaliana. *Plant J.* **2010**, *62*, 302–315. [CrossRef] [PubMed]

18. Ellis, M.H.; Dennis, E.S.; Peacock, W.J. Arabidopsis roots and shoots have different mechanisms for hypoxic stress tolerance. *Plant Physiol.* **1999**, *119*, 57–64. [CrossRef]

19. Mustroph, A.; Boamfa, E.I.; Laarhoven, L.J.; Harren, F.J.; Albrecht, G.; Grimm, B. Organ-specific analysis of the anaerobic primary metabolism in rice and wheat seedlings. I: Dark ethanol production is dominated by the shoots. *Planta* **2006**, *225*, 103–114. [CrossRef]

20. Mustroph, A.; Boamfa, E.I.; Laarhoven, L.J.; Harren, F.J.; Pörs, Y.; Grimm, B. Organ specific analysis of the anaerobic primary metabolism in rice and wheat seedlings II: Light exposure reduces needs for fermentation and extends survival during anaerobiosis. *Planta* **2006**, *225*, 139–149. [CrossRef]

21. Hsu, F.-C.; Chou, M.-Y.; Peng, H.-P.; Chou, S.-J.; Shih, M.-C. Insights into hypoxic systemic responses based on analyses of transcriptional regulation in Arabidopsis. *PLoS ONE* **2011**, *6*, e28888. [CrossRef]

22. Lee, S.C.; Mustroph, A.; Sasidharan, R.; Vashisht, D.; Pedersen, O.; Oosumi, T.; Voesenek, L.A.; Bailey-Serres, J. Molecular characterization of the submergence response of the *Arabidopsis thaliana* ecotype Columbia. *New Phytol.* **2011**, *190*, 457–471. [CrossRef]

23. Chang, R.; Jang, C.J.; Branco-Price, C.; Nghiem, P.; Bailey-Serres, J. Transient MPK6 activation in response to oxygen deprivation and reoxygenation is mediated by mitochondria and aids seedling survival in Arabidopsis. *Plant Mol. Biol.* **2012**, *78*, 109–122. [CrossRef] [PubMed]

24. Mustroph, A.; Barding Jr, G.A.; Kaiser, K.A.; Larive, C.K.; Bailey-Serres, J. Characterization of distinct root and shoot responses to low-oxygen stress in A rabidopsis with a focus on primary C-and N-metabolism. *Plant Cell Environ.* **2014**, *37*, 2366–2380.

25. Irfan, M.; Hayat, S.; Hayat, Q.; Afroz, S.; Ahmad, A. Physiological and biochemical changes in plants under waterlogging. *Protoplasma* **2010**, *241*, 3–17. [CrossRef]

26. Cui, J.; Abadie, C.; Carroll, A.; Lamade, E.; Tcherkez, G. Responses to K deficiency and waterlogging interact via respiratory and nitrogen metabolism. *Plant Cell Environ.* **2019**, *42*, 647–658. [CrossRef] [PubMed]

27. Cui, J.; Davanture, M.; Zivy, M.; Lamade, E.; Tcherkez, G. Metabolic responses to potassium availability and waterlogging reshape respiration and carbon use efficiency in oil palm. *New Phytol.* **2019**, *223*, 310–322. [CrossRef]

28. Cui, J.; Lamade, E.; Fourel, F.; Tcherkez, G. δ^{15}N values in plants is determined by both nitrate assimilation and circulation. *New Phytol.* **2020**, in press. [CrossRef]

29. Christianson, J.A.; Llewellyn, D.J.; Dennis, E.S.; Wilson, I.W. Global gene expression responses to waterlogging in roots and leaves of cotton (*Gossypium hirsutum* L.). *Plant Cell Physiol.* **2009**, *51*, 21–37. [CrossRef]

30. Amarante, L.; Sodek, L. Waterlogging effect on xylem sap glutamine of nodulated soybean. *Biol. Plant.* **2006**, *50*, 405–410. [CrossRef]

31. Puiatti, M.; Sodek, L. Waterlogging affects nitrogen transport in the xylem of soybean. *Plant Physiol. Biochem.* **1999**, *37*, 767–773. [CrossRef]

32. Peuke, A.D.; Gessler, A.; Trumbore, S.; Windt, C.W.; Homan, N.; Gerkema, E.; Van As, H. Phloem flow and sugar transport in *Ricinus communis* L. is inhibited under anoxic conditions of shoot or roots. *Plant Cell Environ.* **2015**, *38*, 433–447. [CrossRef]

33. Schumacher, T.E.; Smucker, A.J.M. Carbon transport and root respiration of split root systems of *Phaseolus vulgaris* subjected to short term localized anoxia. *Plant Physiol.* **1985**, *78*, 359–364. [CrossRef] [PubMed]

34. Saglio, P.H. Effect of path or sink anoxia on sugar translocation in roots of maize seedlings. *Plant Physiol.* **1985**, *77*, 285–290. [CrossRef] [PubMed]

35. Gupta, K.J.; Shah, J.K.; Brotman, Y.; Jahnke, K.; Willmitzer, L.; Kaiser, W.M.; Bauwe, H.; Igamberdiev, A.U. Inhibition of aconitase by nitric oxide leads to induction of the alternative oxidase and to a shift of metabolism towards biosynthesis of amino acids. *J. Exp. Bot.* **2012**, *63*, 1773–1784. [CrossRef] [PubMed]

36. Castro, L.; Rodriguez, M.; Radi, R. Aconitase is readily inactivated by peroxynitrite, but not by its precursor, nitric oxide. *J. Biol. Chem.* **1994**, *269*, 29409–29415. [PubMed]

37. Lloyd, S.J.; Lauble, H.; Prasad, G.S.; Stout, C.D. The mechanism of aconitase: 1.8 Å resolution crystal structure of the S642A:citrate complex. *Protein Sci.* **1999**, *8*, 2655–2662. [CrossRef]

38. Tórtora, V.; Quijano, C.; Freeman, B.; Radi, R.; Castro, L. Mitochondrial aconitase reaction with nitric oxide, S-nitrosoglutathione, and peroxynitrite: Mechanisms and relative contributions to aconitase inactivation. *Free Radic. Biol. Med.* **2007**, *42*, 1075–1088. [CrossRef]

39. Reggiani, R.; Hochkoeppler, A.; Bertani, A. Polyamines in rice seedlings under oxygen-deficit stress. *Plant Physiol.* **1989**, *91*, 1197–1201. [CrossRef]

40. Verma, S.; Mishra, S.N. Putrescine alleviation of growth in salt stressed *Brassica juncea* by inducing antioxidative defense system. *J. Plant Physiol.* **2005**, *162*, 669–677. [CrossRef]

41. Cui, J.; Pottosin, I.; Lamade, E.; Tcherkez, G. What is the role of putrescine accumulated under potassium deficiency? *Plant Cell Environ.* **2020**, in press. [CrossRef]

42. Merchant, A.; Peuke, A.D.; Keitel, C.; Macfarlane, C.; Warren, C.R.; Adams, M.A. Phloem sap and leaf $\delta^{13}C$, carbohydrates, and amino acid concentrations in *Eucalyptus globulus* change systematically according to flooding and water deficit treatment. *J. Exp. Bot.* **2010**, *61*, 1785–1793. [CrossRef] [PubMed]
43. Geiger, D.R.; Sovonick, S.A. Effects of temperature, anoxia and other metabolic inhibitors on translocation. In *Transport in Plants I: Phloem Transport*; Zimmermann, M.H., Milburn, J.A., Eds.; Springer: Berlin/Heidelberg, Germany, 1975; pp. 256–286.

Publisher's Note: MDPI stays neutral with regard to jurisdictional claims in published maps and institutional affiliations.

Article

Eco-Physiological Traits Related to Recovery from Complete Submergence in the Model Legume *Lotus japonicus*

Florencia B. Buraschi [1], Federico P.O. Mollard [1], Agustín A. Grimoldi [2] and Gustavo G. Striker [1,3,*]

[1] IFEVA, CONICET, Cátedra de Fisiología Vegetal, Facultad de Agronomía, Universidad de Buenos Aires, Av. San Martín 4453, Buenos Aires C1417DSE, Argentina; fburaschi@agro.uba.ar (F.B.B.); fmollard@agro.uba.ar (F.P.O.M.)

[2] IFEVA, CONICET, Cátedra de Forrajicultura, Facultad de Agronomía, Universidad de Buenos Aires, Av. San Martín 4453, Buenos Aires C1417DSE, Argentina; grimoldi@agro.uba.ar

[3] UWA, School of Agriculture and Environment, Faculty of Science, The University of Western Australia, 35 Stirling Hwy, Crawley, WA 6009, Australia

* Correspondence: striker@agro.uba.ar

Received: 3 April 2020; Accepted: 17 April 2020; Published: 21 April 2020

Abstract: Submergence is a severe form of stress for most plants. *Lotus japonicus* is a model legume with potential use in assisting breeding programs of closely related forage *Lotus* species. Twelve *L. japonicus* genotypes (10 recombinant inbred lines (RILs) and 2 parental accessions) with different constitutive shoot to root dry mass ratios (S:R) were subjected to 7 days of submergence in clear water and allowed to recover for two weeks post-submergence; a set of non-submerged plants served as controls. Relative growth rate (RGR) was used to indicate the recovery ability of the plants. Leaf relative water content (RWC), stomatal conductance (g_s), greenness of basal and apical leaves, and chlorophyll fluorescence (Fv/Fm, as a measure of photoinhibition) were monitored during recovery, and relationships among these variables and RGR were explored across genotypes. The main results showed (i) variation in recovery ability (RGR) from short-term complete submergence among genotypes, (ii) a trade-off between growth during vs. after the stress indicated by a negative correlation between RGR during submergence and RGR post-submergence, (iii) an inverse relationship between RGR during recovery and S:R upon de-submergence, (iv) positive relationships between RGR at early recovery and RWC and g_s, which were negatively related to S:R, suggesting this parameter as a good estimator of plant water balance post-submergence, (v) chlorophyll retention allowed fast recovery as revealed by the positive relationship between greenness of basal and apical leaves and RGR during the first recovery week, and (vi) full repair of the submergence-damaged photosynthetic apparatus occurred more slowly (second recovery week) than full recovery of plant water relations. The inclusion of these traits contributing to submergence recovery in *L. japonicus* should be considered to speed up the breeding process of the closely related forage *Lotus* spp. used in current agriculture.

Keywords: post-submergence recovery; legumes; plant water relations; shoot to root ratio; *Lotus japonicus*; leaf greenness; leaf desiccation; stomatal conductance

1. Introduction

As a result of the effects of climate change, an increase in the frequency and intensity of flooding events is expected to occur in the coming years [1]. A higher intensity of flooding can easily lead to plant submergence. Complete submergence denotes a condition where floodwaters increase to a level where plant shoots remain fully underwater; it is one of the most stressful scenarios that plants confront in prone-to-flood environments [2]. This situation drastically decreases the direct exchange of gases between the plant and the atmosphere, resulting in reduced O_2 and CO_2 levels [3]. Moreover, complete submergence often reduces the irradiance for photosynthesis depending on the water depth and turbidity. Two plant strategies have been recognized in plant submergence responses: (i) the "escape strategy" and (ii) the "quiescent' strategy" [4–6]. The first one involves shoot elongation to restore leaf contact with the atmosphere and offers plants a better chance to survive under shallow long-term submergence (>1 week). The "quiescent" strategy is based on maintaining steady energy conservation without shoot elongation and it is usually adopted to cope with deep short-term submergence (<1 week), given that shoot emergence represents a high energy cost that could compromise subsequent plant recovery [3,7]. Thus, in scenarios of short, deep submergence with low CO_2 and/or low light, plants rarely grow, but instead aim just to survive until the water recedes and later resume growth. Additionally, intraspecific variability between the escape and quiescent strategies has been found in rice [4]. The resumption of vigorous growth after experiencing complete submergence indicates the recovery ability of the species/genotype/accession [8]. Plant responses during submergence have been extensively studied and documented [4,9–12], while plant responses after water recession (i.e. recovery phase) have seldom been reported (but see [13,14] for *Arabidopsis thaliana*). This study aims at identifying eco-physiological traits facilitating plant recovery after complete submergence.

Forage legumes are important components of pastures and natural grasslands. The *Lotus* genus includes more than 180 species distributed worldwide [15,16], and it is used to improve pastures in stressful environments, where traditional forage legumes, such as lucerne or red clover, fail to grow or survive [17]. *Lotus corniculatus*, *L. pedunculatus*, and *L. tenuis* have been domesticated and improved through breeding for possessing some degree of tolerance to soil waterlogging, with *L. tenuis* being the most tolerant to root oxygen deficiency [6,18,19]. A few years ago, *L. japonicus*, which is taxonomically closely related to these forage species [20], arose as a model for legumes because it is a small plant with short generation time, easy cultivation, and it is amenable to transformation. So, a set of genetic resources was developed, including ecotypes, mutant lines, and recombinant inbred lines (RILs) [21,22]. Previous studies showed that adult plants of this species could tolerate three weeks of waterlogging [23] and up to 12 days of submergence at the seedling stage [6]. So, the characterization of *L. japonicus* responses to submergence and its post-submergence recovery can help support plans intended for the improvement of submergence tolerance of the closely related forage *Lotus* species.

Relevant traits for the recovery of plants from water submergence have scarcely been identified. It is perceived that plant water status is critical to facilitate fast recovery, particularly considering that a diminished and/or damaged root system due to submergence cannot necessarily cope with the water uptake needed to meet shoot transpiration when re-exposed to atmospheric air upon de-submergence ([24] for the grass *Chloris gayana*). In rice cv. IR42, the intolerance to complete submergence is caused by desiccation of leaves (i.e., decreased hydraulic conductivity in the leaf sheath) after de-submergence, which occurs rapidly, provoking wilting and death of the plants [25]. In this regard, shoot to root dry mass ratio upon de-submergence might be roughly related to water balance (i.e., transpiration by the shoot/water uptake by roots) and might influence plant recovery. Carbon assimilation after submergence can also positively impact plant recovery. Pioneer works [26,27] showed that the ability to resume growth after 20 days of submergence in the forb *Alternanthera philoxeroides* and the grass *Hemarthria altissima* resulted from rapid leaf growth and recovery of functionality of the photosynthetic apparatus. In two contrasting accessions for submergence tolerance of *Arabidopsis thaliana* (Bay-0-sensitive vs. Lp2-tolerant), it was shown that quick stomatal aperture, higher chlorophyll retention, and leaf water turgor maintenance were key factors for the tolerant accession to sustain a fast recovery following the removal of the stress [13], these traits also provide a detailed recovery-signaling network for enhanced flooding tolerance in *Arabidopsis*. In this study, we selected 10 RILs of *L. japonicus* (plus both parents) with differential constitutive shoot to root dry mass ratios (S:R) from [23] with the objective to assess leaf water status, stomatal conductance, dark-adapted chlorophyll fluorescence of photosystem II (Fv/Fm), and greenness of basal and apical leaves (as a surrogate to infer N status) and explore how these variables correlate with plant growth resumption (i.e., plant RGR: relative growth rate). We investigated these plant responses and relationships during the first and the second week after de-submergence to unveil traits that aid in early and late recovery. We hypothesized that there is variation among genotypes (i.e., RILs) of *Lotus japonicus* in their ability to recover from complete submergence associated with differences in the S:R ratio, plant water status, and leaf greenness. We expected that genotypes with low S:R present quick recovery of leaf water status, stomatal conductance, and high retention of leaf chlorophyll during the early recovery and, therefore, show high plant RGR post-submergence.

2. Results

2.1. Plant Growth during Submergence

All genotypes (parents MG-20, Gifu, and 10 RILs) showed positive relative growth rates (RGR) during the first week of the experiment when under control conditions (Figure 1a). In contrast, 9 out of 12 genotypes (including the parent MG20) showed negative RGR during the one-week submergence, indicating the inability to accumulate biomass underwater and loss of tissues due to the stress (Figure 1a); Gifu and RILs 6 and 47 presented slight but positive values for RGR. In all cases/genotypes, plants were not able to emerge from the water, so the registered growth responses corresponded to plants that remained underwater for the whole week. Plant dry mass values from which RGR's were calculated are available in Table S1.

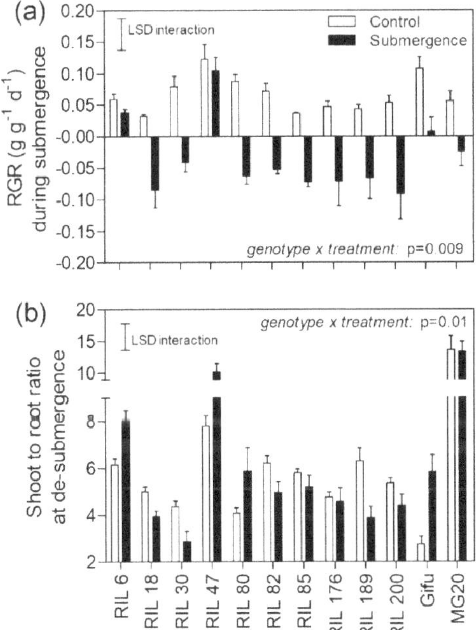

Figure 1. (**a**) Plant relative growth rate (RGR, g g-1 d-1) of 12 genotypes of *Lotus japonicus* (10 recombinant inbred lines and their parents Gifu and MG20) subjected to control and submergence in clear water for one week. (**b**) Shoot to root dry mass ratio (S:R) after the first week of the experiment both at control and de-submergence. The least significant difference (LSD) for the genotype × treatment interaction is shown in (a) (LSD RGR = 0.065) and (b) (LSD S:R = 3.7). Values are means ± standard errors of 6–8 replicates.

Shoot to root dry mass ratio (S:R) of the selected RILs and their parents under control conditions ranged from 2.7 to 13.5 (with a median of 5; Figure 1b), which was wide enough to explore the relationship of this variable with plant recovery from submergence as one of our main objectives. Importantly, differences in S:R ratio among genotypes were constitutive and not related to plant size, as there was no significant relationships between plant dry mass and S:R ($r^2 = 0.34$; $p = 0.23$). Four out of 12 genotypes, RILs 6, 47, 80, and Gifu, increased the S:R in response to complete submergence, while the other three genotypes, RILs 18, 30 and 189, decreased the S:R (Figure 1b). Interestingly, three out of the four genotypes that increased the S:R also showed positive RGR during submergence (RILs 6 and 47, and Gifu).

2.2. Plant Recovery from Submergence

To assess plant recovery, we used plant relative growth rate after de-submergence as an indicator of the ability of the different genotypes to resume growth. RGR values for the first week of recovery were used to infer capacity for an 'early recovery' and RGR values of the second week to infer 'late recovery' responses. In this context, we analyzed how the ability to grow after submergence during these periods (i.e., RGR) correlated across genotypes with several parameters of interest such as (i) RGR during the submergence week per se, (ii) shoot to root dry mass ratio at de-submergence, (iii) leaf relative water content, (iv) stomatal conductance, (v) chlorophyll fluorescence of dark-adapted leaves (Fv/Fm), and (vi) greenness in basal, apical, and new leaves that appeared during the early and late weeks of recovery.

RGR during the first week after submergence (i.e., early recovery) was inversely related to RGR during submergence across genotypes (Figure 2a). During the second week (i.e., late recovery), there was a trend ($p = 0.07$) towards maintenance of a negative relationship between the RGR during submergence and the RGR in the recovery (Figure 2b). So, in general, the genotypes that showed positive RGR during submergence were the poorest performers in terms of RGR during recovery phases (e.g., RILs 6 and 47, and Gifu). In the case of plants growing under control conditions, RGR during the first or second week of recovery (i.e., 2nd and 3rd weeks of the experiment, respectively) were not related to the RGR of these genotypes during the first week of the experiment ($p > 0.29$ in both cases; Figure 2c,d).

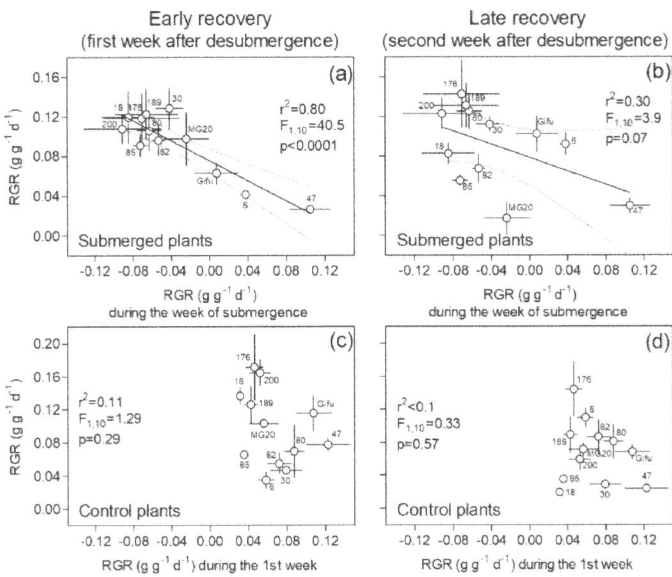

Figure 2. Plant relative growth rate (RGR) during 'early' and 'late' recovery vs. RGR during the first experimental week for one-week submergence (**a,b**) and control (**c,d**) treatments of 12 genotypes of *Lotus japonicus* (10 recombinant inbred lines and their parents Gifu and MG20). The first week after de-submergence was used to explore the relationship between these variables in an 'early recovery' phase (**a,c**) and the second week in a 'late recovery' phase (**b,d**). Values are means ± standard errors of 6–8 replicates.

Shoot to root dry mass ratio (S:R) of submerged plants as an estimator for the potential of transpiration/water uptake under drained conditions showed a negative relationship with RGR during the first week of recovery across genotypes for plants coming from a week of submergence. This means that genotypes with low S:R (e.g., RILs 18, 30, 189) consistently displayed higher RGR during the early recovery (Figure 3a). Even though this negative relation between RGR and S:R persisted during the late recovery (2nd week after submergence), the fitting was weaker than during the early recovery (compare adjustment parameters in Figure 3a vs. Figure 3b). The genotype MG20 was not included in the regressions mentioned above as its S:R was extremely high and was therefore considered an outlier (i.e., S:R$_{MG20}$ > S:R$_{average}$ + 3 standard error). Under control conditions, there was no relation between RGR and S:R ratio either for the 2nd or the 3rd weeks of the experiment ($p > 0.8$ in both cases; Figure 3c,d).

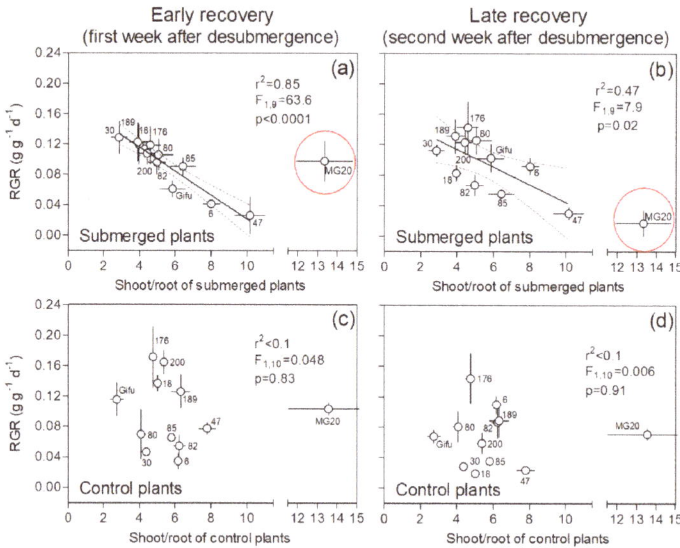

Figure 3. Plant relative growth rate (RGR) during 'early' and 'late' recovery vs. shoot to root dry mass ratio at the end of the first experimental week for one week of submergence (**a**,**b**) and control (**c**,**d**) treatments of 12 genotypes of *Lotus japonicus* (10 recombinant inbred lines and their parents Gifu and MG20). The first week after de-submergence was used to explore the relationship between these variables in an 'early recovery' phase (**a**,**c**) and the second week in a 'late recovery' phase (**b**,**d**). The MG20 genotype was considered an outlier and was not included in the regression parameters shown in (**a**) and (**b**); however, if included, parameters were $r^2 = 0.34$, $F_{1,10} = 5.33$, and $p = 0.043$ and $r^2 = 0.63$, $F_{1,10} = 17.50$, and $p = 0.0019$, respectively. Values are means ± standard errors of 6–8 replicates.

2.3. Leaf Physiological Variables Related to Plant Recovery

Leaf relative water content (leaf RWC), which reflects the balance between water supply to leaf tissue and transpiration rate, showed a positive relationship with RGR of plants upon de-submergence across genotypes during the first week of recovery (Figure 4). This positive relationship, where a higher RWC of the top-most fully expanded leaves corresponded to higher RGR, was true when correlating RGR of the first recovery week with leaf RWC at 2 and 7 days after de-submergence (Figure 4a,b). After ten days of de-submergence most genotypes fully recovered the values of leaf water status, except for RILs 6 and 47 and Gifu (Figure 4c). At the end of the second week of recovery (day 14), all genotypes showed high leaf RWC (Figure 4d), which resulted in similar values to those shown by control plants throughout the experiment (leaf RWC ranged from 89.2% to 92.1%; see Table S2).

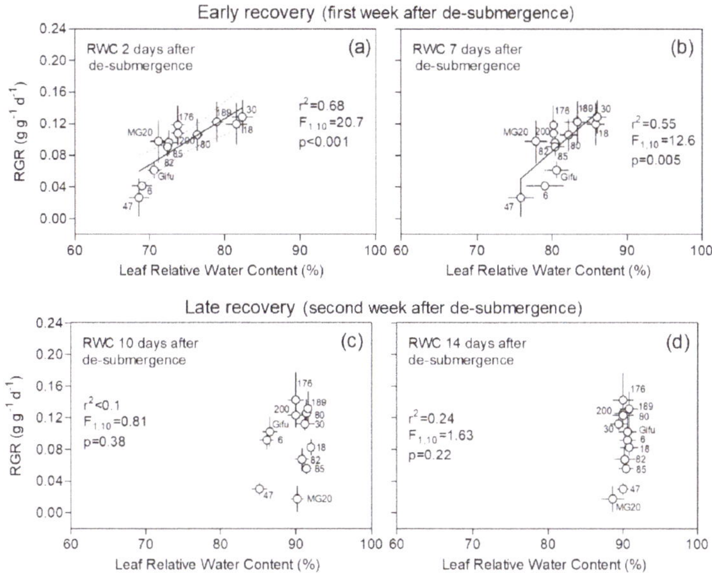

Figure 4. Plant relative growth rate (RGR) during 'early' and 'late' recovery vs. relative water content (RWC, %) on the top-most fully expanded leaves at days 2, 7, 10, and 14 after de-submergence for plants of 12 genotypes of *Lotus japonicus* (10 recombinant inbred lines and their parents Gifu and MG20). The first week after de-submergence was used to explore the relationship between these variables in an 'early recovery' phase (**a,b**) and the second week in a 'late recovery' phase (**c,d**). Values are means ± standard errors (n = 6–8 for RGR and n = 5 for leaf RWC). The leaf RWC values corresponding to the control treatment ranged between 89.2% and 92.4% (all data for each genotype and measurement dates are available in Table S2).

Stomatal conductance (g_s) estimates the rate of gas exchange (uptake of CO_2 favoring photosynthesis) and transpiration (loss of water) through the leaf stomata as determined by the degree of stomatal aperture, which is in turn related to the leaf water status. For this reason, a similar response pattern of g_s and leaf RWC was not unexpected (Figure 5). In this regard, we found a positive relationship between the RGR of plants upon de-submergence across genotypes and g_s on the top-most fully expanded leaves during the first week of recovery (Figure 5). This relationship, where higher g_s values paralleled higher RGRs, was significant when correlating RGR of the first recovery-week with g_s values at 2 and 7 days after de-submergence (Figure 4a,b). Nevertheless, it should be noticed that after reaching one week after de-submergence (day 7), values of g_s were increasing and were close to full recovery (compared to controls; see Table S3), with the exception of genotypes RILs 6 and 47 and Gifu, which remained with the lowest values for this parameter (Figure 5b). At 10 and 14 days after de-submergence (second recovery week), all genotypes presented similar values for g_s as those of the controls for all genotypes (Figure 4c,d). Measured values in control plants of these genotypes through the experiment ranged between 187 and 208 mmoles m^{-2} s^{-1} (Table S3).

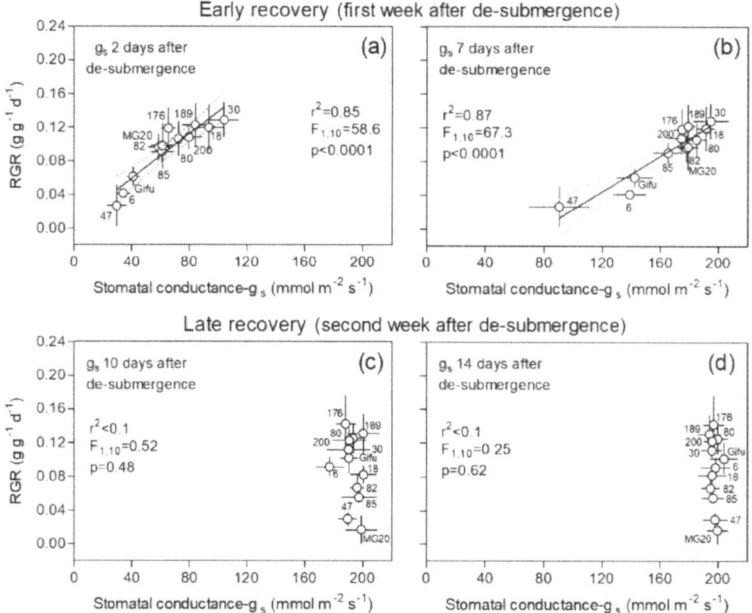

Figure 5. Plant relative growth rate (RGR) during 'early' and 'late' recovery vs. stomatal conductance (g_s, mmol m^{-2} s^{-1}) on the top-most fully expanded leaves at days 2, 7, 10, and 14 after de-submergence for plants of 12 genotypes of *Lotus japonicus* (10 recombinant inbred lines and their parents Gifu and MG20). The first week after de-submergence was used to explore the relationship between these variables in an 'early recovery' phase (**a**,**b**) and the second week in a 'late recovery' phase (**c**,**d**). Values are means ± standard errors (n = 6–8 for RGR and n = 5 for g_s). The stomatal conductance values corresponding to the control treatment ranged between 187.2 and 208.8 mmol m^{-2} s^{-1} (all data for each genotype and measurement dates are available in Table S3).

A decrease in chlorophyll fluorescence of dark-adapted leaves (Fv/Fm) below 0.8 is considered as an indicator of photoinhibition, particularly related to damage to Photosystem II. Interestingly, plants of all genotypes showed low (and similar) values for Fv/Fm (from 0.55 to 0.59) on the top-most fully expanded leaves when measured two days after de-submergence, and no relation with plant RGR during the first recovery week was found across genotypes (Figure 6a). After one week of recovery, Fv/Fm was positively correlated to RGR across genotypes with the poorest performers, namely RILs 6 and 47 and Gifu, still presenting low values indicative of photodamage. During the second recovery week, the possible measurement was made at 11 days after de-submergence, and values for all genotypes showed full recovery for this parameter without differences with their corresponding controls (Table S4). In the case of control plants, the registered values through the experiment ranged between 0.803 and 0.816, considering all genotypes (Table S4).

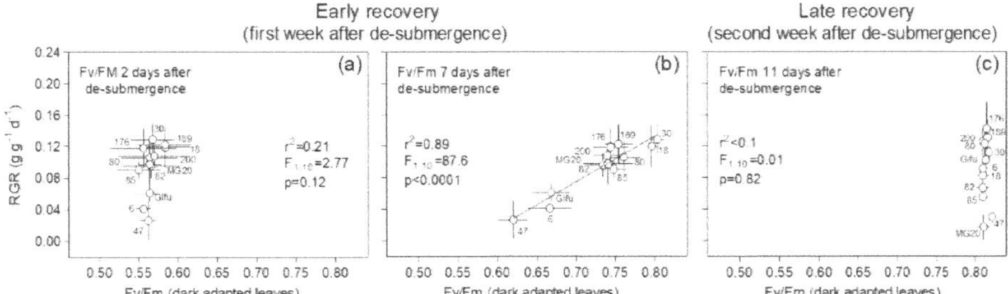

Figure 6. Plant relative growth rate (RGR) during 'early' and 'late' recovery vs. chlorophyll fluorescence on the top-most fully expanded dark-adapted leaves (Fv/Fm) at days 2, 7, and 11 after de-submergence for plants of 12 genotypes of *Lotus japonicus* (10 recombinant inbred lines and their parents Gifu and MG20). The first week after de-submergence was used to explore the relationship between these variables in an 'early recovery' phase (**a**,**b**) and the second week in a 'late recovery' phase (**c**). Values are means ± standard errors (n = 6–8 for RGR and n = 5 for Fv/Fm). The Fv/Fm values corresponding to the control treatment ranged between 0.803 and 0.816 (all data for each genotype and measurement dates are available in Table S4).

Leaf greenness monitoring throughout the experiment after de-submergence in basal and apical positions allows for the estimation of the nitrogen status of leaves, which is in general related to the potential for photosynthesis and, in the case of basal leaves, it allows us to infer senescence and nitrogen remobilization. We found that one week of submergence determined lower greenness values in both basal and apical leaves when compared to controls (see values in Figure 7a,b vs. Table S5). In turn, RGR during the first recovery week was positively related to greenness of both basal and apical leaves across genotypes, both at 2 and 7 days after de-submergence. This means that genotypes that retained more chlorophyll (RILS 18, 30, and 189) showed steadily higher growth (Figure 7a,b). During the second week of recovery, there was no clear relationship between plant RGR and leaf greenness. However, it was noticeable that poor performer genotypes, RILs 6 and 47 and Gifu, showed consistently lower greenness for both basal and apical leaves, without achieving the same full recovery that was reached by the other genotypes (Figure 7c,d).

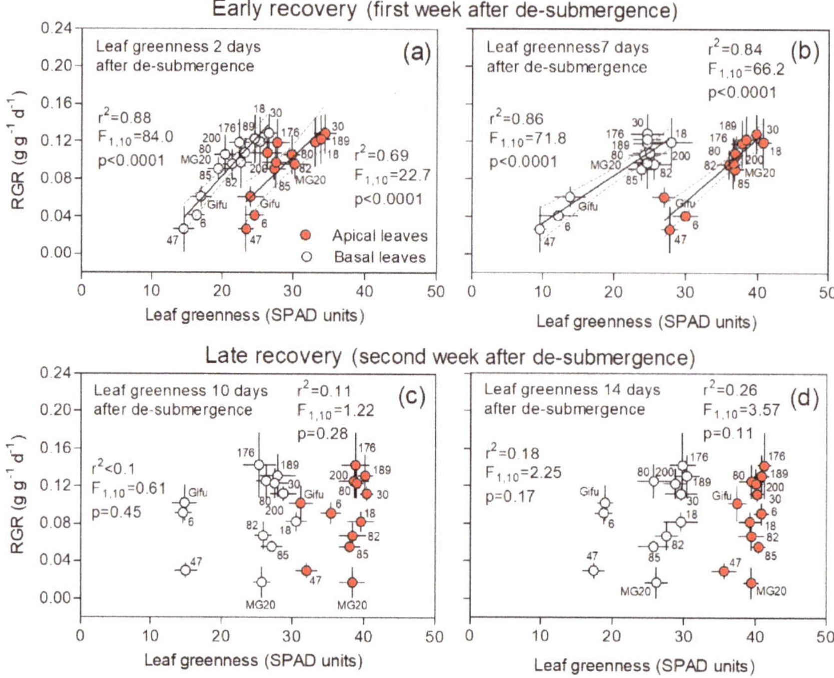

Figure 7. Plant relative growth rate (RGR) during 'early' and 'late' recovery vs. greenness (SPAD units) of basal and apical fully expanded leaves at days 2, 7, 10, and 14 after de-submergence for plants of 12 genotypes of *Lotus japonicus* (10 recombinant inbred lines and their parents Gifu and MG20). The first week after de-submergence was used to explore the relationship between these variables in an 'early recovery' phase (**a,b**) and the second week in a 'late recovery' phase (**c,d**). Values are means ± standard errors (n = 6–8 for RGR and n = 5 for greenness). The leaf greenness values corresponding to the control treatment ranged between 33.6 and 37.6 SPAD units for basal leaves and 33.6 and 37.6 SPAD units for apical young fully expanded leaves (all data for each genotype and measurement dates are available in Table S5).

2.4. Correlations among Traits Aiding Plant Recovery following Submergence

We explored potential correlations among morphophysiological traits associated with RGR of submerged plants as indicative of their recovery ability (Table 1). S:R was negatively correlated to leaf relative water content ($-0.78 < r > -0.69$) and stomatal conductance ($r = -0.62$) during the first week of recovery, meaning that genotypes with high S:R showed poor leaf water status and less open stomata. Moreover, leaf relative water content was positively associated to stomatal conductance ($0.73 > r < 0.94$), Fv/Fm ($0.80 > r < 0.85$), and greenness of both basal ($0.60 > r < 0.86$) and apical leaves ($0.68 > r < 0.92$). Stomatal conductance correlated in a positive way with Fv/Fm ($0.93 > r < 0.95$) and leaf greenness ($0.79 > r < 0.95$). Fv/Fm was positively related to greenness ($0.86 > r < 0.95$) at all leaf positions (basal and apical). The greenness of basal leaves was positively correlated with that of apical leaves ($0.75 > r < 0.91$), meaning that that retention of more chlorophyll upon de-submergence occurred both in young and adult leaves.

Table 1. Pearson correlation coefficients for the morphophysiological traits measured over the recovery phase on 12 genotypes of *Lotus japonicus* (10 recombinant inbred lines and their parents Gifu and MG20) subjected to control (above the diagonal) for three weeks or complete submergence conditions for one week (below the diagonal) and allowed to grow for another 2 weeks. Only correlations for morphophysiological traits that were significantly related to RGR during the recovery from submergence are shown (data from Figures 3–7). Abbreviations are S:R (shoot to root dry mass ratio at de-submergence), LRWC (leaf relative water content), g_s (stomatal conductance), Fv/Fm (chlorophyll fluorescence of dark-adapted leaves). Significant differences: *, $p < 0.05$; **, $p < 0.001$; ns, $p > 0.05$.

Controls ▸ Recovering from sub. ▾	S:R	LRWC 2-d Recovery	LRWC 7-d Recovery	g_s 2-d Recovery	g_s 7-d Recovery	Fv/Fm 7-d Recovery	Greenness Basal Leaves 2-d Recovery	Greenness Apical Leaves 2-d Recovery	Greenness Basal Leaves 7-d Recovery	Greenness Apical Leaves 7-d Recovery
S:R		0.15 ns	0.22 ns	0.18 ns	−0.02 ns	0.50 ns	−0.04 ns	−0.18 ns	−0.21 ns	0.27 ns
LRWC 2-d recovery	−0.69 *		0.31 ns	0.22 ns	0.16 ns	−0.22 ns	0.38 ns	−0.17 ns	−0.20 ns	−0.14 ns
LRWC 7-d recovery	−0.78 **	0.94 **		0.07 ns	0.26 ns	−0.29 ns	0.13 ns	0.33 ns	−0.26 ns	0.25 ns
g_s 2-d recovery	−0.62 *	0.91 **	0.85 **		0.54 ns	0.25 ns	0.31 ns	−0.21 ns	−0.28 ns	0.58 *
g_s 7-d recovery	−0.51 ns	0.75 *	0.73 *	0.85 **		−0.15 ns	0.45 ns	0.30 ns	0.04 ns	0.43 ns
Fv/Fm 7-d recovery	−0.54 ns	0.85 **	0.80 **	0.93 **	0.95 **		−0.03 ns	−0.09 ns	−0.24 ns	0.31 ns
Greenness basal leaves 2-d recovery	−0.51 ns	0.86 **	0.76 *	0.95 **	0.88 **	0.92 **		−0.52 ns	0.22 ns	−0.08 ns
Greenness apical leaves 2-d recovery	−0.44 ns	0.68 *	0.60 *	0.83 **	0.92 **	0.93 **	0.87 **		0.40 ns	−0.72 *
Greenness basal leaves 7-d recovery	−0.58 *	0.92 **	0.85 **	0.88 **	0.79 *	0.86 **	0.87 **	0.75 *		0.26 ns
Greenness apical leaves 7-d recovery	−0.46 ns	0.80 **	0.68 *	0.90 **	0.88 **	0.95 **	0.91 **	0.94 **	0.84 **	

3. Discussion

This research makes five major contributions to the current knowledge of plant recovery from complete submergence by proving the following: First, the existence of variation in the ability to recover from complete submergence in genotypes of *Lotus japonicus* selected for having different shoot to root dry mass ratio [23]. Second, an evident trade-off between growing during the submergence period and the ability to recover indicated by the negative correlation of the RGRs for each period (submergence vs. recovery) across genotypes. Third, the relationship between RGR and shoot to root ratio (S:R) upon de-submergence, where genotypes with low S:R show better plant performance during early days of recovery. Fourth, recovery of plant water relations assessed by leaf relative water content and stomatal conductance appears essential for early recovery and is correlated negatively with S:R, therefore, S:R can be considered as a rough estimator of the water balance between transpiration (shoot) and water uptake (root) in this model legume. Fifth, the importance of leaf chlorophyll retention in supporting a fast recovery from submergence and suggesting the positive correlation between RGR and greenness of basal and apical plant leaves. Additionally, we reported full repair of the photosynthetic apparatus (i.e., all Fv/Fm with values > 0.8) a few days into the second recovery week. In the following paragraphs, we discuss these eco-physiological traits that aid plants in recovery from complete submergence in *L. japonicus*.

Genotypes showing higher RGRs during submergence later presented lower RGRs during the recovery, and the opposite occurred with the genotypes showing low RGRs when underwater (or even negative values due to tissue loss), as these presented the highest RGRs in the post-submergence phase. This trade-off between growing during submergence versus a sit-and-wait response until water recedes is likely a part of the growth strategies to cope with complete submergence already known as "escape" and" quiescence", respectively [3,4]. However, this is one of the first reports demonstrating this trade-off across genotypes for a legume species such as *L. japonicus*, as most studies are available for rice ([28,29], but see [30] for pea genotypes). In this regard, it is possible to speculate that genotypes growing during submergence aim at emerging their leaves above the water using carbon reserves (i.e., starch). In contrast, the ones that do not grow when underwater (see [31] for *L. tenuis* cv. Pampa INTA), use their carbon and nutrient reserves more conservatively (i.e., sustaining basal metabolism) until the stress ends and then resume their growth more vigorously, fueled by stored reserves. Further experiments are needed to quantify the level of reserves in *L. japonicus* materials and its relationships with the eco-physiological variables correlating with plant recovery from complete submergence as identified in this work (S:R, leaf relative water content, stomatal conductance, leaf greenness, Fv/Fm as an indicator of photoinhibition) to corroborate this idea.

An adequate plant water status is critical to resume growth after stress either by lack or excess of water (drought or flooding). So, the balance between transpiration and water uptake is essential to sustain high relative water content in tissues and, therefore, enable fast growth [32,33]. In this regard, the shoot to root ratio can have an important role as a trait by impacting plant water balance. It is alleged that shoot growth and increased shoot to root ratio (S:R) during submergence can induce a better balance between gas transport capacity (oxygen source) and root oxygen consumption (oxygen sink) when plants are able to succeed in the process of emerging leaves above water ("escape strategy"; [34,35]). However, an increased S:R can also constrain plant recovery after de-submergence if water loss by transpiration cannot be replenished by root water uptake, regardless of whether leaf emergence was achieved, as seen in the forage grass *Chloris gayana* subjected to 1 week of full submergence [24]. In our experiment, S:R ratio upon de-submergence was negatively related to the RGR during the recovery phase across *L. japonicus* genotypes. Moreover, we found a negative correlation between S:R and leaf relative water content (RWC) that gives support to the idea that the S:R might have impacted on growth resumption due to an impaired water status of plants. Therefore, any environmental condition experienced by plants in their natural environment that can affect their S:R prior to submergence, such as N, P, or K availability [36,37], can impact the subsequent recovery from complete submergence.

Carbon assimilation and growth after de-submergence are influenced by the degree of stomata aperture enabling diffusion of carbon dioxide towards chloroplasts, the N status of leaves, and the functional degree of the photosynthetic apparatus (if damaged) [13,26,38]. In this study, we registered differences among genotypes in stomatal conductance (g_s), where genotypes with high S:R and low leaf RWC showed lower values of g_s until showing recovery for this parameter during the second week after de-submergence. The g_s was positively related to plant RGRs across genotypes, particularly during the early recovery. Subsequently, all genotypes achieved full recovery during the second week after submergence ceased. On this note, most works dealing with water excess have reported stomatal conductance during/after waterlogging ([19,23,39,40], among others) but rarely after submergence, so opportunities for comparison are minimal (but see [38]). To illustrate, 6-week-old plants of the flood-sensitive grass *Bromus catharticus* were unable to recover the g_s in the same way as controls even after 15 days of recovery from 5 days of complete submergence. In contrast, the flood-tolerant *Phalaris aquatica* did not show, for this parameter, any difference with the controls at any time after de-submergence [38]. In general, *L. japonicus* showed an intermediate behavior compared to the grasses mentioned above, as g_s was reduced by complete submergence during the first week, but it was fully recovered in all genotypes during the second week.

Chlorophyll retention during submergence was positively linked to RGR upon de-submergence both in basal and apical leaves as indicated by the registered leaf greenness dynamics. In this respect, the more tolerant genotypes of *L. japonicus* (RILs 18, 30, and 189) presented lower amounts of (or high sensitivity to) reactive oxygen species (ROS) upon re-aeration (i.e., less oxidative stress), low ethylene formation, and diminished leaf dehydration during recovery compared to the more susceptible ones (RILs 6, 47 and Gifu), similar to what was reported in contrasting accessions of *Arabidopsis thaliana* for submergence tolerance (accessions Lp2-6 vs. Bay-0; [13]). Interestingly, all *L. japonicus* genotypes showed similarly low values for Fv/Fm in young fully expanded leaves, indicating damage to the Photosystem II (PSII) two days after de-submergence. It is known that the re-aeration of leaves (i.e., normoxia) after 7 days in a submergence-induced hypoxic environment can trigger the sudden formation of ROS [41], which ultimately provokes damage to PSII. One week after the stress was released, the genotypes showed a differential ability to recover the Fv/Fm values, and those showing lower values for greenness, g_s, and leaf RWC (RILs 6, 47 and Gifu) were also the poorest performers. Such differential ability to repair a damaged PSII among genotypes could be related to a differential capacity to mitigate effects of ROS (enzymatically and nonenzymatically), or because genotypes generate different amounts of ROS, or a combination of both. Importantly, all genotypes showed full recovery of Fv/Fm (all values> 0.8) at eleven days after submergence, denoting that damage provoked by submergence was not irreversible and that differences among genotypes are related to the time required for recovery. In contrast to our results for *L. japonicus*, the wetland species *Alternantera philoxeroides* and *Hermarthria altissima* are able to recover functionality of the photosynthetic apparatus after de-submergence in just two days as a result of a rapid acclimation to changing oxygen and/or light conditions [26]. It would be interesting to explore further the ability for photosynthetic acclimation that we found in *L. japonicus* and its variability among genotypes as it is an important trait for adaptation to habitats in which the water level often fluctuates.

On top of the eco-physiological traits contributing to plant recovery from submergence reported here, we are now conducting experiments with new genetic resources for *L. japonicus* to specifically address the role of carbon reserves in recovery ability (e.g., mutants of synthesis or degradation of starch; [42]) as well as a quantitative approach of the ROS signaling, scavenging, and homeostasis to provide a comprehensive network of the factors that mediate post-submergence recovery for this model legume. We think that a better understanding of traits and mechanisms contributing to submergence recovery in *L. japonicus* will undoubtedly help to speed up breeding processes of closely related forage *Lotus* spp. used in current agriculture.

4. Materials and Methods

4.1. Species Description

Lotus japonicus L. is a small prostrate perennial legume that flowers profusely, being self-fertile. Although it grows slowly during the first weeks after germination, the adult plant has a bushy growth with many branches up to 30 cm in length, which offers abundant material for biochemical and physiological studies [15]. It has a relatively short generation time of three to four months when grown under glasshouse conditions. *L. japonicus* is taxonomically closely related to *L. corniculatus* (birdsfoot trefoil), *L. uliginosus* (big trefoil), and *L. tenuis* (narrow leaf birdsfoot trefoil), which are used in agriculture as pasture legumes [43]. Importantly, this close relationship among forage *Lotus* species and the model legume *L. japonicus* suggests that identification of tolerance traits to abiotic stresses in the model species, such as submergence in this contribution, will be useful to assist and speed up breeding programs and improvement of plant genotypes that are better adapted to environmentally constrained environments prone to flooding.

4.2. Plant Material and Growing Conditions

A total of 12 genotypes of *L. japonicus* were used in this study: Gifu B-129 (Gifu) and Miyakojima MG-20 (Miyakojima) and 10 RILs that were derived from a cross between these two parent lines and self-pollination to the F_8 generation [44]. The RILs were selected from [23] based on having different shoot to root dry mass ratios as we aimed at investigating the role of this trait (related to the potential transpiration/water uptake ratio) on plant growth when recovering from one week of submergence. The seeds were obtained from the National BioResource Project, Miyazaki University (Miyazaki, Japan).

Seeds of each genotype were scarified by rubbing with a fine sand-paper, surface-sterilized with 0.05% (w/v) sodium hypochlorite, rinsed thoroughly in deionized water, and dark-germinated in an incubator (20 °C) in polystyrene boxes containing absorbent water-saturated white paper. After three days, germinated seeds were transplanted to 2 L plastic pots (2 or 3 seedlings per pot) filled with sand and soil with 3.3% of organic carbon (1:1 v/v). After seeding, pots were transferred to a glasshouse at the Faculty of Agronomy at the University of Buenos Aires and during the first week, the seedlings were thinned to one per pot. In order to avoid nutrient limitation for plant growth, 1.2 g of fertilizer (Nitrofull EmergerR, Argentina: 12% N, 5% P, 15% K, 2% Mg, 8% S, 3% Ca, 0.02% Zn, 0.2% Fe, 0.02% Mn, and 0.015% Bo; % are by weight) was added to every pot. Plants were watered daily before treatment imposition.

4.3. Experimental Design

After a growth period of 45 days, pots were randomized into two groups and subjected to two treatments for one week: (i) control: pots were watered daily and allowed to drain freely, (ii) complete submergence: plants were submerged in clear water at a depth of 45 cm, which corresponded to a water column of *ca.* 15–20 cm above the top of the plants (water column of 1.5–1.8 times the plant height). Submergence water contained 0.50 mM $CaSO_4$, 0.25 mM $MgSO_4$, and 1 mM $KHCO_3$ (at pH *ca.* 7.5) as in [45]. Dissolved oxygen in the submergence water at midday ranged from 3.9 to 4.6 mg L^{-1} (vs. 7.8 mg L^{-1} in air), measured with a DO-5510 dissolved oxygen meter (LT Lutron Electronic Enterprise, Taipei). The photosynthetic photon flux density (PPFD) reaching control plants was 1100 ± 180 μmol m^{-2} s^{-1}, while for submerged plants, it was 490 ± 80 μmol m^{-2} s^{-1} (LI-192 Underwater Quantum Sensor; Li-Cor Inc., Lincoln, NE, USA). The latter is a light environment that allows for underwater photosynthesis in C_3 species [46]. During the first week of the experiment, all plants (both assigned to control and submergence treatments) were placed in plastic containers (0.8 m × 0.5 m × 0.5 m) to facilitate the submergence imposition. After the first experimental week, plants were taken out of the containers and rotated weekly within the glasshouse during a two-week recovery period. The number of replicates per genotype and treatment combination varied from 6 to 8 according to the number of available plants. Average glasshouse relative humidity was 62% ± 12%, and the average air temperature was 24/16 °C (day/night).

4.4. Dry Mass and Relative Growth Rate

Plants of all genotypes were harvested at the beginning of the experiment (day 0), at day 7 (end of the submergence), at day 14 (end of the first recovery week, to infer 'early recovery ability' from submergence), and at day 21 (end of the second recovery week, to infer 'late recovery ability'). The dry mass of plants was obtained after drying at 80 °C for 72 h (i.e. until constant weight). In the second harvest (immediately after submergence; day 7), shoots were separated from roots to calculate the shoot to root dry mass ratio of the different genotypes upon de-submergence.

Relative growth rate of plants (RGR) after submergence was used as an indicator of recovery ability of the genotypes. In this study, the identity of each genotype per se was not particularly relevant as to have sufficient genetic variability for some traits, such as shoot to root dry mass ratio, to potentially influence plant water balance upon de-submergence and return to oxygenated well-drained conditions. The RGR was calculated following the equation by [47]:

$$\text{RGR (g g}^{-1}\text{ d}^{-1}) = [\ln(W_2) - \ln(W_1)] / (t_2 - t_1) \tag{1}$$

where W_2 and W_1 are the plant dry weights from the corresponding treatment at times 2 and 1, respectively, with the value of W_1 being the treatment average, and $t_2 - t_1$ is the number of days between sampling times (i.e., 7 d).

4.5. Leaf Physiological Responses: Relative Water Content, Stomatal Conductance, Chlorophyll Fluorescence, and Greenness

Leaf physiological measurements were made at days 2 and 7 during the first recovery week (days 9 and 14 since the beginning of the experiment), and at days 10 and 14 following de-submergence during the second recovery week (days 17 and 21 of the experiment). These measurements aimed at monitoring the recovery of leaf water status, stomatal functioning, and the impact of one week of full submergence on leaf greenness as indicators of nitrogen status of basal and apical leaves due to damage to the photosynthetic apparatus through measurements of chlorophyll fluorescence of dark-adapted young leaves (i.e., Fv/Fm). For the last parameter, measurements were performed at days 2 and 7 during the first recovery week and at day 11 during the second recovery week. The number of replicates taken for the physiological measurements was five per genotype and treatment combination.

Relative water content of leaves (RWC) was measured according to the methodology proposed by [18]. First, young, fully expanded leaves located at a top position of plants were taken and immediately weighed to obtain their fresh weight. Second, samples were incubated for 12 h in tightly closed tubes containing deionized water to get their turgid weight. Third, the leaves were oven-dried at 70 °C for 48 h to obtain their dry weight (the weights of leaves were latter added to the corresponding plants). Finally, the RWC of leaves was obtained as: [(fresh weight – dry weight) / (turgid weight – dry weight)] × 100.

Stomatal conductance of young, fully expanded leaves at top positions was measured near midday by using a leaf porometer (model SC-1; Decagon Devices, Pullman, WA, USA) to evaluate water loss by transpiration and facilitation of CO_2 diffusion into leaves for photosynthesis. Leaf greenness was measured in young (apical) and old (basal), fully expanded leaves by using a portable chlorophyll meter (SPAD-502; Konica Minolta Sensing, Osaka, Japan). This parameter infers that leaf yellowing is indicative of chlorophyll degradation, nitrogen remobilization, and anticipated leaf senescence as consequences of the impact of submergence stress. Chlorophyll fluorescence (Fv/Fm) was measured on top-most, fully expanded leaves after a dark-adaptation period of 20 min by using leaf clips and the OS-30p portable fluorometer (Opti-Sciences Inc., USA). This parameter indicates the proportion of functional Photosystem II (PSII) reaction centers so that it can be used to quantify the degree of photoinhibition by damage to the PSII [49].

4.6. Statistical Analyses

RGR and shoot to root dry mass ratio responses during submergence were evaluated by two-way ANOVAs, with 'submergence" and "genotype" as main factors. When significant interactions were detected, the least significant difference (LSD) and Fisher's protected tests were used to determine the effect of treatments among genotypes. During the recovery phase, regression analyses were performed to explore relationships between plant RGR and shoot to root dry mass ratio, leaf relative water content, stomatal conductance, chlorophyll fluorescence of dark-adapted leaves (Fv/Fm), and greenness of basal (adult) and apical (young) leaves. These relationships were examined separately for the first and the second recovery weeks to identify traits associated with 'early' and 'late' recovery responses. Additionally, relations between physiological variables that were significantly related to RGR during recovery were explored through Pearson correlations. Normality and the homogeneity of variances were verified before each analysis. All results are presented as means ± standard errors. Statistical analyses were done using Infostat software [50], and graphs were made with GraphPad Prism 5 for Windows (GraphPad Software, San Diego, CA, USA).

5. Conclusions

We found variation in the recovery ability from short-term complete submergence in genotypes of the model legume *Lotus japonicus*. The ability to recover assessed through plant RGR post-submergence negatively correlated with plant RGR displayed during the submergence period, which indicates the existence of a trade-off between growing during vs. after the stress. Several eco-physiological traits and responses correlating to recovery ability across genotypes were identified, which were particularly important in explaining the recovery of plants in the first week after de-submergence. Shoot to root dry mass ratio (S:R) was inversely related to RGR during the recovery. Leaf relative water content and stomatal conductance were positively related to RGR in the early recovery week, and negatively related to S:R ratio, which is why the S:R ratio can be used as an estimator of the potential plant water balance (transpiration in shoot vs. water uptake in root) in post-submergence. Leaf greenness of basal and apical leaves was positively linked to RGR during early days after submergence, indicating that chlorophyll retention is important for a fast recovery. The full repair of the submergence-damaged photosynthetic apparatus occurred later than the recovery of plant water status, and it was completed during the second recovery week. The inclusion of traits contributing to submergence recovery in *L. japonicus* as identified in this work should be considered to speed up breeding processes of closely related forage *Lotus* spp. used in current agriculture.

Supplementary Materials: The following are available online at http://www.mdpi.com/2223-7747/9/4/538/s1, **Table S1**: Plant dry mass (g per plant) of 12 genotypes of *Lotus japonicus* (10 recombinant inbred lines and their parents Gifu and MG20) harvested throughout the experiment. Values are means ± standard errors of 6–8 replicates. **Table S2**: Leaf relative water content (RWC; %) on the top-most, fully expanded leaves of control plants of *Lotus japonicus* recombinant inbred lines (RIL) and the parental MG20 and Gifu lines. Values are means ± standard errors of 5 replicates. For further details, see Materials and Methods Section. **Table S3**: Stomatal conductance, g_s (mmol m^{-2} s^{-1}), of the top-most, fully expanded leaves of control plants of *Lotus japonicus* recombinant inbred lines (RIL) and the parental MG20 and Gifu lines. Values are means ± standard errors of 5 replicates. For further details, see Materials and Methods Section. **Table S4**: Dark-adapted chlorophyll fluorescence (Fv/Fm) on the top-most fully expanded leaves of control plants of *Lotus japonicus* recombinant inbred lines (RIL) and the parental MG20 and Gifu lines. Values are means ± standard errors of 5 replicates. For further details, see Materials and Methods Section. **Table S5**: Greenness (SPAD units) of basal and apical, young, fully expanded leaves of control plants of *Lotus japonicus* recombinant inbred lines (RIL) and their parental MG20 and Gifu lines. Values are means ± standard errors of 5 replicates. For further details, see Materials and Methods Section.

Author Contributions: Conceptualization, F.B.B., F.P.O.M., A.A.G., and G.G.S.; formal analysis, F.B.B. and G.G.S.; writing—original draft preparation, F.B.B. and G.G.S.; writing—review and editing, F.B.B., F.P.O.M., A.A.G., and G.G.S. All authors have read and agreed to the published version of the manuscript.

Funding: This research was funded by the University of Buenos Aires (grant UBACyT20020170100319BA), the National Agency for Scientific and Technological Promotion of Argentina (grant ANPCyT-PICT-2017-0451), and the National Scientific and Technical Research Council of Argentina (grant PIP-CONICET 11220150100041CO).

Acknowledgments: We would like to thank Xiaolin Kuang, Rocío Ploschuk, Eliana Vera, and Andres Michelli for their help in the harvests and assistance during the physiological measurements.

Conflicts of Interest: The authors declare no conflict of interest.

References

1. IPCC. 2014. Climate Change 2014: Synthesis Report. Contribution of Working Groups I. Available online: https://www.ipcc.ch/ (accessed on 20 March 2020).
2. Voesenek, L.A.C.J.; Colmer, T.D.; Pierik, R.; Millenaar, F.F.; Peeters, A.J.M. How plants cope with complete submergence. *New Phytol.* **2006**, *170*, 213–226. [CrossRef]
3. Colmer, T.D.; Voesenek, L.A.C.J. Flooding tolerance: Suites of plant traits in variable environments. *Funct. Plant Biol.* **2009**, *36*, 665–681. [CrossRef]
4. Bailey-Serres, J.; Voesenek, L.A.C.J. Flooding stress: Acclimations and genetic diversity. *Annu. Rev. Plant Biol.* **2008**, *59*, 313–339. [CrossRef]
5. Bailey-Serres, J.; Voesenek, L.A.C.J. Life in the balance: A signaling network controlling survival of flooding. *Curr. Opin. Plant Biol.* **2010**, *13*, 489–494. [CrossRef] [PubMed]
6. Striker, G.G.; Izaguirre, R.F.; Manzur, M.E.; Grimoldi, A.A. Different strategies of *Lotus japonicus, L. corniculatus* and *L. tenuis* to deal with complete submergence at seedling stage. *Plant Biol.* **2012**, *14*, 50–55. [CrossRef] [PubMed]
7. Voesenek, L.A.C.J.; Bailey-Serres, J. Flooding tolerance: O$_2$ sensing and survival strategies. *Curr. Opin. Plant Biol.* **2013**, *16*, 647–653. [CrossRef]
8. Striker, G.G. Time is on our side: The importance of considering a recovery period when assessing flooding tolerance in plants. *Ecol. Res.* **2012**, *27*, 983–987. [CrossRef]
9. Fukao, T.; Xu, K.; Ronald, P.C.; Bailey-Serres, J. A variable cluster of ethylene response factor–like genes regulates metabolic and developmental acclimation responses to submergence in rice. *Plant Cell* **2006**, *18*, 2021–2034. [CrossRef]
10. Hattori, Y.; Nagai, K.; Furukawa, S.; Song, X.J.; Kawano, R.; Sakakibara, H.; Wu, J.; Matsumoto, T.; Yoshimura, A.; Kitano, H.; et al. The ethylene response factors *SNORKEL1* and *SNORKEL2* allow rice to adapt to deep water. *Nature* **2009**, *460*, 1026–1030. [CrossRef]
11. Nagai, K.; Hattori, Y.; Ashikari, M. Stunt or elongate? Two opposite strategies by which rice adapts to floods. *J. Plant Res.* **2010**, *123*, 303–309. [CrossRef]
12. Fukao, T.; Barrera-Figueroa, B.E.; Juntawong, P.; Peña-Castro, J.M. Submergence and waterlogging stress in plants: A review highlighting research opportunities and understudied aspects. *Front. Plant Sci.* **2019**, *10*, 340. [CrossRef] [PubMed]
13. Yeung, E.; van Veen, H.; Vashisht, D.; Paiva, A.L.S.; Hummel, M.; Rankenberg, T.; Steffens, B.; Steffen-Heins, A.; Sauter, M.; de Vries, M.; et al. A stress recovery signaling network for enhanced flooding tolerance in *Arabidopsis thaliana. PNAS* **2018**, *115*, E6085–E6094. [CrossRef]
14. Yeung, E.; Bailey-Serres, J.; Sasidharan, R. After the deluge: Plant revival post-flooding. *Trends Plant Sci.* **2019**, *24*, 443–454. [CrossRef] [PubMed]
15. Pajuelo, E.; Stougaard, J. *Lotus japonicus* as a model system. In *Lotus japonicus handbook*; Stougaard, J., Márquez, A.J., Eds.; Springer: Dordrecht, The Netherlands, 2005; pp. 3–24.
16. Escaray, F.J.; Menendez, A.B.; Gárriz, A.; Pieckenstain, F.L.; Estrella, M.J.; Castagno, L.N.; Carrasco, P.; Sanjuan, J.; Ruiz, O.A. Ecological and agronomic importance of the plant genus *Lotus*. Its application in grassland sustainability and the amelioration of constrained and contaminated soils. *Plant Sci.* **2012**, *182*, 121–133. [CrossRef] [PubMed]
17. Striker, G.G.; Colmer, T.D. Flooding tolerance of forage legumes. *J. Exp. Bot.* **2017**, *68*, 1851–1872. [CrossRef]
18. James, E.K.; Crawford, R.M.M. Effect of oxygen availability on nitrogen fixation by two *Lotus* species under flooded conditions. *J. Exp. Bot.* **1998**, *49*, 599–609. [CrossRef]

19. Striker, G.G.; Insausti, P.; Grimoldi, A.A.; Ploschuk, E.L.; Vasellati, V. Physiological and anatomical basis of differential tolerance to soil flooding of *Lotus corniculatus*, L. and *Lotus glaber Mill*. *Plant Soil* **2005**, *276*, 301–311. [CrossRef]

20. Young, N.D.; Cannon, S.B.; Sato, S.; Kim, D.; Cook, D.R.; Town, C.D.; Roe, B.A.; Tabata, S. Sequencing the genespaces of *Medicago truncatula* and *Lotus japonicus*. *Plant Physiol.* **2005**, *137*, 1174–1181. [CrossRef]

21. Jiang, Q.; Gresshoff, P.M. Classical and molecular genetics of the model legume *Lotus japonicus*. *Mol. Plant Microbe Interact.* **1997**, *10*, 59–68. [CrossRef]

22. Mun, T.; Bachmann, A.; Gupta, V.; Stougaard, J.; Andersen, S.U. Lotus Base: An integrated information portal for the model legume *Lotus japonicus*. *Sci. Rep.* **2016**, *6*, 39447. [CrossRef]

23. Striker, G.G.; Casas, C.; Manzur, M.E.; Ploschuk, R.A.; Casal, J.J. Phenomic networks reveal largely independent root and shoot adjustment in waterlogged plants of *Lotus japonicus*. *Plant Cell Environ.* **2014**, *37*, 2278–2293. [PubMed]

24. Striker, G.G.; Casas, C.; Kuang, X.; Grimoldi, A.A. No escape? Costs and benefits of leaf de-submergence in the pasture grass *Chloris gayana* under different flooding regimes. *Funct. Plant Biol.* **2017**, *44*, 899–906. [CrossRef]

25. Setter, T.L.; Bhekasut, P.; Greenway, H. Desiccation of leaves after de-submergence is one cause for intolerance to complete submergence of the rice cultivar IR 42. *Funct. Plant Biol.* **2010**, *37*, 1096–1104. [CrossRef]

26. Luo, F.L.; Nagel, K.A.; Zeng, B.; Schurr, U.; Matsubara, S. Photosynthetic acclimation is important for post-submergence recovery of photosynthesis and growth in two riparian species. *Ann. Bot.* **2009**, *104*, 1435–1444. [CrossRef] [PubMed]

27. Luo, F.L.; Nagel, K.A.; Scharr, H.; Zeng, B.; Schurr, U.; Matsubara, S. Recovery dynamics of growth, photosynthesis and carbohydrate accumulation after de-submergence: A comparison between two wetland plants showing escape and quiescence strategies. *Ann. Bot.* **2011**, *107*, 49–63. [CrossRef] [PubMed]

28. Setter, T.L.; Laureles, E.V. The beneficial effect of reduced elongation growth on submergence tolerance of rice. *J. Exp. Bot.* **1996**, *47*, 1551–1559. [CrossRef]

29. Hattori, Y.; Nagai, K.; Ashikari, M. Rice growth adapting to deepwater. *Curr. Opin. Plant Biol.* **2011**, *14*, 100–105. [CrossRef]

30. Zaman, M.S.U.; Malik, A.I.; Erskine, W.; Kaur, P. Changes in gene expression during germination reveal pea genotypes with either "quiescence" or "escape" mechanisms of waterlogging tolerance. *Plant Cell Environ.* **2019**, *42*, 245–258. [CrossRef]

31. Manzur, M.E.; Grimoldi, A.A.; Insausti, P.; Striker, G.G. Escape from water or remain quiescent? *Lotus tenuis* changes its strategy depending on depth of submergence. *Ann. Bot.* **2009**, *104*, 1163–1169.

32. Tanguilig, V.C.; Yambao, E.B.; O'toole, J.C.; De Datta, S.K. Water stress effects on leaf elongation, leaf water potential, transpiration, and nutrient uptake of rice, maize, and soybean. *Plant Soil* **1987**, *103*, 155–168. [CrossRef]

33. Aroca, R.; Porcel, R.; Ruiz-Lozano, J.M. Regulation of root water uptake under abiotic stress conditions. *J. Exp. Bot.* **2012**, *63*, 43–57.

34. Herzog, M.; Pedersen, O. Partial versus complete submergence: Snorkelling aids root aeration in *Rumex palustris* but not in *R. acetosa*. *Plant Cell Environ.* **2014**, *37*, 2381–2390.

35. Pan, Y.; Cieraad, E.; Clarkson, B.R.; Colmer, T.D.; Pedersen, O.; Visser, E.J.W.; Voesenek, L.A.C.J.; van Bodegom, P.M. Drivers of plant traits that allow survival in wetlands. *Funct. Ecol.* **2020**. [CrossRef]

36. Cakmak, I.; Hengeler, C.; Marschner, H. Partitioning of shoot and root dry matter and carbohydrates in bean plants suffering from phosphorus, potassium and magnesium deficiency. *J. Exp. Bot.* **1994**, *45*, 1245–1250. [CrossRef]

37. Andrews, M.; Sprent, J.I.; Raven, J.A.; Eady, P.E. Relationships between shoot to root ratio, growth and leaf soluble protein concentration of *Pisum sativum*, *Phaseolus vulgaris* and *Triticum aestivum* under different nutrient deficiencies. *Plant Cell Environ.* **1999**, *22*, 949–958. [CrossRef]

38. Striker, G.G.; Ploschuk, R.A. Recovery from short-term complete submergence in temperate pasture grasses. *Crop Pasture Sci.* **2018**, *69*, 745–753. [CrossRef]

39. Mielke, M.S.; de Almeida, A.A.F.; Gomes, F.P.; Aguilar, M.A.G.; Mangabeira, P.A.O. Leaf gas exchange, chlorophyll fluorescence and growth responses of *Genipa americana* seedlings to soil flooding. *Environ. Exp. Bot.* **2003**, *50*, 221–231. [CrossRef]

40. Ploschuk, R.A.; Grimoldi, A.A.; Ploschuk, E.L.; Striker, G.G. Growth during recovery evidences the waterlogging tolerance of forage grasses. *Crop Pasture Sci.* **2017**, *68*, 574–582. [CrossRef]

41. Blokhina, O.; Virolainen, E.; Fagerstedt, K.V. Antioxidants, oxidative damage and oxygen deprivation stress: A review. *Ann. Bot.* **2003**, *91*, 179–194. [CrossRef]

42. Vriet, C.; Smith, A.M.; Wang, T.L. Root starch reserves are necessary for vigorous re-growth following cutting back in *Lotus japonicus*. *PLoS ONE* **2014**, *9*, e87333. [CrossRef]

43. Swanson, E.B.; Somers, D.A.; Tomes, D.T. Birdsfoot trefoil (*Lotus corniculatus* L.). In *Legumes and Oilseed Crops I*; Springer: Berlin/Heidelberg, Germany, 1990; pp. 323–340.

44. Kawaguchi, M.; Motomura, T.; Imaizumi-Anraku, H.; Akao, S.; Kawasaki, S. Providing the basis for genomics in Lotus japonicus: The accessions Miyakojima and Gifu are appropriate crossing partners for genetic analyses. *Mol Gen. Genom.* **2001**, *266*, 157–166. [CrossRef] [PubMed]

45. Striker, G.G.; Kotula, L.; Colmer, T.D. Tolerance to partial and complete submergence in the forage legume *Melilotus siculus*: An evaluation of 15 accessions for petiole hyponastic response and gas-filled spaces, leaf hydrophobicity and gas films, and root phellem. *Ann. Bot.* **2019**, *123*, 169–180. [CrossRef] [PubMed]

46. Colmer, T.D.; Pedersen, O. Underwater photosynthesis and respiration in leaves of submerged wetland plants: Gas films improve CO_2 and O_2 exchange. *New Phytol.* **2008**, *177*, 918–926. [CrossRef] [PubMed]

47. Hunt, R. Plant growth curves. In *The Functional Approach to Plant Growth Analysis*; Edward Arnold Ltd.: London, UK, 1982.

48. Čatský, J. Determination of water deficit in disks cut out from leaf blades. *Biol. Plant.* **1960**, *2*, 76. [CrossRef]

49. Maxwell, K.; Johnson, G.N. Chlorophyll fluorescence—A practical guide. *J. Exp. Bot.* **2000**, *51*, 659–668. [CrossRef] [PubMed]

50. Di Rienzo, J.A.; Casanoves, F.; Balzarini, M.G.; González, L.; Tablada, M.; Robledo, Y.C. InfoStat versión 2011. *Grupo InfoStat, FCA, Universidad Nacional de Córdoba, Argentina*. Available online: http://www.infostat.com.ar (accessed on 20 March 2020).

Review

Keep Calm and Survive: Adaptation Strategies to Energy Crisis in Fruit Trees under Root Hypoxia

Ariel Salvatierra [1], Guillermo Toro [2], Patricio Mateluna [2], Ismael Opazo [3], Mauricio Ortiz [4] and Paula Pimentel [2,*]

[1]   Plant Genomics Laboratory, Centro de Estudios Avanzados en Fruticultura (CEAF), Rengo 2940000, Chile; asalvatierra@ceaf.cl
[2]   Plant Stress Physiology Laboratory, Centro de Estudios Avanzados en Fruticultura (CEAF), Rengo 2940000, Chile; gtoro@ceaf.cl (G.T.); pmateluna@ceaf.cl (P.M.)
[3]   Plant Breeding Laboratory, Centro de Estudios Avanzados en Fruticultura (CEAF), Rengo 2940000, Chile; iopazo@ceaf.cl
[4]   Agronomy Laboratory, Centro de Estudios Avanzados en Fruticultura (CEAF), Rengo 2940000, Chile; mortiz@ceaf.cl
*   Correspondence: ppimentel@ceaf.cl; Tel.: +56-72-2445009

Received: 28 July 2020; Accepted: 22 August 2020; Published: 27 August 2020

Abstract: Plants are permanently facing challenges imposed by the environment which, in the context of the current scenario of global climate change, implies a constant process of adaptation to survive and even, in the case of crops, at least maintain yield. O_2 deficiency at the rhizosphere level, i.e., root hypoxia, is one of the factors with the greatest impact at whole-plant level. At cellular level, this O_2 deficiency provokes a disturbance in the energy metabolism which has notable consequences on the yield of plant crops. In this sense, although several physiological studies describe processes involved in plant adaptation to root hypoxia in woody fruit trees, with emphasis on the negative impacts on photosynthetic rate, there are very few studies that include -omics strategies for specifically understanding these processes in the roots of such species. Through a de novo assembly approach, a comparative transcriptome study of waterlogged *Prunus* spp. genotypes contrasting in their tolerance to root hypoxia was revisited in order to gain a deeper insight into the reconfiguration of pivotal pathways involved in energy metabolism. This re-analysis describes the classically altered pathways seen in the roots of woody fruit trees under hypoxia, but also routes that link them to pathways involved with nitrogen assimilation and the maintenance of cytoplasmic pH and glycolytic flow. In addition, the effects of root hypoxia on the transcription of genes related to the mitochondrial oxidative phosphorylation system, responsible for providing adenosine triphosphate (ATP) to the cell, are discussed in terms of their roles in the energy balance, reactive oxygen species (ROS) metabolism and aerenchyma formation. This review compiles key findings that help to explain the trait of tolerance to root hypoxia in woody fruit species, giving special attention to their strategies for managing the energy crisis. Finally, research challenges addressing less-explored topics in recovery and stress memory in woody fruit trees are pointed out.

Keywords: hypoxia; waterlogging; fruit trees; *Prunus*; aerenchyma; hypertrophied lenticels; anaerobic fermentation; energy metabolism; root respiration

---

1. Introduction

Plants are aerobic organisms and sensitive to many external conditions that could alter internal homeostasis. O_2 deficiency or deprivation trigger hypoxia or anoxia stress, respectively, depending on the O_2 concentrations. Plants can face hypoxic conditions from different origins: developmental hypoxia, in plant–microbe interactions, or environmental hypoxia (waterlogging/submergence) [1].

Currently, the cultivation surface of fruit trees in Mediterranean and subtropical areas is approximately 49 million hectares. Of this group, 22% corresponds to olive trees, 20% to citrus, 16% to stone fruit, 15% to vines and 13% to pome fruit [2]. Fruit trees of the Mediterranean climate, such as walnut (*Juglans regia* L.), apple (*Malus sylvestris* L.) or sweet cherry (*Prunus avium* L.), show a low tolerance to hypoxia compared to trees from wetland areas [3]. In addition, most of these fruit trees do not have the capacity for acclimatization, or even for the recovery of their physiological and growth parameters while flooded, unlike tree forest species of the wetland areas [4–6]. However, several studies have described some hypoxia–tolerance gradient among fruit trees [4–7].

In the context of fruit production, rootstocks are selected for rooting and grafting capacity, abiotic and biotic stress tolerance, and their ability to beneficially alter scion phenotypes, especially in yielding terms. In perennial and some vegetable crops, grafting is used to join resilient root systems (rootstocks) to shoots (scions) that produce the harvested product [8]. The hydraulic architecture of rootstocks becomes of fundamental importance, since the sustained flow of water controls many plant processes, such as growth, mineral nutrition, scion photosynthesis and transpiration [9]. Species of *Prunus* used as rootstocks are classified as moderately sensitive to root hypoxia, although differences among genotypes regarding their ability to tolerate this stress have been reported [7,10–14]

Mediterranean agriculture has coevolved with harsh environments and changing climate conditions over millennia, generating an extremely rich heritage, but actual climatic change threatens global agriculture especially given the prevalence of highly specialized, low diverse agroecosystems [15]. The impact of global warming plus a human population hovering around 7.7 billion strongly challenges agricultural systems, and the need for a better understanding of how crops and fruit trees can survive and yield under more extreme climatic events becomes of paramount importance, since an increase in precipitation intensity and variability would increase the risks of flooding [16].

Nowadays, as a consequence of global climate change, there are areas of the planet most exposed to intense rains where accumulated amount of water can lead to soil saturation, with a concomitant displacement of O_2 in the rhizosphere. However, although less discussed, it is important to note that there are a number of conditions, beyond excess rainfall, that can configure the establishment of O_2 deficiency at the root level.

2. Edaphic Conditions that Promote O_2 Deficiency

The air capacity of the soil is determined by its texture and structure. Coarse-textured soils can have an air capacity of around 25%, while in fine-textured soils the air capacity can reach 10% [17]. However, when the structure is altered by physicochemical processes or mechanical forces, the macroscopic pores tend to disappear, so a strongly compacted soil may contain less than 5% air by volume at its characteristic field-capacity value of soil moisture [17].

Fine texture and compaction associated with bad irrigation practices are able to generate an excess of water sufficient for establishing an O_2 deficit at rhizosphere level. Silt and clay particles reduce soil aeration because they are tightly packed together, decreasing the air spaces between them and slowing the drainage [18]. This condition creates an O_2 deficient environment for plants by maintaining high moisture on the soil surface [19]. In fine-textured soils, transient soil waterlogging can be generated only by poor irrigation management [20]. Soil compaction occurs primarily when pressure exerted on the soil surface reduces the air spaces between soil particles, and it is associated with agricultural practices, but it can also be the result of natural processes unrelated to the application of compressive forces [19]. This results in a change in the proportion of pores with water and air (mainly loss of coarse pores), and an increase in mechanical resistance to root development [19,21,22]. The soil compaction may be superficial or even reach 100 mm or more [19], which could have an important impact on the perennial fruit tree's development. In almond, a field experiment was performed using a heavy tractor to evaluate the continuous transit of agricultural machinery over the soil simulating multiple applications. The authors concluded that the continuous heavy tractor traffic causes an evident soil compaction (20–40 cm deep) with a drastic decrease in soil porosity reaching up to 11%, and it also produced an increase in bulk density and cone index in subsoil layers [23].

Plants need an adequate supply of air and water in the soil pore space for their suitable development and growth [24]. The rate of the transfer of gases in the air phase is generally much greater than in the water phase; hence soil aeration depends largely on the volume fraction of air-filled pores [17]. Typically, the requirement for plant development is for at least 10% of the soil volume to comprise gas-filled pores at field capacity (air capacity), and for at least 10% of the gas in these pores to be O_2 [25]. Cook and Knight [24] determined that when the air-filled porosity of the soil drops below 12%, the root respiration and the exchange of O_2 and CO_2 between the soil and the atmosphere is hampered. Kawase [26] points out that when it drops below 10%, hypoxia is triggered, and under more drastic conditions it becomes in anoxia.

Hypoxia/anoxia conditions restrict processes such as plant respiration, water and nutrient absorption. Anaerobic conditions in the soil induce a series of physical, chemical and biological processes, such as pH and redox potential, resulting in changes to the soil's elemental profile [27]. O_2 deficiency further affects microbial communities and microbial processes in the soil [28]. Once free O_2 is consumed, nitrate is used by soil microorganisms as an alternative electron acceptor to continue their respiration. The following acceptors are MnO_2, $Fe(OH)_3$, SO_4^{2-} and CO_2 [29]. This generates an increase in the levels of reduced compounds, such as Mn^{2+}, Fe^{2+}, H_2S, NH_4^+ and organic compounds (alkanes, acids, carbonyls, etc.) [30,31]. These solutes can accumulate up to phytotoxic levels and contribute to plant injury [32].

3. Fruit Tree Responses to O_2 Deficiency

3.1. Physiological and Biochemical Response of Fruit Trees under O_2 Deficiency

Under hypoxia stress, it has been observed that the gas exchange parameters are dramatically affected in several fruit trees, such as avocado (*Persea americana* Mill.) [33], kiwi fruits (*Actinidia chinensis* Planch) [34], citrus trees [35–38], pecans (*Carya illinoensis* K. Koch) [39], walnut trees (*Juglans regia* L.) [40], grapevine (*Vitis vinifera* L.) [41,42], pomegranate (*Punica granatum* L.) [43], apple (*Malus* × *domestica* Borkh) [44] and several *Prunus* species [7,45–49]. In general, all these tree species were classified as sensitive to root hypoxia. However, the concept of "relative tolerance" to hypoxia applied to the fruit tree species must be viewed with caution as many factors, such climatic, experimental or edaphic conditions, among others, could influence the responses observed [50]. However, some classifications have been made: extremely tolerant—quince (*Cydonia oblonga* Mill.) and *Pyrus betulaefolia*; very tolerant—pear (*Pyrus* spp.); moderately tolerant—apple (*Malus* × *domestica* Borkh), *Citrus* spp., and plums (*Prunus domestica* L. and *Prunus cerasifera* Ehrh); moderately sensitive—plum (*Prunus salicina* Lind.); very sensitive—cherry (*Prunus avium* L); extremely sensitive, peach (*Prunus persica* Batsch); and most sensitive—almond (*Prunus dulcis* [Mill.] DA Webb) and apricot (*Prunus armeniaca* L.) [50]. In the particular case of *Prunus* species, a tolerance gradient to long-term hypoxia was reported among seven genotypes used as rootstocks, identifying as tolerant to 'Mariana 2624' (*Prunus cerasifera* × *Prunus munsoniana* W. Wight and Hedrick) a plum rootstock, and as the most sensitive to 'Mazzard F12/1' (*P. avium*) a cherry rootstock [7]. 'Mariana 2624' plants survived through 14 days of waterlogging treatment, showing similar stomatal conductance and CO_2 assimilation rate values between waterlogged and control plants, unlike in the hypoxia-sensitive genotype, which showed intense leaf and root damage and a drastic decrease in the gas exchange parameters of the leaves during root hypoxia [7]. The ability to maintain a high photosynthetic rate, such as that observed in hypoxia-tolerant species, would guarantee an adequate supply of carbohydrates from the leaves to the roots. The carbohydrate supply is correlated with the production of highly energetic molecules (ATP), and the level of carbohydrate reserves or the capacity to maintain their transport throughout the plant appears to be a key feature in the tolerance to long-term flooding [48,51–54]. Consequently, maintaining glycolysis by a steady and sufficient supply with carbohydrates seems to be crucial for survival under hypoxia [3].

One of the first responses to O_2 deprivation is a hydraulic adjustment, the purpose of which is to sustain a constant water supply from roots to shoots, which is essential to maintaining gas exchange parameters in hypoxia-tolerant species [3,55]. Root hydraulic conductance is also affected under hypoxia stress, usually decreasing this parameter, but the response would depend on the species, age and even the experimental set-up [56]. There is a huge body of literature about the importance of root water transport in plants under different abiotic stresses (reviewed in [56–58]). In this context, root hypoxia modifies the root water transport in different manners: (1) cellular acidosis and the depletion of ATP affect the phosphorylation of aquaporins and the transport through these water channels is inhibited [59]; (2) hypoxia can alter root structure by inducing suberization (generation of a radial oxygen loss (ROL) barrier), but at the same time, this modification can affect the apoplastic water transport [60]; and (3) massive damage of the root system [3,32]. After 15 days of long-term waterlogging, the hypoxia-tolerant genotype "Mariana 2624" showed similar values of root hydraulic conductance (*Kr*) between normoxic and waterlogged plants. Unlike these, the hypoxia-sensitive genotype showed a strong decrease in *Kr* triggered by hypoxia (Pimentel, unpublished data).

Reactive oxygen species (ROS) are by-products of various metabolic pathways and are generated enzymatically or nonenzymatically [61]. Nonenzymatic ROS production can occur in mitochondria and chloroplast through electron transport chains (ETC) [61–63]. Enzymatic ROS production can occur in peroxisomes, cell walls, plasma membrane and apoplast [64], and also through respiratory burst oxidase homologs (RBOHs), a plasma-membrane-bound NADPH oxidase [62,65]. ROS induce $[Ca^{2+}]_{cyt}$ elevations by activation of the specialized Ca^{2+}-permeable ion channels in the plasma membrane. In addition, NADPH oxidases (RBOHs) are directly activated by cytosolic Ca^{2+}. Both ROS and Ca^{2+} form a self-amplifying loop named "ROS-Ca^{2+} hub" [66]. Elevation of $[Ca^{2+}]_{cyt}$ under hypoxia triggers multiple metabolic events and it is associated with both early and late responses to low oxygen conditions (deeply reviewed in [67]). ROS generated in response to abiotic stresses may be involved in various responses, acting as signaling molecules or triggering ROS-induced cell death [65]. Under hypoxic stress conditions, ROS can be generated due to an impairment of photosynthesis and aerobic respiration processes by inhibiting mETC [54,61,68]. In some cases, re-oxygenation of the soil after prolonged flooding can cause severe oxidative damage to the roots of sensitive trees [36,69]. Indeed, re-oxygenation has been recognized as an abiotic stress that can injure plants post-submergence (reviewed in [36,69])

In a re-analysis of the transcriptome published by Arismendi et al. [70] (commented on in Section 4), it was possible to find three differentially expressed isoforms of the *RBOH* gene, *RBOHA*, *RBOHC* and *RBOHE*. *RBOHA* and *RBOHE* genes showed a similar pattern between the two rootstocks, being that both genes were upregulated in hypoxic conditions. On the other hand, the *RBOHC* gene was downregulated in the hypoxia-tolerant genotype 'Mariana 2624,' but induced in the sensitive one under hypoxia stress (Table 1). Interestingly, the *RBOHC* gene has been reported as principally expressed in roots, where it is related to root hair formation and primary root elongation and development in Arabidopsis [65]. ROS have been described as toxic molecules generated by aerobic respiration that can cause oxidative damage. However, ROS also play a key role in signaling to trigger several processes such as cell proliferation and differentiation [71]. The hypoxia-sensitive genotype response suggests an ROS signaling role in the early stages of O_2 deficiency. Thus, 'Mazzard F12/1' could activate the formation of new roots, possibly replacing the original root system progressively injured in hypoxia.

Table 1. Transcript levels (of genes related to ROS production (*RBOH*) and ROS scavenging (*SOD*, *CAT* and *APX*)) from transcriptomics data from *Prunus* rootstocks under hypoxia.

| | | Log_2 FC | | | | | |
| | | 'Mariana 2624' | | | 'Mazzard F12/1' | | |
Gene		6 h	24 h	72 h	6 h	24 h	72 h
RBOHA	Prupe.6G321500	4.751	4.196	3.828	3.940	3.656	3.112
RBOHC/RHD2	Prupe.1G211000	−0.300	−1.902	−1.240	0.944	0.935	1.138
RBOHE	Prupe.5G107400	0.398	0.309	0.618	0.224	0.429	0.770
Cu Zn SOD1	Prupe.2G269400	−0.451	−1.268	−2.900	−0.454	−0.695	−0.863
Cu Zn SOD2	Prupe.1G347200	0.911	2.752	2.557	−0.713	−1.446	−0.485
Cu Zn SOD3	Prupe.2G262400	1.022	3.112	2.748	0.930	3.460	2.933
Fe SOD1	Prupe.6G042300	0.512	1.369	0.869	1.058	1.402	0.840
CAT1	Prupe.5G011300	0.174	−0.134	−0.818	nd	nd	nd
CAT2	Prupe.5G011400	0.498	−0.115	−0.526	0.376	−0.739	−0.580
APX 1	Prupe.1G481000	0.681	0.724	0.792	0.685	0.873	0.525
APX 2	Prupe.1G493900	−0.383	−0.638	−0.547	−0.017	−0.156	−0.395
APX 3	Prupe.6G091600	0.719	−1.498	−2.527	0.555	−1.351	−2.630
APX 5	Prupe.6G242200	6.538	7.284	6.864	3.323	3.948	4.591
APX 6	Prupe.7G171200	0.718	0.106	−0.252	1.158	0.502	0.740
APX S	Prupe.8G164400	−0.421	−2.284	−2.494	−0.414	−1.564	−2.650

FC: fold change; nd: no detected value.

As ROS accumulates after hypoxic events, the probability of the cell membrane being involved in a lipoperoxidation process is increased. Malondialdehyde (MDA) and electrolyte leakage are widely used as indicators of oxidative damage in plants. Differences in the accumulation of MDA have been reported under hypoxia, which depend on the degree of tolerance of the genotypes evaluated. For instance, in citrus species, the most tolerant genotype showed a delayed accumulation of MDA in leaves and roots in comparison with the other genotypes under hypoxia conditions [72]. Hypoxia treatment dramatically increased MDA content in the roots of two *Malus* species, but higher concentrations of H_2O_2 and superoxide radicals ($O_2^{\bullet-}$) were detected in the hypoxia-sensitive species [73]. In *Prunus* rootstocks under long-term hypoxia, 'Mariana 2624' plants, the hypoxia-tolerant genotype, showed no significant changes in MDA concentration in roots and leaves in comparison with the control plants. Opposingly, the sensitive rootstock 'Mazzard F12/1' showed a higher MDA concentration in roots and leaves after seven days of waterlogging, suggesting less capacity to remove ROS than the tolerant genotype [7,47]. Along with a lower MDA content in the roots of the hypoxia-tolerant genotype, a lower electrolyte leakage rate was also observed. Both parameters evidenced lesser structural damage in the roots, and are related to higher root membrane stability in 'Mariana 2624' [74].

Plants have enzymatic and non-enzymatic antioxidant compounds that participate in ROS detoxification, reducing the phytotoxic effects of radical species at the cellular level [75,76]. In citrus trees, a higher tolerance to O_2 deficiency by flooding is associated with the ability to delay the apparition of oxidative damage caused by a high activity of antioxidant enzymes such as superoxide dismutase (SOD), ascorbate peroxidase (APX), glutathione reductase (GR) and catalase (CAT) [72]. In Citrumelo trees, the induction of SOD, APX and CAT enzyme activities allows them to maintain a transient tolerance under hypoxia [36]. Similar results were observed in *Malus* [73] and *Prunus* species under hypoxia [77]. However, different results were found by Amador et al. (2012), since an increase in CAT activity in short-term waterlogging was found in the hypoxia-sensitive hybrid 'Felinem', but not in the hypoxia-tolerant rootstock 'Myrobalan'. The authors mention that they cannot conclude that antioxidant enzymes are directly involved in the tolerance of the hypoxia-tolerant genotype. The re-analysis of Arismendi et al. [70] evidenced that, in general terms, there are no notable differences in the expression of the *SOD*, *CAT* and *APX* genes between the hypoxia-sensitive and hypoxia-tolerant genotypes, except in two cases. In the first one, the *CuZnSOD2* gene showed an upregulated expression pattern in the hypoxia-tolerant genotype, and a significant downregulation in the hypoxia-sensitive one. In the second case, the *APX5* gene showed upregulation in both genotypes, but with a significantly higher expression in the tolerant one (Table 1). These results could explain the differences in MDA content and electrolyte leakage between the *Prunus* rootstocks, since the belated rise in the MDA concentration and electrolyte leakage in the root hypoxia-sensitive genotype suggests a lesser capacity to remove ROS, implying the occurrence of massive tissue damage which compromises the survival of the plant, as reported by Pimentel et al. [7] and Toro et al. [74].

3.2. Morpho-Anatomical Changes in Fruit Trees under O_2 Deficiency

Roots are the first organ that directly sense and deal with O_2 deficiency in compacted waterlogged or flooded soils. Therefore, morpho-anatomical modifications that allow the maintaining of better root oxygenation in waterlogging conditions are one of the key mechanisms associated with hypoxia-tolerant genotypes [78]. Adventitious roots, hypertrophied lenticels and aerenchyma all contribute to oxygenating the root system under low O_2 conditions [79,80]. Adventitious roots are produced in the replacement of damaged root systems, and are usually thicker and have more intercellular gas-filled spaces than roots growing in well-aerated soil [79]. Hypertrophied lenticels and aerenchyma favor O_2 diffusion to the root tips and rhizosphere, and in addition the hypertrophied lenticels allow the outwards diffusion of potentially toxic compounds like ethanol, acetaldehyde, ethylene and CO_2 [5,80,81]. In annual crops, both hypertrophied lenticels and the formation of aerenchyma generate a snorkel effect that keeps the roots oxygenated under waterlogging conditions [82], and eliminate toxic products generated from lactic and ethanol fermentation [79,80]. Another anatomical modification in roots is the development of the radial oxygen loss (ROL) barrier observed in many wetland plants. This barrier promotes longitudinal O_2 diffusion down roots, restricts the O_2 loss to the soil and could reduce the entry of phytotoxins into the roots in waterlogged soils [83–87]. This root trait has been not identified or studied in hypoxia-tolerant woody and perennial upland fruit trees yet.

The generation of morpho-anatomical modifications has been observed in different fruit tree species. Hypoxia-tolerant apple rootstocks generated adventitious roots in response to intermittent waterlogging events [88]. The development of hypertrophied lenticels as a response to root hypoxia has been reported in Rosaceae species such as *Pyrus* spp. and *Cydonia oblonga* (Mill.) [79]. Pistelli et al. [52] reported the formation of adventitious roots in *Prunus cerasifera* L. ('Mr.S.2/5_rootstock') under waterlogging stress, but they did not report the development of aerenchyma. Pimentel et al. [7] reported that the hypoxia-tolerant *Prunus* rootstock 'Mariana 2624' was able to develop adventitious roots, aerenchyma in adventitious roots and hypertrophied lenticels 10 days after the onset of waterlogging treatment. A large adventitious root system grew from the stem base and just beneath the water surface. In this new root system, aerenchyma tissue was not observed near the root apex, but was widely developed distant from it (at 30 and 55 mm from the root apex) and, at the same time, the development of hypertrophied lenticels on the submerged portion of stems was observed (Figure 3). Hypertrophied lenticels appear in wetland species and in several woody and herbaceous plant species subjected to waterlogging [80,89,90]. In *Prunus* rootstocks, the hypertrophied lenticels were developed only in the hypoxia-tolerant genotypes, and when the submerged stems of the rootstock 'Mariana 2624' were sealed with lanoline to prevent its development, a significant decrease in gas exchange parameters was detected in comparison with the non-lanoline-treated plants [7]. Interestingly, Toro et al. [74] showed that aerenchyma formation also takes place in the existing root system in 'Mariana 2624' at 30 mm from the root apex (Figure 3). In woody plants, all these morpho-anatomical modifications are late responses. In the case of long-term hypoxia treatment, they appear after 6 or 10 days from the beginning of the stress, and they are part of the strategies used to avoid the negative effects of the energetic crisis triggered by an O_2 deficient environment.

4. Transcriptomic Reprogramming of Principal Pathways Involved in Energy Metabolism under O_2 Deficiency

The effect of O_2 deficiency on the rhizosphere of woody plant species has been more widely addressed from a physiological perspective than from a genomic point of view. Despite the massive amount of data provided by next generation sequencing analyses (NGS), which would allow a more integrative view of the plant's response to root hypoxia stress, few transcriptomic studies have been reported in woody species [91–93] and fruit trees such as avocado [94] *Prunus* sp. [49,53,70,95], kiwi fruit [96] and grapevine [41,97].

Using RNA-seq time series data from our collaborative study on *Prunus* rootstocks previously reported [70], a revisited analysis with a de novo assembly approach was carried out in order to gain a deeper insight into the relevant pathways involved in energy metabolism under hypoxic conditions. The de novo assembly evidenced a higher number of differentially expressed genes (DEGs) in the hypoxia-sensitive *Prunus* rootstock 'Mazzard F12/1' at each of the sampling times of the waterlogging treatment, in comparison to the hypoxia-tolerant rootstock 'Mariana 2624' (6 h: 2638 vs. 2356; 24 h: 5819 vs. 5228 and 72 h: 10,255 vs. 6366). Specifically, those RNA-Seq analyses focused on the hypoxia adaptation of root systems [41,53,70,94,95] have evidenced certain metabolic pathways and groups of genes commonly affected by O_2 deficiency.

Maintaining a continuous glycolytic flux is a crucial factor in ensuring the energy pool necessary for the processes involved in the survival of trees under hypoxia [3]. In woody species, it has been reported that hypoxia-sensitive plants deplete their soluble sugars rapidly in flooded conditions, but the tolerant ones keep a higher level of soluble sugars for longer periods [81,98,99]. In this context, the regulation of gene expression involved in primary metabolism and energy homeostasis is essential to avoiding the detrimental effects of energy depletion under hypoxic stress. In the sweet cherry rootstock *Cerasus sachalinensis* (F. Schmidt), the waterlogging treatment upregulated most genes associated with sucrose metabolism. Genes encoding *SUCROSE SYNTHASE* (*SuSy*) were upregulated, however *INVERTASE* (*INV*) genes were downregulated [53]. In 'Mariana 2624' and 'Mazzard F12/1' plants, four *SuSy* isoforms were upregulated under hypoxic conditions. Additionally, in 'Mazzard F12/1' rootstocks, one *INV* isoform was upregulated after 72 h of waterlogging treatment, and four were downregulated at the same time. On the other hand, 'Mariana 2624' rootstocks exhibited three *INV* isoforms downregulated at 24 and 72 h (Figure 1). Both *SuSy* and *INV* can cleave sucrose to release its constituent monosaccharides, although a lower energy cost of generating hexose phosphates for glycolysis is required in the case of sucrose synthase. This latter is a typical feature of sucrose metabolism during O_2 deficiency [100]. Thus, those plants that favor the activity of SuSy would be opting for a more energy efficient way to provide substrates for glycolysis under hypoxic conditions.

Transcriptomic evidences in the roots of waterlogged forestry trees revealed enhanced glycolytic flux and an activation of fermentative pathways in order to maintain the energy supply when mitochondrial respiration is inhibited by O_2 deficiency [81,91]. Alongside the induction of genes related to glycolysis, an absence of transcripts for genes associated to gluconeogenesis, such as *GLUCOSE 6-PHOSPHATASE* and *FRUCTOSE 1,6-BISPHOSPHATASE*, was reported in roots of flooded avocado (*Persea americana* Mill.) [94]. In this sense, the downregulation of *PHOSPHOGLUCOMUTASE* genes reported in flooded grapevine roots also supports the idea of a hampered flux to gluconeogenesis under O_2 deficiency [41]. This pattern was also presented in Myrobalan 'P.2175', another hypoxia-tolerant *Prunus* rootstock, but not in the hypoxia-sensitive 'Felinem' [95]. In the *Prunus* rootstock 'Mariana 2624', two *PHOSPHOGLUCOMUTASE* isoforms were downregulated at 24 and 72 h of waterlogging treatment, but the gene induction of two isoforms detected in 'Mazzard F12/1' plants after 72 h of O_2 deficiency is worth noting (Figure 1). Furthermore, genes encoding for *PHOSPHOENOLPYRUVATE CARBOXYKINASES* (*PEPCK*) repeated the behavior described for *PHOSPHOGLUCOMUTASE* in these *Prunus* rootstocks contrasting in their tolerance to hypoxia (Figure 1). Regarding the above-mentioned, the inhibition of gluconeogenesis appears to be strongly related to hypoxia-tolerant genotypes in *Prunus* spp.

HEXOSE-PHOSPHORYLATING HEXOKINASES (HXK) are the only plant enzymes able to phosphorylate glucose, so they are considered a key factor in glycolysis activation [101]. The hypoxia-tolerant *Prunus* rootstock 'Mariana 2624' evidenced the induction of more *HXK* genes than the hypoxia-sensitive 'Mazzard F12/1', and what is more, the *HEXOKINASE 3* (Prupe.1G366000) was strongly upregulated in the tolerant genotype, but consistently downregulated in the sensitive one during the whole hypoxia treatment (Figure 1) [70].

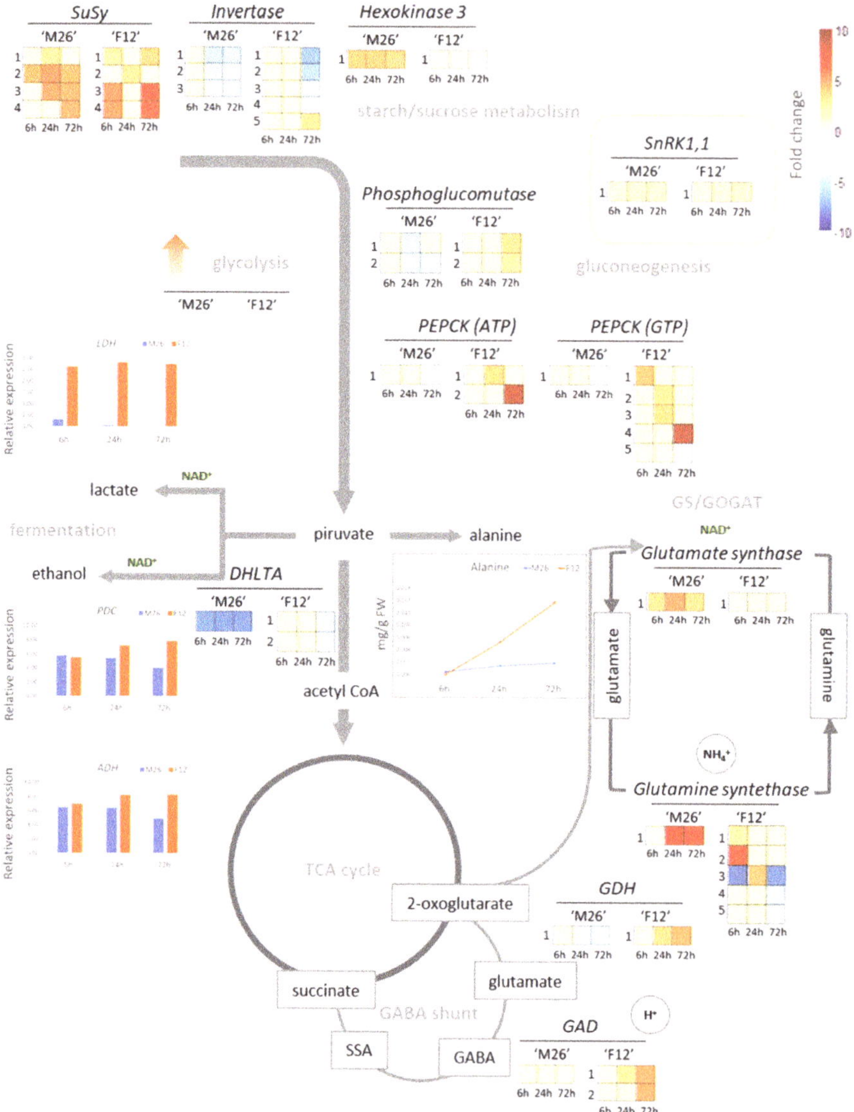

Figure 1. Transcriptomic reconfiguration of pathways involved in energy metabolism in response to O$_2$ deficiency. Through a comparative analysis between two genotypes of *Prunus* rootstocks, the hypoxia-tolerant 'Mariana 2624' ('M26') and the hypoxia-sensitive 'Mazzard F12/1' ('F12'), the changes

in the transcript levels of genes belonging to the routes typically related to energy metabolism (starch/sucrose, glycolysis/gluconeogenesis and lactic/ethanolic fermentations) and that are expressed in a contrasting way between the genotypes are described. Along with these routes, their connections to TCA cycle, GABA shunt and the GS/GOGAT cycle are presented. The differential transcript levels of the genes of the last two pathways show clear differences that point to the favoring of one or another metabolic pathway in a genotype-dependent manner. In addition, the steps where the regeneration of NAD^+, consumption of excess H^+ (regulation of cytoplasmic pH) and assimilation of NH_4^+ take place are shown. The data included in this re-analysis were obtained through a de novo approach via RNAseq data retrieved from NCBI BioProjects PRJNA215360 and PRJNA215068 [70], corresponding to *P. cerasifera* × *P. munsoniana* 'Mariana 2624' and *P. avium* 'Mazzard F12/1', respectively. Quality control of the libraries was performed with FastQC and the AfterQC tool [102]. Consecutively, all libraries were merged into one unique file to perform a de novo assembly with Trinity [103]. The resulting fasta file was depurated to obtain unigenes using Transdecoder with the pFam database [104] and CD-HIT-EST [105,106]. This output was used as a reference to perform an alignment using Hisat2 [107]. Transcript assembly was performed through StringTie [108]. DEGs were obtained using EdgeR from Bioconductor [109]. Genes with adjusted *p*-value < 0.05 and LogFC > 2 & < −2 were used for further analysis. The annotation was performed using GO FEAT [110] and KEGG [111]. DEGs included in this figure are indicated with their homolog loci from *P. persica*: *SuSy*—(1) Prupe.7G192300.4; (2) Prupe.8G264300.1; (3) Prupe.1G131700.1; (4) Prupe.7G192300.4. *Invertase*—(1) Prupe.2G277900.1; (2) Prupe.2G277900.1; (3) Prupe.6G122600.1; (4) Prupe.6G122600.2; (5) Prupe.1G111800.1. *Hexokinase 3*—(1) Prupe.1G366000. *SnRK1,1*—(1) Prupe.3G262900.1. *Phosphoglucomutase*—(1) Prupe.2G286900.1; (2) Prupe.1G330700.1. *PEPCK (ATP)*—(1) Prupe.6G211000.1; (2) Prupe.1G541200.7. *PEPCK (GTP)*—(1) Prupe.1G541200.4; (2) Prupe.6G210900.1; (3) Prupe.6G211000.1; (4) Prupe.1G541200.7; (5) Prupe.4G166400.4. *DHLTA*—(1) Prupe.1G309100.1; (2) Prupe.8G056000.1. *Glutamate synthase*—(1) Prupe.2G311700.2. *Glutamine syntethase*—(1) Prupe.1G148700.1; (2) Prupe.3G166500.3; (3) Prupe.1G346600.1; (4) Prupe.5G236300.1; (5) Prupe.3G166500.2. *GDH*—(1) Prupe.7G004100.1; (2) Prupe.2G269800.1. *GAD*—(1) Prupe.1G339900.1; (2) Prupe.7G252300.1. For more detailed results of the transcriptional patterns of *LDH*, *PDC* and *ADH* (Quantitative Reverse Transcription (qRT)-PCR) and alanine levels (HPLC-DAD) inserted in this figure, refer to [47,112].

As a product of the phosphorylation of glucose by HXK, glucose 6-phosphate (G6P) can block sucrose-non-fermenting-related protein kinase-1 (SnRK1) activity [113]. SnRK1 is a kinase recognized as a metabolic sensor that can decode energy deficiency signals and induce the extensive metabolic reprogramming required for the adaptation to nutrient availability through inhibition of expensive energy processes and growth arrest. Stress conditions, such as hypoxia, can affect photosynthesis and photoassimilates biosynthesis along with respiration triggering a low energy syndrome (Tomé et al., 2014). Transcriptomics data evidenced the induction of *SnRK1* genes majorly associated with the hypoxia-sensitive *Prunus* rootstocks 'Felinem' [95] and 'Mazzard F12/1' (Figure 1, this review). Remarkably, low levels of G6P induce the activity of SnRK1, which triggers the signaling for upregulating several genes such as *PEPCKs*, commented on above. Regarding the genes encoding glycolytic enzymes, these two genotypes exhibited similar transcriptional activation between them, without significant differences in response to waterlogging. In grapevine, an upregulation of *DIPHOSPHATE-DEPENDENT PHOSPHOFRUCTOKINASES*, instead of *ATP DEPENDENT 6-PHOSPHOFRUCTOKINASES 1*, was found, responsible for converting D-fructose 6-phosphate to D-fructose 1,6-bisphosphate, which is another typical feature of the glycolysis in hypoxia that favors energy saving through PPi-dependent rather than ATP-dependent processes [41].

In order to maintain ATP production by the glycolytic pathway, it is necessary to replenish the NAD^+ pool that was reduced during this process. In this context, activation of the fermentative pathways results in the classic and most widely reported response in the anaerobic metabolism. From pyruvate, lactic and ethanolic fermentation generate lactate and ethanol, respectively, and in the process NADH is oxidized, returning NAD^+ to keep glycolysis active [68]. However, both fermentative pathways present eventual disadvantages, since lactate is toxic for the cells, and ethanol diffuses rapidly out of the cells, implying a considerable loss of carbon under hypoxic conditions [114]. The gene induction of *LACTATE DEHYDROGENASE* (*LDH*) in fruit trees under root hypoxia has been reported in avocado [94], *Prunus* spp. [70,112] and *C. sachalinensis* [53], but not in grapevine [41] or the *Prunus* genotypes analyzed by Rubio-Cabetas et al. [95]. In 'Mariana 2624' and 'Mazzard F12/1', root hypoxia triggered a rapid increase in the *LDH1* (Prupe.5 G072700) mRNA levels with a peak in their transcripts at six hours, but such increase was more dramatic and prolonged in the root hypoxia-sensitive genotype (Figure 1) [112]. Consequently, a higher L-lactate content was evidenced in 'Mazzard F12/1' roots with a maximum at 3 and 6 h of waterlogging. An increase in LDH activity in response to waterlogging in three species of *Prunus* (*P. mira*, *P. persica* and *P. amygdalus*) indicated that *P. amygdalus* was the genotype with a greater and more sustained increase in LDH activity. This fact was concomitant with its increased accumulation of lactic acid in the cytoplasm, and its lower tolerance to root hypoxia [115]. Lactate accumulation and cytoplasmic acidosis are determinants of hypoxia-sensitive phenotypes in maize [116]. In addition, in *Limonium* spp., plants capable of removing excesses of lactate from the cytoplasm were more tolerant to hypoxia conditions [117]. In this sense, the increase of *Prunus* spp. *NIP1;1* mRNA, a putative lactic acid transporter, was not linked to a lower lactate content in the roots of 'Mazzard F12/1'. Bioinformatic approaches identified steric hindrances in PruavNIP1;1 given by the residues Phe107 and Trp88 in the NPA region and ar/R filter, respectively, but such blockages were absent in the NIP1;1 of 'Mariana 2624'. The functional characterization of these aquaporins in the yeast strain Δjen1 corroborated the lower efficiency of the lactic acid transport of PruavNIP1;1, which could be related to a higher lactate accumulation and detrimental effects at cell level in 'Mazzard F12/1' roots under hypoxia [112].

The drop in cytoplasmic pH, in part related to the dissociation of lactic acid, inhibits LDH activity and stimulates that of pyruvate decarboxylase (PDC), the first step involved in ethanol fermentation [118]. *PDC* and *ALCOHOL DEHYDROGENASE* (ADH) transcripts were found to be expressed in each transcriptome analyzed. Both *PDC* and *ADH* transcripts were coordinately induced in 'Mariana 2624' and 'Mazzard F12/1' under O_2 deficiency, although the hypoxia-tolerant genotype showed a decreasing trend after 6 h of waterlogging (Figure 1) [47]. This transcriptional finding suggests that 'Mariana 2624' resorts to other adaptation mechanisms, distinct to the fermentation pathways, after the first hours of flooding.

The pyruvate dehydrogenase complex catalyzes the oxidative decarboxylation of pyruvate with the formation of acetyl-CoA, CO_2 and NADH(H$^+$) [119], linking glycolysis to the Tricarboxylic acid (TCA) cycle. Interestingly, transcripts of the subunit E2 (*DIHYDROLIPOYL TRANS-ACETYLASE*, *DHLTA*), a component of this enzymatic complex, were downregulated in both 'Mariana 2624' and 'Mazzard F12/1' roots under hypoxic conditions; however, this repression was earlier and much stronger in the hypoxia-tolerant *Prunus* rootstock (Figure 1). This fact suggests a diminished metabolic flux in the TCA cycle and an accumulation of pyruvate. However, another conclusion for pyruvate as a result of O_2 deficiency, different from the fermentative pathways, is its conversion to alanine by means of the enzyme ALANINE AMINOTRANSFERASE (AlaAT). The accumulation of this amino acid is typically linked to hypoxia in plants [114,120,121]. *AlaAT* mRNAs induced during hypoxia were reported in the roots of avocado [94], grapevine [41] and Myrobalan 'P.2175' [95]. Although *AlaAT* transcripts were not detected in the transcriptome of 'Mariana 2624' or 'Mazzard F12/1', the accumulation of alanine was evidenced in both genotypes during waterlogging, this being more intense in the hypoxia-sensitive genotype (Figure 1) [47]. A hypoxia-induced alanine accumulation is also possible through the activity of GABA-transaminase (GABA-T) that converts GABA into succinic semialdehyde (SSA), releasing alanine (from pyruvate) to the mitochondrial lumen [122,123]. Here, the eventual SSA accumulation will be toxic for the cell. The succinic-semialdehyde dehydrogenase (SSADH) converts SSA into succinate, consuming an NAD$^+$ molecule [124] but this latter would limit the ability of the cell to properly maintain the active glycolysis.

Another alternative means of draining pyruvate to alanine by AlaAT under hypoxic conditions involves the glutamate metabolism. The reductive amination of 2-oxoglutarate by glutamine oxoglutarate amino transferase (GOGAT) regenerates glutamate (substrate of AlaAT to produce alanine) and NAD$^+$ (Diab and Limami, 2016). Interestingly, root hypoxia induced *GOGAT* in 'Mariana 2624' rootstock, but not in the hypoxia-sensitive 'Mazzard F12/1', whose transcript levels remained unchanged under waterlogging (Figure 1). It is possible that a higher GOGAT enzyme activity is related to the higher metabolism of alanine in 'Mariana 2624', which would explain the modest accumulation of this amino acid during waterlogging in this genotype, as opposed to the notorious accumulation shown by the hypoxia-sensitive *Prunus* rootstock (Figure 1) [47]. In flooded grapevine, *GOGAT, GLUTAMINE SYNTHETASE* (*GS*) and *GLUTAMATE DEHYDROGENASE* (*GDH*) were overexpressed [41]. GS catalyzes the ATP-dependent assimilation of NH4$^+$ into glutamine using glutamate as substrate. The GS/GOGAT cycle is the principal route of ammonium assimilation in plants [125]. In tomato (*Solanum lycopersicum* L.), the activity of the ATP-consuming GS was significantly enhanced in roots during prolonged root hypoxia [126]. Here, a striking contrast in the *GS* transcriptional pattern was evidenced between 'Mariana 2624' and 'Mazzard F12/1' roots under hypoxic conditions, as *GS* transcripts were strongly accumulated in the hypoxia-tolerant genotype after 24 h of waterlogging, but consistently downregulated as stress progressed in the hypoxia-sensitive one (Figure 1). This evidence suggests that a more active GS/GOGAT cycle, capable of assimilating nitrogen and regenerating NAD$^+$ to support glycolytic flux under conditions of O$_2$ deficiency, shapes one of the successful metabolic strategies involved in defining the hypoxia-tolerant phenotype seen in 'Mariana 2624'. As in the flooded grapevine, *GDH* transcripts were overexpressed in the roots of waterlogged 'Mazzard F12/1', but clearly downregulated in the hypoxia-tolerant genotype (Figure 1). Another gene involved in the GABA shunt, *GLUTAMATE DECARBOXYLASE* (*GAD*), showed an upregulation only associated with the hypoxia-sensitive genotype (Figure 1). Thus, 'Mazzard F12/1' appears to boost the flux of the GABA shunt by acquiring glutamate from GDH-mediated 2-oxoglutarate amination (instead of from its inhibited GS/GOGAT cycle). This amination generates NAD$^+$, but the detoxification of the SSA consequently generated in this pathway, through SSADH activity, consumes NAD$^+$, so the net gain of this cofactor is 0.

The transcriptomic antecedents compiled from different fruit trees under hypoxic conditions show common alterations of genes involved in starch/sucrose metabolism and glycolysis, together with the inhibition of gluconeogenesis in the case of hypoxia-tolerant genotypes. The activation of, firstly, lactic fermentation (*LDH*), and then ethanolic fermentation (*PDC* and *ADH*), was also evident in all fruit trees. Since ethanol can represent a carbon leak from the plant, a more energy-efficient destination for pyruvate is its conversion to alanine. At this point, the regeneration of the glutamate involved in alanine biosynthesis from the GS/GOGAT cycle instead of from the GDH activity in the context of GABA shunt is postulated as one of the best metabolic strategies for explaining the survival and growth capacity during O$_2$ deficiency in root hypoxia-tolerant genotypes, such as in the case of the *Prunus* rootstock 'Mariana 2624'.

5. Root Respiration under O$_2$ Deficiency

The carbon cycle is a bio-geochemical cycling process of the continuous flowing of organic and inorganic forms of carbon through the biosphere, geosphere and atmosphere, supporting life on Earth [127]. In the biosphere, plants play a key role via autotrophic respiration, which represents an important component of the carbon cycle and corresponds to respiratory processes in the leaf, shoot and root [128]. Respiration involves the participation of different processes responsible for the oxidation of glucose molecules for energy and carbon structures, either in the presence (aerobic) [129,130] or absence (anaerobic) of O$_2$ [131]. Root respiration is a process sensitive to changes in soil conditions, such as chemical composition [132], temperature [133], salinity [134] and water excess (hypoxia/anoxia stress) [74], among others.

Root respiration is highly dependent on the availability of O$_2$ in the root zone, and a lack of this element may lead plants into an imbalance in energy distribution within metabolic processes, whereby the deficit of ATP could range between 3% and 37.5% with respect to well-aerated roots [135].

When the O_2 in the rhizosphere decreases to a point where the formation of ATP by cytochrome oxidase (COX, or complex IV) is hampered, the activation of less efficient metabolic pathways takes place (e.g., fermentative pathways) and plants may enter a state of energy crisis [136]. The growth and maintenance of tissues are two processes that require energy from root respiration, and must coexist coordinately for the correct development of plants. However, energy crisis caused by hypoxia/anoxia stress (O_2 deficiency) induces an energy redistribution either to the maintenance or the growth of new tissue [74,137]. Membrane stability, active ion transport and de novo synthesis of proteins are the most expensive processes whereby the cell metabolism must adjust its energy budget [138]. It has been reported that these processes are controlled by gene regulation at both the transcript and translation level [139], and they strategically determine how plants cope with the energy reduction imposed by hypoxia/anoxia. It is well known that low O_2 levels impair the respiratory metabolism of plant tissues. The damage induced by O_2 depletion, especially on root respiration, could compromise the development and growth of the entire plant, because root respiration drives the energetic support for generating new biomass and/or cellular and structural maintenance [140–142].

As previously commented, there is limited information available about transcriptome analysis in trees or woody species under low O_2 conditions, and even fewer works have been reported that relate the differentiated expressions of genes from transcriptomes and the physiological responses of the respiratory metabolism. The respiratory chain has a principal function of transferring electrons to the terminal oxidases, where O_2 acts as the final electron acceptor, producing high-energy phosphate bonds (ATP) [129,130]. The mitochondrial oxidative phosphorylation system consists of four multi-subunit oxidoreductases involved in the electron transport chain (mETC) (complexes I-IV) and the ATP synthase complex (complex V) [143,144]. In the revisited analysis of the transcriptomic study comparing *Prunus* rootstocks contrasting in their tolerance to hypoxia [70], DEGs encoding for proteins belonging to the mETC, such as *Respiratory Supercomplex Factor 2* (*RCF2*), subunits of complex III (*Cyt$_{bc1(sub8)}$*, *Cyt$_{bo3}$* and *Cyt$_{b\ red}$*) and IV (*COX$_{(sub5b2)}$* and *COX$_{(sub6b2)}$*), *Cytochrome c* (*Cytc*) and *Alternative Oxidase* (*AOX*), were detected (Figure 2). The synchrony of the activity of each protein in the mETC may be altered depending on the O_2 availability in the rhizosphere [145], having as a direct consequence a partially restricted or completely inhibited energy production [131]. As in herbaceous plants, in woody species one of the most dangerous subproducts of the aerobic metabolism is the formation of ROS such as H_2O_2 and O_2^- [62]; however, under an anaerobic condition, other harmful molecules are also formed. Thus, the combination of ROS and nitric oxide (NO) may be extremely detrimental for the cell [70,146].

The processes involved in coping with low O_2 at the root level are quite expensive for the plant's energy budget [137,147], and therefore to scavenge harmful molecules, plants are required to invest a significant amount of energy in synthesizing expensive enzymatic or nonenzymatic molecule scavengers [148]. In the mETC, ROS are formed principally through electron leakage from the protein complexes inserted into the mitochondrial membrane, such as complex I (NDH, NADH dehydrogenase) and III (*Cyt$_{bc1}$*, cytochrome bc1 dehydrogenase) [149], and NO formation is more associated with alternative oxidase (*AOX*), complex III and IV (*COX*, cytochrome *c* oxidase) [150–152]. The revisited analysis of [70] revealed large differences in gene expression related to mETC, which are closely related to proteins of complex III, IV and *AOX* (Figure 2). Regarding complex III, no alterations were found in the expressions of the *Cyt$_{b_red}$* and *Cyt$_{bc1}$* genes in waterlogged 'Mariana 2624' plants, however the gene inductions of these isoforms were evident in the root-hypoxia-sensitive genotype 'Mazzard F12/1' in response to O_2 deficiency (Figure 2). On the other hand, the root-hypoxia-tolerant 'Mariana 2624' repressed the expression of two *Cyt$_{bo3}$* isoforms, but 'Mazzard F12/1' showed a different behavior since the three *Cyt$_{bo3}$* isoforms were induced in response to hypoxia in both early and late stages (Figure 2). With respect to complex IV, 'Mariana 2624' did not present DEGs over time, while in contrast, 'Mazzard F12/1' reduced the expression of *COX$_{(sub5b2)}$* at 72 h, but a strong induction of *COX$_{(sub6b2)}$* was evident from 24 h of root hypoxia (Figure 2). Some of the protein complexes of the mitochondrial membrane may contribute to the scavenging of these harmful molecules. In this sense, NO is formed from nitrite by COX, however, this protein is inhibited by the raised NO, while that *AOX* is an NO-resistant protein [153]. For instance, it has been found in *A. thaliana* that the overexpression of *AOX* may prevent excesses of NO modulating the formation of ONOO$^-$ from interacting with O_2^- [153]. Therefore, despite the low affinity of O_2 with *AOX* and the limiting proton translocation, the activity of *AOX* allows the maintenance of the energy balance under hypoxic conditions [154]. Recently, Vishwakarma et al. [153] found an increase in the haemoglobin–nitric oxide (Hb/NO) cycle under hypoxia, which might be mediated by the *AOX* protein and improves the redox and energy status of the hypoxic cell [136]. This cycle consumes NADH-regenerating NAD$^+$, which would contribute to maintaining the glycotytic flux during O_2 deficiency [155]. In the roots of *Prunus* rootstock under O_2 deficiency, the class 1 *non-symbiotic haemoglobin*-like (*nsHb*) gene showed a higher expression in the hypoxia-tolerant genotype than in the sensitive one [156]. This was also found in roots of hypoxia-tolerant oak genotypes under low O_2 [157]. In addition, the transcriptomic analysis revealed a higher *AOX* gene expression in 'Mariana2624' at 6 h of waterlogging, but no upregulation of this gene was detected in the sensitive genotype (Figures 2 and 3). Thereby, the participation of class 1 *nsHb* and *AOX* genes in hypoxia tolerance genotypes may suggest that the possibility of the participation of the nsHb/NO cycle drives energetic support to the roots of woody plants in the early stages of hypoxia, promoting the electron flow [158] and allowing the prevention of mETC overreduction [159].

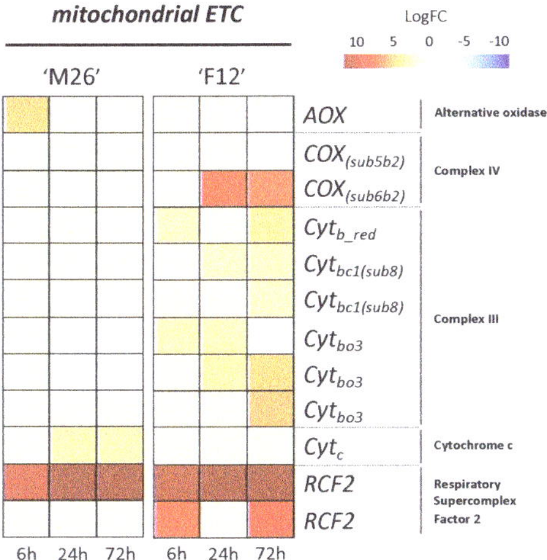

Figure 2. Transcriptomic reconfiguration of pathways involved in mitochondrial electron transport chain in response to O_2 deficiency. The changes in the transcript levels of genes belonging to the mitochondrial electron transport chain typically related to energy metabolism, and that are expressed in a contrasting way between the genotypes, are described. Data obtained from de novo transcriptomic analysis of 'Mariana 2624' ('M26') and 'Mazzard F12/1' ('F12'). *AOX*, alternative oxidase; *Cyt$_c$*, Cytochrome *c*; *RCF2*, Respiratory Supercomplex Factor 2; *Cyt$_{bo3}$*, Cytochrome bo3 ubiquinol oxidase; *Cyt$_{bc1(sub8)}$*, Cytochrome bc1 complex subunit 8; *Cyt$_{b\ red}$*, Cytochrome b reductase; *COX$_{(sub5b2)}$*, Cytochrome c oxidase subunit 5b2; *COX$_{(sub6b2)}$*, Cytochrome c oxidase subunit 5b2.

To generate an optimal response under hypoxia is necessary in order to activate the complete genetic machinery, which could have an extra cost to the root metabolism [160]. Under hypoxia, the control of protein synthesis may be a 'double-edged sword', since this is necessary for proper cell function, but an increase in protein turnover may increase the respiratory costs of maintenance, which might lead to compromised growth [138]. Under optimal conditions, the energy generated in mitochondrial phosphorylation as ATP is used to synthesize new structures in growing plants (growth respiration), and for all processes related to cellular maintenance, such as protein turnover, maintenance of ion gradients and membrane potentials in the cell (maintenance respiration) [141,142,161]. However, environmental changes may alter the distribution of energy. Under waterlogging, the forced entry into a state of energy crisis leads the roots to allocate resources to priority processes, thus modifying the costs associated with root respiration [138]. In *Carex* plants, O_2 deficiency drives the activation of strategies for reducing root growth in order to maximize the respiratory cost imposed on I upon uptake [137]. However, the allocation of O_2 to components of root respiration depends on the species. In the hypoxia-tolerant *Prunus* genotype, 'Mariana M2624', Toro et al. [74] found a greater ability of the root growth to spend less energy in processes related to maintenance, such as protein turnover and membrane integrity. These costs could reach up to 80% of the plant's energy budget [162], and therefore its regulation would be a key factor in tolerance to hypoxia. An EuKaryotic Orthologous Groups (KOG) classification in *Cerasus sachalinensis* roots under short-term waterlogging showed that the largest groups of DEGs were included in the categories of post-translational modification, protein turnover, and chaperones [53]. In addition, the authors found that a high number of transcripts were associated with translation pathways, and also energy metabolism. The revisited transcriptomic analysis showed a remarkable difference between transcripts related to mETC from 'Mariana 2624' and 'Mazzard F12/1' rootstocks in response to hypoxia. Thus, low O_2 in roots increases dramatically the gene expression of the sensitive genotype from 6 to 72 h of waterlogging, whereat one would usually find upregulated genes from subunits of complex III (Cyt_{bo3}, Cyt_{b_red}, and Cyt_{bc1}) and IV ($COX_{(sub6b2)}$), and genes that have control over processes associated with supercomplex formation (*RCF2*) (Figures 2 and 3). According to Arru and Fornaciari [160], protein synthesis would depend on post-transcriptional or post-translational regulation, there being extremely high energy-requiring step at the translational level. A study performed on *Prunus* showed that hypoxia-tolerant rootstocks manifested reduced ATP demands for protein turnover and the maintenance of membrane integrity, which was closely related to the low respiratory costs of maintenance [74].

Into the mitochondrial inner membrane, the respiratory protein complexes associate to form supramolecular assemblies known as supercomplexes [163,164]. The supercomplex formation seems to be crucial for the proper functioning of mETC, and requires the assistance of specific genes to efficiently assemble its constituent proteins [165]. In higher plants, the presence of the supercomplex has been identified in several species such as *Arabidopsis*, bean, potato and barley [164]. However, there is still a lack of information about the genes encoding for proteins involved in supercomplex formation. The *Respiratory Supercomplex Factor 2 (RCF2)* gene is part of the conserved gene family termed *hypoxia-induced gene 1*, which is highly expressed under hypoxia conditions and has been described as necessary for supercomplex formation [166,167].

As for other genes from mETC, there are scarce reports about the *RCF2* genes of woody species or even of higher plants. Recently, Shin et al. [168] identified the *RCF2* gene from *Cucumis melo*, but further efforts are required to identify this gene in a larger number of woody species and evaluate its response under O_2 deficiency. Further, in wheat roots, *RCF2* helps to overcome an energy deficit by enhancing ADP/ATP transfer and, ultimately, improving the supply of ATP [158]. An early expression of the *RCF2* gene in both hypoxia-tolerant and -sensitive *Prunus* rootstocks is evidenced (Figures 2 and 3). However, only the hypoxia-tolerant genotype 'Mariana 2624' showed expression of *AOX*, while the hypoxia-sensitive genotype 'Mazzard F12/1' showed the expression of genes related to complex III, and after 72 h also evidenced the expression of genes related to complex IV (Figure 2). According to Eubel et al. [164], in the mETC, complex III is commonly found, forming a higher (I+III) and lower (III+IV) abundance of the supercomplex, and on the other hand, *AOX* does not seem to form part of a supercomplex, because complex I and III would limit *AOX* activity by reducing substrate ubiquinol. It is has been reported that higher levels of complex I and supercomplex I+III could contribute directly to the maintenance of mitochondrial function under hypoxia [152]. The transcriptomic data of mETC genes from RNAseq could lead to the proposal that the presence of the AOX gene (and consequently AOX protein) would not be limited by the presence of genes related to complex III, which could lead to an over-formation of the supercomplex and reduce the AOX protein activity (Figure 2). In addition, the fact that the 'Mariana 2624' rootstock did not express genes related to complex III or IV, although

RCF2 genes were effectively expressed, would indicate an appropriate regulation of the supercomplex in the mETC in the tolerant genotype. In the hypoxia-sensitive *Prunus* rootstock 'Mazzard F12/1', a higher energy cost related to protein turnover was reported as being triggered by O_2 deficiency, and root tissue injury triggered by waterlogging [74]. As consequence, the hypoxia-sensitive genotype should require greater protein synthesis/breakdown, which is partially observed here in the high differential expression of the complex proteins at different times of waterlogging (Figure 2). mETC have several protein complexes with different roles that guarantee the maintenance of cellular energy both under optimal and stress conditions. Cytochrome c (Cyt_c) corresponds to a small and conserved protein family that is responsible for generating the proton gradient across complexes III and IV, driving ATP synthesis [169] and in addition potentially plays a key role in the development of an adaptative mechanism for tolerating low O_2 [170]. It has been described that the Cyt_c protein may be released (with ROS production) from the mitochondrial membrane into the cytosol, to trigger the key step in the early execution phase of programmed cell death (apoptosis) [171,172].

Generally, the avoiding strategies of plant adaptation to O_2 deficiency involve the formation of aerenchyma structures by apoptosis, in order to support the increase in O_2 diffusion and maintain the aerobic energy supply in root cells [7,74,80]. Scarce information for woody species has been reported regarding the relationship between Cyt_c and hypoxia stress [170]; however, it is widely known that waterlogged woody species are able to develop aerenchyma in the roots through apoptosis [5,7,74,173]. The Cyt_c-dependent aerenchyma formation relies on the redox state of the cell environment, which will depend on the presence of H_2O_2 and/or O_2^- [174]. Under hypoxia, there are relatively high concentrations of H_2O_2 [175], which would lead the cellular environment to an oxidized state, changing the Cyt_c protein structure into its oxidized state, which is capable of triggering apoptosis [174]. After 24 h of waterlogging treatment, a higher induction of the Cyt_c gene was detected in the hypoxia-tolerant genotype 'Mariana 2624', while no or very low expression was observed in 'Mazzard F12/1' (Figures 2 and 3). On the other hand, waterlogged 'Mariana 2624' plants developed aerenchyma in roots, which would combine with the air-filled spaces in supplying the O_2 needed for aerobic metabolic processes [7,74] (Figure 3). Therefore, we suggest that under low O_2 stress, the overexpression of the Cyt_c gene in an oxidized cell environment could increase the synthesis of Cyt_c proteins, generating a higher concentration of proteins in the cytoplasm, which would support the induction of apoptosis and end with aerenchyma formation (Figure 3). Certainly, in the future, more comprehensive studies with woody plants are required to help understand the steps between Cyt_c gene expression and the modulation of protein synthesis, especially under conditions of O_2 deficiency.

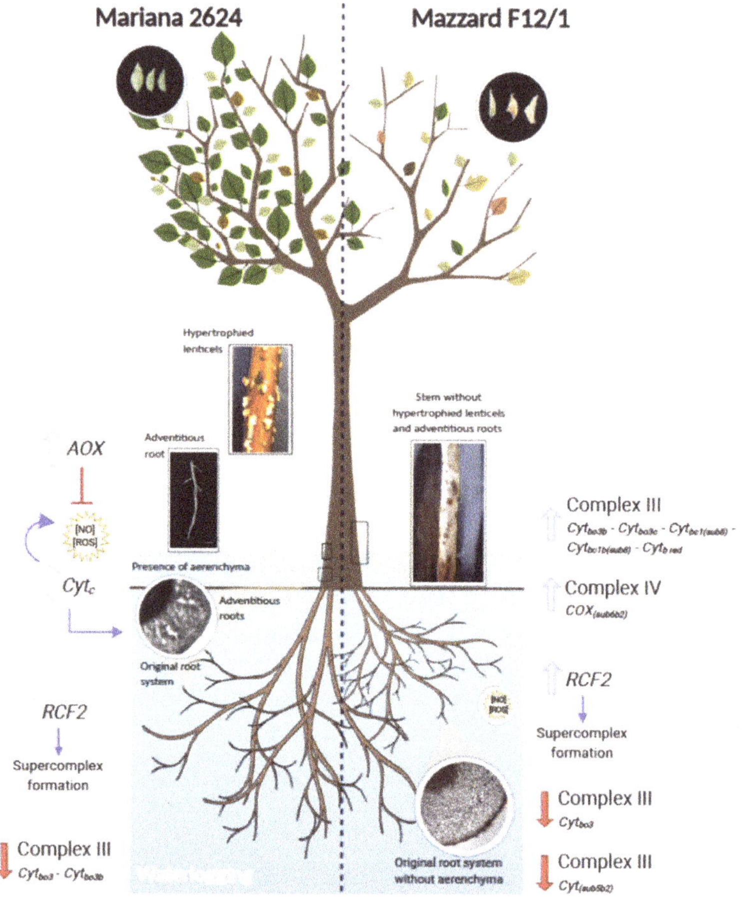

Figure 3. Schematic overview of principal mETC-expressing genes and morpho-anatomical changes in *Prunus* rootstocks under root hypoxia. The figure summarizes the effect of *AOX* and *Cyt*$_c$ on NO and ROS generation. Regarding ROS, the early induction of *AOX* (6 h) and the late induction of *Cyt*$_c$ (24 and 72 h) would lead to a regulation of ROS accumulation in a temporally non-exclusive way in the root-hypoxia-tolerant *Prunus* rootstock ('Mariana M2624'). Thus, at the beginning of stress by O_2 deficiency, this genotype would be preventing the accumulation of ROS in roots, but as stress progresses, in the context of apoptosis, the elevation of ROS levels has been associated with the activity of Cyt$_c$ [171,172]. The generation of aerenchyma mediated by apoptosis present in the original root system, and the generation of adventitious roots with aerenchyma and hypertrophied lenticels, are characteristic features of 'Mariana M2624'. These traits are completely absent in the root-hypoxia-sensitive *Prunus* rootstock ('Mazzard F12/1'). The generation of air-filled spaces results in an avoidance strategy in relation to O_2 deprivation, which is associated with the maintenance of an adequate metabolism and energy supply at the whole plant level. The blue and red arrows represent the over- and down-expression of genes related to mETC. *AOX*, alternative oxidase; *Cyt*$_c$, Cytochrome c; *RCF2*, Respiratory Supercomplex Factor 2; *Cyt*$_{bo3}$, Cytochrome bo3 ubiquinol oxidase; *Cyt*$_{bc1(sub8)}$, Cytochrome bc1 complex subunit 8; *Cyt*$_{b\,red}$, Cytochrome b reductase; *COX*$_{(sub5b2)}$, Cytochrome c oxidase subunit 5b2; *COX*$_{(sub6b2)}$, Cytochrome c oxidase subunit 5b2; [NO], nitric oxide; [ROS], reactive oxygen species. Images of morpho-anatomical changes were obtained from [7,74]. This figure was created using BioRender.

6. Conclusions and Perspectives in Fruit Trees Research

The effects of oxygen deficiency and the adaptive responses of plants have been extensively studied in herbaceous species, mainly with annual life cycles. However, woody fruit tree species, of great economic importance in temperate and sub-tropical zones, have been poorly attended. From a physiological point of view, characterizing the adjustment of photosynthesis to hypoxic conditions is a useful approach to contribute to the definition of the tolerance of these species to such environmental stress. Thus, species more tolerant to root hypoxia, capable of maintaining higher levels of photosynthetic activity, may provide greater carbohydrate reserves for facing the energy crisis triggered by anaerobiosis. On the other hand, transcriptomic studies have become useful for getting a broader view of the metabolic adaptation of fruit trees to hypoxia. Furthermore, together with evaluating the classically described metabolic pathways for plants under hypoxia, transcriptomic analyses allow the investigation of routes or processes less explored in these perennial plant species, such as the genetic determinants of energy sensing or the genes involved in the mETC.

In this review, we focused on analyzing the physiological and molecular aspects of the responses of fruit trees under root hypoxia, with emphasis on a study model that involves two genotypes of *Prunus* rootstocks with contrasting tolerances to O_2 deficiency. In light of the multiple and diverse antecedents evaluated, it seems clear that the tolerant species of fruit trees manage to adapt and survive waterlogging or flooding due to their ability to detect oxygen deficiency more quickly and change their metabolism through a suitable transcriptomic reprogramming.

In this sense, keeping calm, and avoiding the activation of more routes than those that are strictly necessary or less energy efficient in an anaerobic environment, would allow the plant to invest its energy budget precisely, without exhausting it. This would be exemplified by the measured amount of DEGs detected in the root-hypoxia-tolerant *Prunus* rootstock versus the massive transcriptomic reconfiguration evidenced in the sensitive genotype 'Mazzard F12/1', implying a higher energy expenditure for the latter. At the same time, the anatomical and biochemical factors that operate in favor of maintenance processes, which are less energy demanding than those of repair, contribute to the early adaptation to root hypoxia of the tolerant genotype. Subsequently, late morpho-anatomical modifications end up defining a hypoxia-avoidance strategy that allows long-term survival in tolerant genotypes.

Finally, further studies are needed into the effects that re-oxygenation exerts on the fruit trees during the hypoxia recovery phase. In addition, a rising challenge in the study of the adaptive response to root hypoxia of woody fruit trees would involve epigenetic approaches oriented towards analyzing the phenomena of the memory of stress that can be transmitted between rootstocks and scion, or between growing seasons, and how it affects the yielding behavior of the orchards.

Author Contributions: Conceptualization, A.S. and P.P.; transcriptomic data re-analysis, P.M.; writing—original draft preparation, A.S., G.T., I.O., M.O. and P.P.; writing—review and editing, A.S., G.T. and P.P.; visualization, A.S., G.T. and P.P.; supervision, A.S. and P.P.; funding acquisition, A.S., G.T., M.O. and P.P. All authors have read and agreed to the published version of the manuscript.

Funding: This research was funded by AGENCIA NACIONAL DE INVESTIGACION Y DESARROLLO (ANID), grant number R19A10003 and FONDO NACIONAL DE DESARROLLO CIENTÍFICO Y TECNOLÓGICO (FONDECYT), grant number 1190816.

Acknowledgments: We thanks to all CEAF colleagues who have contributed from different areas with their research on root hypoxia and whose work has been part of this review.

Conflicts of Interest: The authors declare no conflict of interest.

References

1. Loreti, E.; Perata, P. The many facets of hypoxia in plants. *Plants* **2020**, *9*, 745. [CrossRef]

2. FAOSTAT. Crops Production in 2018. 2018. Available online: http://faostat.fao.org/ (accessed on 5 June 2019).

3. Kreuzwieser, J.; Rennenberg, H. Molecular and physiological responses of trees to waterlogging stress. *Plant Cell Environ.* **2014**, *37*, 2245–2259. [CrossRef] [PubMed]

4. Herrera, A.; Tezara, W.; Marín, O.; Rengifo, E. Stomatal and non-stomatal limitations of photosynthesis in trees of a tropical seasonally flooded forest. *Physiol. Plant.* **2008**, *134*, 41–48. [CrossRef] [PubMed]

5. Calvo-Polanco, M.; Senorans, J.; Zwiazek, J. Role of adventitious roots in water relations of tamarack (*Larix laricina*) seedlings exposed to flooding. *BMC Plant Biol.* **2012**, *12*, 99. [CrossRef] [PubMed]

6. Herrera, A. Responses to flooding of plant water relations and leaf gas exchange in tropical tolerant trees of a black-water wetland. *Front. Plant Sci.* **2013**, *4*, 1–12. [CrossRef]

7. Pimentel, P.; Almada, R.; Salvatierra, A.; Toro, G.; Arismendi, M.J.; Pino, M.T.; Sagredo, B.; Pinto, M. Physiological and morphological responses of *Prunus* species with different degree of tolerance to long-term root hypoxia. *Sci. Hortic.* **2014**, *180*, 14–23. [CrossRef]

8. Warschefsky, E.J.; Klein, L.L.; Frank, M.H.; Chitwood, D.H.; Londo, J.P.; Von Wettberg, E.J.B.; Miller, A.J. Rootstocks: Diversity, domestication, and impacts on shoot phenotypes. *Trends Plant Sci.* **2016**, *21*, 418–437. [CrossRef]

9. Martínez-Ballesta, M.C.; Alcaraz-López, C.; Muries, B.; Mota-Cadenas, C.; Carvajal, M. Physiological aspects of rootstock–scion interactions. *Sci. Hort.* **2010**, *127*, 112–118. [CrossRef]

10. Mizutani, F.; Yamada, M.; Tomana, T. Differential water tolerance and ethanol accumulation in Prunus species under flooded conditions. *J. Jpn. Soc. Hortic. Sci.* **1982**, *51*, 29–34. [CrossRef]

11. Ranney, T.G. Differential tolerance of eleven *Prunus* taxa to root zone flooding. *J. Environ. Hortic.* **1994**, *12*, 138–141. [CrossRef]

12. Pinochet, J. 'Replantpac' (Rootpac®R), a plum–almond hybrid rootstock for replant situations. *HortScience* **2010**, *45*, 299–301. [CrossRef]

13. Rubio Cabetas, M.J.; Pons, C.; Amador Delgado, M.L.; Marti, C.; Granell, A. Transcriptomic analysis of two *Prunus* genotypes differing in waterlogging response reveals the importance of ANP and hypoxia-associated oxidative response. In Proceedings of the Thirteenth Eucarpia Symposium on Fruit Breeding and Genetics, Warsaw, Poland, 11 September 2011.

14. Iacona, C.; Cirilli, M.; Zega, A.; Frioni, E.; Silvestri, C.; Muleo, R. A somaclonal myrobalan rootstock increases waterlogging tolerance to peach cultivar in controlled conditions. *Sci. Hort.* **2013**, *156*, 1–8. [CrossRef]

15. Aguilera, E.; Díaz-Gaona, C.; García-Laureano, R.; Reyes-Palomo, C.; Guzmán, G.I.; Ortolani, L.; Sánchez-Rodríguez, M.; Rodríguez-Estévez, V. Agroecology for adaptation to climate change and resource depletion in the Mediterranean region. A review. *Agric. Syst.* **2020**, *181*, 102809. [CrossRef]

16. Venkatramanan, V.; Shah, S.; Prasad, R. *Global Climate Change and Environmental Policy*; Springer: Singapore, 2020.

17. Hillel, D. Introduction to Environmental Soil Physics. *Eur. J. Soil Sci.* **2004**, *56*, 684. [CrossRef]

18. Bhattarai, S.; Su, N.; Midmore, D. Oxygation unlocks yield potentials of crops in oxygen-limited soil environments. *Adv. Agron.* **2005**, *88*, 313–377. [CrossRef]

19. Batey, T. Soil compaction and soil management—A Review. *Soil Use Manag.* **2009**, *25*, 335–345. [CrossRef]

20. Morales-Olmedo, M.; Ortiz, M.; Sellés, G. Effects of transient soil waterlogging and its importance for rootstock selection. *Chil. J. Agric. Res.* **2015**, *75*, 45–56. [CrossRef]

21. Ellies, A.; Horn, R.; Smith, R. Effect of management of a volcanic ash soil on structural properties. *Int. Agrophys* **2000**, *14*, 377–384.

22. Seguel, O.; Farías, E.; Luzio, W.; Casanova, M.; Pino, I.; Parada, M.; Videla, X.; Nario, A. Changes in soil physical properties on hillsides vineyard (*Vitis vinifera*). In Proceedings of the ISTRO 18th Triennial conference, Izmir, Turkey, 15–19 June 2009; pp. 15–19.

23. Becerra, A.T.; Botta, G.F.; Bravo, X.L.; Tourn, M.; Melcon, F.B.; Vazquez, J.; Rivero, D.; Linares, P.; Nardon, G. Soil compaction distribution under tractor traffic in almond (Prunus amigdalus L.) orchard in Almería España. *Soil Tillage Res.* **2010**, *107*, 49–56. [CrossRef]

24. Cook, F.J.; Knight, J.H. Oxygen transport to plant roots. *Soil Sci. Soc. Am. J.* **2003**, *67*, 20–31. [CrossRef]

25. Dexter, A.R. Advances in characterization of soil structure. *Soil Tillage Res.* **1988**, *11*, 199–238. [CrossRef]

26. Kawase, M. Anatomical and morphological adaptation of plants to waterlogging. *Hort. Sci.* **1981**, *16*, 8–12.

27. Shabala, S. Physiological and cellular aspects of phytotoxicity tolerance in plants: The role of membrane transporters and implications for crop breeding for waterlogging tolerance. *New Phytol.* **2011**, *190*, 289–298. [CrossRef]

28. Unger, I.M.; Kennedy, A.C.; Muzika, R.-M. Flooding effects on soil microbial communities. *Appl. Soil Ecol.* **2009**, *42*, 1–8. [CrossRef]

29. Vepraskas, M.J.; Faulkner, S.; Richardson, J. Redox chemistry of hydric soils. In *Wetland Soils: Genesis, Hydrology, Landscapes, and Classification*; CRC Press: Boca Raton, FL, USA, 2001; pp. 85–106.

30. Ponnamperuma, F. Flooding and plant growth. Effects of flooding on soils. In *Flooding and Plant Growth*; NRC Research Press: Ottawa, ON, Canada, 1984; pp. 9–45.

31. McKee, W.H., Jr.; McKevlin, M.R. Geochemical processes and nutrient uptake by plants in hydric soils. *Environ. Toxicol. Chem.* **1993**, *12*, 2197–2207. [CrossRef]

32. Drew, M. Oxygen deficiency and root metabolism: Injury and acclimation under hypoxia and anoxia. *Annu. Rev. Plant Physiol. Plant Mol. Biol.* **1997**, *48*, 223–250. [CrossRef]

33. Sanclemente, M.A.; Schaffer, B.; Gil, P.M.; Vargas, A.I.; Davies, F.S. Pruning after flooding hastens recovery of flood-stressed avocado (*Persea americana* Mill.) trees. *Sci. Hort.* **2014**, *169*, 27–35. [CrossRef]

34. Savé, R.; Serrano, L. Some physiological and growth responses of kiwi fruit (*Actinidia chinensis*) to flooding. *Physiol. Plant.* **1986**, *66*, 75–78. [CrossRef]

35. Vu, J.; Yelenosky, G. Photosynthetic responses of *citrus* trees to soil flooding. *Physiol. Plant.* **1991**, *81*, 7–14. [CrossRef]

36. Hossain, Z.; López-Climent, M.F.; Arbona, V.; Pérez-Clemente, R.M.; Gómez-Cadenas, A. Modulation of the antioxidant system in citrus under waterlogging and subsequent drainage. *J. Plant Physiol.* **2009**, *166*, 1391–1404. [CrossRef]

37. Arbona, V.; López-Climent, M.F.; Pérez-Clemente, R.M.; Gómez-Cadenas, A. Maintenance of a high photosynthetic performance is linked to flooding tolerance in citrus. *Environ. Exp. Bot.* **2009**, *66*, 135–142. [CrossRef]

38. García-Sánchez, F.; Syvertsen, J.P.; Gimeno, V.; Botía, P.; Perez-Perez, J.G. Responses to flooding and drought stress by two *citrus* rootstock seedlings with different water-use efficiency. *Physiol. Plant.* **2007**, *130*, 532–542. [CrossRef]

39. Kallestad, J.C.; Sammis, T.W.; Mexala, J.G.; Gutschick, V. The impact of prolonged flood-irrigation on leaf gas exchange in mature pecans in an orchard setting. *Int. J. Plant Prod.* **2012**, *1*, 163–178. [CrossRef]

40. Belloni, V.; Mapelli, S. Effects of drought or flooding stresses on photosynthesis xylem flux and stem radial growth. *Acta Hortic.* **2001**, *544*, 327–333. [CrossRef]

41. Ruperti, B.; Botton, A.; Populin, F.; Eccher, G.; Brilli, M.; Quaggiotti, S.; Trevisan, S.; Cainelli, N.; Guarracino, P.; Schievano, E.; et al. Flooding responses on grapevine: A physiological, transcriptional, and metabolic perspective. *Front. Plant Sci.* **2019**, *10*, 339. [CrossRef]

42. Striegler, R.K.; Howell, G.S.; Flore, J.A. Influence of rootstock on the response of Seyval grapevines to flooding stress. *Am. Soc. Enol. Vitic.* **1993**, *44*, 313–319.

43. Olmo-Vega, A.; García-Sánchez, F.; Simón-Grao, S.; Simón, I.; Lidón, V.; Nieves, M.; Martínez-Nicolás, J.J. Physiological responses of three pomegranate cultivars under flooded conditions. *Sci. Hort.* **2017**, *224*, 171–179. [CrossRef]

44. Bhusal, N.; Kim, H.S.; Han, S.-G.; Yoon, T.-M. Photosynthetic traits and plant–water relations of two apple cultivars grown as bi-leader trees under long-term waterlogging conditions. *Environ. Exp. Bot.* **2020**, *176*, 104111. [CrossRef]

45. Domingo, R.; Pérez-Pastor, A.; Ruiz-Sánchez, M.C. Physiological responses of apricot plants grafted on two different rootstocks to flooding conditions. *J. Plant Physiol.* **2002**, *159*, 725–732. [CrossRef]

46. Xiloyannis, C.; Celano, G.; Vicinanza, L.; Esmenjaud, D.; Gómez-Aparisi, J.; Salesses, G.; Dichio, B. Performance of new selections of *Prunus* rootstocks, resistant to root knot nematodes, in waterlogging conditions. In Proceedings of the I International Symposium on Rootstocks for Deciduous Fruit Tree Species 658, Zaragoza, Spain, 11–14 June 2002; pp. 403–405.

47. Salvatierra, A.; Pimentel, P.; Almada, R.; Hinrichsen, P. Exogenous GABA application transiently improves the tolerance to root hypoxia on a sensitive genotype of *Prunus* rootstock. *Environ. Exp. Bot.* **2016**, *125*, 52–66. [CrossRef]

48. Iacona, C.; Pistelli, L.; Cirilli, M.; Gatti, L.; Mancinelli, R.; Ripa, M.N.; Muleo, R. Day-length is involved in flooding tolerance response in wild type and variant genotypes of rootstock *Prunus cerasifera* L. *Front. Plant Sci.* **2019**, *10*, 546. [CrossRef] [PubMed]

49. Klumb, E.; Rickes, L.; Braga, E.; Bianchi, V. Evaluation of gas exchanges in different *Prunus* spp. rootstocks under drought and flooding stress. *Rev. Bras. Frutic.* **2017**, *39*, 1–8. [CrossRef]

50. Schaffer, B.; Andersen, P.C.; Ploetz, R.C. Responses of fruit crops to flooding. *Hortic. Rev.* **1992**, *13*, 257–313.

51. Parent, C.; Capelli, N.; Berger, A.; Crèvecoeur, M.; Dat, J.F. An overview of plant responses to soil waterlogging. *Plant Stress* **2008**, *2*, 20–27.

52. Pistelli, L.; Iacona, C.; Miano, D.; Cirilli, M.; Colao, M.C.; Mensuali-Sodi, A.; Muleo, R. Novel *Prunus* rootstock somaclonal variants with divergent ability to tolerate waterlogging. *Tree Physiol.* **2012**, *32*, 355–368. [CrossRef]

53. Zhang, P.; Lyu, D.; Jia, L.; He, J.; Qin, S. Physiological and de novo transcriptome analysis of the fermentation mechanism of *Cerasus sachalinensis* roots in response to short-term waterlogging. *BMC Genom.* **2017**, *18*, 649. [CrossRef]

54. Loreti, E.; Valeri, M.C.; Novi, G.; Perata, P. Gene regulation and survival under hypoxia requires starch availability and metabolism. *Plant Physiol.* **2018**, *176*, 1286–1298. [CrossRef]

55. Tan, X.; Xu, H.; Khan, S.; Equiza, M.A.; Lee, S.H.; Vaziriyeganeh, M.; Zwiazek, J.J. Plant water transport and aquaporins in oxygen-deprived environments. *J. Plant Physiol.* **2018**, *227*, 20–30. [CrossRef]

56. Aroca, R.; Porcel, R.; Ruiz-Lozano, J.M. Regulation of root water uptake under abiotic stress conditions. *J. Exp. Bot.* **2011**, *63*, 43–57. [CrossRef]

57. Chaumont, F.; Tyerman, S.D. Aquaporins: Highly regulated channels controlling plant water relations. *Plant Physiol.* **2014**, *164*, 1600. [CrossRef]

58. Pawłowicz, I.; Masajada, K. Aquaporins as a link between water relations and photosynthetic pathway in abiotic stress tolerance in plants. *Gene* **2019**, *687*, 166–172. [CrossRef] [PubMed]

59. Tournaire-Roux, C.; Sutka, M.; Javot, H.; Gout, E.; Gerbeau, P.; Luu, D.-T.; Bligny, R.; Maurel, C. Cytosolic pH regulates root water transport during anoxic stress through gating of aquaporins. *Nature* **2003**, *425*, 393–397. [CrossRef] [PubMed]

60. Tylova, E.; Peckova, E.; Blascheová, Z.; Soukup, A. Casparian bands and suberin lamellae in exodermis of lateral roots: An important trait of roots system response to abiotic stress factors. *Ann. Bot.* **2017**, *120*. [CrossRef] [PubMed]

61. Sasidharan, R.; Hartman, S.; Liu, Z.; Martopawiro, S.; Sajeev, N.; Van Veen, H.; Yeung, E.; Voesenek, L.A.C.J. Signal dynamics and interactions during flooding stress. *Plant Physiol.* **2018**, *176*, 1106–1117. [CrossRef]

62. Møller, M. Plant mitochondria and oxidative stress: Electron transport, NADPH turnover, and metabolism of reactive oxygen species. *Annu. Rev. Plant Physiol. Plant Mol. Biol.* **2001**, *52*, 561–591. [CrossRef]

63. Asada, K. Production and scavenging of reactive oxygen species in chloroplasts and their functions. *Plant Physiol.* **2006**, *141*, 391–396. [CrossRef]

64. Mignolet-Spruyt, L.; Xu, E.; Idänheimo, N.; Hoeberichts, F.A.; Mühlenbock, P.; Brosché, M.; Van Breusegem, F.; Kangasjärvi, J. Spreading the news: Subcellular and organellar reactive oxygen species production and signalling. *J. Exp. Bot.* **2016**, *67*, 3831–3844. [CrossRef]

65. Chapman, J.M.; Muhlemann, J.K.; Gayomba, S.R.; Muday, G.K. RBOH-dependent ROS synthesis and ROS scavenging by plant specialized metabolites to modulate plant development and stress responses. *Chem. Res. Toxicol.* **2019**, *32*, 370–396. [CrossRef]

66. Demidchik, V.; Shabala, S. Mechanisms of cytosolic calcium elevation in plants: The role of ion channels, calcium extrusion systems and NADPH oxidase-mediated 'ROS-Ca^{2+} Hub'. *Funct. Plant Biol.* **2017**, *45*, 9–27. [CrossRef]

67. Igamberdiev, A.U.; Hill, R.D. Elevation of cytosolic Ca^{2+} in response to energy deficiency in plants: The general mechanism of adaptation to low oxygen stress. *Biochem. J.* **2018**, *475*, 1411–1425. [CrossRef]

68. Bailey-Serres, J.; Voesenek, L.A.C.J. Flooding stress: Acclimations and genetic diversity. *Annu. Rev. Plant. Biol.* **2008**, *59*, 313–339. [CrossRef] [PubMed]

69. Fukao, T.; Barrera-Figueroa, B.E.; Juntawong, P.; Peña-Castro, J.M. Submergence and waterlogging stress in plants: A review highlighting research opportunities and understudied aspects. *Front. Plant Sci.* **2019**, *10*. [CrossRef] [PubMed]

70. Arismendi, M.J.; Almada, R.; Pimentel, P.; Bastias, A.; Salvatierra, A.; Rojas, P.; Hinrichsen, P.; Pinto, M.; Di Genova, A.; Travisany, D.; et al. Transcriptome sequencing of *Prunus* sp. rootstocks roots to identify candidate genes involved in the response to root hypoxia. *Tree Genet. Genomes* **2015**, *11*, 1–16. [CrossRef]

71. Mittler, R. ROS are good. *Trends Plant Sci.* **2017**, *22*, 11–19. [CrossRef] [PubMed]

72. Arbona, V.; Hossain, Z.; López-Climent, M.F.; Pérez-Clemente, R.M.; Gómez-Cadenas, A. Antioxidant enzymatic activity is linked to waterlogging stress tolerance in citrus. *Physiol. Plant.* **2008**, *132*, 452–466. [CrossRef]

73. Bai, T.; Li, C.; Ma, F.; Feng, F.; Shu, H. Responses of growth and antioxidant system to root-zone hypoxia stress in two *Malus* species. *Plant Soil* **2010**, *327*, 95–105. [CrossRef]

74. Toro, G.; Pinto, M.; Pimentel, P. Root respiratory components of *Prunus* spp. rootstocks under low oxygen: Regulation of growth, maintenance, and ion uptake respiration. *Sci. Hortic.* **2018**, *239*, 259–268. [CrossRef]

75. Mittler, R. Oxidative stress, antioxidants and stress tolerance. *Trends Plant Sci.* **2002**, *7*, 405–410. [CrossRef]

76. Mittler, R.; Vanderauwera, S.; Gollery, M.; Van Breusegem, F. Reactive oxygen gene network of plants. *Trends Plant Sci.* **2004**, *9*. [CrossRef]

77. Radmann, E.; Klumb, E.; Deuner, S.; Bianchi, V. Antioxidant capacity in leaf and root tissues of *Prunus* spp. under flooding. *J. Exp. Agric. Int.* **2018**, *26*, 1–10. [CrossRef]

78. Sauter, M. Root responses to flooding. *Curr. Opin. Plant Biol.* **2013**, *16*, 282–286. [CrossRef]

79. Kozlowski, T.T. Responses of woody plants to flooding and salinity. *Tree Physiol.* **1997**, *17*, 490. [CrossRef]

80. Yamauchi, T.; Shimamura, S.; Nakazono, M.; Mochizuki, T. Aerenchyma formation in crop species: A review. *Field Crops Res.* **2013**, *152*, 8–16. [CrossRef]

81. Le Provost, G.; Sulmon, C.; Frigerio, J.-M.; Bodénès, C.; Kremer, A.; Plomion, C. Role of waterlogging-responsive genes in shaping interspecific differentiation between two sympatric oak species. *Tree Physiol.* **2011**, *32*, 119–134. [CrossRef] [PubMed]

82. Shimamura, S.; Yamamoto, R.; Nakamura, T.; Shimada, S.; Komatsu, S. Stem hypertrophic lenticels and secondary aerenchyma enable oxygen transport to roots of soybean in flooded soil. *Ann. Bot.* **2010**, *106*, 277–284. [CrossRef] [PubMed]

83. Armstrong, W. Aeration in higher plants. In *Advances in Botanical Research*; Woolhouse, H.W., Ed.; Academic Press: London, UK, 1979; Volume 7, pp. 225–332.

84. Colmer, T.D. Long-distance transport of gases in plants: A perspective on internal aeration and radial oxygen loss from roots. *Plant Cell Environ.* **2003**, *26*, 17–36. [CrossRef]

85. Armstrong, J.; Armstrong, W. Rice: Sulfide-induced barriers to root radial oxygen loss, Fe^{2+} and water uptake, and lateral root emergence. *Ann. Bot.* **2005**, *96*, 625–638. [CrossRef] [PubMed]

86. Colmer, T.D.; Voesenek, L.A.C.J. Flooding tolerance: Suites of plant traits in variable environments. *Funct. Plant Biol.* **2009**, *36*, 665–681. [CrossRef]

87. Yamauchi, T.; Colmer, T.D.; Pedersen, O.; Nakazono, M. Regulation of root traits for internal aeration and tolerance to soil waterlogging-flooding stress. *Plant Physiol.* **2018**, *176*, 1118–1130. [CrossRef]

88. Marchioretto, L.D.R.; Rossi, A.D.; Amaral, L.O.d.; Ribeiro, A.M.A.d.S. Tolerance of apple rootstocks to short-term waterlogging. *Ciência Rural* **2018**, *48*. [CrossRef]

89. Andersen, P.C.; Lombard, P.B.; Westwood, M.N. Effect of root anaerobiosis on the water relations of several *Pyrus* species. *Physiol. Plant.* **1984**, *62*, 245–252. [CrossRef]

90. Vartapetian, B.B.; Andreeva, I.N.; Generozova, I.P.; Polyakova, L.I.; Maslova, I.P.; Dolgikh, Y.I.; Stepanova, A.Y. Functional electron microscopy in studies of plant response and adaptation to anaerobic stress. *Ann. Bot.* **2003**, *91*, 155–172. [CrossRef] [PubMed]

91. Kreuzwieser, J.; Hauberg, J.; Howell, K.A.; Carroll, A.; Rennenberg, H.; Millar, A.H.; Whelan, J. Differential response of gray poplar leaves and roots underpins stress adaptation during hypoxia. *Plant Physiol.* **2009**, *149*, 461–473. [CrossRef] [PubMed]

92. Qi, B.; Yang, Y.; Yin, Y.; Xu, M.; Li, H. De novo sequencing, assembly, and analysis of the Taxodium 'Zhongshansa' roots and shoots transcriptome in response to short-term waterlogging. *BMC Plant Biol.* **2014**, *14*, 201. [CrossRef] [PubMed]

93. Le Provost, G.; Lesur, I.; Lalanne, C.; Da Silva, C.; Labadie, K.; Aury, J.M.; Leple, J.C.; Plomion, C. Implication of the suberin pathway in adaptation to waterlogging and hypertrophied lenticels formation in pedunculate oak (*Quercus robur* L.). *Tree Physiol.* **2016**, *36*, 1330–1342. [CrossRef] [PubMed]

94. Reeksting, B.J.; Coetzer, N.; Mahomed, W.; Engelbrecht, J.; Van den Berg, N. De novo sequencing, assembly, and analysis of the root transcriptome of *Persea americana* (Mill.) in response to *Phytophthora cinnamomi* and flooding. *PLoS ONE* **2014**, *9*, e86399. [CrossRef]

95. Rubio-Cabetas, M.J.; Pons, C.; Bielsa, B.; Amador, M.L.; Marti, C.; Granell, A. Preformed and induced mechanisms underlies the differential responses of *Prunus* rootstock to hypoxia. *J. Plant Physiol.* **2018**, *228*, 134–149. [CrossRef]

96. Zhang, J.-Y.; Huang, S.-N.; Mo, Z.-H.; Xuan, J.-P.; Jia, X.-D.; Wang, G.; Guo, Z.-R. De novo transcriptome sequencing and comparative analysis of differentially expressed genes in kiwifruit under waterlogging stress. *Mol. Breed.* **2015**, *35*, 208. [CrossRef]

97. Zhu, X.; Li, X.; Jiu, S.; Zhang, K.; Wang, C.; Fang, J. Analysis of the regulation networks in grapevine reveals response to waterlogging stress and candidate gene-marker selection for damage severity. *R. Soc. Open Sci.* **2018**, *5*, 172253. [CrossRef]

98. Ferner, E.; Rennenberg, H.; Kreuzwieser, J. Effect of flooding on C metabolism of flood-tolerant (*Quercus robur*) and non-tolerant (*Fagus sylvatica*) tree species. *Tree Physiol.* **2012**, *32*, 135–145. [CrossRef] [PubMed]

99. Martínez-Alcántara, B.; Jover, S.; Quiñones, A.; Forner-Giner, M.Á.; Rodríguez-Gamir, J.; Legaz, F.; Primo-Millo, E.; Iglesias, D.J. Flooding affects uptake and distribution of carbon and nitrogen in citrus seedlings. *J. Plant Physiol.* **2012**, *169*, 1150–1157. [CrossRef]

100. Fukao, T.; Bailey-Serres, J. Plant responses to hypoxia—Is survival a balancing act? *Trends Plant Sci.* **2004**, *9*, 449–456. [CrossRef] [PubMed]

101. Granot, D.; Kelly, G.; Stein, O.; David-Schwartz, R. Substantial roles of hexokinase and fructokinase in the effects of sugars on plant physiology and development. *J. Exp. Bot.* **2013**, *65*, 809–819. [CrossRef] [PubMed]

102. Chen, S.; Huang, T.; Zhou, Y.; Han, Y.; Xu, M.; Gu, J. AfterQC: Automatic filtering, trimming, error removing and quality control for fastq data. *BMC Bioinform.* **2017**, *18*, 80. [CrossRef] [PubMed]

103. Haas, B.J.; Papanicolaou, A.; Yassour, M.; Grabherr, M.; Blood, P.D.; Bowden, J.; Couger, M.B.; Eccles, D.; Li, B.; Lieber, M.; et al. De novo transcript sequence reconstruction from RNA-seq using the Trinity platform for reference generation and analysis. *Nat. Protocols* **2013**, *8*, 1494–1512. [CrossRef]

104. El-Gebali, S.; Mistry, J.; Bateman, A.; Eddy, S.R.; Luciani, A.; Potter, S.C.; Qureshi, M.; Richardson, L.J.; Salazar, G.A.; Smart, A.; et al. The Pfam protein families database in 2019. *Nucleic Acids Res.* **2018**, *47*, D427–D432. [CrossRef]

105. Fu, L.; Niu, B.; Zhu, Z.; Wu, S.; Li, W. CD-HIT: Accelerated for clustering the next-generation sequencing data. *Bioinformatics* **2012**, *28*, 3150–3152. [CrossRef]

106. Li, W.; Godzik, A. Cd-Hit: A Fast Program for Clustering and Comparing Large Sets of Protein or Nucleotide Sequences. *Bioinformatics* **2006**, *22*, 1658–1659. [CrossRef]

107. Kim, D.; Paggi, J.M.; Park, C.; Bennett, C.; Salzberg, S.L. Graph-based genome alignment and genotyping with HISAT2 and HISAT-genotype. *Nat. Biotechnol.* **2019**, *37*, 907–915. [CrossRef]

108. Kovaka, S.; Zimin, A.V.; Pertea, G.M.; Razaghi, R.; Salzberg, S.L.; Pertea, M. Transcriptome assembly from long-read RNA-seq alignments with StringTie2. *Genome Biol.* **2019**, *20*, 278. [CrossRef]

109. Robinson, M.D.; McCarthy, D.J.; Smyth, G.K. edgeR: A Bioconductor package for differential expression analysis of digital gene expression data. *Bioinformatics* **2009**, *26*, 139–140. [CrossRef] [PubMed]

110. Araujo, F.A.; Barh, D.; Silva, A.; Guimarães, L.; Ramos, R.T.J. GO FEAT: A rapid web-based functional annotation tool for genomic and transcriptomic data. *Sci. Rep.* **2018**, *8*, 1794. [CrossRef] [PubMed]

111. Kanehisa, M.; Goto, S. KEGG: Kyoto Encyclopedia of Genes and Genomes. *Nucleic Acids Res.* **2000**, *28*, 27–30. [CrossRef] [PubMed]

112. Mateluna, P.; Salvatierra, A.; Solis, S.; Nuñez, G.; Pimentel, P. Involvement of aquaporin NIP1;1 in the contrasting tolerance response to root hypoxia in *Prunus* rootstocks. *J. Plant Physiol.* **2018**, *228*, 19–28. [CrossRef] [PubMed]

113. Toroser, D.; Plaut, Z.; Huber, S.C. Regulation of a plant SNF1-related protein kinase by glucose-6-Phosphate. *Plant Physiol.* **2000**, *123*, 403–412. [CrossRef]

114. Rocha, M.; Licausi, F.; Araújo, W.L.; Nunes-Nesi, A.; Sodek, L.; Fernie, A.R.; Van Dongen, J.T. Glycolysis and the tricarboxylic acid cycle are linked by alanine aminotransferase during hypoxia induced by waterlogging of *Lotus japonicus*. *Plant Physiol.* **2010**, *152*, 1501–1513. [CrossRef]

115. Zhou, C.; Bai, T.; Wang, Y.; Wu, T.; Zhang, X.; Xu, X.; Han, Z. Morpholoical and enzymatic responses to waterlogging in three *Prunus* species. *Sci. Hort.* **2017**, *221*, 62–67. [CrossRef]

116. Xia, J.-H.; Saglio, P.H. Lactic acid efflux as a mechanism of hypoxic acclimation of maize root tips to anoxia. *Plant Physiol.* **1992**, *100*, 40–46. [CrossRef]

117. Rivoal, J.; Hanson, A.D. Evidence for a large and sustained glycolytic flux to lactate in anoxic roots of some members of the halophytic genus *Limonium*. *Plant Physiol.* **1993**, *101*, 553–560. [CrossRef]

118. O'Carra, P.; Mulcahy, P. Plant lactate dehydrogenase: NADH kinetics and inhibition by ATP. *Phytochemistry* **1997**, *45*, 897–902. [CrossRef]

119. Patel, M.S.; Roche, T.E. Molecular biology and biochemistry of pyruvate dehydrogenase complexes1. *FASEB J.* **1990**, *4*, 3224–3233. [CrossRef] [PubMed]

120. De Sousa, C.A.F.; Sodek, L. Alanine metabolism and alanine aminotransferase activity in soybean (*Glycine max*) during hypoxia of the root system and subsequent return to normoxia. *Environ. Exp. Bot.* **2003**, *50*, 1–8. [CrossRef]

121. Limami, A.M.; Glévarec, G.; Ricoult, C.; Cliquet, J.-B.; Planchet, E. Concerted modulation of alanine and glutamate metabolism in young *Medicago truncatula* seedlings under hypoxic stress. *J. Exp. Bot.* **2008**, *59*, 2325–2335. [CrossRef] [PubMed]

122. Clark, S.M.; Di Leo, R.; Dhanoa, P.K.; Van Cauwenberghe, O.R.; Mullen, R.T.; Shelp, B.J. Biochemical characterization, mitochondrial localization, expression, and potential functions for an Arabidopsis γ-aminobutyrate transaminase that utilizes both pyruvate and glyoxylate. *J. Exp. Bot.* **2009**, *60*, 1743–1757. [CrossRef]

123. Clark, S.M.; Di Leo, R.; Van Cauwenberghe, O.R.; Mullen, R.T.; Shelp, B.J. Subcellular localization and expression of multiple tomato γ-aminobutyrate transaminases that utilize both pyruvate and glyoxylate. *J. Exp. Bot.* **2009**, *60*, 3255–3267. [CrossRef]

124. Michaeli, S.; Fromm, H. Closing the loop on the GABA shunt in plants: Are GABA metabolism and signaling entwined? *Front. Plant Sci.* **2015**, *6*. [CrossRef]

125. Lea, P.J.; Miflin, B.J. Glutamate synthase and the synthesis of glutamate in plants. *Plant Physiol. Biochem.* **2003**, *41*, 555–564. [CrossRef]

126. Horchani, F.; Aschi-Smiti, S. Prolonged root hypoxia effects on enzymes involved in nitrogen assimilation pathway in tomato plants. *Plant Signal. Behav.* **2010**, *5*, 1583–1589. [CrossRef]

127. Canuel, E.A.; Hardison, A.K. Carbon Cycle. In *Encyclopedia of Geochemistry: A Comprehensive Reference Source on the Chemistry of the Earth*; White, W.M., Ed.; Springer International Publishing: Cham, Switzerland, 2018; pp. 191–194. [CrossRef]

128. Gifford, R.M. Plant respiration in productivity models: Conceptualisation, representation and issues for global terrestrial carbon-cycle research. *Funct. Plant Biol.* **2003**, *30*, 171–186. [CrossRef]

129. Van Dongen, J.T.; Gupta, K.J.; Ramírez-Aguilar, S.J.; Araújo, W.L.; Nunez-Nesi, A.; Fernie, A.R. Regulation of respiration in plants: A role for alternative metabolic pathways. *J. Plant Physiol.* **2011**, *168*, 1434–1443. [CrossRef]

130. Millar, A.H.; Whelan, J.; Soole, K.L.; Day, D.A. Organization and Regulation of Mitochondrial Respiration in Plants. *Ann. Rev. Plant Biol.* **2011**, *62*, 79–104. [CrossRef] [PubMed]

131. Gupta, K.J.; Zabalza, A.; Van Dongen, J.T. Regulation of respiration when the oxygen availability changes. *Physiol. Plant.* **2009**, *137*, 383–391. [CrossRef]

132. Martínez, F.; Lazo, Y.O.; Fernández-Galiano, J.M.; Merino, J.A. Chemical composition and construction cost for roots of Mediterranean trees, shrub species and grassland communities. *Plant Cell Environ.* **2002**, *25*, 601–608. [CrossRef]

133. Rachmilevitch, S.; Lambers, H.; Huang, B. Root respiratory characteristics associated with plant adaptation to high soil temperature for geothermal and turf-type *Agrostis* species. *J. Exp. Bot.* **2006**, *57*, 623–631. [CrossRef]

134. Rewald, B.; Shelef, O.; Ephrath, J.E.; Rachmilevitch, S. Adaptive Plasticity of Salt-Stressed Root Systems. In *Ecophysiology and Responses of Plants under Salt Stress*; Ahmad, P., Azooz, M.M., Prasad, M.N.V., Eds.; Springer: New York, NY, USA, 2013; pp. 169–201. [CrossRef]

135. Gibbs, J.; Greenway, H. *Review*: Mechanisms of anoxia tolerance in plants. I. Growth, survival and anaerobic catabolism. *Funct. Plant Biol.* **2003**, *30*, 1–47. [CrossRef] [PubMed]

136. Armstrong, W.; Beckett, P.M.; Colmer, T.D.; Setter, T.L.; Greenway, H. Tolerance of roots to low oxygen: 'Anoxic' cores, the phytoglobin-nitric oxide cycle, and energy or oxygen sensing. *J. Plant Physiol.* **2019**, *239*, 92–108. [CrossRef] [PubMed]

137. Nakamura, T.; Nakamura, M. Root respiratory costs of ion uptake, root growth, and root maintenance in wetland plants: Efficiency and strategy of O_2 use for adaptation to hypoxia. *Oecologia* **2016**, *182*, 667–678. [CrossRef]

138. Greenway, H.; Gibbs, J. Review: Mechanisms of anoxia tolerance in plants. II. Energy requirements for maintenance and energy distribution to essential processes. *Funct. Plant Biol.* **2003**, *30*, 999–1036. [CrossRef]

139. Sorenson, R.; Bailey-Serres, J. Selective mRNA Translation Tailors Low Oxygen Energetics. In *Low-Oxygen Stress in Plants: Oxygen Sensing and Adaptive Responses to Hypoxia*; Van Dongen, J.T., Licausi, F., Eds.; Springer: Vienna, Austria, 2014; pp. 95–115. [CrossRef]

140. Thornley, J.H. Plant growth and respiration re-visited: Maintenance respiration defined—It is an emergent property of, not a separate process within, the system - and why the respiration: Photosynthesis ratio is conservative. *Ann. Bot.* **2011**, *108*, 1365–1380. [CrossRef]

141. Penning de Vries, F.W.T. The Cost of Maintenance Processes in Plant Cells. *Ann. Bot.* **1975**, *39*, 77–92. [CrossRef]

142. Penning de Vries, F.W.T.; Brunsting, A.H.; Van Laar, H.H. Products, requirements and efficiency of biosynthesis: A quantitative approach. *J. Theor. Biol.* **1974**, *45*, 339–377. [CrossRef]

143. Noctor, G.; De Paepe, R.; Foyer, C.H. Mitochondrial redox biology and homeostasis in plants. *Trends Plant Sci.* **2007**, *12*, 125–134. [CrossRef]

144. Dudkina, N.V.; Heinemeyer, J.; Sunderhaus, S.; Boekema, E.J.; Braun, H.-P. Respiratory chain supercomplexes in the plant mitochondrial membrane. *Trends Plant Sci.* **2006**, *11*, 232–240. [CrossRef] [PubMed]

145. Bailey-Serres, J.; Fukao, T.; Gibbs, D.J.; Holdsworth, M.J.; Lee, S.C.; Licausi, F.; Perata, P.; Voesenek, L.A.C.J.; Van Dongen, J.T. Making sense of low oxygen sensing. *Trends Plant Sci.* **2012**, *17*, 129–138. [CrossRef] [PubMed]

146. Blokhina, O.; Virolainen, E.; Fagerstedt, K.V. Antioxidants, Oxidative Damage and Oxygen Deprivation Stress: A Review. *Ann. Bot.* **2003**, *91*, 179–194. [CrossRef] [PubMed]

147. Toro, G.; Pinto, M. Plant respiration under low oxygen. *Chil. J. Agric. Res.* **2015**, *75*, 57–70. [CrossRef]

148. Caretto, S.; Linsalata, V.; Colella, G.; Mita, G.; Lattanzio, V. Carbon Fluxes between Primary Metabolism and Phenolic Pathway in Plant Tissues under Stress. *Int. J. Mol. Sci.* **2015**, *16*, 26378–26394. [CrossRef]

149. Møller, I.M.; Jensen, P.E.; Hansson, A. Oxidative Modifications to Cellular Components in Plants. *Annu. Rev. Plant Biol.* **2007**, *58*, 459–481. [CrossRef]

150. Gupta, K.J.; Mur, L.A.J.; Wany, A.; Kumari, A.; Fernie, A.R.; Ratcliffe, R.G. The role of nitrite and nitric oxide under low oxygen conditions in plants. *New Phytol.* **2020**, *225*, 1143–1151. [CrossRef]

151. Igamberdiev, A.U.; Bykova, N.V.; Shah, J.K.; Hill, R.D. Anoxic nitric oxide cycling in plants: Participating reactions and possible mechanisms. *Physiol. Plant.* **2010**, *138*, 393–404. [CrossRef]

152. Gupta, K.J.; Lee, C.P.; Ratcliffe, R.G. Nitrite Protects Mitochondrial Structure and Function under Hypoxia. *Plant Cell Physiol.* **2017**, *58*, 175–183. [CrossRef] [PubMed]

153. Vishwakarma, A.; Kumari, A.; Mur, L.A.J.; Gupta, K.J. A discrete role for alternative oxidase under hypoxia to increase nitric oxide and drive energy production. *Free Radic. Biol. Med.* **2018**, *122*, 40–51. [CrossRef]

154. Jayawardhane, J.; Cochrane, D.W.; Vyas, P.; Bykova, N.V.; Vanlerberghe, G.C.; Igamberdiev, A.U. Roles for Plant Mitochondrial Alternative Oxidase Under Normoxia, Hypoxia, and Reoxygenation Conditions. *Front. Plant Sci.* **2020**, *11*. [CrossRef] [PubMed]

155. Stoimenova, M.; Igamberdiev, A.; Gupta, K.; Hill, R. Nitrite-driven anaerobic ATP synthesis in barley and rice root mitochondria. *Planta* **2007**, *226*, 465–474. [CrossRef]

156. Almada, R.; Arismendi, M.J.; Pimentel, P.; Rojas, P.; Hinrichsen, P.; Pinto, M.; Sagredo, B. Class 1 non-symbiotic and class 3 truncated hemoglobin-like genes are differentially expressed in stone fruit rootstocks (*Prunus* L.) with different degrees of tolerance to root hypoxia. *Tree Genet. Genomes* **2013**, *9*, 1051–1063. [CrossRef]

157. Parent, C.; Crévecoeur, M.; Capelli, N.; Dat, J.F. Contrasting growth and adaptive responses of two oak species to flooding stress: Role of non-symbiotic haemoglobin. *Plant Cell Environ.* **2011**, *34*, 1113–1126. [CrossRef]

158. Gazizova, N.; Rakhmatullina, D.; Minibayeva, F. Effect of respiratory inhibitors on mitochondrial complexes and ADP/ATP translocators in the *Triticum aestivum* roots. *Plant Physiol. Biochem.* **2020**, *151*, 601–607. [CrossRef]

159. Millenaar, F.F.; Benschop, J.J.; Wagner, A.M.; Lambers, H. The role of the alternative oxidase in stabilizing the in vivo reduction state of the ubiquinone pool and the activation state of the alternative oxidase. *Plant Physiol.* **1998**, *118*, 599–607. [CrossRef]

160. Arru, L.; Fornaciari, S. Root Oxygen Deprivation and Leaf Biochemistry in Trees. In *Waterlogging Signalling and Tolerance in Plants*; Mancuso, S., Shabala, S., Eds.; Springer: Berlin/Heidelberg, Germany, 2010; pp. 181–195. [CrossRef]

161. Veen, B.W. Energy cost of ion transport. In *Genetic Engineering of Osmoregulation. Impact on Plant Productivity for Food, Chemicals and Energy*; Rains, D.W., Valentine, R.C., Hollander, A., Eds.; Springer: New York, NY, USA, 1980; pp. 187–195.

162. Bouma, T.J.; De Visser, R.; Janssen, J.; De Kock, M.; Van Leeuwen, P.; Lambers, H. Respiratory energy requirements and rate of protein turnover in vivo determined by the use of an inhibitor of protein synthesis and a probe to assess its effect. *Physiol. Plant.* **1994**, *92*, 585–594. [CrossRef]

163. Hirst, J. Open questions: Respiratory chain supercomplexes—Why are they there and what do they do? *BMC Biol.* **2018**, *16*, 111. [CrossRef] [PubMed]

164. Eubel, H.; Jänsch, L.; Braun, H.-P. New insights into the respiratory chain of plant mitochondria. Supercomplexes and a unique composition of complex II. *Plant Physiol.* **2003**, *133*, 274–286. [CrossRef] [PubMed]

165. Römpler, K.; Müller, T.; Juris, L.; Wissel, M.; Vukotic, M.; Hofmann, K.; Deckers, M. Overlapping Role of Respiratory Supercomplex Factor Rcf2 and Its N-terminal Homolog Rcf3 in Saccharomyces cerevisiae. *J. Biol. Chem.* **2016**, *291*, 23769–23778. [CrossRef]

166. Lobo-Jarne, T.; Ugalde, C. Respiratory chain supercomplexes: Structures, function and biogenesis. *Semin. Cell Dev. Biol.* **2018**, *76*, 179–190. [CrossRef] [PubMed]

167. Strogolova, V.; Furness, A.; Robb-McGrath, M.; Garlich, J.; Stuart, R.A. Rcf1 and Rcf2, members of the hypoxia-induced gene 1 protein family, are critical components of the mitochondrial cytochrome bc1-cytochrome c oxidase supercomplex. *Mol. Cell. Biol.* **2012**, *32*, 1363–1373. [CrossRef] [PubMed]

168. Shin, A.-Y.; Koo, N.; Kim, S.; Sim, Y.M.; Choi, D.; Kim, Y.-M.; Kwon, S.-Y. Draft genome sequences of two oriental melons, *Cucumis melo* L. var. makuwa. *Sci. Data* **2019**, *6*, 220. [CrossRef]

169. Schertl, P.; Braun, H.-P. Respiratory electron transfer pathways in plant mitochondria. *Front. Plant Sci.* **2014**, *5*, 163. [CrossRef]

170. Virolainen, E.; Blokhina, O.; Fagerstedt, K. Ca^{2+}-induced High Amplitude Swelling and Cytochrome c Release from Wheat (*Triticum aestivum* L.) Mitochondria Under Anoxic Stress. *Ann. Bot.* **2002**, *90*, 509–516. [CrossRef]

171. Bernardi, P.; Scorrano, L.; Colonna, R.; Petronilli, V.; Di Lisa, F. Mitochondria and cell death. *Eur. J. Biochem.* **1999**, *264*, 687–701. [CrossRef]

172. Bouranis, D.L.; Chorianopoulou, S.N.; Siyiannis, V.F.; Protonotarios, V.E.; Hawkesford, M.J. Lysigenous aerenchyma development in roots–triggers and cross-talks for a cell elimination program. *Int. J. Plant. Dev. Biol.* **2007**, *1*, 127–140.

173. Pimenta, J.A.; Bianchini, E.; Medri, M.E. Adaptations to flooding by tropical trees: Morphological and anatomical modifications. *Oecol. Aust.* **2010**, *4*, 157–176. [CrossRef]

174. Hancock, J.T.; Desikan, R.; Neill, S.J. Cytochrome c, Glutathione, and the Possible Role of Redox Potentials in Apoptosis. *Ann. N. Y. Acad. Sci.* **2003**, *1010*, 446–448. [CrossRef] [PubMed]

175. Rhoads, D.M.; Umbach, A.L.; Subbaiah, C.C.; Siedow, J.N. Mitochondrial Reactive Oxygen Species. Contribution to Oxidative Stress and Interorganellar Signaling. *Plant Physiol.* **2006**, *141*, 357. [CrossRef] [PubMed]

Article

Differential Expression of Maize and Teosinte microRNAs under Submergence, Drought, and Alternated Stress

Edgar Baldemar Sepúlveda-García [1], José Francisco Pulido-Barajas [2], Ariana Arlene Huerta-Heredia [3], Julián Mario Peña-Castro [1], Renyi Liu [4] and Blanca Estela Barrera-Figueroa [1,*]

[1] Laboratorio de Biotecnología Vegetal, Instituto de Biotecnología, Universidad del Papaloapan, Tuxtepec 68301, Mexico; edgarbal31@gmail.com (E.B.S.-G.); julianpc@unpa.edu.mx (J.M.P.-C.)
[2] Division de Estudios de Posgrado, Universidad del Papaloapan, Tuxtepec 68301, Mexico; jfpb.leg@gmail.com
[3] Cátedras CONACyT-UNPA, Universidad del Papaloapan, Tuxtepec 68301, Mexico; arianaahuertah@hotmail.com
[4] Center for Agroforestry Mega Data Science, Haixia Institute of Science and Technology, Fujian Agriculture and Forestry University, Fuzhou 350002, China; ryliu@fafu.edu.cn
*   Correspondence: bbarrera@unpa.edu.mx

Received: 15 August 2020; Accepted: 11 October 2020; Published: 15 October 2020

Abstract: Submergence and drought stresses are the main constraints to crop production worldwide. MicroRNAs (miRNAs) are known to play a major role in plant response to various stresses. In this study, we analyzed the expression of maize and teosinte miRNAs by high-throughput sequencing of small RNA libraries in maize and its ancestor teosinte (*Zea mays* ssp. *parviglumis*), under submergence, drought, and alternated stress. We found that the expression patterns of 67 miRNA sequences representing 23 miRNA families in maize and other plants were regulated by submergence or drought. miR159a, miR166b, miR167c, and miR169c were downregulated by submergence in both plants but more severely in maize. miR156k and miR164e were upregulated by drought in teosinte but downregulated in maize. Small RNA profiling of teosinte subject to alternate treatments with drought and submergence revealed that submergence as the first stress attenuated the response to drought, while drought being the first stress did not alter the response to submergence. The miRNAs identified herein, and their potential targets, indicate that control of development, growth, and response to oxidative stress could be crucial for adaptation and that there exists evolutionary divergence between these two subspecies in miRNA response to abiotic stresses.

Keywords: hypoxia; submergence; drought; alternated stress; maize; teosinte; microRNAs

---

1. Introduction

As a consequence of global warming, hydrological fluctuation events such as excessive rainfall and droughts are common and projected to continue in the future, affecting economic activities [1]. Alterations in water availability in the field, either caused by water surplus or deficit, produce water stress in plants that negatively impacts the growth and productivity of crops worldwide [2].

Flooding affects the properties of soil and the composition of associated microbial communities [3] and reduces the availability of nutrients and flux of oxygen to the plant. This leads to hypoxic stress at the cellular level, especially when the column of water exceeds the length of the stem (i.e., submergence) [4,5]. In response to flooding stress, plants express genes known as hypoxia core genes (HCG) that promote anaerobic metabolism [6]. Additionally, some plants use the escape strategy by redirecting growth to stem elongation to overpass the column of water and maintain the oxygen flux. Plants may also delay growth, flowering, and other processes that are energetically expensive, as part of a strategy known as quiescence to save resources useful to reestablish normal growth when the water level recedes [7]. On the contrary, when water availability is limited, as in the case of drought, plants undergo a similar series of changes to adapt to the adverse conditions, which include stomatal closure with limitations in CO_2 uptake and reduction in photosynthetic activity, increase in root elongation rate, early or delayed flowering time, and other physiological and morphological changes [8].

Cellular dehydration caused by drought, and hypoxia driven by submergence, trigger a cascade of adaptive responses that are regulated at the molecular level in plants and are directed to maintain vital functions and protect structures from damage. Since cellular protection is a priority during stress, it has been shown that some responses to dehydration and excess water have common molecular effectors acting as nodes in the crosstalk between responses to drought and submergence [9–11]. This may have allowed wild plants to evolve tolerance mechanisms to cope with alternate drought and flooding events, which commonly occur consecutively in some regions of the world [12,13].

Maize (*Zea mays* ssp. *mays*) is one of the main dietary cereals in the world [14]. In the United States, the world's main maize producer, extreme drought and excessive rainfall are the first and second major causes of losses in maize production, respectively, with both affecting yields to a comparable extent [15]. According to archeological, botanical, and genetic evidence, maize was domesticated from its single ancestor teosinte (*Zea mays* ssp. *parviglumis*) around 10,000 years ago [16,17]. It is estimated that maize retained 83% of the nucleotide diversity from its ancestor [18]. During domestication and breeding, maize has been subject to selection for traits that are beneficial to improve feeding quality and yield, while other traits like tolerance to stress could have been lost. For this reason, teosinte is considered a living reservoir of genes and mechanisms that could be of importance to improve tolerance to stress in maize [19,20].

MicroRNAs (miRNAs) have been proposed as promising targets for developing plants with improved tolerance to multiple abiotic stresses [21]. miRNAs are small RNA molecules with pivotal roles in the response to environmental stresses in plants that usually act as negative regulators of gene expression via silencing target genes, which are recognized by sequence complementarity and subsequently targeted for degradation and/or translation inhibition [22]. With the development of new platforms for the high-throughput sequencing of small RNAs, several miRNAs involved in the response to drought or submergence have been described in plants such as *Arabidopsis thaliana* [23,24], *Brachypodium distachyon* [25,26], *Oryza sativa* [27], *Nelumbo nucifera*, [28] and *Zea mays* [29,30].

According to miRBase, 325 mature miRNAs from 174 miRNA precursors have been identified in maize [31], and some of them have been reported to be responsive to drought [32,33], submergence, [34] or waterlogging [35]. For example, miR159, miR164, miR167, miR393, miR408, and miR528 are upregulated by short-term waterlogging in roots of maize lines with high tolerance and suggest the involvement of hormonal control in the response mediated by miRNAs [30]. Another study of two maize lines, with contrasting tolerance to drought, observed the downregulation of miR164 and upregulation of miR159, miR390, and miR398 in the tolerant line compared with the sensitive line [29]. However, it is not clear how miRNAs are regulated in teosinte under submergence or drought stress and whether miRNA response to these stresses is conserved during maize domestication and breeding.

Therefore, this study aimed to identify microRNAs expressed in maize and teosinte under submergence and drought stress in order to analyze the conserved and differential mechanisms of response in these plants. For this, we used high-throughput small RNA sequencing to profile maize and teosinte under submergence, drought, and consecutive drought and submergence stresses. We found a group of miRNAs that were regulated by drought or submergence, including some miRNAs that were differentially regulated in the two species. Analyses of small RNA data from teosinte subject to alternate treatments of drought and submergence indicated that miRNA response to drought was attenuated when submergence was the first stress treatment, but miRNA response to submergence did not change when drought stress was applied first. The potential targets of these miRNAs indicate that control of development, growth, and response to oxidative stress could be crucial for understanding the conservation and divergence of stress tolerance between maize and teosinte.

2. Results

2.1. Physiological Response of Maize and Teosinte to Submergence and Drought

With the objective to determine a time period of treatment for RNA collection, maize seedlings with two vegetative leaves and 14 days after sowing (stage V2, 14 DAS) were exposed to submergence for 2, 4, and 6 days (Figure 1A). All treatments induced growth reduction in maize, but plants submerged for 4 or 6 days developed necrotic spots on the leaves, and plants treated for 6 days were not able to recover from the stress. Therefore, a 2 day submergence treatment was selected for subsequent assays. Similarly, drought assays were performed on maize seedlings that were deprived of water until they reached half their initial weight and were maintained under this level of water limitation for 2, 4, 6, and 8 days (Figure 1B).

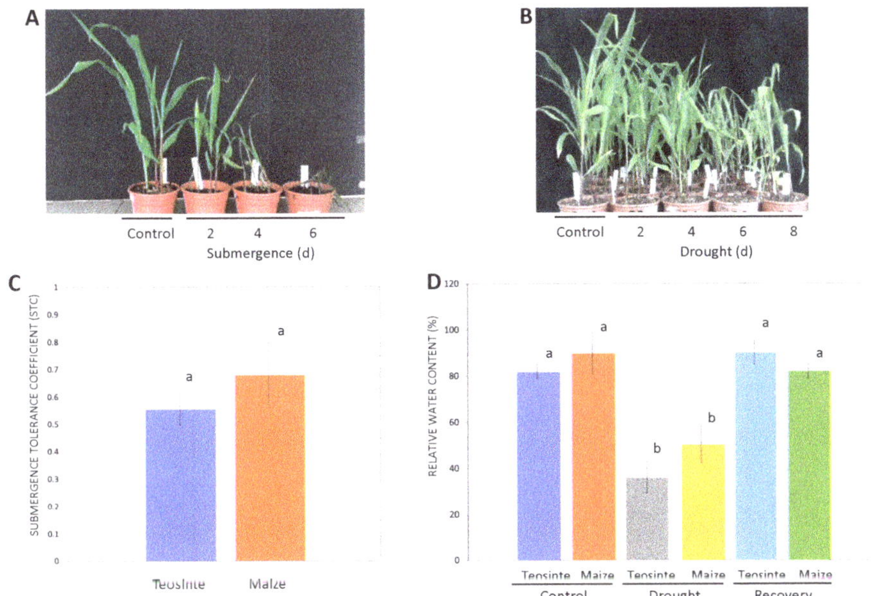

Figure 1. Effects of submergence and drought on maize and teosinte plants. (**A**) Plants exposed to submergence for 2, 4, and 6 days, and (**B**) plants exposed to drought for 2, 4, 6, and 8 days. Response of teosinte and maize plants to (**C**) submergence for 48 h, expressed as submergence tolerance coefficient (STC), and (**D**) drought for 6 days and recovery 24 h after rehydration, expressed as relative water content (RWC) in the leaves. Data are means, with error bars representing ±SD of three biological replicates, each replicate consisting of six plants. Different letters above the error bars indicate statistically significant differences between samples ($p < 0.05$) in a Student's *t*-test.

Drought caused a reduction in the growth of maize plants when compared with control plants, but this reduction was stabilized, which may be explained by the fact that pots were maintained at a constant, intermediate water deficit condition, and not exposed to progressive drought stress. Based on these observations, maize and teosinte plants were exposed to submergence for 2 days and drought for 8 days, and the submergence tolerance coefficient (STC) and relative water content (RWC) were registered for the respective treatments. Submergence affected the growth of maize and teosinte seedlings to a similar extent, since STC was not significantly different (Figure 1C). There were also no significant differences in RWC between maize and teosinte in control, drought, and 24 h of recovery after the stress (Figure 1D). These results suggest that both plants are able to sense and respond to stress by adjusting water status and growth to a similar extent at this stage. This is expected because both maize CML496 and teosinte grow in conditions prevalent in the tropics, where drought, high temperature, and extreme seasonal variations in monthly rainfall are commonly observed. Therefore, this constitutes a solid system to investigate the intra-specific conservation and divergence in the response to stress between maize and its direct ancestor that lead to adaptation in both plants.

2.2. Overall Analysis of Small RNA Sequencing Data

The alterations in growth, STC, and RWC observed in maize and teosinte seedlings in the assays suggest that the times and stress intensities can help to study molecular responses in both plants. Accordingly, maize and teosinte seedlings exposed to 2 days of submergence or 6 days of constant drought were selected for profiling the microRNA populations involved in the corresponding responses. In addition, teosinte seedlings exposed to alternated stress of drought-submergence and submergence-drought were also selected.

High-throughput sequencing of small RNA libraries prepared from teosinte and maize under submergence, drought, and alternated stress rendered from 16 to 46 million raw reads (Table 1), with PHRED score above 30 (Figure S1). After removing low-quality reads and adapters, the libraries were analyzed for size distribution of reads, revealing an enrichment of 24 nt sRNAs across the libraries (Figure S2). Reads less than 18 or longer than 35 nt were removed, resulting in a number of clean reads ranging from 14.8 to 41.9 million (Table 1).

Table 1. Summary of small RNA sequencing data from eight maize and teosinte libraries. Distribution of small RNAs in the corresponding non-coding RNA category.

	Maize			Teosinte				
	Control	Submerg	Drought	Control	Submerg	Drought	Drought-Submerg	Submerg-Drought
Raw	16,139,354	26,621,448	31,642,210	20,672,413	28,656,756	29,063,588	17,997,037	46,522,229
Clean	14,897,058	24,568,439	29,744,161	18,955,892	26,296,921	26,960,313	15,760,258	41,915,673
Maize	14,870,947	24,488,359	29,696,341	18,927,434	26,217,672	26,912,777	15,727,757	41,761,974
	99.82%	99.67%	99.84%	99.85%	99.70%	99.82%	99.79%	99.63%
mt	87,401	106,625	134,420	114,741	73,980	105,999	106,451	145,709
	0.59%	0.43%	0.45%	0.61%	0.28%	0.39%	0.68%	0.35%
cp	182,414	452,301	459,718	475,121	444,434	835,675	973,949	1,289,213
	1.22%	1.84%	1.55%	2.51%	1.69%	3.10%	6.18%	3.08%
mRNA	195,418	258,447	342,833	207,230	202,191	256,739	238,774	309,551
	1.31%	1.05%	1.15%	1.09%	0.77%	0.95%	1.52%	0.74%
rRNA	288,257	601,629	778,148	687,038	591,868	588,819	1,056,243	1,240,514
	1.94%	2.45%	2.62%	3.62%	2.25%	2.18%	6.71%	2.97%
tRNA	103,451	161,288	200,294	569,294	480,975	553,939	953,794	1,426,872
	0.70%	0.66%	0.67%	3.00%	1.83%	2.05%	6.06%	3.41%
snRNA	516,999	838,151	1,071,157	576,299	868,349	890,657	421,404	1,220,748
	3.47%	3.42%	3.60%	3.04%	3.31%	3.30%	2.67%	2.92%
snoRNA	105,518	168,088	228,389	141,006	174,443	190,297	111,528	256,675
	0.70%	68.00%	0.77%	0.74%	0.66%	0.70%	0.70%	0.61%
lncRNA	258,027	416,896	519,449	362,527	530,505	528,963	231,375	719,402
	1.73%	1.70%	1.74%	1.91%	2.02%	1.96%	1.47%	1.72%
repeats	2,219,656	3,623,199	4,820,455	5,064,798	5,305,481	5,343,314	4,734,879	9,267,896
	14.92%	14.79%	16.23%	26.75%	20.23%	19.85%	30%	22.19%
Maize	397,473	248,193	774,680	180,091	98,923	310,478	47,836	350,678
miRNAs	2.67%	1.01%	2.60%	0.95%	0.38%	1.15%	0.30%	0.84%
Other	24,837	21,526	39,108	6498	5862	7393	6470	13,887
miRNAs	0.17%	0.09%	0.13%	0.03%	0.02%	0.03%	0.04%	0.03%

Submerg: Submergence; mt: mitochondria; cp: chloroplast; mRNA: messenger RNA; rRNA: ribosomal RNA; tRNA: transfer RNA; snRNA: small nuclear RNA; snoRNA: small nucleolar RNA; lncRNA: long non-coding RNA.

On average, 99.7% of the clean reads were mapped to the maize genome. From the reads mapped to the maize genome, less than 3% represented plant miRNAs reported in the miRBase (Table 1). Interestingly, libraries from teosinte showed the lowest percentage of miRNAs (approximately 0.74% for teosinte and approximately 2.19% for maize), suggesting a potential underrepresentation effect caused by the absence of teosinte miRNAs in the miRBase. However, the percentage of teosinte reads that were mapped to the maize genome (approximately 99.75%) was the same as that for maize (approximately 99.77%), indicating that differences are likely due to lower levels of miRNA expression in teosinte. In addition, a reduced abundance of plant miRNAs under submergence stress was observed in both maize and teosinte, and in teosinte treated with drought-submergence stress compared with the corresponding controls (Table 1). After removing low frequency reads and normalization to tags per million (TPM), 178 miRNA sequences in total (118 mapping to maize and 60 to miRNAs of other plants) were expressed across the libraries (Table S1).

2.3. Global Changes in Expression of miRNA Sequences in Response to Stress

Further analysis revealed that 164 miRNA sequences (110 from maize and 54 annotated in other plants) were differentially expressed in maize and/or teosinte in response to at least one of the treatments (Table S1).

Regarding submergence stress, maize and teosinte showed common expression profiles of 10 upregulated and 72 downregulated miRNA sequences (Figure 2A). A total of 42 miRNA sequences (14 upregulated and 28 downregulated) were regulated exclusively in maize, while 27 miRNA sequences (4 upregulated and 23 downregulated) were exclusive of teosinte (Figure 2A). A similar trend in the number of miRNAs responsive to stress, with lesser miRNAs being regulated in teosinte than in maize, was observed for drought treatments (Figure 2B). The analysis revealed 70 miRNA sequences (21 upregulated, 49 downregulated) regulated exclusively in maize, while 27 were differentially expressed only in teosinte (17 upregulated, 10 downregulated).

The overlap between submergence and drought responses in maize and teosinte showed that in maize, most of the miRNAs (68 sequences) were downregulated in both submergence and drought (Figure 2C), whereas in teosinte, the number of sequences downregulated exclusively in submergence was the highest (69 sequences; Figure 2D).

In addition, small RNA libraries from teosinte exposed to alternated water stress were analyzed to assess the effects of consecutive treatments of drought and submergence on miRNA expression. Figure 2E shows that 18 miRNA sequences (5 upregulated and 13 downregulated) were differentially expressed in a similar pattern in response to submergence, drought, and alternated drought-submergence treatments. The number of miRNA sequences (69) overlapping between drought-submergence and submergence (representing 63% of total sequences in submergence) was higher compared with only 22 miRNA sequences between drought-submergence and drought (39% of total sequences in drought; Figure 2E), indicating that when submergence is the second stress, it has the strongest effect on miRNA expression.

When submergence is followed by drought, 21 out of 56 miRNA sequences regulated under drought overlapped between submergence-drought and drought treatments (37% of sequences in drought), indicating that when submergence is the first stress, it affects the subsequent response to drought in teosinte (Figure 2F).

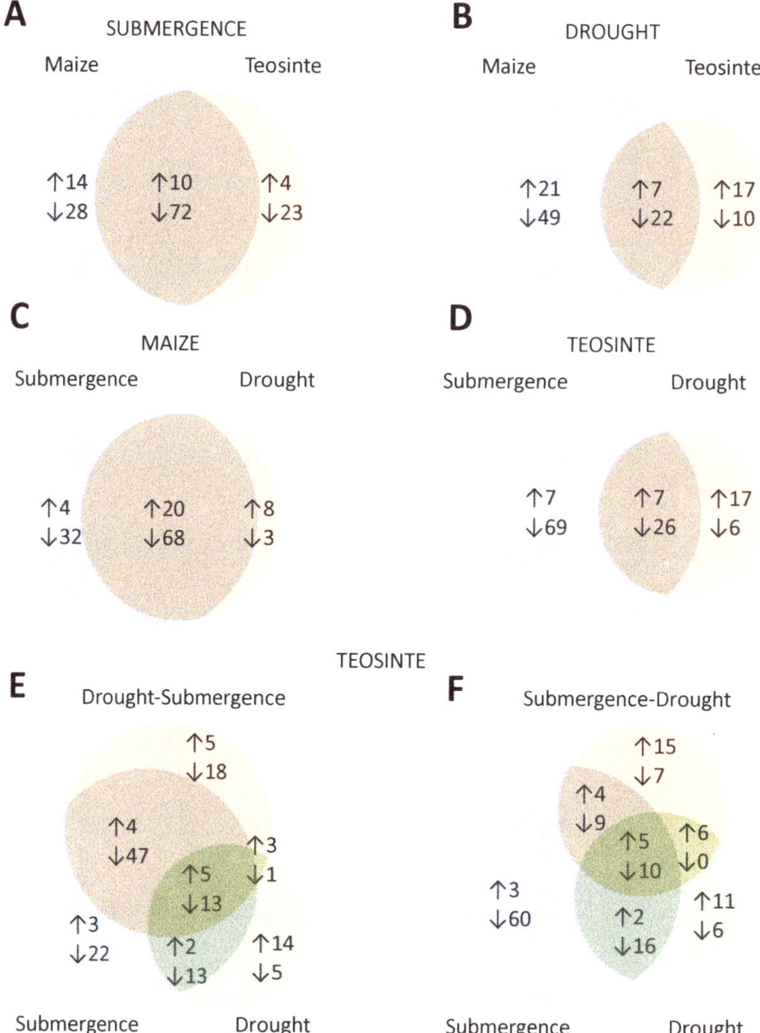

Figure 2. Venn diagrams showing differentially expressed miRNA reads that overlap in maize and teosinte under (**A**) submergence and (**B**) drought treatments. Plant miRNAs overlapping in submergence and drought treatments in (**C**) maize, and (**D**) teosinte. Plant miRNAs overlapping in teosinte under single and alternated treatments of (**E**) drought followed by submergence, and (**F**) submergence followed by drought. Numbers are based on fold change (FC) ($\log_2$) values (treatment/control). Upward arrows represent upregulated miRNAs (FC $\geq$ 1) and downward arrows represent downregulated miRNAs (FC $\leq$ −1).

2.4. Redundancy of miRNA Reads Annotated to miRBASE and Grouping of Representative Sequences

The data showed a high level of redundancy of reads in the libraries annotated to miRBASE, mainly caused by positional variants of mature miRNAs and other sequences matching miRNA precursors. Regrouping of mature miRNAs, positional variants derived from the same mature sequence, and sequences that exactly match those from other plants but with a close similarity to known maize miRNAs (i.e., differing by one or two nucleotides in the 5′ or 3′ ends), rendered 67 representative sequences belonging to 23 miRNA families (Table S2).

2.5. Search for Novel miRNAs

In order to identify new variants of known miRNAs in teosinte, a search was performed against the miRBASE, allowing up to two mismatches in the mature or precursor sequences. This rendered 102 polymorphic variants that showed the same expression profiles as those of their corresponding exact sequences (Table S3). However, these sequences failed to provide a perfect hit to the reference genome, which did not allow the analysis of a putative precursor sequence. We also searched for potential novel miRNAs in our datasets and generated several candidates, but after a detailed inspection, they did not comply with the criteria for annotation of novel miRNAs in terms of secondary structure and/or distribution of reads along the potential precursor [36].

2.6. Analysis of miRNAs Responsive to Submergence and Drought in Maize and Teosinte

In the group of 67 representative non-redundant miRNA sequences, 13 were identified as responsive to at least one of the stress conditions, with fold change values ranging from −4.39 for the most downregulated miRNA (miR166bd) to 2.17 for the most upregulated miRNA (miR319b) (Table 2). Besides these miRNAs, others showing differential expression such as miR1511, miR2916, miR482, and miR4995 (Table S2), were discarded from the analysis due to failure to hit the maize genome or originating from ribosomal RNA or transposons.

Table 2. Representative plant miRNAs responsive to submergence and drought in maize and teosinte.

miRNA Name	Fold Change (FC) [1]			
	Maize		Teosinte	
	Submergence	Drought	Submergence	Drought
miR156k	−2.94	−2.99	−0.86	1.09
miR159ab	−2.17	0.80	−1.46	1.06
miR164e	−4.24	−1.49	−0.29	1.20
miR166bd	−4.39	−3.79	−2.77	−1.49
miR167cdeg	−2.99	−2.48	−1.62	−0.33
miR169cr	−3.13	−0.95	−1.05	0.11
miR319b	−0.75	0.80	−0.66	2.17
miR396cd	−1.04	−0.17	−0.39	1.08
miR398ab	−0.38	−0.29	−1.09	0.05
miR398b	−2.83	−0.94	−0.69	0.29
miR408	−1.34	−0.51	−1.78	−1.76
miR408b	−2.14	−0.89	−2.57	0.17
miR528ab	−1.30	−1.65	−2.22	−0.57
Constitutive				
miR166c	0.00	−0.01	−0.21	−0.19

[1] Fold change (treatment/control) values ≥ 1 indicate upregulated miRNAs (orange), and values ≤ −1 indicate downregulated miRNAs (green).

Overall, the grouping of miRNAs presented a clear picture of the stress response. Changes in miRNA expression were more marked in maize than in teosinte for most miRNAs downregulated under submergence, suggesting that maize plants could be more sensitive or reactive to submergence than teosinte in terms of miRNA responses. miRNAs that were specifically downregulated in maize were miR159ab, miR164e, miR166bd, miR167cdeg, miR169cr, miR396cd, and miR398b (Table 2).

In maize, drought stress repressed the expression of five miRNAs (miR156k, miR164e, miR166bd, miR167cdeg, and miR528a). In teosinte, only miR166bd and miR408 were downregulated, while five miRNAs were upregulated under drought (miR156k, miR159ab, miR164e, miR319b, and miR396cd) (Table 2).

2.7. Analysis of miRNAs Responsive to Alternated Stress in Teosinte

When plants were exposed to drought-submergence conditions, the miRNA expression profile resembled that of submergence, suggesting that at the end of the drought treatment, plants were not compromised to respond to a submergence event (Table 3). The changes between drought-submergence and submergence stress were limited only to the expression of miR408. This miRNA target gene encodes plastocyanin (PLC) and laccase (LAC) involved in copper homeostasis and the formation of lignin.

Table 3. Representative plant miRNAs responsive to single (submergence or drought) and alternated treatments (drought followed by submergence and submergence followed by drought) in teosinte plants.

miRNA Name	Fold Change (FC) [1]			
	Submergence	Drought	Drought-Submergence	Submergence-Drought
miR156k	−0.86	1.09	−0.54	1.17
miR159ab	−1.46	1.06	−1.43	0.55
miR164e	−0.29	1.20	0.06	1.09
miR166bd	−2.77	−1.49	−1.95	−0.77
miR167cdeg	−1.62	−0.33	−1.87	0.71
miR169cr	−1.05	0.11	−1.60	0.11
miR319b	−0.66	2.17	−0.41	0.16
miR396cd	−0.39	1.08	−0.94	0.53
miR398ab	−1.09	0.05	−1.56	−0.58
miR398b	−0.69	0.29	−1.92	−0.57
miR408	−1.78	−1.76	−0.88	0.55
miR408b	−2.57	0.17	−0.43	−0.87
miR528ab	−2.22	−0.57	−3.16	−0.99
Constitutive				
miR166c	−0.21	−0.19	−0.27	−0.19

[1] Fold change (treatment/control) values ≥ 1 indicate upregulated miRNAs (orange), and values ≤ −1 indicate downregulated miRNAs (green).

When plants were exposed to submergence-drought conditions, most miRNAs were regulated in a way similar to that in plants under drought alone, but the response was attenuated. For example, miR159ab, miR319b, and miR396cd were upregulated in drought as a single stress, but when drought was preceded by submergence, regulation of these miRNAs was maintained within the same trend but at low levels (Table 3). These results suggest that plants are plastic in response to environmental cues and have the capacity to oscillate to adapt to different stresses.

2.8. Assessment and Validation of miRNA Expression by Quantitative RT-PCR

Quantitative RT-PCR assays were performed using a modified SL-RTPCR method for seven miRNAs regulated by stress. Based on the fold change values of miRNAs expressed in the libraries, miR166c-5P was selected as a constitutive control for qRT-PCR assays, since it showed intermediate abundance levels and stable regulation in single and alternated stress conditions (Tables 2 and 3). SL-RTPCR assays showed changes in expression similar to those observed by analysis of sequencing data for miR156k, miR159ab, miR167cdeg, miR396cd, miR398ab, miR408b, and miR528ab (Figure 3A–G).

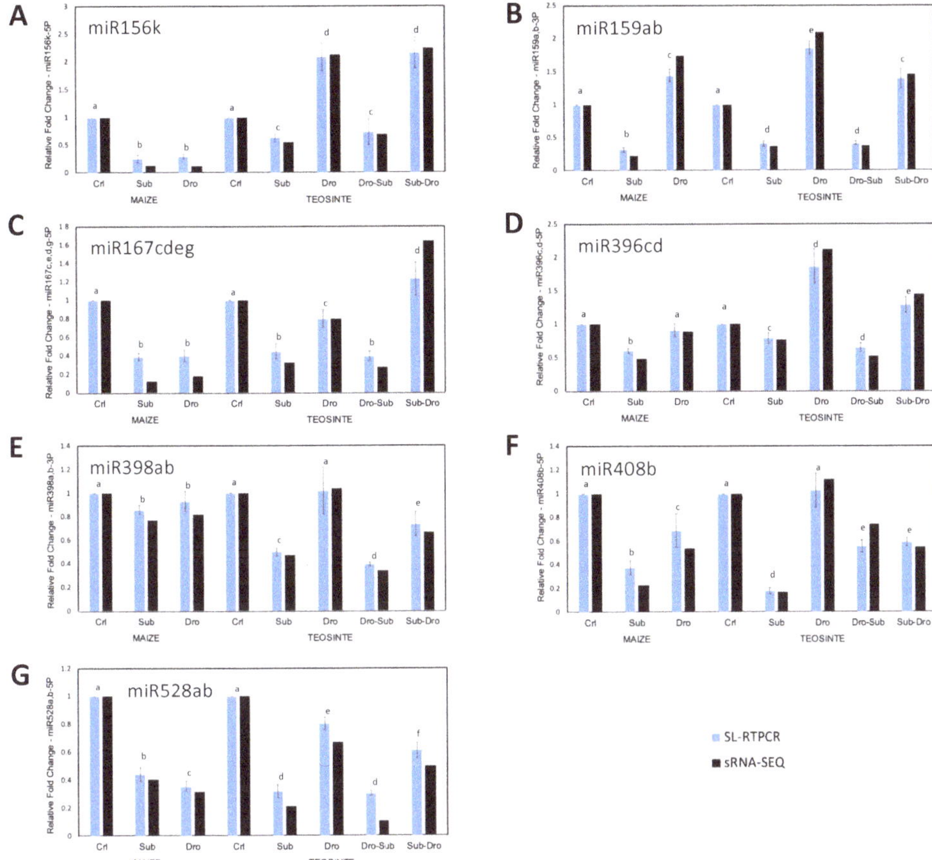

Figure 3. qRT-PCR validation of miRNAs response to submergence, drought, and alternated stress by SL-RTPCR for (**A**) miR156, (**B**) miR159, (**C**) miR167, (**D**) miR396, (**E**) miR398, (**F**) miR408, and (**G**) miR528. Relative fold change values were calculated as $\Delta\Delta Ct$ for SL-RTPCR (blue bars) using miR166c as the constitutive control or as the treatment/control ratio from normalized reads for sRNA-SEQ data (black bars). Results are mean with error bars representing ±SD of three biological replicates, each replicate consisting of a pool of six plants and three technical replicates. Different letters above the error bars indicate statistically significant differences between samples ($p < 0.05$).

For example, based on sequencing data, miR167 had a fold-change value of 0.12 and 0.17 in maize under submergence and drought, respectively. The values obtained by SL-RTPCR were 0.39 (±0.03) and 0.40 (±0.06), respectively, which confirms that miR167 is downregulated by these treatments and thus the expression trends for upregulated or downregulated miRNAs were maintained in both methods.

2.9. Other Small RNAs

Mitochondria and chloroplasts are the main compartments for energy production that are affected by low water and oxygen availability and may be a source of signals to integrate the cellular response [37,38]. A search was performed in the libraries with the objective of identifying small RNAs potentially originating from maize mitochondria or plastids. For both mitochondria and plastids, the highest number of reads mapped to the subunit 2 of NADH dehydrogenase (Table S4). The reads were derived from a region spanning 53 bp of NADH dehydrogenase subunit 2 with an abundance of up to 820 TPM, comparable to miR408 abundance (Table S2). The second highest number of reads mapped to the trnL-trnF intergenic spacer and tRNA-Phe (trnF) gene from the maize chloroplast. Interestingly, these reads derived from a 32 bp region of the gene did not show a clear expression trend in response to stress; however, a high accumulation of reads in teosinte compared with that in maize was evident, with up to 2273 TPM, comparable to miR398 abundance (Tables S2 and S5). The causes and implications of the existence of these small RNA signatures need to be investigated in the future.

3. Discussion

Drought and submergence represent complex conditions that can be dissected into several components such as hypoxia, nutrient, light, osmotic, temperature, and oxidative stress [39]. Understanding how the different components act to integrate the response to stress at all levels of regulation is fundamental to underpin the improvement of tolerance to drought and flooding stress in crops.

miRNAs are a class of small RNAs that act as regulators of gene expression at the transcriptional level. In this study, high-throughput sequencing and analysis of small RNA populations from maize and teosinte under submergence and drought revealed that submergence caused a global reduction in the representation of miRNA expression compared with control or drought conditions. This was observed in roots of wild tomato treated with hypoxia, where only 1.45% of total reads were expressed in hypoxia-treated roots against 2.45% expressed in control roots [40].

Further exploration of the expression of specific miRNA family members allowed the identification of 13 miRNAs with differential expression profiles in response to submergence and drought. The analysis of their expression patterns and functions of their predicted, or previously confirmed targets, allowed the recognition of two main components of the responses to submergence and drought in maize: transcriptional regulation and antioxidant activity.

3.1. Transcriptional Regulation

The growth and development processes are driven by hormonal signaling pathways that are fundamental in the response to submergence and drought in plants [41], where transcription factors and miRNAs participate. Several differentially expressed miRNAs identified in this work are known to act over target genes encoding transcription factors such as miR156 (squamosa promoter-binding protein like, SPL) [42–44], miR159 (giberellic acid-MYB transcription factor, GAMYB [28,29]), miR164 (NAM-ATAF-CUC domain transcription factor NAC, MYB [45]), miR166 (basic leucine zipper transcription factor, HD-ZIP III [46,47]), miR167 (auxin-responsive factor, ARF [30,35,48]), miR169 (nuclear transcription factor Y subunit alpha, NFYA [49–52]), miR319 (teosinte branched1/cycloidea/proliferating cell nuclear antigen factor, TCP [53]), and miR396 (growth-regulating factor, GRF [54,55]). Other target genes predicted in this study are listed in Table S6.

Most miRNAs in this category have been related to responses to abiotic stress. miR156 controls developmental transitions and flowering through the activity of their targets (SPL) [42–44] which is upregulated in *Arabidopsis* roots under hypoxia and in lotus under submergence [23,28]. Other miRNAs such as miR166 were upregulated in maize roots in response to waterlogging [34] and downregulated in *Arabidopsis* roots in response to hypoxia [23]. miR164 participates in the development of lateral roots [45] and the response to waterlogging in maize roots [30]. miR159 controls petiole elongation and flowering [56] and is upregulated in maize roots exposed to waterlogging [29]. miR167 is upregulated by waterlogging in maize roots [30,34], and downregulated in submerged lotus seedlings [28].

In this study, miRNAs in the category of transcriptional regulation overlapped between maize and teosinte in response to submergence and between submergence and drought in maize (Figure 4), indicating that the control of growth is of central importance for adaptation in maize and teosinte. Interestingly, all these miRNAs were downregulated in the overlaps. Downregulation of this set of miRNAs means that their targets may be released from post transcriptional control, which may activate hormone-responsive genes to initiate elongation (miR159), cell elongation and differentiation (miR167), transition to flowering (miR156), and other responses. In addition, it is possible that changes observed in miRNA expression at the transcriptional level are not only directed to increase growth, but also to mediate feedback in hormonal pathways, or to prepare for reoxygenation or rehydration after stress. Interestingly, the response to drought in teosinte was regulated by five miRNAs that were upregulated (miR156, miR164, miR159, miR319, and miR396). The upregulation of these miRNAs suggests that growth processes may be restricted in teosinte during drought stress (Figure 4).

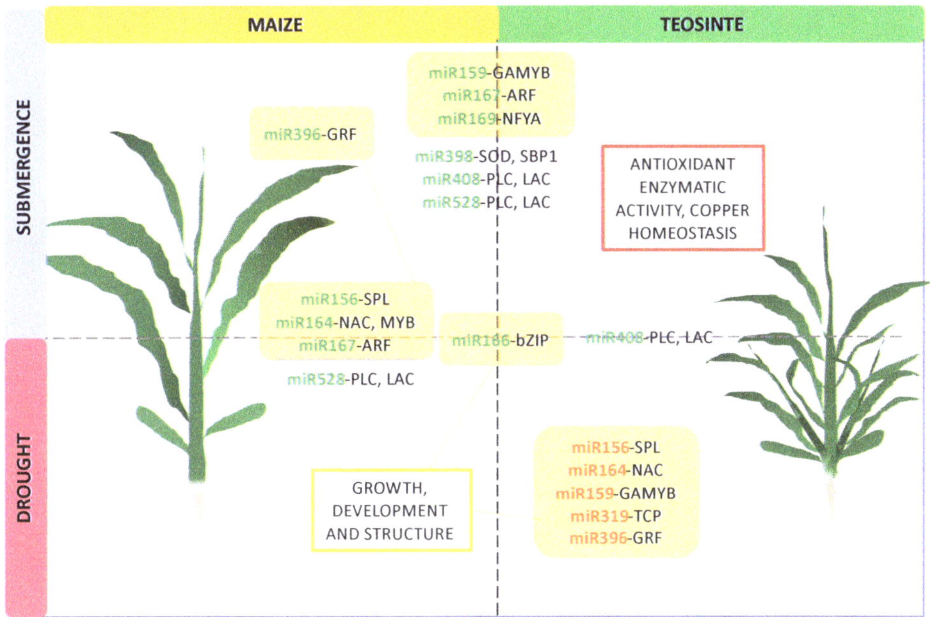

Figure 4. Hypothetical model of the regulation of miRNAs in the response to submergence and drought in maize and teosinte. The importance of antioxidant enzymatic activity, copper homeostasis, and control of growth and development as the main lines of response regulated by miRNAs in both plants is highlighted. Interrupted lines represent overlap between plants and/or treatments. Green, downregulated miRNAs; orange, upregulated miRNAs. GRF: growth-regulating factor; GAMYB: gibberellic acid-MYB transcription factor; ARF: auxin-responsive factor; NFYA: nuclear transcription factor Y subunit alpha; PLC: plastocyanin; LAC: laccase; SOD: superoxide dismutase; SBP1: selenium-binding protein1; bZIP: basic leucine zipper transcription factor; SPL: squamosa promoter-binding protein like; NAC: NAM-ATAF-CUC domain transcription factor; TCP: teosinte branched1/cycloidea/proliferating cell nuclear antigen factor; SBP1: selenium-binding protein 1.

3.2. Antioxidant Activity

Submergence and drought stress result in excessive production of reactive oxygen species (ROS) and efficient mechanisms to cope with oxidative injury are determinants of tolerance [57]. miR398, miR408, and miR528 are included in this category. miR398 target genes encode Cu/Zn superoxide dismutases (CSD1 and CSD2) that act in the defense against toxic ROS in *Arabidopsis* [58]. In *Phaseolus vulgaris*, miR398 is downregulated under drought and submergence [59] and is also downregulated in response to submergence in *Arabidopsis* [60]. We predicted other miR398 targets including a selenium-binding protein (SBP) (Table S6). An SBP was recently reported to be induced in *Arabidopsis* in response to submergence [60] and confirmed as a miR398 target by degradome analysis in maize [61]. In this study, miR398 was downregulated in maize and teosinte under submergence, thereby implicating a potential increase in antioxidant activity.

Other members, such as miR408 and miR528, target genes encoding the cupredoxins PLC and LAC [62–64], and other genes controlling circadian clock and flowering time [65,66]. PLC functions as an electron transporter and LAC participates in the formation of lignin by oxidation. miR408 and miR528 have been previously reported to respond to submergence in Lotus seedlings and maize roots [28,30]. In this study, miR408 and miR528 were downregulated in maize and teosinte under submergence, suggesting the maintenance of electron flux, oxidation homeostasis, and lignin synthesis (Figure 4).

3.3. Comparison of miRNA Expression in Maize and Teosinte

Previous work demonstrated a high level of conservation of miRNA mature sequences between maize and teosinte [67]. Thus, divergence in tolerance to stress is most likely the result of differences in gene expression between these plants. Overall, in this work maize and teosinte showed similar responses to submergence, with a clear trend toward downregulation of miRNAs involved in the control of growth, development, antioxidant response, and copper homeostasis. However, the response to submergence was more intense in maize than in teosinte for most of the miRNAs responsive to stress, especially for those miRNAs controlling growth. This suggests that maize is more reactive to submergence and responds by releasing the control on their target genes, which in turn may accelerate growth to escape from stress. In teosinte, these responses were less dramatic, suggesting that a finer tuning of expression could be a key for tolerance in this plant. A similar pattern was observed in a study comparing miRNA expression in inbred maize lines with different levels of tolerance to waterlogging, where the stress sensitive line reacted by downregulating most of the miRNAs and induced target genes to accelerate growth. Instead, the tolerant line responded with a moderate level of downregulation, or even upregulation of some miRNAs to repress growth under stress [30].

In the case of drought, maize responded in a similar way as that in submergence by downregulating miRNAs to induce growth, while teosinte upregulated miRNAs to restrict growth and probably save energy resources. Regarding defense against oxidative stress, oxidation homeostasis, and reinforcement of cell structure, both plants maintained the downregulation of miRNAs, suggesting the possibility that their target genes were actively protecting the plants from the oxidative damage caused by stress.

3.4. Effects of Alternated Stress on Teosinte

Alternated events of drought and submergence are common in some regions of the world where plants experience the succession of these events throughout their life cycle. In this study, teosinte plants treated with alternated stress revealed that drought as the first stress had no effect on the capacity to respond to a subsequent event of submergence (Figure 5). In fact, plants treated with alternated stress, and plants treated only with submergence had resembling replicates for most of the miRNAs. This suggests that teosinte was able to rapidly reverse the effects of drought and redirect the response towards submergence, even though drought had induced clear symptoms of stress in plants. The only miRNA that did not respond to submergence as the second stress was miR408, but a functional overlap with miR528 may have supplied the activity of LAC and PLC.

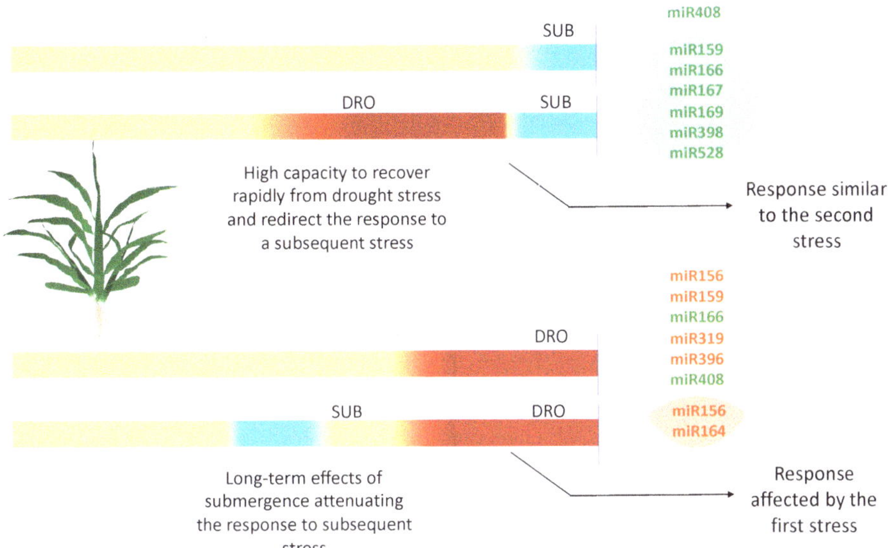

Figure 5. Hypothetical model of the response to alternated stress in teosinte. The influence of the first stress and the consequences in miRNA responsiveness to the second stress are indicated. Green, downregulated miRNAs; orange, upregulated miRNAs.

In the opposite order, submergence followed by drought produced an attenuated response to drought. The effects of submergence as the first stress had a long-term influence, considering that between the end of submergence and the beginning of drought, there was a gap of 8 days. This effect also suggests the plastic nature of plants to adapt to adverse factors and oscillate between different stresses.

Research has shown that recovery from submergence represents an additional stress by itself. Upon de-submergence, plants suffer reoxygenation stress conducive to the accumulation of ROS, reillumination stress, dehydration, and senescence [68]. During these processes, hormonal signals, ROS, protective proteins, and other molecular effectors could be induced in plants over a period of time that could be useful to survive any other event of stress such as drought, establishing an overlap between submergence and drought tolerance, as described in rice harboring the SUB1A gene [9].

Several aspects in the involvement of miRNAs in the responses to stress still need to be addressed for constructing a solid system that improves stress tolerance in plants. These include a detailed study of expression dynamics between miRNA and target genes during exposure to combined or alternated stress and during recovery in lines with contrasting tolerance, implications of other factors in the response such as intensity and quality of light during stress, circadian cycle on miRNA expression, plant age, the function of other sRNAs involved, and additional layers of regulation of gene expression. The present study constitutes a base for information that could be useful in further efforts to broaden the knowledge on the roles of miRNAs in stress response and explore miRNAs as tools for improving tolerance to multiple stresses.

4. Materials and Methods

4.1. Plant Materials

Maize seeds (*Zea mays ssp. mays*) of tropical homozygous line CML496, and teosinte (*Zea mays ssp. parviglumis*) were obtained from Centro de Investigación y de Estudios Avanzados (CINVESTAV) and Instituto Nacional de Investigaciones Forestales, Agrícolas y Pecuarias (INIFAP) seed collections, respectively. The seeds were germinated directly in pots with Sunshine Mix 3 (N: 36 ppm from

NO$_3$/9 ppm from NH$_4$; P: 7 ppm; K: 49 ppm; Ca: 45 ppm; Mg: 30 ppm; S: 39 ppm; B: 0.046 ppm; Cu: 0.01 ppm; Fe: 0.478 ppm; Mn: 0.251 ppm; Mo: 0.035 ppm; Na: 10 ppm; Cl: 8 ppm; Al: 0.48 ppm) (Sun Gro Horticulture Distribution Inc., Agawam, MA, USA), three seeds per pot. In the case of teosinte, the seeds were mechanically scarified to remove the coat before sowing.

4.2. Growth Conditions and Treatments

Maize and teosinte plants were grown in a plant growth room at 23 °C and 16 h/8 h light/dark cycle in order to lengthen the vegetative growth phase. The drought stress applied in this study was set to maintain plants at a constant intermediate intensity of drought rather than a progressive drought stress. For drought stress, pots with plants in vegetative stage V2, 12 days after sowing (12 DAS) were water-deprived for 6 days by interrupting water supply until half of their initial weight (water deficit) was reached. Then, water was added in small amounts on a daily basis to constantly maintain the plants in a water deficit condition at half their weight for 6 days. Submergence stress was applied on plants at 26 DAS, with pots in empty 32 cm deep submergence tanks that were filled with filtered tap water to increase the water column by 6 cm every 2 h for 10 h, until the maximum capacity was reached, and plants were fully submerged. The plants were maintained in this condition for 48 h after the first addition of water. For all the treatments, growth conditions were maintained at 23 °C and 16 h/8 h light/dark cycle.

For drought followed by submergence (drought-submergence), teosinte plants (14 DAS) were water-deprived for 6 days followed by 6 days under constant drought stress and then submerged for 48 h as previously described. For submergence followed by drought (submergence-drought), teosinte plants (12 DAS) were submerged for 48 h and removed from the water tank followed by depriving them of water until the plants reached half their initial weight (8 days). Then, small amounts of water were added to maintain a constant water deficit condition for 6 days. The single and alternated stress experiments with maize and teosinte were designed to end simultaneously at 28 DAS. At the end of the experiments, the shoots were immediately collected, frozen in liquid nitrogen, and stored at −80 °C for further processing. All the assays, including control and treatments, were performed with three biological replicates, with each replicate containing six plants.

4.3. Assessment of Physiological and Morphological Effects of Stress in Maize and Teosinte

For drought stress, the relative water content (RWC) was measured in the second leaf of maize plants under drought stress and control conditions, as previously described [69]. Briefly, square leaf sections (2 × 2 cm) were excised from plants and weighed to obtain the fresh weight, then submerged in deionized water for 24 h to obtain the turgid weight, and finally dried at 70 °C until constant dry weight was achieved. Then, RWC was calculated as [(fresh weight-dry weight)/(turgid weight-dry weight) × 100].

For submergence stress, the submergence tolerance coefficient (STC) was adapted from the coefficient applied for waterlogging [70]. Briefly, the length of shoots and roots of plants exposed to submergence or control conditions were noted at the end of experiments. The plants were then dried at 70 °C until a constant dry weight was reached. The sums of length and dry weight of plants exposed to the stress and control conditions were used to calculate the STC as [(sum of treated plants)/(sum of control plants)]. RWC and STC measurements were performed in three biological replicates, each consisting of six plants per treatment.

4.4. Total RNA Extraction and Construction of Small RNA Libraries

Total RNA was extracted from treated and control samples (three biological replicates, each consisting of a pool of six plants) with TRIzol reagent (Life Technologies, Carlsbad, CA, USA) according to the manufacturer's instructions. Total RNA was quantified using a Nanodrop ND-1000 Spectrophotometer (Thermo Fisher Scientific Inc., Waltham, MA, USA) and visualized by 1% agarose gel electrophoresis under denaturing conditions. Total RNA from triplicates of the same treatment was pooled (4 µg from each biological replicate in a total of 12 µg per treatment). The small RNA fraction (20–30 nt) was purified from total RNA in a denaturing 15% acrylamide gel. Next, a pre-adenylated adapter (linker1, Integrated DNA Technologies, San Diego, CA, USA) was ligated to the 3′ end of small RNAs with T4 RNA ligase II (New England Biolabs, Ipswich, MA, USA) in the absence of ATP. Then, the RNA adapter Illumina RA5 was ligated to the 5′ end of small RNAs with T4 RNA ligase 1 (Ambion, Austin, TX, USA). The ligated products were used as templates in a cDNA synthesis reaction with the primer RT Bridge and the retrotranscriptase MMuLV (Thermo Fisher Scientific Inc., Waltham, MA, USA), followed by a 15-cycle PCR reaction with Phusion High-Fidelity DNA polymerase (New England Biolabs). The PCR products were purified, quantified, and sequenced in the Illumina HiSEQ2500 at the Genomics Core of the University of California-Riverside (Riverside, CA, USA). Adapters and primers are listed in Table S7.

4.5. Bioinformatics Analysis of Small RNA Libraries

Raw sequencing datasets from small RNA libraries were processed with the CLC Genomics Workbench 12 (QIAGEN, Hilden, Germany) to remove adapters, low-quality reads, and reads shorter than 18 nt and longer than 35 nt. Clean reads were mapped to the B73 RefGen_v5 maize genome [71] and classified into the different non-coding RNA categories (rRNA, tRNA, snRNA, snoRNA, lncRNA, and repeats). In addition, clean reads were mapped to the maize chloroplast and mitochondria DNA (B73 RefGen_v4). The clean reads were searched against the plant miRNAs deposited in the miRBASE (Release 22.1) [31], allowing (1) zero mismatches, or (2) up to two mismatches to detect polymorphisms. miRNA sequence counts were normalized as TPM relative to the total clean read counts in the corresponding library. Individual miRNA sequences were then analyzed for differential expression. Fold change values were calculated as $\log_2$ (normalized counts in treatment/normalized counts in control) (Table S1). Only miRNA sequences with raw counts ≥ 50 in at least one of the libraries were considered for differential expression analysis, and fold change values ≥1 or ≤−1 were considered as upregulated or downregulated, respectively.

miRNA sequences with a high level of similarity were then grouped. All sequences that differed from the mature sequence by one or two nucleotides in the 5′ or 3′ ends were considered positional variants and were then grouped with the mature sequence to constitute a representative miRNA group. The sum of reads and fold change values were recalculated for grouped miRNA sequences (Table S2).

Small RNA sequencing data were deposited into the NCBI/GEO database with accession number GSE155050.

4.6. Assessment of miRNA Differential Expression by Quantitative RT-PCR (qRT-PCR)

A modified stem-loop method based on that previously described [72,73] was used to assess miRNA expression. Briefly, total RNA from treated and control plants were treated with DNase I (Thermo Fisher Scientific, Inc, Waltham, MA, USA.). Subsequently, 0.1 µg of treated RNA was used in 10 µL-cDNA synthesis reactions using the TaqMan miRNA reverse transcription kit (Applied Biosystems, Foster City, CA, USA) and miRNA specific stem-loop primers. For quantitative PCR, 20 µL-PCR reactions were prepared using Maxima Probe qPCR Master Mix (Thermo Fisher Scientific Inc, Waltham, MA, USA) with a miRNA-specific direct primer, PCR stem-loop reverse primer, and a universal fluorescein amidite (FAM)-labeled probe, in a PikoReal real-time thermal cycler (Thermo Fisher Scientific Inc.). Differential expression values were calculated using the ΔΔCt method [74] with the use of miR166c as an internal control to normalize the relative abundance of each miRNA. All primers and probes used are listed in Table S7.

4.7. Prediction of Target Genes

The psRNATarget analysis server (http://plantgrn.noble.org/psRNATarget) was used for identification of target genes of miRNAs responsive to submergence and drought [75]. DPMIND [76] was used to search for miRNA targets in nine maize degradomes.

Supplementary Materials: The following are available online at http://www.mdpi.com/2223-7747/9/10/1367/s1, Figure S1: PHRED score distribution of small RNA reads in maize and teosinte libraries; Figure S2: Size distribution of small RNA reads in maize and teosinte libraries; Table S1: Known miRNA sequences expressed in maize and teosinte libraries; Table S2: Grouped sequences of known microRNAs expressed in maize and teosinte libraries; Table S3: Sequence variants of microRNAs expressed in maize and teosinte libraries; Table S4: The most abundant reads that mapped to NADH dehydrogenase subunit 2 from the *Zea mays* organelle DNA; Table S5: The most abundant reads that mapped to the trnL-trnF intergenic spacer and tRNA-Phe (trnF) gene from the maize chloroplast; Table S6: List of predicted targets of submergence and drought-responsive miRNAs; Table S7: List of adapters and primers used in this study.

Author Contributions: Conceptualization, R.L. and B.E.B.-F.; formal analysis, E.B.S.-G., J.F.P.-B. and B.E.B.-F.; funding acquisition, J.M.P.-C., R.L. and B.E.B.-F.; investigation, E.B.S.-G., J.M.P.-C., R.L. and B.E.B.-F.; methodology, E.B.S.-G., J.F.P.-B. and A.A.H.-H.; resources, J.M.P.-C. and B.E.B.-F.; supervision, B.E.B.-F.; writing—original draft, B.E.B.-F.; writing—review & editing, E.B.S.-G., A.A.H.-H., J.M.P.-C. and R.L. All authors have read and agreed to the published version of the manuscript.

Funding: This material is based upon work supported by a grant from the University of California Institute for Mexico and the United States (UC MEXUS) and the Consejo Nacional de Ciencia y Tecnología de México (CONACYT) to B.E.B.-F. and R.L., grants from Consejo Nacional de Ciencia y Tecnología CONACYT—México (Ciencia Básica 287137) to J.M.P.-C., and (Ciencia Básica 169619) to B.E.B.-F., CONACYT postdoctoral fellowship to E.B.S.-G. and A.A.H.-H., and CONACYT postgraduate fellowship to J.F.P.-B.

Acknowledgments: We wish to thank Axel Thiessen-Favier (CINVESTAV-Irapuato; *in memoriam*) for providing maize seeds; Flavio Aragon Cuevas (INIFAP Oaxaca Valles Centrales) for providing teosinte seeds; Mikael Roose (UCR) for academic advice and administrative assistance; Glenn Hicks and John Robert Weger (Genomics Core, UCR) for technical advice and performing sRNA sequencing; Fabiola Hernández, L.E. María de Jesús Hernández, L.C. Yenny Reyes Roque and Francisco Campos (Universidad del Papaloapan) for administrative assistance.

Conflicts of Interest: The authors declare no conflict of interest. The sponsors had no role in the design, execution, interpretation, or writing of the study.

References

1. He, X.; Pan, M.; Wei, Z.; Wood, E.F.; Sheffield, J. A Global Drought and Flood Catalogue from 1950 to 2016. *Bull. Am. Meteorol. Soc.* **2020**, *101*, E508–E535. [CrossRef]

2. Lesk, C.; Rowhani, P.; Ramankutty, N. Influence of Extreme Weather Disasters on Global Crop Production. *Nature* **2016**, *529*, 84–87. [CrossRef]

3. Nguyen, L.T.T.; Osanai, Y.; Anderson, I.C.; Bange, M.P.; Tissue, D.T.; Singh, B.K. Flooding and Prolonged Drought Have Differential Legacy Impacts on Soil Nitrogen Cycling, Microbial Communities and Plant Productivity. *Plant Soil* **2018**, *431*, 371–387. [CrossRef]

4. Voesenek, L.A.C.J.; Bailey-Serres, J. Flood Adaptive Traits and Processes: An Overview. *New Phytol.* **2015**, *206*, 57–73. [CrossRef] [PubMed]

5. Fukao, T.; Barrera-Figueroa, B.E.; Juntawong, P.; Peña-Castro, J.M. Submergence and Waterlogging Stress in Plants: A Review Highlighting Research Opportunities and Understudied Aspects. *Front. Plant Sci.* **2019**, *10*, 340. [CrossRef] [PubMed]

6. Loreti, E.; Valeri, M.C.; Novi, G.; Perata, P. Gene Regulation and Survival under Hypoxia Requires Starch Availability and Metabolism. *Plant Physiol.* **2018**, *176*, 1286–1298. [CrossRef] [PubMed]

7. Bailey-Serres, J.; Lee, S.C.; Brinton, E. Waterproofing Crops: Effective Flooding Survival Strategies. *Plant Physiol.* **2012**, *160*, 1698–1709. [CrossRef]

8. Osakabe, Y.; Osakabe, K.; Shinozaki, K.; Tran, L.-S.P. Response of Plants to Water Stress. *Front. Plant Sci.* **2014**, *5*. [CrossRef]

9. Fukao, T.; Yeung, E.; Bailey-Serres, J. The Submergence Tolerance Regulator SUB1A Mediates Crosstalk between Submergence and Drought Tolerance in Rice. *Plant Cell* **2011**, *23*, 412–427. [CrossRef]

10. Liang, S.; Xiong, W.; Yin, C.; Xie, X.; Jin, Y.; Zhang, S.; Yang, B.; Ye, G.; Chen, S.; Luan, W. Overexpression of OsARD1 Improves Submergence, Drought, and Salt Tolerances of Seedling Through the Enhancement of Ethylene Synthesis in Rice. *Front. Plant Sci.* **2019**, *10*, 1088. [CrossRef] [PubMed]

11. Bin Rahman, A.N.M.R.; Zhang, J. Flood and Drought Tolerance in Rice: Opposite but May Coexist. *Food Energy Secur.* **2016**, *5*, 76–88. [CrossRef]

12. Amissah, L.; Mohren, G.M.J.; Kyereh, B.; Agyeman, V.K.; Poorter, L. Rainfall Seasonality and Drought Performance Shape the Distribution of Tropical Tree Species in Ghana. *Ecol. Evol.* **2018**, *8*, 8582–8597. [CrossRef] [PubMed]

13. He, X.; Sheffield, J. Lagged Compound Occurrence of Droughts and Pluvials Globally over the Past Seven Decades. *Geophys. Res. Lett.* **2020**. [CrossRef]

14. Kim, W.; Iizumi, T.; Nishimori, M. Global Patterns of Crop Production Losses Associated with Droughts from 1983 to 2009. *J. Appl. Meteorol. Climatol.* **2019**, *58*, 1233–1244. [CrossRef]

15. Li, Y.; Guan, K.; Schnitkey, G.D.; DeLucia, E.; Peng, B. Excessive Rainfall Leads to Maize Yield Loss of a Comparable Magnitude to Extreme Drought in the United States. *Glob. Chang. Biol.* **2019**, *25*, 2325–2337. [CrossRef]

16. González, J.D.J.S.; Corral, J.A.R.; García, G.M.; Ojeda, G.R.; Larios, L.D.L.C.; Holland, J.B.; Medrano, R.M.; Romero, G.E.G. Ecogeography of Teosinte. *PLoS ONE* **2018**, *13*, e0192676. [CrossRef]

17. Matsuoka, Y.; Vigouroux, Y.; Goodman, M.M.; Sanchez, G.J.; Buckler, E.; Doebley, J. A Single Domestication for Maize Shown by Multilocus Microsatellite Genotyping. *Proc. Natl. Acad. Sci. USA* **2002**, *99*, 6080–6084. [CrossRef]

18. Hufford, M.B.; Xu, X.; van Heerwaarden, J.; Pyhäjärvi, T.; Chia, J.-M.; Cartwright, R.A.; Elshire, R.J.; Glaubitz, J.C.; Guill, K.E.; Kaeppler, S.M.; et al. Comparative Population Genomics of Maize Domestication and Improvement. *Nat. Genet.* **2012**, *44*, 808–811. [CrossRef]

19. Liu, J.; Fernie, A.R.; Yan, J. The Past, Present, and Future of Maize Improvement: Domestication, Genomics, and Functional Genomic Routes toward Crop Enhancement. *Plant Commun.* **2020**, *1*, 100010. [CrossRef]

20. Mano, Y.; Omori, F. Breeding for Flooding Tolerant Maize Using "Teosinte" as a Germplasm Resource. *Plant Roots* **2007**, *1*, 17–21. [CrossRef]

21. Shriram, V.; Kumar, V.; Devarumath, R.M.; Khare, T.S.; Wani, S.H. MicroRNAs As Potential Targets for Abiotic Stress Tolerance in Plants. *Front. Plant Sci.* **2016**, *7*, 817. [CrossRef] [PubMed]

22. Song, X.; Li, Y.; Cao, X.; Qi, Y. MicroRNAs and Their Regulatory Roles in Plant–Environment Interactions. *Annu. Rev. Plant Biol.* **2019**, *70*, 489–525. [CrossRef] [PubMed]

23. Moldovan, D.; Spriggs, A.; Yang, J.; Pogson, B.J.; Dennis, E.S.; Wilson, I.W. Hypoxia-Responsive MicroRNAs and Trans-Acting Small Interfering RNAs in Arabidopsis. *J. Exp. Bot.* **2010**, *61*, 165–177. [CrossRef] [PubMed]

24. Pegler, J.; Oultram, J.; Grof, C.; Eamens, A. Profiling the Abiotic Stress Responsive MicroRNA Landscape of Arabidopsis Thaliana. *Plants* **2019**, *8*, 58. [CrossRef]

25. Jeong, D.-H.; Schmidt, S.A.; Rymarquis, L.A.; Park, S.; Ganssmann, M.; German, M.A.; Accerbi, M.; Zhai, J.; Fahlgren, N.; Fox, S.E.; et al. Parallel Analysis of RNA Ends Enhances Global Investigation of MicroRNAs and Target RNAs of Brachypodium Distachyon. *Genome Biol.* **2013**, *14*, R145. [CrossRef]

26. Bertolini, E.; Verelst, W.; Horner, D.S.; Gianfranceschi, L.; Piccolo, V.; Inzé, D.; Pè, M.E.; Mica, E. Addressing the Role of MicroRNAs in Reprogramming Leaf Growth during Drought Stress in Brachypodium Distachyon. *Mol. Plant* **2013**, *6*, 423–443. [CrossRef]

27. Barrera-Figueroa, B.E.; Gao, L.; Wu, Z.; Zhou, X.; Zhu, J.; Jin, H.; Liu, R.; Zhu, J.-K. High Throughput Sequencing Reveals Novel and Abiotic Stress-Regulated MicroRNAs in the Inflorescences of Rice. *BMC Plant Biol.* **2012**, *12*, 132. [CrossRef]

28. Jin, Q.; Xu, Y.; Mattson, N.; Li, X.; Wang, B.; Zhang, X.; Jiang, H.; Liu, X.; Wang, Y.; Yao, D. Identification of Submergence-Responsive MicroRNAs and Their Targets Reveals Complex MiRNA-Mediated Regulatory Networks in Lotus (Nelumbo Nucifera Gaertn). *Front. Plant Sci.* **2017**, *8*. [CrossRef] [PubMed]

29. Liu, X.; Zhang, X.; Sun, B.; Hao, L.; Liu, C.; Zhang, D.; Tang, H.; Li, C.; Li, Y.; Shi, Y.; et al. Genome-Wide Identification and Comparative Analysis of Drought-Related MicroRNAs in Two Maize Inbred Lines with Contrasting Drought Tolerance by Deep Sequencing. *PLoS ONE* **2019**, *14*, e0219176. [CrossRef] [PubMed]

30. Liu, Z.; Kumari, S.; Zhang, L.; Zheng, Y.; Ware, D. Characterization of MiRNAs in Response to Short-Term Waterlogging in Three Inbred Lines of Zea Mays. *PLoS ONE* **2012**, *7*, e39786. [CrossRef]

31. Kozomara, A.; Birgaoanu, M.; Griffiths-Jones, S. MiRBase: From MicroRNA Sequences to Function. *Nucleic Acids Res.* **2019**, *47*, D155–D162. [CrossRef] [PubMed]

32. Li, J.; Fu, F.; An, M.; Zhou, S.; She, Y.; Li, W. Differential Expression of MicroRNAs in Response to Drought Stress in Maize. *J. Integr. Agric.* **2013**, *12*, 1414–1422. [CrossRef]

33. Seeve, C.M.; Sunkar, R.; Zheng, Y.; Liu, L.; Liu, Z.; McMullen, M.; Nelson, S.; Sharp, R.E.; Oliver, M.J. Water-Deficit Responsive MicroRNAs in the Primary Root Growth Zone of Maize. *BMC Plant Biol.* **2019**, *19*, 447. [CrossRef]

34. Zhang, Z.; Wei, L.; Zou, X.; Tao, Y.; Liu, Z.; Zheng, Y. Submergence-Responsive MicroRNAs Are Potentially Involved in the Regulation of Morphological and Metabolic Adaptations in Maize Root Cells. *Ann. Bot.* **2008**, *102*, 509–519. [CrossRef] [PubMed]

35. Zhai, L.; Liu, Z.; Zou, X.; Jiang, Y.; Qiu, F.; Zheng, Y.; Zhang, Z. Genome-Wide Identification and Analysis of MicroRNA Responding to Long-Term Waterlogging in Crown Roots of Maize Seedlings. *Physiol. Plant.* **2013**, *147*, 181–193. [CrossRef]

36. Axtell, M.J.; Meyers, B.C. Revisiting Criteria for Plant MicroRNA Annotation in the Era of Big Data. *Plant Cell* **2018**, *30*, 272–284. [CrossRef]

37. Meng, X.; Li, L.; Narsai, R.; De Clercq, I.; Whelan, J.; Berkowitz, O. Mitochondrial Signalling Is Critical for Acclimation and Adaptation to Flooding in *Arabidopsis Thaliana*. *Plant J.* **2020**, *103*, 227–247. [CrossRef]

38. Mommer, L.; Pons, T.L.; Wolters-Arts, M.; Venema, J.H.; Visser, E.J.W. Submergence-Induced Morphological, Anatomical, and Biochemical Responses in a Terrestrial Species Affect Gas Diffusion Resistance and Photosynthetic Performance. *Plant Physiol.* **2005**, *139*, 497–508. [CrossRef]

39. Schmidt, R.R.; Weits, D.A.; Feulner, C.F.J.; van Dongen, J.T. Oxygen Sensing and Integrative Stress Signaling in Plants. *Plant Physiol.* **2018**, *176*, 1131–1142. [CrossRef]

40. Hou, Y.; Jiang, F.; Zheng, X.; Wu, Z. Identification and Analysis of Oxygen Responsive MicroRNAs in the Root of Wild Tomato (S. Habrochaites). *BMC Plant Biol.* **2019**, *19*, 100. [CrossRef]

41. Voesenek, L.A.C.J. Interactions Between Plant Hormones Regulate Submergence-Induced Shoot Elongation in the Flooding-Tolerant Dicot Rumex Palustris. *Ann. Bot.* **2003**, *91*, 205–211. [CrossRef]

42. Chuck, G.S.; Tobias, C.; Sun, L.; Kraemer, F.; Li, C.; Dibble, D.; Arora, R.; Bragg, J.N.; Vogel, J.P.; Singh, S.; et al. Overexpression of the Maize Corngrass1 MicroRNA Prevents Flowering, Improves Digestibility, and Increases Starch Content of Switchgrass. *Proc. Natl. Acad. Sci. USA* **2011**, *108*, 17550–17555. [CrossRef] [PubMed]

43. Mao, H.-D.; Yu, L.-J.; Li, Z.-J.; Yan, Y.; Han, R.; Liu, H.; Ma, M. Genome-Wide Analysis of the SPL Family Transcription Factors and Their Responses to Abiotic Stresses in Maize. *Plant Gene* **2016**, *6*, 1–12. [CrossRef]

44. Wu, G. Temporal Regulation of Shoot Development in Arabidopsis Thaliana by MiR156 and Its Target SPL3. *Development* **2006**, *133*, 3539–3547. [CrossRef]

45. Guo, H.-S.; Xie, Q.; Fei, J.-F.; Chua, N.-H. MicroRNA Directs MRNA Cleavage of the Transcription Factor *NAC1* to Downregulate Auxin Signals for Arabidopsis Lateral Root Development. *Plant Cell* **2005**, *17*, 1376–1386. [CrossRef] [PubMed]

46. Zhang, J.; Long, Y.; Xue, M.; Xiao, X.; Pei, X. Identification of MicroRNAs in Response to Drought in Common Wild Rice (Oryza Rufipogon Griff.) Shoots and Roots. *PLoS ONE* **2017**, *12*, e0170330. [CrossRef] [PubMed]

47. Singh, A.; Roy, S.; Singh, S.; Das, S.S.; Gautam, V.; Yadav, S.; Kumar, A.; Singh, A.; Samantha, S.; Sarkar, A.K. Phytohormonal Crosstalk Modulates the Expression of MiR166/165s, Target Class III HD-ZIPs, and KANADI Genes during Root Growth in Arabidopsis Thaliana. *Sci. Rep.* **2017**, *7*, 3408. [CrossRef] [PubMed]

48. Li, S.-B.; Xie, Z.-Z.; Hu, C.-G.; Zhang, J.-Z. A Review of Auxin Response Factors (ARFs) in Plants. *Front. Plant Sci.* **2016**, *7*. [CrossRef]

49. Li, W.-X.; Oono, Y.; Zhu, J.; He, X.-J.; Wu, J.-M.; Iida, K.; Lu, X.-Y.; Cui, X.; Jin, H.; Zhu, J.-K. The *Arabidopsis* NFYA5 Transcription Factor Is Regulated Transcriptionally and Posttranscriptionally to Promote Drought Resistance. *Plant Cell* **2008**, *20*, 2238–2251. [CrossRef]

50. Luan, M.; Xu, M.; Lu, Y.; Zhang, Q.; Zhang, L.; Zhang, C.; Fan, Y.; Lang, Z.; Wang, L. Family-Wide Survey of MiR169s and NF-YAs and Their Expression Profiles Response to Abiotic Stress in Maize Roots. *PLoS ONE* **2014**, *9*, e91369. [CrossRef]

51. Xu, M.Y.; Zhang, L.; Li, W.W.; Hu, X.L.; Wang, M.-B.; Fan, Y.L.; Zhang, C.Y.; Wang, L. Stress-Induced Early Flowering Is Mediated by MiR169 in Arabidopsis Thaliana. *J. Exp. Bot.* **2014**, *65*, 89–101. [CrossRef]

52. Sorin, C.; Declerck, M.; Christ, A.; Blein, T.; Ma, L.; Lelandais-Brière, C.; Njo, M.F.; Beeckman, T.; Crespi, M.; Hartmann, C. A MiR169 Isoform Regulates Specific NF-YA Targets and Root Architecture in Arabidopsis. *New Phytol.* **2014**, *202*, 1197–1211. [CrossRef]

53. Zhou, M.; Luo, H. Role of MicroRNA319 in Creeping Bentgrass Salinity and Drought Stress Response. *Plant Signal. Behav.* **2014**, *9*, e28700. [CrossRef]

54. Wang, L.; Gu, X.; Xu, D.; Wang, W.; Wang, H.; Zeng, M.; Chang, Z.; Huang, H.; Cui, X. MiR396-Targeted AtGRF Transcription Factors Are Required for Coordination of Cell Division and Differentiation during Leaf Development in Arabidopsis. *J. Exp. Bot.* **2011**, *62*, 761–773. [CrossRef] [PubMed]

55. Beltramino, M.; Ercoli, M.F.; Debernardi, J.M.; Goldy, C.; Rojas, A.M.L.; Nota, F.; Alvarez, M.E.; Vercruyssen, L.; Inzé, D.; Palatnik, J.F.; et al. Robust Increase of Leaf Size by Arabidopsis Thaliana GRF3-like Transcription Factors under Different Growth Conditions. *Sci. Rep.* **2018**, *8*, 13447. [CrossRef] [PubMed]

56. Gocal, G.F.W.; Sheldon, C.C.; Gubler, F.; Moritz, T.; Bagnall, D.J.; MacMillan, C.P.; Li, S.F.; Parish, R.W.; Dennis, E.S.; Weigel, D.; et al. *GAMYB-like* Genes, Flowering, and Gibberellin Signaling in Arabidopsis. *Plant Physiol.* **2001**, *127*, 1682–1693. [CrossRef]

57. Rivera-Contreras, I.K.; Zamora-Hernández, T.; Huerta-Heredia, A.A.; Capataz-Tafur, J.; Barrera-Figueroa, B.E.; Juntawong, P.; Peña-Castro, J.M. Transcriptomic Analysis of Submergence-Tolerant and Sensitive Brachypodium Distachyon Ecotypes Reveals Oxidative Stress as a Major Tolerance Factor. *Sci. Rep.* **2016**, *6*, 27686. [CrossRef]

58. Sunkar, R.; Kapoor, A.; Zhu, J.-K. Posttranscriptional Induction of Two Cu/Zn Superoxide Dismutase Genes in *Arabidopsis* Is Mediated by Downregulation of MiR398 and Important for Oxidative Stress Tolerance. *Plant Cell* **2006**, *18*, 2051–2065. [CrossRef] [PubMed]

59. De la Rosa, C.; Covarrubias, A.A.; Reyes, J.L. A Dicistronic Precursor Encoding MiR398 and the Legume-Specific MiR2119 Coregulates CSD1 and ADH1 MRNAs in Response to Water Deficit: MiR398 and MiR2119 Co-Regulate CSD1 and ADH1 MRNAs. *Plant Cell Environ.* **2019**, *42*, 133–144. [CrossRef]

60. Loreti, E.; Betti, F.; Ladera-Carmona, M.J.; Fontana, F.; Novi, G.; Valeri, M.C.; Perata, P. ARGONAUTE1 and ARGONAUTE4 Regulate Gene Expression and Hypoxia Tolerance. *Plant Physiol.* **2020**, *182*, 287–300. [CrossRef]

61. Gong, S.; Ding, Y.; Huang, S.; Zhu, C. Identification of MiRNAs and Their Target Genes Associated with Sweet Corn Seed Vigor by Combined Small RNA and Degradome Sequencing. *J. Agric. Food Chem.* **2015**, *63*, 5485–5491. [CrossRef]

62. Abdel-Ghany, S.E.; Pilon, M. MicroRNA-Mediated Systemic Down-Regulation of Copper Protein Expression in Response to Low Copper Availability in *Arabidopsis*. *J. Biol. Chem.* **2008**, *283*, 15932–15945. [CrossRef]

63. Sun, Q.; Liu, X.; Yang, J.; Liu, W.; Du, Q.; Wang, H.; Fu, C.; Li, W.-X. MicroRNA528 Affects Lodging Resistance of Maize by Regulating Lignin Biosynthesis under Nitrogen-Luxury Conditions. *Mol. Plant* **2018**, *11*, 806–814. [CrossRef]

64. Zhang, H.; Zhao, X.; Li, J.; Cai, H.; Deng, X.W.; Li, L. MicroRNA408 Is Critical for the *HY5-SPL7* Gene Network That Mediates the Coordinated Response to Light and Copper. *Plant Cell* **2014**, *26*, 4933–4953. [CrossRef]

65. Zhao, X.Y.; Hong, P.; Wu, J.Y.; Chen, X.B.; Ye, X.G.; Pan, Y.Y.; Wang, J.; Zhang, X.S. The Tae-MiR408-Mediated Control of *TaTOC1* Genes Transcription Is Required for the Regulation of Heading Time in Wheat. *Plant Physiol.* **2016**, *170*, 1578–1594. [CrossRef] [PubMed]

66. Yang, R.; Li, P.; Mei, H.; Wang, D.; Sun, J.; Yang, C.; Hao, L.; Cao, S.; Chu, C.; Hu, S.; et al. Fine-Tuning of MiR528 Accumulation Modulates Flowering Time in Rice. *Mol. Plant* **2019**, *12*, 1103–1113. [CrossRef]

67. Zhang, L.; Chia, J.-M.; Kumari, S.; Stein, J.C.; Liu, Z.; Narechania, A.; Maher, C.A.; Guill, K.; McMullen, M.D.; Ware, D. A Genome-Wide Characterization of MicroRNA Genes in Maize. *PLoS Genet.* **2009**, *5*, e1000716. [CrossRef] [PubMed]

68. Yeung, E.; Bailey-Serres, J.; Sasidharan, R. After The Deluge: Plant Revival Post-Flooding. *Trends Plant. Sci.* **2019**, *24*, 443–454. [CrossRef] [PubMed]

69. Barrs, H.; Weatherley, P. A Re-Examination of the Relative Turgidity Technique for Estimating Water Deficits in Leaves. *Aust. J. Biol. Sci.* **1962**, *15*, 413. [CrossRef]

70. Liu, Y.; Tang, B.; Zheng, Y.; Ma, K.; Xu, S.; Qiu, F. Screening Methods for Waterlogging Tolerance at Maize (*Zea mays* L.) Seedling Stage. *Agric. Sci. China* **2010**, *9*, 362–369. [CrossRef]

71. Portwood, J.L.; Woodhouse, M.R.; Cannon, E.K.; Gardiner, J.M.; Harper, L.C.; Schaeffer, M.L.; Walsh, J.R.; Sen, T.Z.; Cho, K.T.; Schott, D.A.; et al. MaizeGDB 2018: The Maize Multi-Genome Genetics and Genomics Database. *Nucleic Acids Res.* **2019**, *47*, D1146–D1154. [CrossRef]

72. Chen, C. Real-Time Quantification of MicroRNAs by Stem-Loop RT-PCR. *Nucleic Acids Res.* **2005**, *33*, e179. [CrossRef]

73. Jung, U.; Jiang, X.; Kaufmann, S.H.E.; Patzel, V. A Universal TaqMan-Based RT-PCR Protocol for Cost-Efficient Detection of Small Noncoding RNA. *RNA* **2013**, *19*, 1864–1873. [CrossRef] [PubMed]

74. Livak, K.J.; Schmittgen, T.D. Analysis of Relative Gene Expression Data Using Real-Time Quantitative PCR and the $2^{-\Delta\Delta CT}$ Method. *Methods* **2001**, *25*, 402–408. [CrossRef]

75. Dai, X.; Zhuang, Z.; Zhao, P.X. PsRNATarget: A Plant Small RNA Target Analysis Server (2017 Release). *Nucleic Acids Res.* **2018**, *46*, W49–W54. [CrossRef] [PubMed]

76. Fei, Y.; Wang, R.; Li, H.; Liu, S.; Zhang, H.; Huang, J. DPMIND: Degradome-Based Plant MiRNA–Target Interaction and Network Database. *Bioinformatics* **2018**, *34*, 1618–1620. [CrossRef] [PubMed]

Publisher's Note: MDPI stays neutral with regard to jurisdictional claims in published maps and institutional affiliations.

Article

Foliar Glycine Betaine or Hydrogen Peroxide Sprays Ameliorate Waterlogging Stress in Cape Gooseberry

Nicolas E. Castro-Duque, Cristhian C. Chávez-Arias * and Hermann Restrepo-Díaz

Departamento de Agronomía, Facultad de Ciencias Agrarias, Universidad Nacional de Colombia, Bogotá 111321, Colombia; necastrod@unal.edu.co (N.E.C.-D.); hrestrepod@unal.edu.co (H.R.-D.)
* Correspondence: ccchaveza@unal.edu.co; Tel.: +57-1-316-5000 (ext. 19018)

Received: 20 April 2020; Accepted: 13 May 2020; Published: 19 May 2020

Abstract: Exogenous glycine betaine (GB) or hydrogen peroxide (H_2O_2) application has not been explored to mitigate waterlogging stress in Andean fruit trees. The objective of this study was to evaluate foliar GB or H_2O_2 application on the physiological behavior of Cape gooseberry plants under waterlogging. Two separate experiments were carried out. In the first trial, the treatment groups were: (1) plants without waterlogging and with no foliar applications, (2) plants with waterlogging and without foliar applications, and (3) waterlogged plants with 25, 50, or 100 mM of H_2O_2 or GB, respectively. The treatments in the second trial were: (1) plants without waterlogging and with no foliar applications, (2) plants with waterlogging and without foliar applications, and (3) waterlogged plants with 100 mM of H_2O_2 or GB, respectively. In the first experiment, plants with waterlogging and with exogenous GB or H_2O_2 applications at a dose of 100 mM showed higher leaf water potential (−0.5 Mpa), dry weight (1.0 g), and stomatal conductance (95 mmol·m^{-2}·s^{-1}) values. In the second experiment, exogenously supplied GB or H_2O_2 also increased the relative growth rate, and leaf photosynthesis mitigating waterlogging stress. These results show that short-term GB or H_2O_2 supply can be a tool in managing waterlogging in Cape gooseberry.

Keywords: hypoxia; leaf gas exchange; waterlogging tolerance; organic compound; plant growth; *Physalis peruviana* L.

---

1. Introduction

Cape gooseberry (*Physalis peruviana* L.) is a plant that belongs to the *Solanaceae* family and its center of origin is located in the Andes, specifically in Peru, from where it expanded to various areas of the tropics and subtropics [1–3]. In Colombia, the production of this crop was 16,445 t, occupying 1312 ha during 2018 [4].

Climate change and variability alter the normal rainfall cycle causing floods of agricultural land and affecting crop production [5]. In Colombia, climate variability phenomena, such as the "La Niña" phenomenon, are characterized by an increase in rainfall that enhances the probability of floods [6,7]. In 2010, La Niña phenomenon produced an increase in rainfall, exceeding historical averages and causing a decrease in agricultural production from 7888 to 1515 t in Cundinamarca, one of the main producer departments of the country [4,7,8].

It has been reported that there is a high susceptibility of cultivated plants to waterlogging stress, affecting their growth, development, yield, and finally their survival [9,10]. One of the main effects of waterlogging is on plant growth. In this regard, several authors have observed that moderate or prolonged periods of O_2 deficit in the soil cause a low leaf area [11], a reduction in plant height [12], and an alteration in stem diameter [13]. In Cape gooseberry, short periods of waterlogging stress (6 days) also cause a decrease in plant height, leaf area, and stem diameter [14,15].

A reduction of growth parameters due to waterlogging may be associated with an impairment of the leaf gas exchange properties (stomatal conductance), chlorophyll content, and efficiency of photosystem II (PSII) [11,16–18]. Plants susceptible to waterlogging have been reported to show stomatal closure 24 h after the exposure to stress [19]. On the other hand, the leaf chlorophyll content can drop due to imbalances in the nutrient uptake or increased ethylene synthesis, causing impairment or decrease in the efficiency of PSII [19–21]. A previous experiment has also shown low stomatal conductance, leaf chlorophyll content, and F_v/F_m ratio in Cape gooseberry plants under moderate waterlogging periods (4 days) [15].

Waterlogging alters the plant water status due to stomatal closure [22]. The negative effects of periods of oxygen deprivation on the leaf water potential have been reported in cacao [17], bean [16], and tomato [23]. Likewise, the relative water content (RWC) has been widely used to describe the plant water status and has been correlated with the level of soil moisture [24]. In this regard, the RWC is a reliable variable to measure the susceptibility of plants to waterlogging [17,18]. These variables have also been useful to evaluate the susceptibility or efficiency of management techniques to O_2 deficit conditions in the soil in Andean fruit trees such as Lulo or Cape gooseberry [15,20,25].

Exogenous applications of compounds such as glycine betaine (GB) or hydrogen peroxide (H_2O_2) can help tolerate or lessen negative effects on plants under abiotic stress conditions by activating defense mechanisms or aiding plant growth, development, and productivity [26,27]. Some authors have reported that physiological parameters such as leaf gas exchange properties (photosynthesis), efficiency of PSII, water relations (water potential), growth, and antioxidant activity are favored by these compounds under waterlogging stress in different cultivated species [13,28]. Glycine betaine helps plants under abiotic stress conditions by acting as an osmolyte that protects cells [29], increases cell water retention [30], reduces levels of reactive oxygen species (ROS) and helps in the protection of the plasma membrane [31]. Regarding waterlogging stress, the exogenous application of this molecule has been little studied; however, Rasheed et al. [28] reported that GB applications caused an increase in plant biomass, leaf total chlorophyll, and K^+ concentration compared to fully waterlogged plants.

Hydrogen peroxide is a molecule that has also been studied to mitigate the effects of abiotic stresses in crops such as potato [32], tomato [33], bean [34,35], rice [36], maize [37] and soybean [13]. Different studies have concluded that exogenous H_2O_2 application helps leaf gas exchange properties (stomatal conductance and photosynthesis) [13], dry matter accumulation [35], leaf relative water content, and water potential [33,34], and plant height under different abiotic stresses [34,35]. Finally, the use of H_2O_2 has been little studied under waterlogging conditions. However, Andrade et al. [13] reported that pretreatments with H_2O_2 favored the increase in plant biomass, stomatal conductance, and net photosynthetic rate in soybean.

Increases in the intensity and frequency of rainfall in Colombia are estimated for the coming years [6,38]. For this reason, studies on the acclimatization response of Andean fruit trees to waterlogging scenarios have recently gained importance [14,20,21]. However, research on agronomic strategies to mitigate the negative impact of waterlogging with foliar GB and H_2O_2 sprays on Andean fruit trees has yet to be explored. Rasheed et al. [28] and Andrade et al. [13] mention the positive effect of these molecules on tolerance to waterlogging stress. For this reason, the objective of this study was to evaluate the exogenous application of different doses of GB or H_2O_2 on the physiological behavior of Cape gooseberry plants ecotype Colombia subjected to waterlogging, to determine the best molecule and dose to use to mitigate this stress.

2. Results

2.1. First Experiment: Evaluation of Different Doses of Glycine Betaine (GB) or Hydrogen Peroxide (H_2O_2) under a Waterlogging Period

Table 1 summarizes the effect of foliar GB and H_2O_2 sprays on the growth parameters of Cape gooseberry plants. Control plants without waterlogging (CWoW) (not exposed to waterlogging) generally showed the highest growth parameter values throughout the experiment compared to the other treatments. In this sense, foliar GB or H_2O_2 sprays mainly contributed to a greater stem length in Cape gooseberry plants under waterlogging conditions at 4 Days After Waterlogging (DAW), with approximate stem length values of 21 cm, while plants with waterlogging and without any foliar compound sprays (control with waterlogging, CWW) had a height of 16.70 cm. At 4 DAW, it was also observed that the foliar applications of both compounds at their different doses favored the leaf area, stem diameter, and shoot dry weight of waterlogged plants. At 13 DAW, the obtained results of plant growth showed that foliar GB applications at a concentration of 100 mM caused an increase mainly on stem diameter (0.53 cm), leaf area (222.56 cm^2), and shoot dry weight (1.06 g) in waterlogged plants compared to the CWW (0.42 cm, 116.14 cm^2, and 0.39 g, respectively). Regarding foliar H_2O_2 applications, this compound directly affected the plant height (22.10 cm) of waterlogged plants, while the CWW showed values of 16.56 cm. Table 2 shows how foliar GB or H_2O_2 applications at their different doses influenced physiological variables such as leaf temperature, stomatal conductance (g_s), efficiency of PSII (F_v/F_m), and water potential (Ψ_{wf}) in Cape gooseberry leaves at 4 and 13 DAW, respectively. It is observed that waterlogging causes a higher leaf temperature (26.89 and 28.49 °C) and lower g_s (157.10 and 42.76 mmol $CO_2 \cdot m^{-2} \cdot s^{-1}$) in plants at both sampling points. Foliar GB or H_2O_2 applications, mainly at a dose of 100 mM, caused a reduction in leaf temperature (22.27 and 24.71 °C, respectively) and an increase in g_s (180.70 and 191.78 mmol $CO_2 \cdot m^{-2} \cdot s^{-1}$, respectively) at 4 DAW, with similar values to the ones recorded for plants without waterlogging (18.85 °C and 194.42 mmol $CO_2 \cdot m^{-2} \cdot s^{-1}$). Similar trends were also observed for the variables previously described at 13 DAW. On the other hand, the F_v/F_m ratio was also conditioned by the treatments at both points, with the lowest ratio being obtained in the CWW treatment (around 0.6). Furthermore, foliar applications of these compounds helped to increase this ratio (~0.77). Finally, the Ψ_{wf} was higher in the control without waterlogging and in the GB treatment at a dose of 100 mM, compared to the other treatments in both samples. The Waterlogging Tolerance Coefficient (WTC) was obtained only at 13 DAW (Figure 1A), observing that the foliar GB application at 100 mM caused greater tolerance to waterlogging (0.52) compared to the rest of the treatments. Then, the correlation between leaf area and WTC (r^2 = 0.96) also confirmed that the foliar GB or H_2O_2 sprays at a concentration of 100 mM were the best at conferring tolerance to a waterlogging condition (Figure 1B).

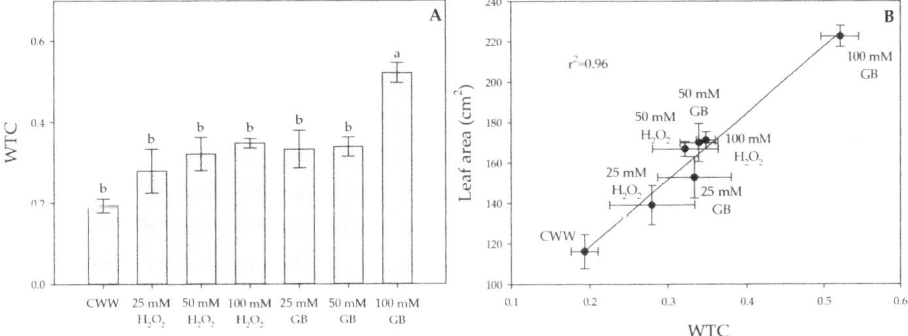

Figure 1. (**A**) Waterlogging Tolerance Coefficient (WTC) at 13 Days After Waterlogging (DAW) and (**B**) correlation between the leaf area and the WTC of Cape gooseberry (*Physalis peruviana* L.) plants ecotype Colombia subjected to a waterlogging period with treatments of control with waterlogging (CWW), 25 mM of hydrogen peroxide (H_2O_2), 50 mM of hydrogen peroxide (H_2O_2), 100 mM of hydrogen peroxide (H_2O_2), 25 mM of glycine betaine (GB), 50 mM of glycine betaine (GB) and 100 mM of glycine betaine (GB). Evaluated 13 days after waterlogging (DAW). Data represent the average of ten plants ± standard error per treatment (*n* = 5). Bars followed by different letters indicate statistically significant differences according to the Tukey test (*p* ≤ 0.05).

Table 1. Stem length, stem diameter, leaf area, and shoot dry weight of Cape gooseberry (*Physalis peruviana* L.) plants ecotype Colombia subjected to a waterlogging period and with exogenous applications of 25, 50, and 100 mM of hydrogen peroxide (H_2O_2) or glycine betaine (GB), respectively. Control without waterlogging (CWoW) and control with waterlogging (CWW). Evaluated at 4 and 13 days after waterlogging (DAW).

Treatment	4 DAW				13 DAW			
	Stem Length (cm)	Stem Diameter (cm)	Leaf Area (cm^2)	Shoot Dry Weight (g)	Stem Length (cm)	Stem Diameter (cm)	Leaf Area (cm^2)	Shoot Dry Weight (g)
CWoW	24.62 a [1]	0.664 a	369.74 a	1.45 a	25.50 a	0.72 a	376.58 a	2.06 a
CWW	16.70 c	0.428 b	156.37 d	0.48 e	16.56 d	0.42 c	116.14 d	0.39 d
Waterlogging + 25 mM H_2O_2	20.82 abc	0.524 ab	254.84 c	0.65 de	18.10 cd	0.55 bc	139.02 cd	0.57 cd
Waterlogging + 50 mM H_2O_2	19.50 bc	0.500 b	255.60 c	0.64 de	19.30 c	0.50 bc	166.86 c	0.65 cd
Waterlogging + 100 mM H_2O_2	22.40 ab	0.556 ab	311.18 b	0.99 c	22.10 b	0.55 b	171.18 c	0.71 c
Waterlogging + 25 mM GB	22.60 ab	0.470 b	240.54 c	0.74 d	18.40 cd	0.49 bc	152.78 c	0.67 cd
Waterlogging + 50 mM GB	19.98 bc	0.490 b	266.36 c	1.00 c	18.84 cd	0.52 bc	169.98 c	0.69 cd
Waterlogging + 100 mM GB	20.90 ab	0.484 b	318.51 b	1.25 b	18.20 cd	0.53 bc	222.56 b	1.06 b
Significance (*p* value)	0.0001	0.0013	0.0000	0.0000	0.0000	0.0000	0.0000	0.0000
CV (%) [2]	9.67	14.45	6.35	9.26	5.99	12.17	9.33	18.07

[1] Values ($n = 5$) within a column followed by different letters are significantly different from $p \leq 0.05$ according to the Tukey test. [2] CV: Coefficient of variation.

Table 2. Estimation of physiological parameters: Leaf temperature, stomatal conductance, efficiency of photosystem II (PSII) (Fv/Fm), and leaf water potential in Cape gooseberry (*Physalis peruviana* L.) plants ecotype Colombia subjected to a waterlogging period with exogenous applications of 25, 50, and 100 mM of hydrogen peroxide (H_2O_2) or glycine betaine (GB), respectively. Control without waterlogging (CWoW) and control with waterlogging (CWW). Evaluated at 4 and 13 Days After Waterlogging (DAW).

Treatment	4 DAW				13 DAW			
	Leaf Temperature (°C)	Stomatal Conductance (mmol $CO_2 \cdot m^{-2} \cdot s^{-1}$)	Efficiency of PSII (F_v/F_m)	Leaf Water Potential (−Mpa)	Leaf Temperature (°C)	Stomatal Conductance (mmol $CO_2 \cdot m^{-2} \cdot s^{-1}$)	Efficiency of PSII (F_v/F_m)	Leaf Water Potential (−Mpa)
CWoW	18.85 a [1]	194.42 a	0.82 a	0.23 a	18.93 a	190.26 a	0.81 ab	0.30 a
CWW	26.89 d	157.10 c	0.57 c	0.54 c	28.49 d	42.76 d	0.64 ab	0.66 c
Waterlogging + 25 mM H_2O_2	24.52 c	189.66 a	0.63 bc	0.49 bc	26.98 cd	66.16 c	0.77 ab	0.62 c
Waterlogging + 50 mM H_2O_2	24.16 c	149.00 b	0.70 abc	0.45 bc	26.92 cd	93.40 b	0.84 a	0.59 bc
Waterlogging + 100 mM H_2O_2	24.71 c	191.78 a	0.70 abc	0.420 bc	24.16 b	94.56 b	0.78 ab	0.53 bc
Waterlogging + 25 mM GB	22.12 b	176.78 b	0.74 ab	0.49 bc	26.49 c	66.62 c	0.63 b	0.54 bc
Waterlogging + 50 mM GB	23.87 c	156.90 b	0.81 a	0.48 bc	26.49 c	93.36 b	0.74 ab	0.57 bc
Waterlogging + 100 mM GB	22.27 b	180.70 a	0.80 a	0.37 ab	24.07 b	95.64 b	0.75 ab	0.45 ab
Significance (*p* value)	0.0000	0.0000	0.0000	0.0001	0.0000	0.0000	0.0270	0.0175
CV (%) [2]	1.86	6.52	4.67	18.74	3.52	10.96	8.6	18.97

[1] Values (*n* = 5) within a column followed by different letters are significantly different from $p \leq 0.05$ according to the Tukey test. [2] CV: Coefficient of variation.

2.2. Experiment 2: Evaluation of the Most Efficient Doses of Glycine Betaine (GB) and Hydrogen Peroxide (H₂O₂) in Plants Exposed to Two Waterlogging Periods

Growth parameters (stem diameter, shoot dry weight, and leaf area) showed differences ($p \leq 0.05$) between treatments throughout experiment 2 (Table 3). Regarding stem diameter, it was observed that foliar GB or H_2O_2 applications began to cause an increase in this variable under stress conditions from 4 DAW, maintaining this trend during the experiment. At the end of the trial (36 DAW), higher stem diameter (0.49 cm) values were observed in plants without waterlogging and with no foliar applications compared to waterlogged plants without any foliar application (0.31 cm). Exogenously supplied GB or H_2O_2 promoted an increase in stem diameter in waterlogged plants (0.39 cm for GB and 0.35 for H_2O_2). On the other hand, the shoot dry weight was considerably higher in plants without waterlogging (9.90 g) than in waterlogged plants with or without foliar applications (~1.6 g). Finally, foliar GB or H_2O_2 sprays caused an increase in leaf area (95.53 cm^2 for GB and 91.54 cm^2 for H_2O_2) compared to only waterlogged plants (25.82 cm^2). However, plants under waterlogging with foliar applications did not reach the values obtained in plants without conditions of hypoxia in the soil or foliar sprays (1236.10 cm^2) at the end of the trial (36 DAW).

The stomatal conductance (g_s), leaf relative chlorophyll content (soil plant analysis development (SPAD) readings), and efficiency of PSII (F_v/F_m) were significantly affected by the treatments ($p \leq 0.05$) (Table 4). Regarding g_s, it is observed that the group of waterlogged plants with and without foliar applications of the compounds always showed lower values (between 9.90 and 40.63 mmol $CO_2 \cdot m^{-2} \cdot s^{-1}$) throughout the experiment (at 6, 12, 18, and 36 DAW, respectively) compared to control plants without waterlogging (CWoW) (between 118 and 216 mmol $CO_2 \cdot m^{-2} \cdot s^{-1}$ at the different sampling points). However, it was observed that GB or H_2O_2 sprays favored this variable in the group of waterlogged plants, observing significant differences ($p \leq 0.05$) (31.78 and 40.63 mmol $CO_2 \cdot m^{-2} \cdot s^{-1}$, respectively) compared to CWW (9.90 mmol $CO_2 \cdot m^{-2} \cdot s^{-1}$) at the end of the experiment (36 DAW).

Similar results were also observed for SPAD readings, with the highest values for the CWoW. It is important to note that the treatments with foliar GB or H_2O_2 application began to show higher values compared to the waterlogged control from 18 DAW to the end of the experiment (36 DAW). Finally, the efficiency of PSII (F_v/F_m) was also affected by the waterlogging conditions, with the greatest negative effects registered at 36 DAW. At this point, the treatments with waterlogging and foliar GB or H_2O_2 application had a positive effect on the F_v/F_m ratio (0.42 for both treatments) compared to CWW (0.28), whereas control plants and without waterlogging showed higher values (0.74) throughout the experiment.

The Relative Water Content (RWC) showed significant differences between the treatments from 18 DAW (Figure 2A). The best water status throughout the experiment was observed in plants without waterlogging with an RWC of 80%. Therefore, foliar GB or H_2O_2 applications favored the RWC of Cape gooseberry plants under waterlogging conditions throughout the experiment. Plants with foliar GB applications showed a higher RWC than waterlogged plants treated with H_2O_2 (43.97%) and plants with only waterlogging (34.18%) at 18 DAW. Between 18 and 36 DAW, it was observed that the waterlogging conditions continued to decrease the RWC mainly in the groups of waterlogged plants treated with H_2O_2 (36.37%) and CWW (34.18%). Significant differences ($p \leq 0.05$) were only obtained on the Relative Growth Rate (RGR) (Figure 2B) at 36 DAW. The treatment with foliar GB application at a concentration of 100 mM (0.022 cm) obtained a higher RGR compared to the other plant groups (0.017 cm for control without waterlogging, 0.016 cm for 100 mM, and 0.002 cm for control with waterlogging).

Table 3. Growth parameters (leaf area, shoot dry weight and stem diameter) of Cape gooseberry (*Physalis peruviana* L.) plants ecotype Colombia subjected to two waterlogging periods with control treatments without waterlogging (CWoW), control with waterlogging (CWW), 100 mM of hydrogen peroxide (H_2O_2) and 100 mM of glycine betaine (GB). Evaluated at 6, 18, 24, and 36 days after waterlogging (DAW).

Treatment	Stem Diameter (cm)				Shoot Dry Weight (g)				Foliar Area (cm^2)			
	DAW				DAW				DAW			
	6	18	24	36	6	18	24	36	6	18	24	36
CWoW	0.29 a [1]	0.34 a	0.43 a	0.49 a	2.61 a	6.14 a	5.95 a	9.90 a	584.7 a	882.2 a	1163.2 a	1236.1 a
CWW	0.21 b	0.24 c	0.29 c	0.31 c	0.93 b	2.22 b	2.17 b	1.59 b	104.2 c	100.9 b	106.8 c	25.8 c
Waterlogging + 100 mM H_2O_2	0.25 ab	0.27 bc	0.32 bc	0.35 b	1.53 ab	2.02 b	2.83 b	1.99 b	154.3 b	142.1 b	202.4 bc	91.5 b
Waterlogging + 100 mM GB	0.30 a	0.31 ab	0.34 b	0.39 b	2.21 ab	2.33 b	3.06 b	1.75 b	135.4 ab	129.8 b	221.9 b	95.5 b
Significance (*p* value)	0.006	0.000	0.000	0.000	0.046	0.000	0.000	0.000	0.000	0.000	0.000	0.000
CV (%) [2]	11.69	6.56	6.12	5.53	43.11	20.71	22.76	60.06	6.41	15.14	12.08	7.84

[1] Values (*n* = 5) within a column followed by different letters are significantly different from $p \leq 0.05$ according to the Tukey test. [2] CV: Coefficient of variation.

Table 4. Physiological parameters (stomatal conductance, soil plant analysis development (SPAD) chlorophylls, and efficiency of PSII) of Cape gooseberry (*Physalis peruviana* L.) plants ecotype Colombia subjected to two waterlogging periods with treatments of control without waterlogging (CWoW), control with waterlogging (CWW), 100 mM of hydrogen peroxide (H_2O_2) and 100 mM of glycine betaine (GB). Evaluated at 6, 18, 24, and 36 days after waterlogging (DAW).

Treatment	Stomatal Conductance (mmol $CO_2 \cdot m^{-2} \cdot s^{-1}$)				SPAD Chlorophylls				Efficiency of PSII (F_v/F_m)			
	DAW				DAW				DAW			
	6	18	24	36	6	18	24	36	6	18	24	36
CWoW	195.8 a [1]	129.1 a	215.8 a	117.7 a	38.3 a	32.8 a	36.5 a	42.2 a	0.83 a	0.82 a	0.75 a	0.74 a
CWW	18.9 b	14.1 b	13.9 b	9.90 c	31.3 ab	19.6 b	18.7 b	11.0 c	0.67 ab	0.68 a	0.58 a	0.28 c
Waterlogging + 100 mM H_2O_2	27.6 b	24.9 b	29.1 b	40.6 b	33.6 ab	23.6 ab	24.7 ab	17.5 b	0.65 b	0.77 a	0.45 a	0.42 b
Waterlogging + 100 mM GB	32.4 b	20.6 b	33.2 b	31.8 b	29.8 b	23.1 ab	28.6 ab	12.0 bc	0.78 ab	0.73 a	0.69 a	0.42 b
Significance (*p* value)	0.000	0.000	0.000	0.000	0.020	0.012	0.008	0.000	0.025	0.099	0.154	0.000
CV (%) [2]	16.53	46.02	22.3	12.94	10.25	19.28	22.04	13.75	10.47	9.74	30.14	11.69

[1] Values (*n* = 5) within a column followed by different letters are significantly different from $p \leq 0.05$ according to the Tukey test. [2] CV: Coefficient of variation.

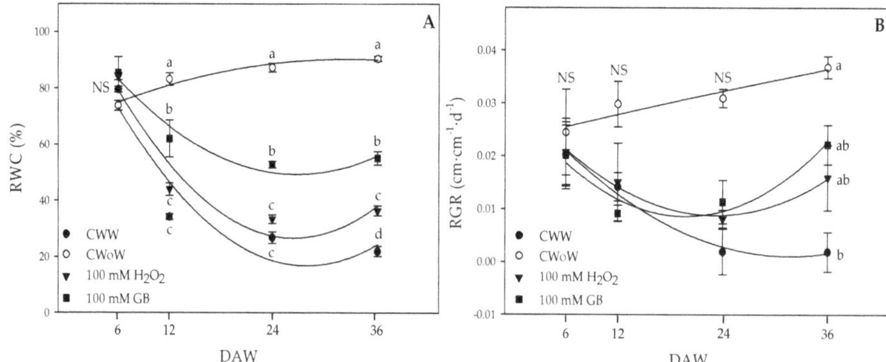

Figure 2. (**A**) Relative Water Content (RWC) and (**B**) Relative Growth Rate (RGR) of Cape gooseberry (*Physalis peruviana* L.) plants ecotype Colombia subjected to two waterlogging periods. Filled circle: control treatment subjected to waterlogging, circle without filling: control treatment without waterlogging, filled triangle: treatment with a concentration of 100 mM of H_2O_2, filled square: treatment with a concentration of 100 mM of GB. Evaluated at 6, 12, 24, and 36 days after waterlogging (DAW). Data represent the mean of five plants ± standard error per treatment ($n = 5$). Points followed by different letters indicate statistically significant differences according to the Tukey test ($p \leq 0.05$).

Photosynthesis (Figure 3A) and Canopy Temperature Index (CTI) (Figure 3B) were calculated at 36 DAW. The highest photosynthesis value was obtained in CWoW plants (8.37 mmol·m^{-2}·s^{-1}). Waterlogging conditions were observed to cause a reduction in the photosynthesis rate of 79% (1.75 mmol·m^{-2}·s^{-1}). However, photosynthesis under oxygen deficiency in the soil was stimulated by foliar application of GB (4.99 mmol·m^{-2}·s^{-1}) and H_2O_2 (2.65 mmol· m^{-2}·s^{-1}) (Figure 3A). Similar trends were observed in the CTI where the treatment with foliar GB application (0.76) favored the canopy temperature of Cape gooseberry plants with waterlogging, obtaining higher values than those recorded in plants subjected to waterlogging (0.30) (Figure 3B).

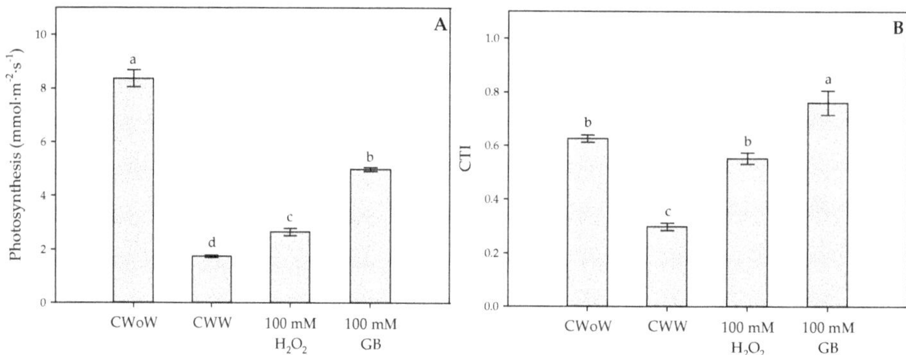

Figure 3. (**A**) Photosynthesis and (**B**) canopy temperature index (CTI) of Cape gooseberry (*Physalis peruviana* L.) plants ecotype Colombia subjected to two waterlogging periods with treatments of control without waterlogging (CWoW), control with waterlogging (CWW), 100 mM of hydrogen peroxide (H_2O_2) and 100 mM of glycine betaine (GB). Evaluated at 36 Days After Waterlogging (DAW). Each point represents the mean of the four values. Data represent the average of ten plants ± standard error per treatment ($n = 5$). Bars followed by different letters indicate statistically significant differences according to the Tukey test ($p \leq 0.05$).

3. Discussion

Waterlogging stress causes adverse effects on the physiological and biochemical parameters of cultivated plants [39]. Rao and Li [19] point out that periods of oxygen deficit in the soil greater than 24 h cause a decrease in the leaf gas exchange properties (photosynthesis and stomatal conductance), leaf chlorophyll content, plant growth, and water status. Likewise, stomatal closure and low plant water status caused by waterlogging generate an increase in leaf temperature [40,41]. These responses, induced by waterlogging conditions in the soil, may be associated with physiological dysfunctions, such as impaired water and nutrient uptake caused by a reduction in root hydraulic conductance or root cell death [42,43], restricted CO_2 entry due to stomatal closure [44,45], low Rubisco activation during CO_2 assimilation [19], oxidative damage on photosystem II caused by reactive oxygen species (ROS) [44,46], and increased chlorophyllase activity (chlorophyll degradation) and ethylene synthesis [19,47]. High ethylene production in plants under conditions of anoxia or hypoxia is caused by fermentative enzymes (fructose-1,6-bisphosphate aldolase (ALD), enolase (ENO), pyruvate decarboxylase (PDC), and alcohol dehydrogenase 2 (ADH2)) as an adaptive response to oxygen deficit in the soil [48]. Based on the above, the treatment with only waterlogged plants showed physiological affectations such as a reduction of the leaf gas exchange properties, low plant growth, water status, and leaf chlorophyll content, and an increase in leaf temperature in both experiments (Tables 1–4; Figures 2 and 3). Similar observations have also been reported in Cape gooseberry plants under oxygen deficit conditions in the soil [14,15,25].

Foliar GB or H_2O_2 applications (mainly at a dose of 100 mM of each compound) helped to mitigate the negative effects caused by waterlogging conditions in the soil by favoring the evaluated variables in both experiments (Tables 1–4; Figures 1–3). The results obtained in this study confirm similar observations found in other cultivated species in which the exogenously supplied GB and H_2O_2 helped to alleviate the negative impact caused by waterlogging conditions on plant physiology. The beneficial role of foliar applications of these compounds on the physiological and biochemical parameters of plants under moderate (approximately 15 days) waterlogging periods has been documented for soybean (exogenous H_2O_2 supply also increased leaf gas exchange parameters and growth) [13] and tomato (foliar GB application also enhanced growth and chlorophyll concentration) [28]. Glycine betaine (GB) is a low molecular weight ammonium compound easily absorbed by roots or leaves. It helps to alleviate abiotic stress conditions in plants since it can participate in different physiological processes, such as osmoregulation, stabilization of the quaternary structure of proteins, enzymes (e.g., Rubisco) and membranes, protection of the photosynthetic apparatus, maintenance of the electron flow in the thylakoid membranes and regulation of antioxidant enzymes activity [49,50]. On the other hand, hydrogen peroxide (H_2O_2) has a signaling role in the mediation of physiological processes in the plant during the process of acclimatization to different types of abiotic stress. Some of these processes are antioxidant defense, stomatal behavior, regulation of the photosynthesis rate, and promotion of biosynthesis of compatible osmolytes to maintain leaf water content [51–53].

Foliar GB and H_2O_2 applications favored growth variables (stem length and diameter, leaf area, shoot dry weight, and relative growth rate) and physiological variables such as leaf temperature, stomatal conductance (g_s), maximum photochemical efficiency of PSII (F_v/F_m), leaf water potential (Ψ_{wf}), relative water content (RWC), chlorophyll content, and net photosynthesis (Pn) in Cape gooseberry plants under waterlogging conditions (Tables 1–4; Figures 2 and 3). It was demonstrated that exogenous applications of H_2O_2 increase biomass in cucumber (*Cucumis sativus* L.) plants under water stress conditions compared to plants without H_2O_2 [52]. Stem diameter is another important trait to determine the plant's adaptive response and evaluate the efficiency of a treatment to alleviate stress caused by hypoxia conditions in the soil [20,21]. Soybean (*Glycine max* L.) seeds treated exogenously with H_2O_2 increased the stem diameter of seedlings exposed to a prolonged waterlogging period (32 days) compared to seedlings under conditions of oxygen deficit in the soil and without the application of this compound [13]. It has also been confirmed that exogenously supplied H_2O_2 stimulates aerenchyma formation in plants, such as rice [54].

In this study, it was observed that foliar H_2O_2 sprays increase stem diameter by promoting lysigenous aerenchyma (internal gas space) formation for adaptation to waterlogging conditions [54]. Finally, the positive effect of foliar GB applications on growth parameters was also reported in tomato (*Solanum lycopersicum* L.) plants favoring dry matter accumulation in the shoot and root under waterlogging conditions [28].

An increase in leaf gas exchange parameters (Pn and g_s), chlorophyll content, F_v/F_m and plant water relations (Ψ_{wf} and RWC) after exogenous application of H_2O_2 and GB has also been reported by several authors under abiotic stress conditions. For example, Andrade et al. [13] showed that exogenous H_2O_2 applications increased the shoot and root dry matter, root volume, stem diameter, Pn, g_s, transpiration, leaf chlorophyll content and activity of antioxidative enzymes in soybean plants under waterlogging stress. On the other hand, Sorwong and Sakhonwasee [55] also observed a positive effect on leaf gas exchange parameters (Pn and g_s) and F_v/F_m ratio in Mexican marigold (*Tagetes erecta* L.) plants subjected to abiotic stress after foliar treatment with GB. Likewise, GB application caused a significant increase in the content of photosynthetic pigments (chlorophyll and carotenoids) in tomato plants exposed to conditions of oxygen deficiency in the soil [28].

In this study, H_2O_2 was found to mitigate the negative effects of waterlogging by improving plant growth. This response may be associated with the fact that H_2O_2 can activate signaling pathways to stimulate cell proliferation and also increase leaf area [56], cell differentiation [57], and plant elongation [58]. Additionally, this compound enhanced the gas exchange parameters, efficiency of PSII, and leaf chlorophyll content in plants under conditions of oxygen deficit in the soil. The positive results of exogenous H_2O_2 application could be related to the induction of plant tolerance to different abiotic stress conditions by modulating processes involved in ROS detoxification, the increase in glutathione (GSH) and ascorbate (AsA) contents, the biosynthesis of proline to protect the photosynthetic machinery, and the regulation of stomatal conductance [13,59]. Finally, H_2O_2 sprays favored the accumulation of compatible solutes such as proline, maintaining the leaf RWC [51,52].

Applications of GB to Cape gooseberry plants under waterlogging conditions showed a biostimulant effect on plant growth and the evaluated physiological parameters. It has been reported that foliar GB sprays can quickly penetrate the leaf surface and be easily transported to other plant organs, where it would contribute to improving tolerance to different types of abiotic stress [60]. This compound may be involved in the inhibition of ROS accumulation, protection of photosynthetic machinery, accumulation of compatible solutes to maintain turgidity in cells, activation of some stress-related genes, and protection of the cell membrane and quaternary structure of proteins [61,62]. Likewise, GB regulates stomatal movements by prolonging their opening under a condition of abiotic stress as a consequence of increased osmoprotective activity [63]. Finally, an increase in the activity of antioxidant enzymes such as catalases (CAT) and peroxidases (POD) has been reported in response to GB application by regulating the oxidative stress generated by waterlogging conditions in the soil [13]. Leaf temperature has been used as an indicator of stress in plants [64]. In the present study, exogenous H_2O_2 and GB applications favored leaf temperature regulation by improving the CTI (Figure 3 and Table 2). As mentioned above, these molecules have an osmoprotective or stomatal regulation effect that directly favors the plant water status, resulting in lower leaf temperature. Finally, it was observed that WTC and CTI can be associated with physiological variables such as leaf gas exchange properties or plant growth, which can be considered to evaluate the effectiveness of treatments to mitigate the effects of waterlogging [65].

The most inexpensive way to deal with short-term waterlogging periods is improving plant tolerance to low oxygen conditions in the soil [66]. In recent years, studies have focused on the evaluation of the physiological and biochemical responses to different periods of waterlogging, finding that Cape gooseberry plants are highly susceptible to short-term periods of this stress [14,15,67]. In this regard, the analysis of easy-to-use, economic, and profitable techniques is necessary to mitigate the effects of short or intermittent periods of waterlogging. It has been observed that the use of foliar applications of organic compounds or elicitors is an important tool to help plants acquire tolerance to short periods of waterlogging [19,66]. The results of present study demonstrated that exogenously supplied GB or H_2O_2 may be an appropriate agronomic practice to counteract the negative effects of non-prolonged waterlogging periods, since these treatments increased the tolerance of a susceptible plant (expressed as WTI) such as Cape gooseberry. Exogenous application of elicitors (hormones) or organic compounds (botanical extracts) have also favored plant tolerance to a short period of stress [25]. However, the novelty of this research was that foliar application of the evaluated compounds increased Cape gooseberry tolerance to intermittent periods of waterlogging.

There is still a lack of knowledge to understand the advantages or disadvantages of several techniques to manage Andean fruit crops under different waterlogging periods. The obtained results indicate a series of advantages from the physiological and crop management points of view; it is important to highlight that the exogenous application of organic compounds or elicitors is a technique that can help plants to cope with the unfavorable condition of waterlogging relatively fast when soil conditions are limiting. These observations are in agreement with other studies that report the foliar application of organic or nitrogenous compounds, elicitors, or phytohormones as an efficient technique to reduce the negative effects of short or moderate periods of waterlogging [19,20,25]. Another advantage of this study is that the analysis of responses such as growth, stem diameter, plant water status, leaf gas exchange, and plant temperature has recently helped to understand the acclimatization mechanisms of plants of the *Solanaceae* family, mainly Andean fruit trees, to conditions of oxygen deprivation in the soil in tropical countries. The obtained information has provided support for the evaluation and development of efficient crop management techniques [21,68,69]. However, it has been reported that foliar applications of compounds such as growth regulators, nutrients, or elicitors are not effective for prolonged periods of waterlogging. Thus, the combination of techniques such as soil drainage and crop management practices (foliar applications) is still required [66].

4. Materials and Methods

4.1. Plant Material and General Growth Conditions of the Experiments

Two experiments were carried out separately between March and July 2019 under greenhouse conditions at the Faculty of Agricultural Sciences of the Universidad Nacional de Colombia, (4°35′56″ and 74°04′51″), Bogotá campus. The general growth conditions in the greenhouse in both experiments were: average temperature of 25/15 °C, 60%–80% relative humidity, and a natural photoperiod of 12 h. For both experiments, 2-month-old Cape gooseberry (*Physalis peruviana* L.) ecotype Colombia seedlings were transplanted into 2 L capacity plastic pots with a mixture of peat and sand (3:1 *v/v*) as substrate. The plants were watered every day from the transplant until the beginning of treatments with 50 mL of a nutritive solution prepared with a complete liquid fertilizer (Nutriponic®, Walco SA, Colombia) at a concentration of 5 mL·L^{-1} H_2O. The final concentration of the nutrient solution was as follows: 2.08 mM Ca (NO$_3$)$_2$·4H$_2$O, 1.99 mM MgSO$_4$·7 H$_2$O, 2.00 mM NH$_4$H$_2$PO$_4$, 10.09 mM KNO$_3$, 46.26 nM H$_3$BO$_3$, 0.45 nM Na$_2$MoO$_4$·2H$_2$O, 0.32 nM CuSO$_4$·5H$_2$O, 9.19 nM MnCl$_2$·4H$_2$O, 0.76 nM ZnSO$_4$·7H$_2$O, and 19.75 nM FeSO$_4$·H$_2$O. The volume of irrigated water was obtained using the technique described by Hainaut et al. [70], in which daily evapotranspiration was estimated gravimetrically. The treatments in each experiment were established from 45 days after transplanting (DAT) and these are detailed below:

4.1.1. Experiment 1—Evaluation of Different Glycine Betaine (GB) or Hydrogen Peroxide Doses (H_2O_2) under a Waterlogging Period

At 45 DAT, when plants reached five fully expanded leaves, eight treatment groups were established to estimate the effect of foliar GB and H_2O_2 sprays under waterlogging conditions. The treatment groups were as follows: (1) plants without waterlogging and with no GB or H_2O_2 application (absolute control), (2) waterlogged plants with no GB or H_2O_2 application, (waterlogged control), (3) waterlogged plants with foliar H_2O_2 sprays at concentrations of 25, 50 or 100 mM (JGB SA, Cali, Colombia), (4) waterlogged plants with foliar GB sprays at concentrations of 25, 50 or 100 mM

(VivaGrow, Westminster, CO, USA). Waterlogging treatments consisted in placing the plants in plastic containers with dimensions of 53 × 53 × 30 cm and a capacity of 120 L, filled until reaching a 5 cm water level on the root neck. The plants were subjected to a 6-day waterlogging period (between 45 and 51 DAT) since previous studies showed that this period caused damage to Cape gooseberry plants [14,15,67]. Foliar GB or H_2O_2 applications were performed at 0, 6, and 10 DAW using an application volume of 20 mL, wetting both the upper and lower surfaces of leaves using a 1.8 L manual spray pump (Royal Condor Garden®, Soacha, Colombia). All foliar applications were carried out between 07:00 and 09:00 h (with sunrise at 06:00 h). Each treatment group consisted of 10 plants (five replicates per sampling point), for a total of 80 plants in this experiment, which were arranged in a completely randomized design (CRD) in the greenhouse. Finally, the experiment lasted approximately 60 days.

4.1.2. Experiment 2—Evaluation of the Most Efficient Doses of Glycine Betaine (GB) and Hydrogen Peroxide (H_2O_2) (Experiment 1) in Plants Exposed to Two Waterlogging Periods

Based on the first trial, the dose of the two chemical compounds (GB and H_2O_2) that showed the best response to mitigate the waterlogging stress was selected (Figure 1). Four groups of treatments were also established when plants reached five fully mature leaves at 45 DAT. The treatments are described as follows: (1) plants without waterlogging and with no GB or H_2O_2 sprays (absolute control), (2) waterlogged plants with no GB or H_2O_2 sprays (waterlogged control), (3) waterlogged plants with foliar H_2O_2 sprays at a concentration of 100 mM, and (4) waterlogged plants with foliar GB sprays at a concentration of 100 mM. Waterlogging treatments were also carried out by placing the plants in plastic containers with dimensions of 53 × 53 × 30 cm and a capacity of 120 L, which were filled until reaching a water level of 5 cm above the root neck. In this experiment, plants were subjected to two different waterlogging periods to quantify the effect of chemicals on mitigation under two short time (6 days) stress conditions. The first stress period was established between 45 and 51 DAT, while the second period was between 63 and 69 DAT. Between each waterlogging period, the plants were removed from each plastic container to allow water to drain until reaching the field capacity of the substrate. Subsequently, plants were watered during the recovery period (between 52 and 62 DAT) according to the evapotranspiration rate for 12 days. Foliar GB or H_2O_2 applications were also performed at 0, 3, 6, and 9 DAW between 07:00 and 09:00 h, wetting the upper and lower surfaces of leaves using the manual spray pump. Each treatment group consisted of 20 plants (five replicates per sampling point), for a total of 80 plants in this experiment, which were arranged in a completely randomized design (CRD) in the greenhouse. Finally, the experiment lasted 85 days.

4.2. Stomatal Conductance, Relative Chlorophyll Content, and Efficiency of PSII (F_v/F_m)

Stomatal conductance (gs) was estimated using a portable porometer (SC-1, Decagon Devices Inc., Pullman, WA, USA) between 10:00 and 13:00 h on a fully expanded leaf in the middle portion of the canopy. The relative chlorophyll content was then measured with a chlorophyll meter (AtLeaf, FT Green LLC Wilmington, USA), also on the same leaves used for gs readings in both experiments. Finally, the leaves used for gs and SPAD readings were dark-adapted with clips for 20 min to determine the maximum efficiency of PSII (F_v/F_m) by using a modulated fluorometer (MINI-PAM, Walz, Effeltrich, Germany) with an actinic light pulse of up to 2600 $\mu mol \cdot m^{-2} \cdot s^{-1}$ on their surface.

4.3. Plant Growth Parameters (Stem Diameter and Height, Leaf Area, Shoot Dry Weights and Relative Growth Rate)

Plant height and root neck diameter were recorded weekly with a ruler and vernier caliper, respectively. Then plants were harvested and separated into each of the shoot organs (leaves and stems). The leaf area of each plant was estimated by taking a photograph of the leaves of the plant canopy and, subsequently, the digital images were analyzed with a Java image-processing program (Image J; National Institute of Mental Health, Bethesda, MD, USA). Finally, the harvested organs were dried for 72 h in an oven at 70 °C to determine their respective dry weight (DW). In general, measurements of the above variables were performed at 4 and 13 DAW for experiment 1, and at 6, 18, 24, and 36 DAW for experiment 2, respectively.

On the other hand, RGR was determined for experiment 2. It was indirectly calculated using the length of the stem regarding the different sampling days. RGR was calculated using the following Equation (1):

$$\text{RGR} = \frac{\text{Ln } SL\ T2 - \text{Ln } SL\ T1}{\text{DS2} - \text{DS1}} \tag{1}$$

where $SL\ T2$ means stem length at time 2, $SL\ T1$ means stem length at time 1, DS2—DS1 means the difference in the number of days between sample 2 and sample 1.

4.4. Leaf Temperature and Canopy Temperature Index (CTI)

In experiment 1, the same leaves used for gs, leaf relative chlorophyll content, and F_v/F_m readings were also selected to determine leaf temperature using an infrared thermometer (Cole Parmer Instruments, Vernon Hills, IL 60061, USA). The readings were taken at 4 and 13 DAW.

In experiment 2, CTI was determined by means of a thermal camera (FLIR C2, FLIR Systems, Wilsonville, OR, USA), using Equation (2) described by Jones [71]:

$$\text{CTI} = \frac{\text{PT} - \text{Twl}}{\text{Tdl} - \text{Twl}} \tag{2}$$

where PT is the plant temperature, Twl is the temperature of the wet leaf, and Tdl is the temperature of the dry leaf. PT is determined by taking a thermal photograph at a distance of 0.9 m from the entire plant. Likewise, Tdl was estimated by collecting a leaf and Twl was obtained by wetting the leaf with a mixture of water and an agricultural adjuvant (Agrotin, Bayer CropScience, Bogotá, Colombia). Both leaves were placed on a white Styrofoam surface at the base of the plant pot. Plant canopy and reference leaves (Tdl and Twl) temperatures were analyzed using the software FLIR® Tools Plus 3.1.13080.1002 (FLIR® Systems, Wilsonville, OR, US). Thermal images were taken at noon at 36 DAW.

4.5. Water Relations (Leaf Water Potential and Relative Water Content (RWC))

After estimating the leaf temperature in experiment 1, the leaves were cut at the petiole with a scalpel to immediately record the leaf water potential (Ψ_{wf}) with a Scholander pressure chamber (PMS Instruments, Albany, OR, USA). The chamber was then sealed and gradually pressurized with nitrogen. The water potential was recorded by observing the expulsion of the sap from the xylem system out of the cut edge of the leaf petiole with an X15 magnifying glass. The Ψ_{wf} was obtained at noon at 4 and 13 DAW. In experiment 2, a leaf was taken from the middle third of the plant to determine the (RWC). Five 25 mm diameter discs were cut and their fresh weight (FW) was obtained. Subsequently, the discs were placed in a Petri dish with water for 24 h at laboratory temperature to determine the turgid weight (TW). Finally, the discs were dried for 72 h in an oven at 70 °C and their DW was determined. The RWC was calculated at 6, 18, 24, and 36 DAW using the following Equation (3):

$$\text{RWC} = \frac{(\text{FW} - \text{DW})}{(\text{TW} - \text{DW})} \times 100 \tag{3}$$

4.6. Waterlogging Tolerance Coefficient (WTC)

The WTC was indirectly calculated to determine the tolerance of treatments to waterlogging, using the shoot dry weight of the waterlogged treatments in relation to the control treatment without waterlogging [65]. The WTC was obtained at 13 DAW in experiment 1 using the following Equation (4):

$$\text{WTC} = \frac{\text{DW}_{\text{WT}}}{\text{DW}_{\text{CT}}} \tag{4}$$

where DW_{WT} is the shoot dry weight of waterlogged treatments with or without foliar GB or H_2O_2 sprays, whereas DW_{CT} is the shoot dry weight of the absolute control treatment (plants without waterlogging and foliar sprays).

4.7. Photosynthesis

Photosynthesis was estimated using a portable photosynthesis meter (LICOR 6200, Lincoln, NE, USA) on a leaf from the middle third of the plant in experiment 2. Measurements were taken on completely sunny days between 11:00 and 13:00 h at 36 DAW. During photosynthesis measurements, the conditions inside the chamber were as follows: photosynthetically active radiation (PAR) greater than 800 μmol m^{-2}·s^{-1}, leaf temperature of 27 $\pm$ 5 °C, and leaf to air water vapor pressure difference of 1.8 $\pm$ 0.5 kPa.

4.8. Experimental Design and Data Analysis

The data obtained from the first and second experiments were analyzed using a completely randomized design (CRD) in which each treatment had five plants as replicates. All percentage values were transformed using the arcsine transformation before analysis. Likewise, a correlation analysis was performed between the WTC and the leaf area to determine the best dose of the products used in experiment 1. When the analysis of variance (ANOVA) showed significant differences ($p \leq 0.05$), a Tukey post hoc test was used for mean comparison. The data were analyzed with the Statistix v 9.0 software (Analytical Software, Tallahassee, FL, USA) and the graphs were developed in SigmaPlot 12.0 (Systat Software, San Jose, CA, USA).

5. Conclusions

In summary, the present study continued to show the susceptibility of Cape gooseberry plants to waterlogging conditions, since this plant species shows a reduction in growth mainly associated with a low photosynthetic rate and water status under conditions of oxygen deficiency in the soil. However, foliar H_2O_2 or GB applications at a concentration of 100 mM helped to lessen the waterlogging conditions on Cape gooseberry plants and favored their physiological response. The foregoing allows us to conclude that the use of H_2O_2 or GB can be a viable tool in managing stress conditions due to moderate waterlogging in Cape gooseberry crops when periods of heavy rainfall are expected.

Author Contributions: Conceptualization, C.C.C.-A. and H.R.-D.; methodology, N.E.C.-D.; software, N.E.C.-D. and C.C.C.-A.; validation, N.E.C.-D., C.C.C.-A., and H.R.-D.; formal analysis, N.E.C.-D.; investigation, N.E.C.-D.; resources, H.R.-D.; data curation, N.E.C.-D. and C.C.C.-A.; writing—original draft preparation, N.E.C.-D.; writing—review and editing, H.R.-D. and C.C.C.-A.; visualization, C.C.C.-A.; supervision, H.R.-D.; project administration, H.R.-D.; funding acquisition, H.R.-D. All authors have read and agreed to the published version of the manuscript.

Funding: This research received no external funding.

Conflicts of Interest: The authors declare no conflict of interest.

References

1. Ramírez, M.; Roveda, G.; Bonilla, R.; Cabra, L.; Peñaranda, A.; López, M.; Serralde, D.P.; Tamayo, A.; Navas, G.E.; Díaz, C.A. *Uso y Manejo de Biofertilizantes en el Cultivo de la Uchuva*; Produmedios: Bogotá, Colombia, 2008; p. 56.

2. Fischer, G.; Miranda, D. Uchuva (*Physalis peruviana* L.). In *Manual Para el Cultivo de Frutales en el Trópico*; Fischer, G., Ed.; Produmedios: Bogotá, Colombia, 2012; pp. 851–873.

3. Álvarez-Flórez, F.; López-Cristoffanini, C.; Jáuregui, O.; Melgarejo, L.M.; López-Carbonell, M. Changes in ABA, IAA and JA levels during calyx, fruit and leaves development in cape gooseberry plants (*Physalis peruviana* L.). *Plant Physiol. Bioch.* **2017**, *115*, 174–182. [CrossRef] [PubMed]

4. Agronet. Available online: http://www.agronet.gov.co/estadistica/Paginas/default.aspx (accessed on 16 March 2020).

5. Kreuzwieser, J.; Gessler, A. Global climate change and tree nutrition: Influence of water availability. *Tree Physiol.* **2010**, *30*, 1221–1234. [CrossRef] [PubMed]

6. Ramirez-Villegas, J.; Salazar, M.; Jarvis, A.; Navarro-Racines, C.E. A way forward on adaptation to climate change in Colombian agriculture: Perspectives towards 2050. *Clim. Chang.* **2012**, *115*, 611–628. [CrossRef]

7. Vargas, G.; Hernández, Y.; Pabón, J.D. La Niña event 2010–2011: Hydroclimatic effects and socioeconomic impacts in Colombia. In *Climate Change, Extreme Events and Disaster Risk Reduction*; Mal, S., Singh, R., Huggel, C., Eds.; Springer: Cham, Switzerland, 2018; pp. 217–232.

8. Euscateguí, C.; Hurtado, G. *Análisis del Impacto del Fenómeno "La Niña" 2010–2011 en la Hidroclimatología del País*; Instituto de Hidrología, Meteorología y Estudios Ambientales: Bogotá, Colombia, 2011; p. 32.

9. Ou, L.J.; Dai, X.Z.; Zhang, Z.Q.; Zou, X.X. Responses of pepper to waterlogging stress. *Photosynthetica* **2011**, *49*, 339. [CrossRef]

10. Lekshmy, S.; Jha, S.K.; Sairam, R.K. Physiological and molecular mechanisms of flooding tolerance in plants. In *Elucidation of Abiotic Stress Signaling in Plants*; Pandey, G.K., Ed.; Springer: New York, NY, USA, 2015; Volume 2, pp. 227–242.

11. Issarakraisila, M.; Ma, Q.; Turner, D.W. Photosynthetic and growth responses of juvenile Chinese kale (*Brassica oleracea* var. *alboglabra*) and Caisin (*Brassica rapa* subsp. *parachinensis*) to waterlogging and water deficit. *Sci. Hortic.* **2007**, *111*, 107–113. [CrossRef]

12. Elkelish, A.A.; Alhaithloul, H.A.S.; Qari, S.H.; Soliman, M.H.; Hasanuzzaman, M. Pretreatment with *Trichoderma harzianum* alleviates waterlogging-induced growth alterations in tomato seedlings by modulating physiological, biochemical, and molecular mechanisms. *Environ. Exp. Bot.* **2020**, *171*, 103946. [CrossRef]

13. Andrade, C.A.; de Souza, K.R.D.; de Oliveira Santos, M.; da Silva, D.M.; Alves, J.D. Hydrogen peroxide promotes the tolerance of soybeans to waterlogging. *Sci. Hortic.* **2018**, *232*, 40–45. [CrossRef]

14. Aldana, F.; García, P.N.; Fischer, G. Effect of waterlogging stress on the growth, development and symptomatology of cape gooseberry (*Physalis peruviana* L.) plants. *Rev. Acad. Colomb. Cienc. Exact. Fis. Nat.* **2014**, *38*, 393–400. [CrossRef]

15. Chávez-Arias, C.C.; Gómez-Caro, S.; Restrepo-Díaz, H. Physiological, Biochemical and Chlorophyll Fluorescence Parameters of *Physalis Peruviana* L. Seedlings Exposed to Different Short-Term Waterlogging Periods and Fusarium Wilt Infection. *Agronomy* **2019**, *9*, 213. [CrossRef]

16. Ahmed, S.; Nawata, E.; Sakuratani, T. Changes of endogenous ABA and ACC, and their correlations to photosynthesis and water relations in mungbean (*Vigna radiata* (L.) Wilczak cv. KPS1) during waterlogging. *Environ. Exp. Bot.* **2006**, *57*, 278–284. [CrossRef]

17. De Almeida, J.; Tezara, W.; Herrera, A. Physiological responses to drought and experimental water deficit and waterlogging of four clones of cacao (*Theobroma cacao* L.) selected for cultivation in Venezuela. *Agric. Water Manag.* **2016**, *171*, 80–88. [CrossRef]

18. Jurczyk, B.; Pociecha, E.; Kościelniak, J.; Rapacz, M. Different photosynthetic acclimation mechanisms are activated under waterlogging in two contrasting Lolium perenne genotypes. *Funct. Plant Biol.* **2016**, *43*, 931–938. [CrossRef]

19. Rao, R.; Li, Y. Management of flooding effects on growth of vegetable and selected field crops. *HortTechnology* **2003**, *13*, 610–616. [CrossRef]

20. Flórez-Velasco, N.; Balaguera-López, H.E.; Restrepo-Díaz, H. Effects of foliar urea application on lulo (*Solanum quitoense* cv. *septentrionale*) plants grown under different waterlogging and nitrogen conditions. *Sci. Hortic.* **2015**, *186*, 154–162. [CrossRef]

21. Betancourt-Osorio, J.; Sanchez-Canro, D.; Restrepo-Díaz, H. Effect of Nitrogen Nutritional Statuses and Waterlogging Conditions on Growth Parameters, Nitrogen Use Efficiency and Chlorophyll Fluorescence in Tamarillo Seedlings. *Not. Bot. Horti. Agrobo.* **2016**, *44*, 375–381. [CrossRef]

22. Irfan, M.; Hayat, S.; Hayat, Q.; Afroz, S.; Ahmad, A. Physiological and biochemical changes in plants under waterlogging. *Protoplasma* **2010**, *241*, 3–17. [CrossRef]

23. Perez, O.; Dell'Amica, J.M.; Reynaldo, I. Performance of tomato seedlings under soil flooding. *Cultiv. Trop.* **1999**, *20*, 41–44.

24. Tanentzap, F.M.; Stempel, A.; Ryser, P. Reliability of leaf relative water content (RWC) measurements after storage: Consequences for in situ measurements. *Botany* **2015**, *93*, 535–541. [CrossRef]

25. Chávez-Arias, C.C.; Gómez-Caro, S.; Restrepo-Díaz, H. Mitigation of the Impact of Vascular Wilt and Soil Hypoxia on Cape Gooseberry Plants by Foliar Application of Synthetic Elicitors. *HortScience* **2020**, *55*, 121–132. [CrossRef]

26. Cha-um, S.; Samphumphuang, T.; Kirdmanee, C. Glycinebetaine alleviates water deficit stress in indica rice using proline accumulation, photosynthetic efficiencies, growth performances and yield attributes. *Aust. J. Crop Sci.* **2013**, *7*, 213–218.

27. Ashfaque, F.; Khan, M.I.R.; Khan, N.A. Exogenously applied H_2O_2 promotes proline accumulation, water relations, photosynthetic efficiency and growth of wheat (*Triticum aestivum* L.) under salt stress. *Annu. Res. Rev. Biol.* **2014**, *4*, 105–120. [CrossRef]

28. Rasheed, R.; Iqbal, M.; Ashraf, M.A.; Hussain, I.; Shafiq, F.; Yousaf, A.; Zaheer, A. Glycine betaine counteracts the inhibitory effects of waterlogging on growth, photosynthetic pigments, oxidative defence system, nutrient composition, and fruit quality in tomato. *J. Hortic. Sci. Biotech.* **2018**, *93*, 385–391. [CrossRef]

29. Craig, S.A. Betaine in human nutrition. *Am. J. Clin. Nutr.* **2004**, *80*, 539–549. [CrossRef] [PubMed]

30. Mäkelä, P. Agro-industrial uses of glycinebetaine. *Sugar Tech* **2004**, *6*, 207–212. [CrossRef]

31. Chen, T.H.; Murata, N. Glycinebetaine: An effective protectant against abiotic stress in plants. *Trends Plant Sci.* **2008**, *13*, 499–505. [CrossRef]

32. López-Delgado, H.A.; Martínez-Gutiérrez, R.; Mora-Herrera, M.E.; Torres-Valdés, Y. Induction of freezing tolerance by the application of hydrogen peroxide and salicylic acid as tuber-dip or canopy spraying in *Solanum tuberosum* L. plants. *Potato Res.* **2018**, *61*, 195–206. [CrossRef]

33. İşeri, Ö.D.; Körpe, D.A.; Sahin, F.I.; Haberal, M. Hydrogen peroxide pretreatment of roots enhanced oxidative stress response of tomato under cold stress. *Acta Physiol. Plant.* **2013**, *35*, 1905–1913. [CrossRef]

34. Khan, T.A.; Yusuf, M.; Fariduddin, Q. Seed treatment with H_2O_2 modifies net photosynthetic rate and antioxidant system in mung bean (*Vigna radiata* L. Wilczek) plants. *Isr. J. Plant Sci.* **2015**, *62*, 167–175. [CrossRef]

35. Hasan, S.A.; Irfan, M.; Masrahi, Y.S.; Khalaf, M.A.; Hayat, S. Growth, photosynthesis, and antioxidant responses of *Vigna unguiculata* L. treated with hydrogen peroxide. *Cogent Food Agric.* **2016**, *2*, 1155331. [CrossRef]

36. Bhattacharjee, S. An inductive pulse of hydrogen peroxide pretreatment restores redox-homeostasis and oxidative membrane damage under extremes of temperature in two rice cultivars. *Plant Growth Regul.* **2012**, *68*, 395–410. [CrossRef]

37. Guzel, S.; Terzi, R. Exogenous hydrogen peroxide increases dry matter production, mineral content and level of osmotic solutes in young maize leaves and alleviates deleterious effects of copper stress. *Bot. Stud.* **2013**, *54*, 26. [CrossRef] [PubMed]

38. Ocampo, O. El cambio climático y su impacto en el agro. *Rev. Ing.* **2011**, *33*, 115–123.

39. Ashraf, M.; Arfan, M. Gas exchange characteristics and water relations in two cultivars of *Hibiscus esculentus* under waterlogging. *Biol. Plant.* **2005**, *49*, 459–462. [CrossRef]

40. Nickum, M.T.; Crane, J.H.; Schaffer, B.; Davies, F.S. Reponses of mamey sapote (*Pouteria sapota*) trees to continuous and cyclical flooding in calcareous soil. *Sci. Hortic.* **2010**, *123*, 402–411. [CrossRef]

41. Nemeskéri, E.; Helyes, L. Physiological Responses of Selected Vegetable Crop Species to Water Stress. *Agronomy* **2019**, *9*, 447. [CrossRef]

42. Barickman, T.C.; Simpson, C.R.; Sams, C.E. Waterlogging Causes Early Modification in the Physiological Performance, Carotenoids, Chlorophylls, Proline, and Soluble Sugars of Cucumber Plants. *Plants* **2019**, *8*, 160. [CrossRef]

43. Drew, M.C. Oxygen deficiency and root metabolism: Injury and acclimation under hypoxia and anoxia. *Annu. Rev. Plant Mol. Biol.* **1997**, *48*, 223–250. [CrossRef]

44. Striker, G.G. Flooding stress on plants: Anatomical, morphological and physiological responses. In *Botany*; Mworia, J., Ed.; InTech: Rijeka, Croatia, 2012; pp. 3–28.

45. Bashar, K.K.; Tareq, M.Z.; Amin, M.R.; Honi, U.; Tahjib-Ul-Arif, M.; Sadat, M.A.; Hossen, Q.M.M. Phytohormone-Mediated Stomatal Response, Escape and Quiescence Strategies in Plants under Flooding Stress. *Agronomy* **2019**, *9*, 43. [CrossRef]

46. Anee, T.I.; Nahar, K.; Rahman, A.; Mahmud, J.A.; Bhuiyan, T.F.; Alam, M.U.; Fujita, M.; Hasanuzzaman, M. Oxidative Damage and Antioxidant Defense in *Sesamum indicum* after Different Waterlogging Durations. *Plants* **2019**, *8*, 196. [CrossRef]

47. Bansal, R.; Srivastava, J.P. Effect of waterlogging on photosynthetic and biochemical parameters in pigeonpea. *Russ. J. Plant Physiol.* **2015**, *62*, 322–327. [CrossRef]

48. Andrews, D.L.; MacAlpine, D.M.; Johnson, J.R.; Kelley, P.M.; Cobb, B.G.; Drew, M.C. Differential induction of mRNAs for the glycolytic and ethanolic fermentative pathways by hypoxia and anoxia in maize seedlings. *Plant Physiol.* **1994**, *106*, 1575–1582. [CrossRef] [PubMed]

49. Gupta, N.; Thind, S.K.; Bains, N.S. Glycine betaine application modifies biochemical attributes of osmotic adjustment in drought stressed wheat. *Plant Growth Regul.* **2014**, *72*, 221–228. [CrossRef]

50. Hasanuzzaman, M.; Fujita, M.; Oku, H.; Islam, M.T. *Plant Tolerance to Environmental Stress: Role of Phytoprotectants*, 1st ed.; CRC Press: Boca Raton, FL, USA, 2019; p. 448.

51. Hossain, M.A.; Bhattacharjee, S.; Armin, S.M.; Qian, P.; Xin, W.; Li, H.Y.; Burritt, D.J.; Fujita, M.; Tran, L.S.P. Hydrogen peroxide priming modulates abiotic oxidative stress tolerance: Insights from ROS detoxification and scavenging. *Front. Plant Sci.* **2015**, *6*, 420. [CrossRef] [PubMed]

52. Sun, Y.; Wang, H.; Liu, S.; Peng, X. Exogenous application of hydrogen peroxide alleviates drought stress in cucumber seedlings. *S. Afr. J. Bot.* **2016**, *106*, 23–28. [CrossRef]

53. Khan, T.A.; Yusuf, M.; Fariduddin, Q. Hydrogen peroxide in regulation of plant metabolism: Signalling and its effect under abiotic stress. *Photosynthetica* **2018**, *56*, 1237–1248. [CrossRef]

54. Steffens, B.; Geske, T.; Sauter, M. Aerenchyma formation in the rice stem and its promotion by H_2O_2. *New Phytol.* **2011**, *190*, 369–378. [CrossRef]

55. Sorwong, A.; Sakhonwasee, S. Foliar application of glycine betaine mitigates the effect of heat stress in three marigold (*Tagetes erecta*) cultivars. *Hort. J.* **2015**, *84*, 161–171. [CrossRef]

56. Foreman, J.; Demidchik, V.; Bothwell, J.H.F.; Mylona, P.; Miedema, H.; Torres, M.A.; Linstead, P.; Costa, S.; Brownlee, C.; Jones, J.D.G.; et al. Reactive oxygen species produced by NADPH oxidase regulate plant cell growth. *Nature* **2003**, *422*, 442–446. [CrossRef]

57. Niu, L.; Liao, W. Hydrogen Peroxide Signaling in Plant Development and Abiotic Responses: Crosstalk with Nitric Oxide and Calcium. *Front. Plant Sci.* **2016**, *7*, 230. [CrossRef]

58. Barba-Espin, G.; Diaz-Vivancos, P.; Clemente-Moreno, M.J.; Albacete, A.; Faize, L.; Faize, M.; Pérez-Alfocea, F.; Hernández, J.A. Interaction between hydrogen peroxide and plant hormones during germination and the early growth of pea seedlings. *Plant Cell Environ.* **2010**, *33*, 981–994. [CrossRef]

59. Liao, W.B.; Huang, G.B.; Yu, J.H.; Zhang, M.L. Nitric oxide and hydrogen peroxide alleviate drought stress in marigold explants and promote its adventitious root development. *Plant Physiol. Bioch.* **2012**, *58*, 6–15. [CrossRef] [PubMed]

60. Mäkelä, P.; Jokinen, K.; Kontturi, M.; Peltonen-Sainio, P.; Pehu, E.; Somersalo, S. Foliar application of glycinebetaine—A novel product from sugar beet—As an approach to increase tomato yield. *Ind. Crop. Prod.* **1998**, *7*, 139–148. [CrossRef]

61. Chimenti, C.A.; Pearson, J.; Hall, A.J. Osmotic adjustment and yield maintenance under drought in sunflower. *Field Crop. Res.* **2002**, *75*, 235–246. [CrossRef]

62. Giri, J. Glycinebetaine and abiotic stress tolerance in plants. *Plant Signal. Behav.* **2011**, *6*, 1746–1751. [CrossRef] [PubMed]

63. Korkmaz, A.; Değer, Ö.; Kocaçınar, F. Alleviation of water stress effects on pepper seedlings by foliar application of glycinebetaine. *N. Z. J. Crop Hortic. Sci.* **2015**, *43*, 18–31. [CrossRef]

64. Martynenko, A.; Shotton, K.; Astatkie, T.; Petrash, G.; Fowler, C.; Neily, W.; Critchley, A.T. Thermal imaging of soybean response to drought stress: The effect of *Ascophyllum nodosum* seaweed extract. *SpringerPlus* **2016**, *5*, 1–14. [CrossRef]

65. Velasco, N.F.; Ligarreto, G.A.; Díaz, H.R.; Fonseca, L.P.M. Photosynthetic responses and tolerance to root-zone hypoxia stress of five bean cultivars (*Phaseolus vulgaris* L.). *S. Afr. J. Bot.* **2019**, *123*, 200–207. [CrossRef]

66. Manik, S.N.; Pengilley, G.; Dean, G.; Field, B.; Shabala, S.; Zhou, M. Soil and crop management practices to minimize the impact of waterlogging on crop productivity. *Front. Plant Sci.* **2019**, *10*, 14010. [CrossRef]

67. Villarreal-Navarrete, A.; Fischer, G.; Melgarejo, L.M.; Correa, G.; Hoyos-Carvajal, L. Growth response of the cape gooseberry (*Physalis peruviana* L.) to waterlogging stress and *Fusarium oxysporum* infection. *Acta Hortic.* **2017**, *1178*, 161–168. [CrossRef]

68. Cardona, W.A.; Bautista-Montealegre, L.G.; Flórez-Velasco, N.; Fischer, G. Desarrollo de la biomasa y raíz en plantas de lulo (*Solanum quitoense* var. *septentrionale*) en respuesta al sombrío y anegamiento. *Rev. Colomb. Cienc. Horticolas* **2016**, *10*, 53–65. [CrossRef]

69. Sánchez-Reinoso, A.D.; Jiménez-Pulido, Y.; Martínez-Pérez, J.P.; Pinilla, C.S.; Fischer, G. Chlorophyll fluorescence and other physiological parameters as indicators of waterlogging and shadow stress in lulo (*Solanum quitoense* var. *septentrionale*) seedlings. *Rev. Colomb. Cienc. Hortícolas* **2019**, *13*. [CrossRef]
70. Hainaut, P.; Remacle, T.; Decamps, C.; Lambert, R.; Sadok, W. Higher forage yields under temperate drought explained by lower transpiration rates under increasing evaporative demand. *Eur. J. Agron.* **2016**, *72*, 91–98. [CrossRef]
71. Jones, H.G. Use of infrared thermometry for estimation of stomatal conductance as a possible aid to irrigation scheduling. *Agric. Forest Meteorol.* **1999**, *95*, 139–149. [CrossRef]

Article

Hypoxic Treatment Decreases the Physiological Action of the Herbicide Imazamox on *Pisum sativum* Roots

Miriam Gil-Monreal, Mercedes Royuela and Ana Zabalza *

Institute for Multidisciplinary Research in Applied Biology (IMAB), Universidad Publica de Navarra, Campus Arrosadia s/n, 31006 Pamplona, Spain; mirian.gil@unavarra.es (M.G.-M.); royuela@unavarra.es (M.R.)
* Correspondence: ana.zabalza@unavarra.es; Tel.: +34-948-169118

Received: 30 June 2020; Accepted: 30 July 2020; Published: 3 August 2020

Abstract: The inhibition of acetolactate synthase (ALS; EC 2.2.1.6), an enzyme located in the biosynthetic pathway of branched-chain amino acids, is the target site of the herbicide imazamox. One of the physiological effects triggered after ALS inhibition is the induction of aerobic ethanol fermentation. The objective of this study was to unravel if fermentation induction is related to the toxicity of the herbicide or if it is a plant defense mechanism. Pea plants were exposed to two different times of hypoxia before herbicide application in order to induce the ethanol fermentation pathway, and the physiological response after herbicide application was evaluated at the level of carbohydrates and amino acid profile. The effects of the herbicide on total soluble sugars and starch accumulation, and changes in specific amino acids (branched-chain, amide, and acidic) were attenuated if plants were subjected to hypoxia before herbicide application. These results suggest that fermentation is a plant defense mechanism that decreases the herbicidal effect.

Keywords: acetolactate synthase; ethanol fermentation; imidazolinones; mode of action; aerobic fermentation

---

1. Introduction

Imazamox (IMX) is an imidazolinone herbicide that inhibits acetolactate synthase (ALS), also known as acetohydroxy acid synthase (EC 2.2.1.6), the enzyme that catalyzes the condensation of either two molecules of pyruvate to form acetolactate or one molecule of pyruvate and one molecule of 2-ketobutyrate to form 2-aceto-2-hydroxybutyrate in the biosynthetic pathway of branched-chain amino acids (BCAAs): leucine, isoleucine, and valine [1].

ALS-inhibiting herbicides emerged in the 1980s and they have been demonstrated as potent, selective, broad-spectrum herbicides. These chemicals are the largest site-of-action group on the market, with more than 50 active ingredients belonging to five classes (imidazolinones, sulfonylureas, triazolopyrimidines, pyrimidinyl(thio)benzoates, and sulfonylamino-carbonyl triazolinones). Altogether, ALS-inhibiting herbicides accounted for approximately 15% of the total herbicide market in 2015 [2].

Although the biochemical mechanisms underlying the blocking of ALS activity through ALS inhibitors has been studied [3], less is known regarding their modes of action, which are the physiological processes underlying plant death resulting from inactivated ALS. To understand the physiological effects involved in the lethal process after herbicide treatment is important as it can lead to their more rational use, and because it can help in the development of new compounds with similar herbicidal activities, but with different enzyme inhibition targets to avoid the evolution of weed resistance.

Previous findings showed that ALS inhibitors cause growth arrest followed by the slow death of treated plants [4,5], along with changes in the free amino acid content [6,7] and impairment of carbon metabolism, such as carbohydrate accumulation [8–10]. Another effect on the roots of plants treated with ALS inhibitors is the induction of aerobic fermentation and the alternative respiratory pathway; both of which are low-ATP producing pathways [9,11–15]. All these metabolic impairments indicate that the effect of these herbicides on primary plant metabolism has broader physiological consequences than a lack of certain amino acids alone.

Fermentation is an essential pathway activated in plants exposed to hypoxia. Plant hypoxia often originates because of variations in the environment, such as flooding or severe rainfall [16]. Under these circumstances, plants have to adapt their metabolism in order to avoid energy shortage [17]. Induction of the fermentative metabolism allows the plant to use glycolysis for ATP production by recycling NAD^+ [18]. In the ethanol fermentation pathway, pyruvate decarboxylase (PDC, EC 4.1.1.1) catalyzes the conversion of pyruvate to acetaldehyde, which is then converted to ethanol by the action of the enzyme alcohol dehydrogenase (ADH, EC 1.1.1.1), with the concomitant regeneration of NAD^+ [18]. Besides its role during oxygen deprivation, ethanol fermentation is induced in response to a number of stressful aerobic conditions, such as osmotic stress or low-temperature conditions [19–23]. It has been suggested that fermentation improves cold stress tolerance [24] and that this pathway might act as an overflow regulating carbohydrate metabolism [25].

Although fermentation in plants exposed to low-oxygen conditions has been studied in depth, its role in plants exposed to other stresses when oxygen is not limited has yet to be fully explained; therefore, the role of fermentation induction in the mode of action of ALS inhibitors remains to be elucidated. Two, non-contradictory explanations can be considered. Firstly, the induction of this pathway could be a plant defense mechanism that promotes better tolerance of the herbicide. Secondly, it could contribute to the chemical's toxicity, since ethanol and lactate, metabolites shown to be toxic for the plants, are produced during fermentation [18].

The objective of this study was to unravel the importance of ethanol fermentation in plant responses to ALS inhibitors, trying to outline if fermentation induction is related to the toxicity of the herbicides or if it is a plant defense mechanism that alleviates the herbicidal effect. An original experimental design was employed to achieve this aim. Pea plants were exposed to two different times of hypoxia before herbicide application in order to induce the ethanol fermentation pathway prior to ALS inhibition. Then, IMX was applied to the nutrient solution and the characteristic physiological effects triggered by ALS inhibitors were evaluated (carbohydrates and amino acid profile).

2. Results

Pea plants were exposed to low-oxygen conditions before the herbicide was applied, thus when the plants were treated with the herbicide, they presented an enhanced fermentative metabolism. Using this approach, the effects of the herbicide could be compared between the plants that demonstrated an induced fermentative metabolism before ALS inhibition, and those in which the fermentation pathway was not activated when the herbicide was applied. Specifically, one group of plants was exposed to 48 h of hypoxia (the Hypoxia-48h group), another group was exposed to 24 h of hypoxia (the Hypoxia-24h group), and a third group was maintained in normal oxygen conditions (the No-Hypoxia group). IMX was applied to half of the plants in each group, while the others were not treated with herbicides and were used as the control for the corresponding group. The different treatments were named with a combination of two codes: the first code refers to the hours of hypoxia (0, 24 h, or 48 h), and the second one indicates if IMX was applied (IMX) or if they were controls (C) (see Table 1 for details): 0-C, 0-IMX, 24h-C, 24h-IMX, 48h-C, 48h-IMX.

Table 1. Summary of the treatments.

Group	Abbreviation	Treatment Description
No-Hypoxia	0-C	No treatment was applied.
	0-IMX	Application of 5 mg L^{-1} of imazamox at day 0.
Hypoxia-24h	24h-C	The aeration was removed for 24 h before the day 0. Aeration was again placed at day 0 until the end of the experiment. No herbicide was applied.
	24h-IMX	The aeration was removed for 24 h before the day 0. The aeration was again placed at day 0 until the end of the experiment. At day 0 imazamox was applied at a final concentration of 5 mg L^{-1}.
Hypoxia-48h	48h-C	The aeration was removed for 48 h before the day 0. Aeration was again placed at day 0 until the end of the experiment. No herbicide was applied.
	48h-IMX	The aeration was removed for 48 h before the day 0. Aeration was again placed at day 0 until the end of the experiment. The day 0 imazamox was applied at a final concentration of 5 mg L^{-1}.

2.1. Validation of the Experiment

To ascertain that the wanted low-oxygen conditions were obtained, the oxygen concentration in the nutrient solution was monitored (Figure 1). When the aeration was removed (indicated with a black arrow for the Hypoxia-48h group and grey arrow for the Hypoxia-24h group in Figure 1), the oxygen concentration in the nutrient solution drastically decreased. At day 0, the oxygen concentration present in the nutrient solution of the tanks from the Hypoxia-48h and Hypoxia-24h groups was about 30–40%. By contrast, in the tanks that were continuously aerated (the No-Hypoxia group), the oxygen concentration remained at around 100%. These results indicate that the desired conditions to carry out the experiment were obtained.

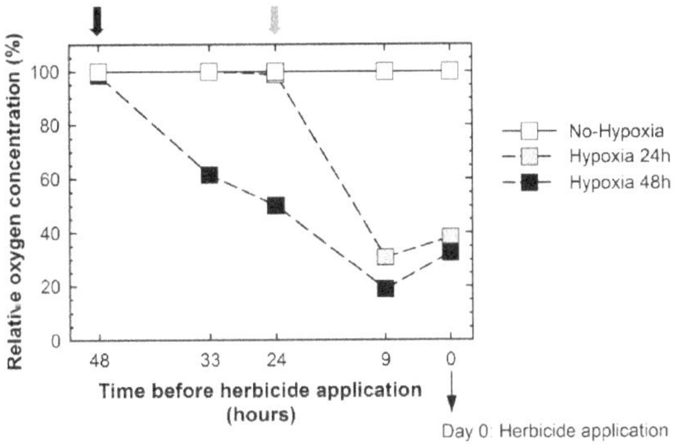

Figure 1. Relative oxygen concentration (%) in the nutrient solution before herbicide application. Black and grey arrows above the graph indicate aeration removal 48 h and 24 h before herbicide application, respectively.

The in vitro enzymatic activities of PDC and ADH (Figure 2A) were measured in the roots of the plants from the six studied treatments (Table 1). The herbicide application increased the activity of PDC in the roots of the plants of the No-Hypoxia group. At day 0, the PDC activity detected in the roots of the plants that were exposed to hypoxia was much higher than the activity found in the roots of the plants from the No-Hypoxia group. Once the plants were again aerated, the PDC activity decreased to control values in the plants that were not treated with IMX, while it remained higher in the IMX-treated plants with respect to their controls in the Hypoxia-24h and Hypoxia-48h groups. Similar to the PDC activity, at the beginning of the experiment, the ADH activity was much higher in the groups that were exposed to hypoxia before herbicide application. Once the plants were again aerated, the activity of ADH decreased to control values. In this case, the herbicidal effect in all groups was not very pronounced.

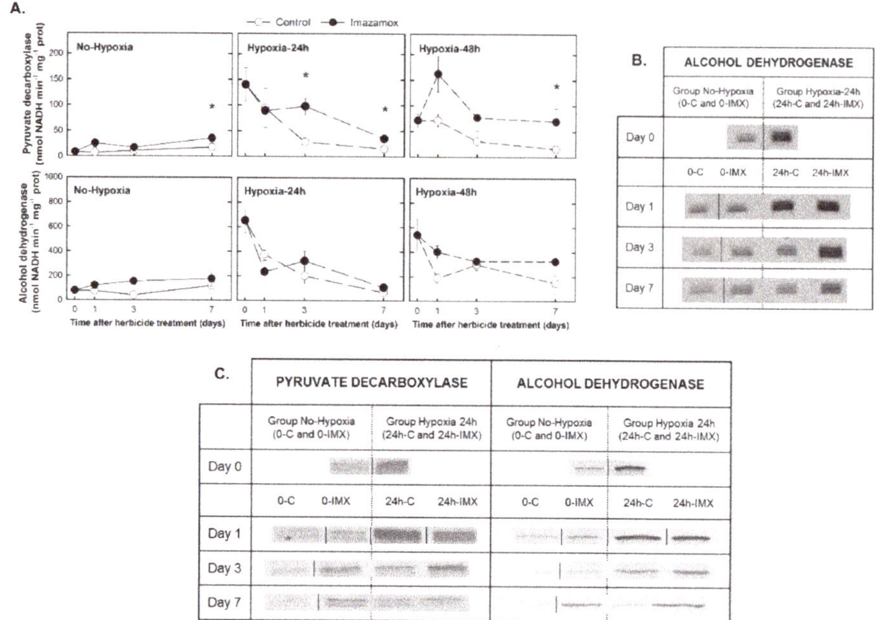

Figure 2. Pyruvate decarboxylase (PDC) and alcohol dehydrogenase (ADH) in the roots of pea plants. Pea plants were not treated with hypoxia before herbicide application (the No-Hypoxia group) or were treated with hypoxia for 24 h or 48 h before herbicide application (the Hypoxia-24h and Hypoxia-48h groups). Half of the plants from each group were treated with imazamox (5 mg L^{-1}, black symbols). The others were not treated with herbicide and were used as controls (white symbols). The different treatments were named with a combination of two codes: the first code refers to the hours of hypoxia (0, 24 h, or 48 h), and the second one indicates if plants were treated with imazamox (IMX) or were controls (C): 0-C, 0-IMX, 24h-C, 24h-IMX, 48h-C, 48h-IMX. (**A**) In vitro activities of PDC and ADH in the roots of pea plants. Values represent the mean ± SE (n = 4 biological replicates). Significant variations are marked with * for differences between control and imazamox-treated plants (t-Test, $p < 0.05$) on a given day. See Supporting Information Table S1 for the two-way ANOVA results. (**B**) Native PAGE for ADH activity in pea roots. Each lane contained 1.75 µg of protein. (**C**) Immunoblot detection of PDC and ADH on days 0, 1, 3, and 7 after imazamox application. Each lane contained 30 µg of protein. In blots and gels, each vertical dividing line indicates lane or lanes removed from the original image. See supporting Information Figures S1, S2 and S3 for full gel images.

The in gel enzymatic activity of ADH (Figure 2B) and the protein content of the enzymes PDC and ADH (Figure 2C) were determined in the No-Hypoxia and Hypoxia-24h groups. The three ADH isoenzymes described in peas were detected in the native PAGE for ADH activity (from top to bottom: ADH1-ADH1, ADH1-ADH2, and ADH2-ADH2) [26], and the pattern of band intensity (Figure S1) was similar to the in vitro activity. In the No-Hypoxia group, IMX application increased the activity of the ADH1-ADH2 and ADH2-ADH2 isoenzymes from day 3. The PDC and ADH protein content (Figure 2C) showed a similar pattern as the in vitro activity, showing that the increase in activity is due, at least in part, to an increase in the amount of fermentative enzymes.

2.2. Growth Parameters

In order to study the effect of IMX on the growth of the different studied groups, the shoot and root lengths were measured (Figure 3). Figure 3A shows the aspect of the plants 7 days after herbicide application. The roots of the IMX-treated plants became brownish while the roots of the control plants were white. The growth of the secondary roots was inhibited and the growth of the shoot and the principal root was also arrested by the herbicides.

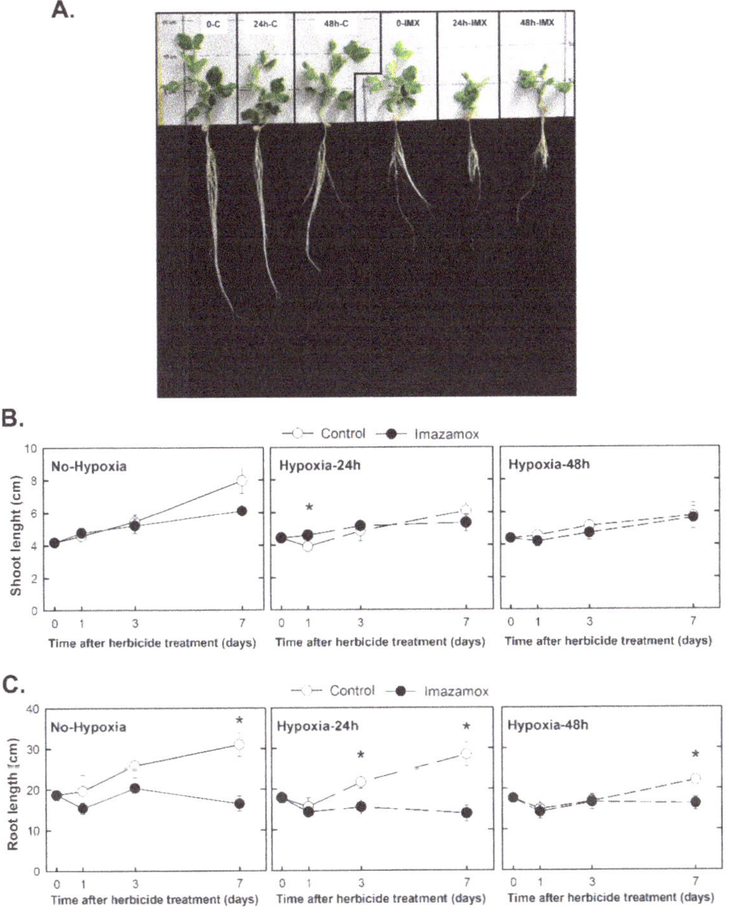

Figure 3. Effect of imazamox on the growth of pea plants. Pea plants were not treated with hypoxia before herbicide application (the No-Hypoxia group) or were treated with hypoxia for 24 h or 48 h before herbicide application (the Hypoxia-24h and Hypoxia-48h groups). Half of the plants from each group were treated with imazamox (5 mg L^{-1}, black symbols) and the others were not treated with herbicide and were used as controls (white symbols). The different treatments were named with a combination of two codes: the first code refers to the hours of hypoxia (0, 24 h, or 48 h), and the second one indicates if plants were treated with imazamox (IMX) or they were controls (C): 0-C, 0-IMX, 24h-C, 24h-IMX, 48h-C, 48h-IMX. (**A**) Aspect of the plants 7 days after herbicide application. (**B**) Shoot length (**C**) Root length. Values represent the mean ± SE (*n* = 4 biological replicates). Significant variations are marked with * for differences between control and imazamox-treated plants (*t*-Test, *p* < 0.05) on a given day. See Supporting Information Table S1 for the two-way ANOVA results.

The shoot of the control plants from the Hypoxia-24h and Hypoxia-48h groups grew less than the plants from the No-Hypoxia group (Figure 3B), indicating that hypoxia had an effect on shoot growth. The effect of IMX on the shoot growth of the treated plants was similar in the three groups, regardless of whether they were previously exposed to hypoxia. Root growth was arrested in the No-Hypoxia group as a consequence of herbicide application after 3 and 7 days of treatment (Figure 3C). The effect of the herbicide in the Hypoxia-48h group was not so evident because the roots of the control plants (48h-C) practically did not grow until the third day.

2.3. Carbohydrate Content

The sum of sucrose, fructose, glucose (Figure 4A), and starch (Figure 4B) contents were measured in the roots of the pea plants. Total soluble sugars were accumulated in the herbicide-treated plants from all the studied groups. While the effect of the herbicide on the non-hypoxic plants was significant as early as day 1 of the IMX treatment, the significant effect on the 24h-IMX and 48h-IMX plants was delayed until the third day of treatment.

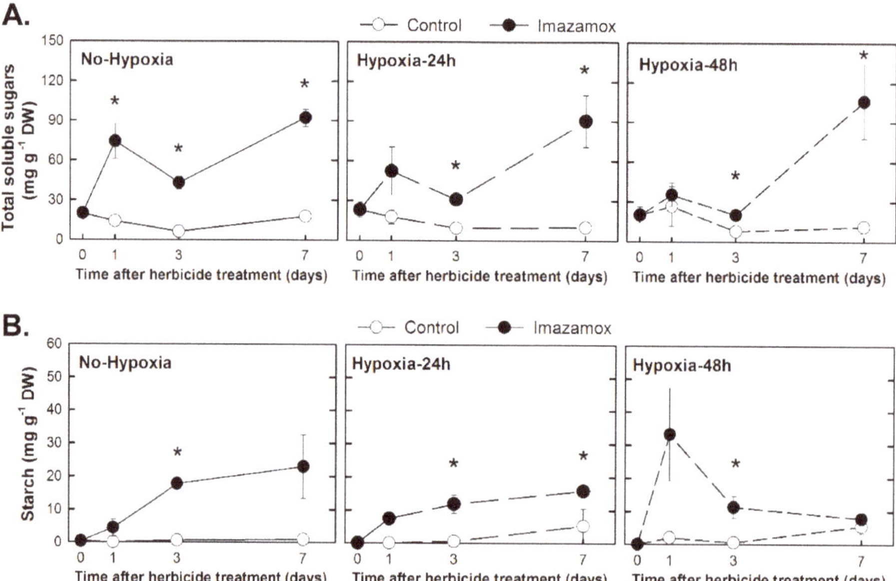

Figure 4. Total soluble sugars (**A**) and starch (**B**) content in the root of pea plants. Pea plants were not treated with hypoxia before herbicide application (the No-Hypoxia group) or were treated with hypoxia for 24 h or 48 h before herbicide application (the Hypoxia-24h and Hypoxia-48h groups). Half of the plants from each group were treated with imazamox (5 mg L^{-1}, black symbols). In each group, the others were not treated with herbicide and were used as the controls (white symbols) for the imazamox-treated plants. Values represent the mean $\pm$ SE ($n = 4$ biological replicates). Significant variations are marked with * for differences between control and imazamox-treated plants (*t*-Test, $p < 0.05$) on a given day. See Supporting Information Table S1 for the two-way ANOVA results.

Starch was also accumulated in the roots of the plants of the three studied groups as a consequence of IMX application (Figure 4B). The plants exposed to hypoxia before herbicide application showed lower starch accumulation at day 7 than the plants that were not exposed to hypoxia (Figure 4B).

2.4. Free Amino Acid Profile

The effects on free amino acid profile were studied in the No-Hypoxia and Hypoxia-24h groups (Figure 5) by monitoring five physiological parameters previously described to be affected by ALS inhibitors: free amino acid content, BCAA, aromatic, acidic, and amide amino acid contents [6,10,14,27]. As ALS -inhibitors cause a general increase in the content of free amino acids that could mask the specific changes in each absolute content [27,28], BCAA, aromatic, acidic and amide amino acids are shown as their relative content in terms of percentage of total amino acid content (Figure 5).

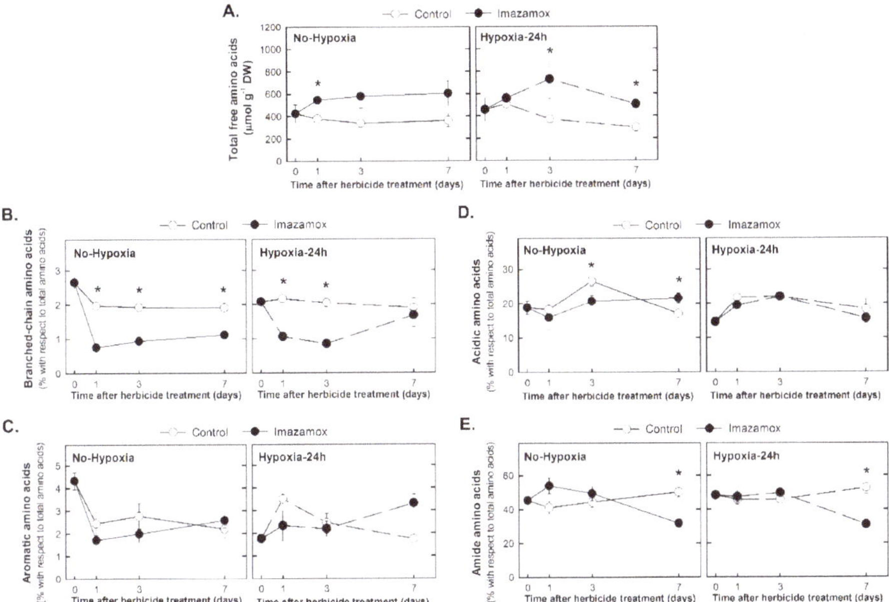

Figure 5. Free amino acid profile in the root of pea plants. Pea plants were not treated with hypoxia before herbicide application (Group No-Hypoxia) or were treated with hypoxia for 24 h before herbicide application (Group Hypoxia-24h). Half of the plants from each group were treated with imazamox (5 mg L⁻¹, black symbols). In each group, the other halves were not treated with herbicide and were used as the controls (white symbols) for the Imazamox-treated plants. Total free amino acids (**A**) branched-chain (**B**) aromatic (**C**) acidic (**D**) and amide (**E**) amino acid contents. Values represent the mean ± SE (*n* = 4 biological replicates). Significant variations are marked with * for differences between control and imazamox-treated plants (*t*-Test, $p < 0.05$) at a given day. See Supporting Information Table S1 for the two-way ANOVA results.

As expected, IMX treatment provoked an increase in the free amino acid pool in both groups of plants (Figure 5A). Nevertheless, the increase was significant earlier (day 1) when plants were not subjected to the hypoxic treatment. The relative content of BCAA was lower in plants treated with the herbicide in both groups, but this effect was abolished after 7 days of treatment if plants were subjected to the hypoxic treatment (Figure 5B). Aromatic amino acid content (sum of phenylalanine, tyrosine, and tryptophan) was not significantly affected by the herbicide or the hypoxic treatment (Figure 5C). Acidic amino acid content (the sum of glutamic and aspartic acids) decreased after 3 days of IMX treatment only when plants were not subjected to hypoxia before herbicide application (Figure 5D). Contrary to expectations, a general increase in the amide amino acid content was not detected (the sum of glutamine and asparagine) (Figure 5E).

Alanine and γ-aminobutyric acid (GABA) are two amino acids that usually accumulate under low-oxygen conditions. Their content was evaluated in both the No-Hypoxia and Hypoxia-24h groups (Figure 6). After 24 h of hypoxia, the content of GABA and alanine increased, thus, plants from the Hypoxia-24h group presented high levels of alanine and GABA at the onset of the herbicide treatment. Exposure to hypoxia did not change the effect of the herbicide, with the pattern of Alanine and GABA contents being similar in both groups: alanine content was accumulated in herbicide-treated plants over the entire time of study and GABA content was not modified by the herbicide.

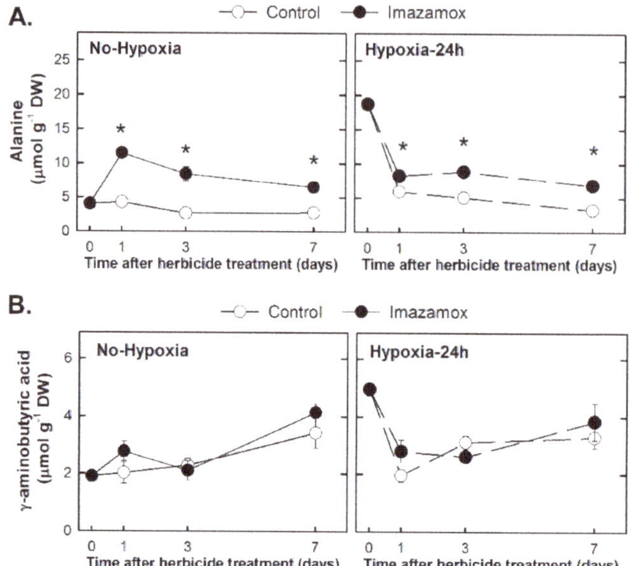

Figure 6. Alanine (**A**) and γ-aminobutyric acid (GABA; **B**) contents in the root of pea plants. Pea plants were not treated with hypoxia before herbicide application (Group No-Hypoxia) or were treated with hypoxia for 24 h before herbicide application (Group Hypoxia-24h). Half of the plants from each group were treated with imazamox (5 mg L^{-1}, black symbols). In each group, the other halves were not treated with the herbicide and were used as the controls (white symbols) for the imazamox-treated plants. Values represent the mean ± SE (n = 4 biological replicates). Significant variations are marked with * for differences between control and imazamox-treated plants (t-Test, $p < 0.05$) at a given day. See Supporting Information Table S1 for the two-way ANOVA results.

3. Discussion

The oxygen concentration in the nutrient solution measurements and the evaluation of the enzymes involved in the ethanol fermentation validated the experimental design since the desired conditions were obtained. The oxygen concentration in the nutrient solution of the plants exposed to hypoxia before herbicide application drastically decreased: it was about 30–40% the day of herbicide application (day 0) (Figure 1). In addition, alanine and GABA were accumulated in these plants (Figure S1), which is a typical response found in plants exposed to hypoxia. On the other hand, at day 0, the activities and protein contents of PDC and ADH of the plants from the Hypoxia-24h and Hypoxia-48h groups were much higher than the activities in the non-hypoxic plants, confirming that ethanol fermentation in the roots was already induced when the herbicide was applied.

The previously described effects on ethanol fermentation, triggered as a consequence of ALS inhibition [11,13–15], were observed in the IMX-treated plants: increased in vitro activity of PDC (Figure 2A), higher ADH activity detected by native gel (Figure 2B), and accumulation of both PDC and ADH proteins (Figure 2C). In addition, native gel showed that the anodic isozyme demonstrated less induction than the other two bands after IMX treatment, as was described before for other ALS inhibitors [13]. Different induction pattern was described for anaerobically stressed cotton, from which the most anodic isozyme (ADH2-ADH2) was induced more than the other two [29].

In concordance with the growth arrest previously described in several plant species (e.g., soybean, pea, rice) as a consequence of ALS inhibition [4,12,30,31], the results obtained in this experiment also evidenced that IMX provoked a decrease in both shoot and root elongation (Figure 3). Oxygen availability also limits the growth of the tissues [18], and root growth of the plants subjected for 48 h to hypoxia was restricted for 3 days after returning to aerobic conditions (Figure 3C). It was not possible to detect a different effect of the herbicide on root growth depending if plants were exposed to hypoxia or not, because IMX arrested the root growth completely in the three groups.

Carbohydrates were accumulated in the roots as a consequence of ALS inhibition (Figure 4) as has been described before [8–11,14,32,33]. The accumulation in roots was due to a lack of use of available sugars because growth was arrested (Figure 3). This carbohydrate accumulation has not been previously related to a change in total respiration rates or cytochrome respiratory capacity, but an increase in AOX capacity was observed in a previous experiment with the same plant material but with other ALS inhibitors [9]. These results suggest that ALS inhibition induce the less-efficient ATP-producing pathways of fermentation and alternative respiration, while growth is arrested and carbohydrates are not consumed.

Although carbohydrate accumulation was detected after IMX treatment in hypoxic plants (groups Hypoxia-24h and Hypoxia-48h), the total soluble accumulation was detected later than in the non-hypoxic plants and the starch accumulation was lower than in the non-hypoxic plants (Figure 4). Although it can be argued that hypoxia incurred a carbon deficit which delayed the later carbohydrate accumulation due to herbicide, this was not true, as previous findings revealed that carbohydrate shortage was not detected in the pea roots subjected to hypoxia to 1 or 3 days (Figure 4 in [34]). So, the lower and later carbohydrate accumulation induced by the herbicide in hypoxic plants compared to non-hypoxic suggests an alleviation of this physiological effect on carbon metabolism if plants are subjected to hypoxia before herbicide application.

Moreover, regarding the nitrogen metabolism, an increase in the total free amino acid content and several changes in individual amino acids were detected after IMX treatment (Figure 5). Free amino acid accumulation has been previously observed in plants treated with different ALS inhibitors [6,7,35] and it has been proposed to occur due to an increased protein turnover [36], which can be related to an altered, proteolytic profile [7]. Contrary to previous studies [10,27], aromatic amino acid content was not affected by the ALS-inhibitors.

BCAA relative content was decreased after IMX treatment during all the period of study while in previous studies the decrease was transitory, abolished after 4 days [27,37,38] and followed by an increase. The initial decrease of the BCAA and the following increase agree with the proteolysis discussed above.

Acidic and amide amino acid contents showed a tendency to the pattern expected, increase of amides and decrease of acidic amino acids [10,27], although the changes were not consistent during all the time of study. In general, the proportion of the two acidic amino acids was lower in herbicide-treated plants than in the control, and the proportion of their amides was higher in herbicide-treated plants than in control plants (Figure 5). This behaviour shows that the plant develops a progress in nitrogen storage from amino acids in the acidic form to the amide form. These coordinated variations are difficult to explain, and it is not possible finally to establish if there are key enzymes under common control, if such modulation involves parameters such as glutamine/glutamate [39] or if there is a general control of C/N metabolism just by controlling the 'glutamine plus asparagine/glutamate plus aspartate' proportion.

Interestingly, several of these effects described after the herbicide on the amino acid profile were attenuated if ethanol fermentation was induced before herbicide application: BCAA decrease after 7 days was abolished and amide amino acid increase and acidic amino acid decrease were attenuated.

Alanine was accumulated at the onset of the herbicide treatments in plants subjected to hypoxia (Figure 6A). The production of alanine from pyruvate through alanine amino-transferase is another quite important pathway related to fermentation and lack of oxygen [40]. Alanine was accumulated in herbicide treated plants both in hypoxic and non-hypoxic plants, as was detected in Arabidopsis plants treated with foramsulfuron, [37] or pea roots treated with imazethapyr [34], other two ALS inhibitors and which can be related to the previously detected induction of alanine amino-transferease [11,14]. The non-protein amino acid GABA (Figure 6B) was accumulated after 48 h of hypoxia. It is a metabolite that is usually accumulated under stress situations, such as anoxia [41,42].

Plant death following ALS inhibition, has been associated with an impairment of carbon and nitrogen metabolism, only detected after ALS inhibition and not detected after inhibition of other enzymes of the BCAA pathway. Thus, the different herbicidal efficacy observed could be associated with a different carbon/nitrogen metabolism imbalance [43]. The results of this study evidence that hypoxia alleviates the impairment of carbon and nitrogen metabolism induced by ALS inhibitors, suggesting that an induced fermentative metabolism alleviates the herbicidal effects on the physiology of treated plants.

Several other approaches have tried to examine the significance of the induction of fermentation in response to the ALS-inhibiting herbicides to evaluate if it is part of the plant defense against herbicides or by contrast, it contributes to their chemical toxicity. Unfortunately, until now, divergent results have been reported. On one hand, alleviation of the effects of ALS inhibitor herbicides on different parameters (carbohydrate and amino acid accumulation) was observed in the *adh1* mutant supporting that fermentation contributes to the herbicide action [14]. On the other hand, mutants with reduced fermentative potential exhibited higher sensitivity to herbicide treatments [44] and plants overexpressing the PDH bypass were less sensitive to ALS-inhibitors [45], supporting the hypothesis that induction of PDC contributes to herbicide stress tolerance. The present study shows that fermentation induction at the moment of herbicide application decreases the herbicide physiological action and supports the second of the two hypothesis proposed about the role of fermentation in the mode of action of ALS-inhibitors: the induction of ethanol fermentation in treated plants is part of the plant defense mechanism after herbicide.

4. Materials and Methods

4.1. Plant Material and Treatment Application

Pisum sativum L. cv. snap sugar boys where surface sterilized and grown under controlled conditions [9]. To prevent roots from hypoxia, the nutrient solution was continuously aerated. When the plants were 12-days-old, they were separated in three groups (Table 1). In order to obtain low-oxygen conditions for the induction of fermentation, the aeration was removed in one of the groups for 48 h (named Hypoxia-48h), and for 24 h in another group (named Hypoxia-24h). After the 48 h or 24 h the tanks were again aerated, this day was considered as the day 0. The other group was maintained continuously aerated (named No-Hypoxia). The nutrient solution was replaced the day 0 in all the tanks and the herbicide was applied to half of the tanks from the group Hypoxia-48h (named 48h-IMX), half of the plants from the group Hypoxia-24h (named 24h-IMX), and to half of the tanks from the group No-Hypoxia (named 0-IMX). The herbicide was applied to the nutrient solution as commercial formulation at a final concentration of 5 mg active ingredient L^{-1} (16.33 µM) of IMX (Pulsar®40, BASF Española SA, Barcelona, Spain). The other half of the plants from the groups Hypoxia-48h, Hypoxia-24h and No-Hypoxia were not treated with herbicide (named as 48h-C, 24h-C and 0-C, respectively) and were used as the control for the comparison with the herbicide-treated plants of the corresponding group. The experiment was performed in duplicate. The applied herbicide doses provoked plant death in 20 days.

For the analytical measurements, intact root samples were taken at day 0, before herbicide application, and at days 1, 3 and 7 after IMX application, this time points were chosen in order to allow us to evaluate physiological and biochemical plant responses induced by the herbicide, but not directly resulting from cell death.

Plant material was immediately frozen in liquid nitrogen and stored at −80 °C for further analysis. Later, frozen samples were ground under liquid nitrogen using a Retsch mixer mill (MM200, Retsch®, Haan, Germany), the required amount of tissue for each analysis was separated and stored at −80 °C. Some fresh material was dried for 48 h at 80 °C in order to obtain the fresh weight/dry weight ratio.

4.2. PDC and ADH Enzymatic Activities

Briefly, PDC and ADH activities were measured in a spectrophotometer monitoring NADH consumption or formation at 340 nm, respectively, as described before [11].

In gel, ADH activity was measured in desalted extracts obtained from pea roots in a ratio of 0.1 g FW/ 2.5 mL extraction buffer. Native electrophoresis was run in a 12.5% polyacrylamide gel (Phast Gel® Homogeneous 12.5% in 0.112 M Acetate, 0.112 M Tris (pH 6.4)) at 4 °C in a Phast SystemTM (Pharmacia, LKB, Biotechnology AB, Uppsala, Sweden). Phast Gel® Buffer Strips Native (0.88 M L-Alanine, 0.25 M Tris (pH 8.8)) were used for the electrophoresis. In each line, 1.95 µg of protein was loaded.

ADH specific staining was performed as described in a previous study [26] with minor modifications. The gel was incubated in darkness for 15 min in a solution composed of 25 mM Tris-Cl (pH 8), 0.8% (*v/v*) ethanol, 0.144 mM nitro blue tetrazolium, 0.65 mM phenazine methosulfate, and 0.24 mM NAD+.

4.3. PDC and ADH Immunoblotting

PDC and ADH protein immunoblot assay was performed according to standard techniques, as described in a previous study [46]. PDC and ADH antibodies were used at dilutions of 1:1000 and 1:500, respectively. Goat anti-rabbit IgG alkaline phosphatase (Sigma-Aldrich) was used as the secondary antibody at a dilution of 1:20,000, and cross-reacting protein bands were visualized using the Amplified Alkaline Phosphatase Goat Anti-Rabbit Immun-Blot® Assay Kit (Bio-Rad Inc., Hercules, CA, USA), according to the manufacturer's instructions. The intensity of the bands was quantified using a GS-800 densitometer (Bio-Rad Inc., Hercules, CA, USA).

4.4. Metabolites Determination

4.4.1. Amino Acids

The extraction of amino acids from pea roots was performed in HCl. After protein precipitation, amino acid concentrations were measured in the supernatant using capillary electrophoresis equipped with a laser-induced fluorescence detector, as previously described [27].

4.4.2. Total Soluble Sugars and Starch

The glucose, fructose, and sucrose (total soluble sugars) concentrations were determined in ethanol-soluble extracts, and the ethanol-insoluble residue was extracted for starch analysis. Starch and soluble sugar concentrations were determined using high-performance capillary electrophoresis, as previously described [8].

4.5. Statistical Analysis

All analyses were performed using four biological replicates from two independent experiments. The mean was used as a measure of central tendency and the standard error (SE) as a measure of dispersion.

First, the data of the three different groups (No-Hypoxia, Hypoxia-24h, and Hypoxia-48h) were compared independently. The herbicide-treated plants were compared with their respective controls for each sampling day using the Student's *t*-Test for the Significance of the Difference between the Means of Two Independent Samples. Second, for all the studied parameters, a two-way analysis of variance (ANOVA) was performed to examine the influence of the studied variables (hypoxia and herbicide application) and their possible interaction. When the results were expressed in percentages, the data were previously transformed according to the following formula: $arcsin \sqrt{x}/100$. The statistical analysis was conducted at a significance level of 5% ($p < 0.05$) and the statistical program used was IBM SPSS Statistics (v.22).

5. Conclusions

In this study, an original approach was used to unravel the role of fermentation induction after ALS-inhibiting herbicides, by evaluating the effect of applying hypoxia before the herbicide on the typical physiological markers in herbicide-treated plants.

There are several physiological effects widely described after ALS inhibition, that have been also detected in pea roots in this study: carbohydrate accumulation, decrease in BCAA and acidic amino acids and increase in amide amino acids. The changes in these physiological markers was lower or later if the plants were subjected to hypoxia before treatment application. This decrease of the herbicide action with hypoxia supports the hypothesis that fermentation induction by ALS inhibitors is a plant defense mechanism counteracting the physiological effect of the herbicide.

Supplementary Materials: The following are available online at http://www.mdpi.com/2223-7747/9/8/981/s1, Figure S1: Original blots and band intensity of the Native PAGE for alcohol dehydrogenase, Figure S2: Original blots and band intensity of the immunoblot detection of pyruvate decarboxylase, Figure S3: Original blots and band intensity of the immunoblot detection of alcohol dehydrogenase, Table S1: Results of the two-way analysis of variance.

Author Contributions: Conceptualization, M.R.; methodology, M.G.-M.; software, M.G.-M.; validation, A.Z.; formal analysis, A.Z. and M.R.; investigation, M.G.-M. and A.Z.; resources, M.R.; data curation, M.G.-M. and A.Z.; writing—original draft preparation, A.Z.; writing—review and editing, M.G.-M. and M.R.; visualization, A.Z.; supervision, M.R.; project administration, M.R.; funding acquisition, M.R. All authors have read and agreed to the published version of the manuscript.

Funding: This work was funded by the Spanish Ministry of Economy and Competitiveness (AGL2016-77531-R).

Acknowledgments: We thank Gustavo Garijo and Micaela García for technical assistance.

Conflicts of Interest: The authors declare no conflict of interest.

References

1. Singh, B.K. Biosynthesis of valine, leucine and isoleucine. In *Plant Amino Acids: Biochemistry and Biotechnology*; Singh, B.K., Ed.; Marcel Dekker: New York, NY, USA, 1999; pp. 227–247.
2. Peters, B.; Strek, H.J. Herbicide discovery in light of rapidly spreading resistance and ever-increasing regulatory hurdles. *Pest Manag. Sci.* **2018**, *74*, 2211–2215. [CrossRef]
3. Duggleby, R.G.; McCourt, J.A.; Guddat, L.W. Structure and mechanism of inhibition of plant acetohydroxyacid synthase. *Plant Physiol. Biochem.* **2008**, *46*, 309–324. [CrossRef]
4. Wittenbach, V.; Abell, L.M. Inhibition of valine, leucine and isoleucine biosynthesis. In *Plant Amino Acids: Biochemistry and Biotechnology*; Singh, B.K., Ed.; Marcel Dekker: New York, NY, USA, 1999; pp. 385–416.
5. Cobb, A.; Reade, J. The inhibition of amino acid biosynthesis. In *Herbicides and Plant Physiology*; Wiley-Blackwell, Ed.; Wiley-Blackwell: Oxford, UK, 2010; pp. 176–199, ISBN 978-1-4051-2935-0.
6. Shaner, D.L.; Reider, M.L. Physiological responses of corn (*Zea mays*) to AC 243,997 in combination with valine, leucine, and isoleucine. *Pestic. Biochem. Physiol.* **1986**, *25*, 248–257. [CrossRef]
7. Zulet, A.; Gil-Monreal, M.; Villamor, J.G.; Zabalza, A.; van der Hoorn, R.A.L.; Royuela, M. Proteolytic pathways induced by herbicides that inhibit amino acid biosynthesis. *PLoS ONE* **2013**, *8*, e73847. [CrossRef]

8. Zabalza, A.; Orcaray, L.; Gaston, S.; Royuela, M. Carbohydrate accumulation in leaves of plants treated with the herbicide chlorsulfuron or imazethapyr is due to a decrease in sink strength. *J. Agric. Food Chem.* **2004**, *52*, 7601–7606. [CrossRef]

9. Armendáriz, O.; Gil-Monreal, M.; Zulet, A.; Zabalza, A.; Royuela, M. Both foliar and residual applications of herbicides that inhibit amino acid biosynthesis induce alternative respiration and aerobic fermentation in pea roots. *Plant Biol.* **2016**, *18*, 382–390. [CrossRef]

10. Fernández-Escalada, M.; Zulet-González, A.; Gil-Monreal, M.; Royuela, M.; Zabalza, A. Physiological performance of glyphosate and imazamox mixtures on *Amaranthus palmeri* sensitive and resistant to glyphosate. *Sci. Rep.* **2019**, *9*, 18225. [CrossRef]

11. Gaston, S.; Zabalza, A.; Gonzalez, E.M.; Arrese-Igor, C.; Aparicio-Tejo, P.M.; Royuela, M. Imazethapyr, an inhibitor of the branched-chain amino acid biosynthesis, induces aerobic fermentation in pea plants. *Physiol. Plant.* **2002**, *114*, 524–532. [CrossRef]

12. Gaston, S.; Ribas-Carbo, M.; Busquets, S.; Berry, J.A.; Zabalza, A.; Royuela, M. Changes in mitochondrial electron partitioning in response to herbicides inhibiting branched-chain amino acid biosynthesis in soybean. *Plant Physiol.* **2003**, *133*, 1351–1359. [CrossRef]

13. Zabalza, A.; González, E.M.; Arrese-Igor, C.; Royuela, M. Fermentative metabolism is induced by inhibiting different enzymes of the branched-chain amino acid biosynthesis pathway in pea plants. *J. Agric. Food Chem.* **2005**, *53*, 7486–7493. [CrossRef]

14. Zulet, A.; Gil-Monreal, M.; Zabalza, A.; van Dongen, J.T.; Royuela, M. Fermentation and alternative oxidase contribute to the action of amino acid biosynthesis-inhibiting herbicides. *J. Plant Physiol.* **2015**, *175*, 102–112. [CrossRef] [PubMed]

15. Gil-Monreal, M.; Fernandez-Escalada, M.; Royuela, M.; Zabalza, A. An aerated axenic hydroponic system for the application of root treatments: Exogenous pyruvate as a practical case. *Plant Methods* **2018**, *14*, 48. [CrossRef] [PubMed]

16. Loreti, E.; van Veen, H.; Perata, P. Plant responses to flooding stress. *Curr. Opin. Plant Biol.* **2016**, *33*, 64–71. [CrossRef] [PubMed]

17. van Dongen, J.T.; Licausi, F. Oxygen sensing and signaling. *Annu. Rev. Plant Biol.* **2015**, *66*, 345–367. [CrossRef]

18. Perata, P.; Alpi, A. Plant responses to anaerobiosis. *Plant Sci.* **1993**, *93*, 1–17. [CrossRef]

19. Christie, P.J.; Hahn, M.; Walbot, V. Low-temperature accumulation of alcohol dehydrogenase-1 mRNA and protein activity in maize and rice seedlings. *Plant Physiol.* **1991**, *95*, 699–706. [CrossRef]

20. Kato-Noguchi, H. Osmotic stress increases alcohol dehydrogenase activity in maize seedlings. *Biol. Plant.* **2000**, *43*, 621–623. [CrossRef]

21. Dolferus, R.; Jacobs, M.; Peacock, W.J.; Dennis, E.S. Differential interactions of promoter elements in stress responses of the Arabidopsis *Adh* gene. *Plant Physiol.* **1994**, *105*, 1075–1087. [CrossRef]

22. Minhas, D.; Grover, A. Transcript levels of genes encoding various glycolytic and fermentation enzymes change in response to abiotic stresses. *Plant Sci.* **1999**, *146*, 41–51. [CrossRef]

23. Kürsteiner, O.; Dupuis, I.; Kuhlemeier, C. The *Pyruvate decarboxylase1* gene of Arabidopsis is required during anoxia but not other environmental stresses. *Plant Physiol.* **2003**, *132*, 968–978. [CrossRef]

24. Peters, J.S.; Frenkel, C. Relationship between alcohol dehydrogenase activity and low-temperature in two maize genotypes, Silverado *F1* and *Adh1-Adh2* - doubly null. *Plant Physiol. Biochem.* **2004**, *42*, 841–846. [CrossRef] [PubMed]

25. Tadege, M.; Dupuis, I.; Kuhlemeier, C. Ethanolic fermentation: New functions for an old pathway. *Trends Plant Sci.* **1999**, *4*, 320–325. [CrossRef]

26. Schwartz, D.; Endo, T. Alcohol dehydrogenase polymorphism in maize-simple and compound loci. *Genetics* **1966**, *53*, 709–715. [PubMed]

27. Orcaray, L.; Igal, M.; Marino, D.; Zabalza, A.; Royuela, M. The possible role of quinate in the mode of action of glyphosate and acetolactate synthase inhibitors. *Pest Manag. Sci.* **2010**, *66*, 262–269. [CrossRef]

28. Fernández-Escalada, M.; Gil-Monreal, M.; Zabalza, A.; Royuela, M. Characterization of the *Amaranthus palmeri* physiological response to glyphosate in susceptible and resistant populations. *J. Agric. Food Chem.* **2016**, *64*, 95–106. [CrossRef]

29. Millar, A.A.; Olive, M.R.; Dennis, E.S. The expression and anaerobic induction of alcohol-dehydrogenase in cotton. *Biochem. Genet.* **1994**, *32*, 279–300. [CrossRef]

30. Zabalza, A.; Gaston, S.; Sandalio, L.M.; del Río, L.A.; Royuela, M. Oxidative stress is not related to the mode of action of herbicides that inhibit acetolactate synthase. *Environ. Exp. Bot.* **2007**, *59*, 150–159. [CrossRef]

31. Qian, H.; Hu, H.; Mao, Y.; Ma, J.; Zhang, A.; Liu, W.; Fu, Z. Enantioselective phytotoxicity of the herbicide imazethapyr in rice. *Chemosphere* **2009**, *76*, 885–892. [CrossRef]

32. Wang, Q.; Ge, L.; Zhao, N.; Zhang, L.; You, L.; Wang, D.; Liu, W.; Wang, J. A Trp-574-Leu mutation in the acetolactate synthase (ALS) gene of *Lithospermum arvense* L. confers broad-spectrum resistance to ALS inhibitors. *Pestic. Biochem. Physiol.* **2019**, *158*, 12–17. [CrossRef]

33. Qian, H.; Wang, R.; Hu, H.; Lu, T.; Chen, X.; Ye, H.; Liu, W.; Fu, Z. Enantioselective phytotoxicity of the herbicide imazethapyr and its effect on rice physiology and gene transcription. *Environ. Sci. Technol.* **2011**, *45*, 7036–7043. [CrossRef]

34. Zabalza, A.; Orcaray, L.; Igal, M.; Schauer, N.; Fernie, A.R.; Geigenberger, P.; van Dongen, J.T.; Royuela, M. Unraveling the role of fermentation in the mode of action of acetolactate synthase inhibitors by metabolic profiling. *J. Plant Physiol.* **2011**, *168*, 1568–1575. [CrossRef] [PubMed]

35. Zabalza, A.; Gaston, S.; Ribas-Carbo, M.; Orcaray, L.; Igal, M.; Royuela, M. Nitrogen assimilation studies using ^{15}N in soybean plants treated with imazethapyr, an inhibitor of branched-chain amino acid biosynthesis. *J. Agric. Food Chem.* **2006**, *54*, 8818–8823. [CrossRef] [PubMed]

36. Rhodes, D.; Hogan, A.L.; Deal, L.; Jamieson, G.C.; Haworth, P. Amino acid metabolism of *Lemna minor* L.: II. Responses to chlorsulfuron. *Plant Physiol.* **1987**, *84*, 775–780. [CrossRef]

37. Trenkamp, S.; Eckes, P.; Busch, M.; Fernie, A.R. Temporally resolved GC-MS-based metabolic profiling of herbicide treated plants treated reveals that changes in polar primary metabolites alone can distinguish herbicides of differing mode of action. *Metabolomics* **2009**, *5*, 277–291. [CrossRef]

38. Less, H.; Angelovici, R.; Tzin, V.; Galili, G. Principal transcriptional regulation and genome-wide system interactions of the Asp-family and aromatic amino acid networks of amino acid metabolism in plants. *Amino Acids* **2010**, *39*, 1023–1028. [CrossRef] [PubMed]

39. Foyer, C.H.; Parry, M.; Noctor, G. Markers and signals associated with nitrogen assimilation in higher plants. *J. Exp. Bot.* **2003**, *54*, 585–593. [CrossRef]

40. Rocha, M.; Licausi, F.; Araújo, W.L.; Nunes-Nesi, A.; Sodek, L.; Fernie, A.R.; van Dongen, J.T. Glycolysis and the tricarboxylic acid cycle are linked by alanine aminotransferase during hypoxia induced by waterlogging of Lotus japonicus. *Plant Physiol.* **2010**, *152*, 1501–1513. [CrossRef]

41. Shelp, B.J.; Bown, A.W.; Mclean, M.D. Metabolism and functions of gamma-aminobutyric acid. *Trends Plant Sci.* **1999**, *4*, 446–452. [CrossRef]

42. Narsai, R.; Rocha, M.; Geigenberger, P.; Whelan, J.; van Dongen, J.T. Comparative analysis between plant species of transcriptional and metabolic responses to hypoxia. *New Phytol.* **2011**, *190*, 472–487. [CrossRef]

43. Zabalza, A.; Zulet, A.; Gil-Monreal, M.; Igal, M.; Royuela, M. Branched-chain amino acid biosynthesis inhibitors: Herbicide efficacy is associated with an induced carbon-nitrogen imbalance. *J. Plant Physiol.* **2013**, *170*, 814–821. [CrossRef]

44. Gil-Monreal, M.; Giuntoli, B.; Zabalza, A.; Licausi, F.; Royuela, M. ERF-VII transcription factors induce ethanol fermentation in response to amino acid biosynthesis-inhibiting herbicides. *J. Exp. Bot.* **2019**, *70*, 5839–5851. [CrossRef] [PubMed]

45. Gil-Monreal, M.; Zabalza, A.; Missihoun, T.D.; Dörmann, P.; Bartels, D.; Royuela, M. Induction of the PDH bypass and upregulation of the *ALDH7B4* in plants treated with herbicides inhibiting amino acid biosynthesis. *Plant Sci.* **2017**, *264*, 16–28. [CrossRef] [PubMed]

46. Zabalza, A.; van Dongen, J.T.; Froehlich, A.; Oliver, S.N.; Faix, B.; Gupta, K.J.; Schmälzlin, E.; Igal, M.; Orcaray, L.; Royuela, M.; et al. Regulation of respiration and fermentation to control the plant internal oxygen concentration. *Plant Physiol.* **2009**, *149*, 1087–1098. [CrossRef] [PubMed]

MDPI

St. Alban-Anlage 66

4052 Basel

Switzerland

Tel. +41 61 683 77 34

Fax +41 61 302 89 18

www.mdpi.com

Plants Editorial Office

E-mail: plants@mdpi.com

www.mdpi.com/journal/plants

CPSIA information can be obtained
at www.ICGtesting.com
Printed in the USA
LVHW071302120521
686547LV00053B/1789